Current Challenges on Large Supramolecular Assemblies

NATO Science Series

A Series presenting the results of activities sponsored by the NATO Science Committee. The Series is published by IOS Press and Kluwer Academic Publishers, in conjunction with the NATO Scientific Affairs Division.

General Sub-Series

A. Life Sciences	IOS Press
B. Physics	Kluwer Academic Publishers
C. Mathematical and Physical Sciences	Kluwer Academic Publishers
D. Behavioural and Social Sciences	Kluwer Academic Publishers
E. Applied Sciences	Kluwer Academic Publishers
F. Computer and Systems Sciences	IOS Press

Partnership Sub-Series

1. Disarmament Technologies	Kluwer Academic Publishers
2. Environmental Security	Kluwer Academic Publishers
3. High Technology	Kluwer Academic Publishers
4. Science and Technology Policy	IOS Press
5. Computer Networking	IOS Press

The Partnership Sub-Series incorporates activities undertaken in collaboration with NATO's Partners in the Euro-Atlantic Partnership Council – countries of the CIS and Central and Eastern Europe – in Priority Areas of concern to those countries.

NATO-PCO-DATA BASE

The NATO Science Series continues the series of books published formerly in the NATO ASI Series. An electronic index to the NATO ASI Series provides full bibliographical references (with keywords and/or abstracts) to more than 50000 contributions from international scientists published in all sections of the NATO ASI Series.
Access to the NATO-PCO-DATA BASE is possible via CD-ROM "NATO-PCO-DATA BASE" with user-friendly retrieval software in English, French and German (© WTV GmbH and DATAWARE Technologies Inc. 1989).

The CD-ROM of the NATO ASI Series can be ordered from: PCO, Overijse, Belgium.

Series C: Mathematical and Physical Sciences - Volume 519

Current Challenges on Large Supramolecular Assemblies

edited by

George Tsoucaris
CNRS - Université Paris-Sud,
Centre Pharmaceutique,
Châtenay-Malabry,
France

Springer-Science+Business Media, B.V.

Proceedings of the NATO Advanced Research Workshop on
Current Challenges on Large Supramolecular Assemblies
Saint George-Lycabettus Hotel, Athens, Greece
31 October – 5 November 1997

A C.I.P. Catalogue record for this book is available from the Library of Congress.

ISBN 978-94-010-6224-4 ISBN 978-94-011-5284-6 (eBook)
DOI 10.1007/978-94-011-5284-6

Printed on acid-free paper

This book contains the proceedings of a NATO Advanced Research Workshop held within the programme of activities of the NATO Special Programme on Supramolecular Chemistry as part of the activities of the NATO Science Committee.

Other books previously published as a result of the activities of the Special Programme are:

WIPFF, G. (Ed.), *Computational Approaches in Supramolecular Chemistry.* (ASIC 426) 1994. ISBN 0-7923-2767-5

FLEISCHAKER, G.R., COLONNA, S. and LUISI, P.L. (Eds.), *Self-Production of Supramolecular Structures.* From Synthetic Structures to Models of Minimal Living Systems. (ASIC 446) 1994. ISBN 0-7923-3163-X

FABBRIZZI, L., POGGI, A. (Eds.), *Transition Metals in Supramolecular Chemistry.* (ASIC 448) 1994. ISBN 0-7923-3196-6

BECHER, J. and SCHAUMBURG, K. (Eds.), *Molecular Engineering for Advanced Materials.* (ASIC 456) 1995. ISBN 0-7923-3347-0

LA MAR, G.N. (Ed.), *Nuclear Magnetic Resonance of Paramagnetic Macromolecules.* (ASIC 457) 1995. ISBN 0-7923-3348-9

SIEGEL, JAY S. (Ed.), *Supramolecular Stereochemistry.* (ASIC 473) 1995. ISBN 0-7923-3702-6

WILCOX, C.S. and HAMILTON A.D. (Eds.), *Molecular Design and Bioorganic Catalysis.* (ASIC 478) 1996. ISBN 0-7923-4024-8

MEUNIER, B. (Ed.), *DNA and RNA Cleavers and Chemotherapy of Cancer and Viral Diseases.* (ASIC 479) 1996. ISBN 0-7923-4025-6

KAHN, O. (Ed.), *Magnetism: A Supramolecular Function.* (ASIC 484) 1996. ISBN 0-7923-4153-8

ECHEGOYEN, L., KAIFER ANGEL E. (Eds.), *Physical Supramolecular Chemistry.* (ASIC 485) 1996. ISBN 0-7923-4181-3

DESVERGNE J.P., CZARNIK A.W. (Eds.), *Chemosensors of Ion and Molecule Recognition.* (ASIC 492) 1997. ISBN 0-7923-4555-X

MICHL J., *Modular Chemistry.* (ASIC 499) ISBN 0-7923-4730-7

This Book contains the Proceedings of a NATO Advanced Research Workshop held within the Supramolecular Chemistry Programme, Scientific and Environmental Affairs Division.

Organising Committee:

J.L. ATWOOD	Dep. of Chemistry, University of Missouri, Columbia, Missouri, USA
J. LIPKOWSKI	Institute of Phys. Chem. Polish Academy. of Science, Warsaw, Poland
1. MAVRIDIS	Institute of Phys.Chem., NCSR "Demokritos", Agia Paraskevi, Greece
G. TSOUCARIS	C.N.R.S.-Univ. Paris-Sud, Châtenay-Malabry, France

The following topics are recorded in this book in virtually the same order as the oral contributions presented at the Workshop:

Crystallography and Design Principles
Synthesis, Reactivity, Molecular Recognition
From 2D to 3D Assemblies
Molecular Recognition and Crystal State
Upgrading Methodologies
Natural and semi-natural systems
Assembly and Self Assembly

The Organising Committee expresses its gratitude to the following sponsors for generous support:

NATO Supramolecular Chemistry Programme

International Union of Crystallography
European Commission, DG XII, Central and Eastern Europe
Office of Naval Research, European Office, U. S. A.
Ministère des Affaires Etrangères, France
General Secretariat of Research and Technology, Greece
NCSR "Democritos"; Greece
Agricultural University of Athens, Greece
Ionian Bank, Greece

The Organising Committee warmly thanks these persons who devoted their time and energy for an excellent local organisation, including several cultural activities:

Dr I. Mavridis, Dr. D. Yanakopoulou
Prof. D. Mentzafos, Prof. E. Eliopoulos, Prof. A. Hountas
Julia Argyroglou, Stella Makedonopoulou, Kyriaki Eliadou,
Demetrios Zouvelekis.

TABLE OF CONTENTS

developments to cope with difficult analytical and structural problems pertaining to large Assemblies:
Direct Methods for crystal structures with 200-1000 atoms; computer simulations at interfaces and large clusters (nucleation); Xe NMR; Electrospray Mass Spectroscopy; conformation changes from Single Molecule Force Spectrometry; fabrication of SM assemblies by positioning individual molecules; Time Resolved Diffraction (picosecond and sub-picosecond range).
Remarkable progress pertains to the direct observation and, more importantly, to the direct manipulation of a single molecule or supermolecule, or of a single electron.

The last part of the Workshop dealt with specific SM families.
Some natural systems most likely to offer insights into the SM organisation as a basis for a given function have been selected: peptide macrocycles; catalytic antibodies; an antibiotic repressor and its complex with operator DNA providing a clue to antibiotic resistance. New specific cyclodextrins and their complexes were designed and synthesised, endowing these semi-natural compounds with a remarkable selectivity.
In a break from the usual process of SM synthesis, a sort of "inverse approach" to creating a perfectly ordered SM Assembly has been achieved by the preparation of nanocrystals (5 nm) with an almost constant "molecular weight", i.e. number of elementary units within one nanocrystal.
Finally, it is remarkable that supermolecules with a repetition motif (ladders, grids..., reaching several nm) have been elegantly synthesised by using the chemical approach of self-assembly of components endowed with adequately imprinted information. Clearly, the latter is a decisive progress towards bridging the supramolecular area with that of nanostructures.

Summarising, the meeting has contributed to further bridging areas of research or methods of investigation: sophisticated molecular components are being synthesised; crystals with predesigned structural characteristics are obtained; more components are assembled to build large SM systems with increasing versatility in applications.
These new compounds can now be the object of further studies in other disciplines: Physics, Materials Sciences, Pharmaceutics.
Even if the complexity of the biological assemblies and functions is still beyond that of the SM assemblies, many of the presented contributions show that the diversity inherent to Chemistry is a powerful tool for creating new structures endowed with ever increasing capabilities and functions.

G. Tsoucaris

QUANTUM CRYSTALLOGRAPHY: FEATURES AND APPLICATION

J. KARLE,* L. HUANG*† and L. MASSA‡
**Laboratory for the Structure of Matter, Naval Research Laboratory, Washington, D.C. 20175-5341, USA*
‡Department of Chemistry, Hunter College, New York, NY 1002, USA

1. Introduction

Quantum crystallography (QCr) is an area of research that arises from the fact that experimental X-ray diffraction data obtained from crystals can also be readily described theoretically by the use of quantum mechanical modeling. The intimate connection between experiment and theory arises from the fact that X-rays are scattered by electrons whose distributions are represented in the experimental data and models of electron density distributions are given by quantum mechanics (Q.M.). An objective of this type of research is to obtain a quantum mechanical model that is consistent with the crystallographic data, thus affording the opportunity to calculate numerous properties of interest, for example, various energies, electron distributions, atomic charges and electrostatic potentials. Our approach to quantum crystallography is based on the use of a single, idempotent density matrix (a projector matrix) [1]. In the initial stages of the process of optimizing the fit of a quantum mechanical model to X-ray diffraction data, it is valuable to have a projector matrix that is as close as possible to the one that results from the fitting process. Such a matrix is obtainable from *ab initio* calculations. The fitting process involves the adjustment of the values of the elements in the projector matrix and certain other parameters while preserving the idempotency of the matrix and its normalized trace. These properties will be described later on.

Structures of interest may be too large for *ab initio* calculations to be made all at once. We have developed a type of fragment calculation in which the individual *ab initio* fragment calculations can be combined to give a very accurate equivalent to the results that would be obtained if the desired *ab inito* calculation could be made all at once [2]. The fragment calculation scales essentially linearly with complexity. Fragment calculations are, evidently, purely quantum mechanical. The fragment calculations can generate wave functions for rather complex molecules in a free state or in a crystalline environment. The theoretical basis for the fragment calculation may be seen from examination of the theoretical basis for quantum crystallography.

†Permanent address: Geo-Genters, Inc., 10903 Indianhead Highway, Ft. Washington, MD 20744-4018,USA

G. Tsoucaris (ed.), Current Challenges on Large Supramolecular Assemblies, 1–5.

As part of the development of the practical aspects of quantum crystallography, there are special treatments for the experimental data. They concern, for example, calculations to remove vibrational effects from the data and a statistical method for detecting and correcting systematic errors.

QCr proceeds by optimizing the agreement between the structure factor magnitudes from experiment and the structure factor magnitudes obtained from a quantum mechanical model. The only crystallographic information included in the quantum mechanical model concerns the use of atomic positions known from experiment. The fitting process allows for the possibility that the atomic positions may incur some minor adjustments. As noted previously, the Q.M. model has adjustable parameters including those of a projector matrix P that must obey strict conditions, namely,

Idempotency: $\mathbf{P}^2 = \mathbf{P}$

and Normalization: trace $\mathbf{P} = \mathrm{N}$

where N is the number of doubly occupied orbitals. These conditions ensure single determinant N-representability. N-representability ensures quantum mechanical validity.

We have presented a general formulation of quantum crystallography which is valid for all 230 space groups, accounting specifically for the symmetry [1]. It can be applied with atomic orbital or symmetry orbital bases. The key features of the presentation concern the quantum mechanical description of an electron density distribution, the definition of a projector matrix which is part of the description and a general expression for the Fourier transform of the electron density distribution that yields the theoretical values for the structure factors. Thus, the theory is in place for making applications of interest. Descriptions of our studies concerning applications of quantum crystallography are being prepared for publication.

2. **Fragment Calculations**

In the course of developing the mathematical theory for quantum crystallography, the following equations appeared

$$\mathbf{P} = \mathbf{R}\,\mathbf{S}$$

$$\mathbf{S} = \int \psi\psi^{\dagger}\, d\mathbf{r}$$

where $\mathbf{P}$ is a projector matrix, a key component of the quantum mechanical model of the electron density, $\mathbf{R}$ is a matrix product $\mathbf{C}^{\dagger}\mathbf{C}$ where $\mathbf{C}$ is a matrix of LCAO coefficients, $\mathbf{S}$ is the overlap matrix composed of integrals over all space of products of pairs of orbitals. The use of fragments is based on the fact that the overlap of pairs of orbitals drops off greatly with increase in distance between the orbital centers.

A structure that has been used to test the accuracy of *ab initio* fragment calculations is leu^1-zervamicin [2], an antibiotic peptide consisting of 16 amino acid residues. Leu^1-zervamicin is a membrane active peptide that suggests a model for ion-transport through a membrane with an associated gating mechanism on the atomic level [3]. With solvent, the structure investigated consisted of 295 atoms, a number that may be currently handled *in toto* by the *ab initio* programs offered by Gaussian 94 [4]. The combined results of the fragment calculations and the results of the *in toto* calculation were readily compared. In pursuing the fragment calculation, a set of moieties is selected which, when combined, comprise the entire structure. Nineteen in all were selected for the leu^1-zervamicin structure, sixteen were based on the 16 amino acid residues and three more were based on the solvent molecules. A fragment consists of two parts, a kernel of atoms and their neighborhood. The neighborhood of a kernel consists of all the atoms that fall within a preassigned distance from the atoms in the kernel. The greater the distance, the greater the potential accuracy. A value of about 5 Å appears to be suitable. Certain adjustments need to be made to the fragments to assure an even number of electrons. This has been achieved by adding hydrogen atoms or additional atoms from the structure that extend beyond the 5 Å limit. The details of this calculation have been described [2]. By calculating difference isodensity surfaces to compare the *in toto* calculation with the fragment calculation, it was found that the maximum errors, which occurred in only minute volumes, was about 1.2×10^{-3} electrons per cubic angstrom. This calculation involved the use of the simple basis functions, STO3g. The more sophisticated bases should give similar results. With development of programs for *ab initio* quantum mechanical calculations on parallel computers, it should be readily possible to carry out *ab initio* calculations for macromolecules by use of fragments.

3. **Nitrocubanes**

Ab initio quantum mechanical calculations on a series of nitrocubanes have been carried out as a preliminary step before the application of quantum crystallographic adjustments in order to optimize the fit of the quantum mechanical models to the experimental data. The molecules of interest consisted of cubane, and the 1,4-dinitro, 1,3,5,7-tetranitro, 1,2,3,4,6,7-hexanitro, 1,2,3,5,6,7-hexanitro, and octanitrocubanes. Three types of quantum mechanical calculations were made to compare with the average C-N and C-C interatomic distances found in crystallographic structure determinations. The 1,2,3,4,6,7-hexanitrocubane and octanitrocubane have not been synthesized and quantum mechanical calculations on them were based on starting structures suggested by those that were known. The purely quantum mechanical calculations involved Hartree-Fock

SCF with an STO3g basis, Hartree-Fock SCF with an STO6-31g(d,p) basis and density functional theory with BLYPSTO6-31g(d,p). For the average C-N distances, the HFSTO3g calculations were far off, being about 0.05 Å too large, the BLYPSTO6-31g(d,p) calculations were about 0.02 Å too large and the HFSTO6-31g(d,p) calculations were about 0.015 Å too small. For the C-C distances, the STO3g calculations were about 0.01 Å too large, the BLYPSTO6-31g(d,p) calculations ranged from 0.02-0.03 Å too large and the HFSTO6-31g(d,p) ranged from about 0.003 Å too large to 0.009 Å too small. The structures for the theoretical calculations were obtained by determining the atomic coordinates that were consistent with energy optimization for the various models. Although the Hartree-Fock calculation was in best agreement with the experimental values for the interatomic distances, the density functional theory gave lower energies in the energy optimization calculations. Agreement with the interatomic distances may be a useful clue for choosing a starting projector matrix for the application of quantum crystallographic adjustments to the experimental structure factor magnitudes.

A number of purely quantum mechanical calculations were made for the various substituted cubanes, based on the results of the energy optimizations. They included the average number of electrons associated with each type of atom present, from which estimates of the charges on the atoms readily follow. Also calculated were total molecular energies, energy differences between molecules with successively increasing nitro substitutions, and energy densities of the crystalline forms of dinitro- and tetranitrocubane. This work is currently being prepared for publication.

4. Maleic Anhydride

Maleic anhydride has played a valuable role in facilitating our development of the details for carrying out quantum crystallographic calculations.

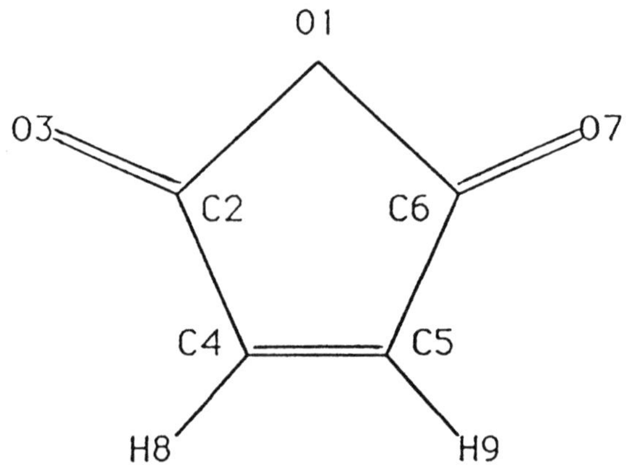

A particularly useful feature is the plane of symmetry that cuts through the molecule in such a way that the charges on both sides of the plane should be equivalent, except, perhaps, for rather minor differences arising from the influences of surrounding molecules. Achieving this in the quantum crystallographic calculations was a valuable learning experience and led to an appreciation of the distinctions between various quantum mechanical models, the importance of having high quality experimental data, the need to examine diffraction data for systematic errors and a statistical method for correcting them when they occur.

The optimization of the calculations on maleic anhydride were guided by the measures of agreement of the fit between the experimental structure factor magnitudes and those obtained from the quantum mechanical model. Maleic anhydride crystallizes in space group $P2_12_12_1$ with four molecules in the unit cell. The experimental data included 507 independent structure factor magnitudes which were corrected for thermal effects and systematic errors. The quantum mechanical model involved the initial use of the basis, 6-31g(d,p). The quantum crystallographic calculations are characterized by a least-squares procedure for optimizing the fit between the experimental data and the model. As a consequence of these calculations, the standard constants in the Gaussian functions that form the orbitals are, in general, altered, the values of the elements of the projector matrix may change and the atomic coordinates also may change. For maleic anhydride, the resulting R-factor as a measure of the agreement between experimental and theoretical structure factor magnitudes is approximately 1%.

It should be noted that the end product of quantum crystallographic calculations are static wave functions. The calculations with maleic anhydride indicate that such wave functions can be obtained in good agreement with crystallographic data. They can be used for making a variety of quantum mechanical calculations that concern, for example, various energies and charge distributions. The details of the study of maleic anhydride are being prepared for publication. (Research supported by the Office of Naval Research.)

References

1. L. Massa, L. Huang and J. Karle (1995) Quantum Crystallography and the Use of Kernel Projector Matrices, *International Journal of Quantum Chemistry: Quantum Chemistry Symposium* **29**, 371-384.

2. L. Huang, L. Massa and J. Karle (1996) Kernel Projector Matrices for Leu[1]-Zervamicin, *International Journal of Quantum Chemistry, Quantum Chemistry Symposium* **30**, 1691-1700.

3. I. L. Karle, J. L. Flippen-Anderson, S. Agarwalla and P. Balaram (1994) Conformation of the Flexible Bent Helix of Leu[1]-Zervamicin in Crystal C and a Possible Gating Action for Ion Passage, *Biopolymers* **34**, 721-735.

4. M. J. Frisch, G. W. Trucks, H. B. Schlegel, P. M. W. Gill, B. G. Johnson, M. A. Robb, J. R. Cheeseman, T. Keith, G. A. Petersson, J. A. Montgomery, K. Raghavachari, M. A. Al-Lanham, V. G. Zakrzewski, J. V. Ortiz, J. B. Foresman, J. Cioslowski, B. B. Stefanov, A. Nanayakkara, M. Challacombe, C. Y. Peng, P. Y. Ayala, W. Chen, M. W. Wong, J. L. Andres, E. S. Replogle, R. Gomperts, R. L Martin, D. J. Fox, J. S. Binkley, D. J. Defrees, J. Baker, J. P. Stewart, M. Head-Gordon, C. Gonzalez, and J. A. Pople (1995) *Gaussian 94, Revision C.4*, Gaussian, Inc., Pittsburgh, PA.

LARGE MOLECULAR ASSEMBLIES HELD TOGETHER BY NON-COVALENT BONDS

J. L. Atwood, L. R. MacGillivray, K. N. Rose,
L. J. Barbour, K. T. Holman, and G. W. Orr
Department of Chemistry
University of Missouri-Columbia
Columbia, MO 65211

One of the goals of supramolecular chemistry is the construction of very large assemblies, assemblies which may shed light on those found in the life processes. Herein we describe two quite large assemblies, but those which are amenable to characterization in solution by NMR techniques and in the solid state by X-ray crystallography. This relative ease of characterization is due to the extraordinally high symmetry, parcticularly of the larger assembly.

Spontaneous self-assembly processes leading to discrete spherical arrays are common in Nature.[1, 2] An important feature of these shell-like architectures is their ability to encapsulate neutral and/or charged guests whose size, shape, and chemical exteriors complement their inner surfaces.[3,4] Indeed, the field of supramolecular chemistry has recently witnessed the emergence of monomeric[5,6,7] and supramolecular dimeric molecular capsules,[8,9] many of which have been based upon the head-to-head alignment of bowl-shaped polyaromatic macrocycles such as calix[4]arenes.[5,7,9] However, structural mimicry of frameworks akin to viruses and fullerenes, which are based upon the self-assembly of $n > 3$ subunits where surface curvature is supplied by edge sharing of regular polygons, has remained elusive. Here we discuss the first example of such a system, a chiral spherical molecular assembly held together by 60 hydrogen bonds, **1**.[10] We demonstrate the ability of **1**, which consists of six calix[4]resorcinarenes, **2,** and eight water molecules, to maintain its structure in apolar media and to encapsulate guest species within a well-defined cavity that possesses an internal volume of ca. 1375 Å^3. A single crystal X-ray crystallographic analysis has revealed that the topology of **1** resembles a spherical virus[1] and conforms to the structure of a snub cube, one of the thirteen Archimedean solids.[11] Further, we reveal the non-covalent synthesis of dimeric entity **3**,[12] a structure held together by sixteen hydrogen bonds and one which mimics a carcerand.[5]

Interest in **2** lies in its ability to function as a multiple hydrogen bond donor in the solid state.[13] During experiments aimed at co-crystallizing *C*-methylcalix[4]resorcinarene, **2a**, with hydrogen bond acceptors in aromatic solvents, we have discovered the ability of **2a** to self-assemble as a spherical hexamer, along with adventitious water molecules, to form **1**. Subsequent solution studies have revealed the ability of *C*-undecylcalix[4]resorcinarene, **2b**, to maintain the structure of the spheroid in apolar organic solvents. We discuss here the synthesis, X-ray crystal structure, and solution behaviour of **1**.

A cross sectional view of the X-ray crystal structure of **1** is shown in Fig. 1. The assembly consists of six molecules of **2a** and eight

G. Tsoucaris (ed.), Current Challenges on Large Supramolecular Assemblies, 7–16.

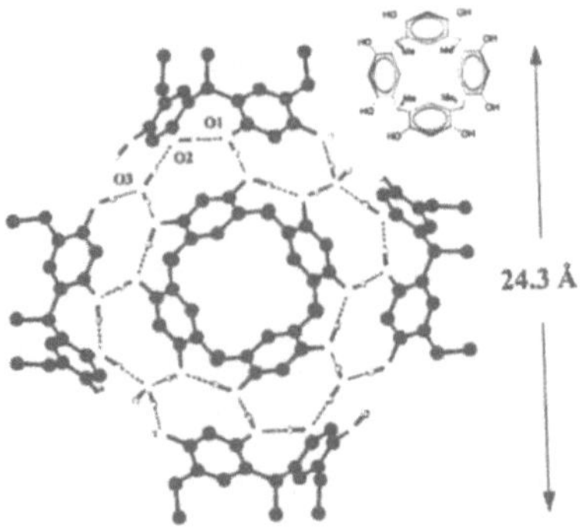

Figure 1. The structure of **1** in cross-section (inset: structural formula for **2a**).

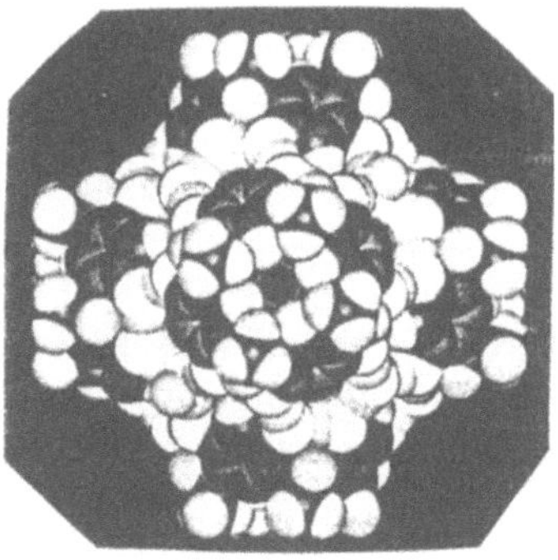

Figure 2. The structure of **1** viewed along the crystallographic 4-fold rotation axis.

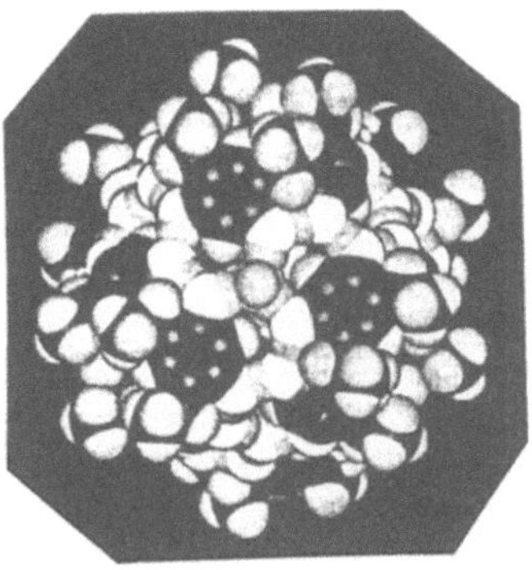

Figure 3. The structure of **1** viewed along the crystallographic 3-fold rotation axis.

molecules of water, assembled *via* 60 O-H...O hydrogen bonds, to form a shell-like octahedral cubic spheroid. The calixarenes point their hydroxyl groups along the periphery of **1** and form two hydrogen bonds to two neighbouring calixarenes (O1...O1', 2.73(1) Å) while the water molecules are embedded along the surface of **1** such that they lie on the vertices of a cube (edge length = 9.00 Å) and participate in three hydrogen bonds with three different calixarenes (O2...O3, 2.73(1) Å). Each calixarene also exhibits four intramolecular hydrogen bonds, one at each of its corners (O1...O2, 2.64(1) Å), which impart stability to its bowl-like conformation. As a result, **1** possesses *4 3 2* symmetry (Figs. 2-4), ignoring all hydroxyl hydrogen atoms, and a well-defined central cavity with a maximum diameter of 17.7 Å and an internal volume of *ca.* 1375 $Å^3$ (Fig. 5).[14] Indeed, the cavity of **1** is vast, being more than 4.5 times larger than the cavity of the largest molecular capsule reported to date (*ca.* 300 $Å^3$).[8] Interestingly, the calixarenes of **1** are twisted by 22.5° with respect to the faces of the water "cuboid" which, as a consequence, makes **1** chiral.

A view displaying the solid state packing of **1** reveals that neighbouring spheroids fall on their four-fold axes and their methyl groups lie staggered, owing to the twisting displayed by the macrocycles (Fig. 6). This gives rise to an interpenetrating body-centred cubic lattice that exhibits interstices between **1** that are occupied by disordered water molecules, each of which participates in two hydrogen bonds with two water molecules from adjacent spheroids, and nitrobenzene molecules, each of which participates in face-to-face π-π interactions with two benzene moieties from two different calixarenes. Despite being able to locate electron density corresponding to guest species within the interior of **1**, it is not possible to determine their identity from the X-ray experiment presumably due to the high symmetry imposed by **1** and high thermal motion within the cavity.[3]

Consultation of polyhedron models has revealed the structure of **1a** to conform to a snub cube (Fig. 7), one of the thirteen Archimedean solids, in which the vertices of the square faces correspond to the corners of the calixarenes and the centroids of the eight triangles that adjoin three squares correspond to the eight water molecules.[10,11] Note that **1** is the first example of a closed surface supramolecular spheroid that possesses $n > 3$ subunits and, owing to the fit exhibited by its components, possesses a topology that agrees with the theory of virus shell structure which states that such octahedral systems must contain 24 asymmetric units and possess *4 3 2* symmetry.[1] Moreover, such observations illustrate that in order to design related spherical hosts, one must consider the limited number of possibilities available in space for such frameworks, those being the five Platonic (regular) solids and the thirteen Archimedean (semi-regular) solids.[11] The ramifications of these observations are important since they suggest that it should be possible to select appropriately sized, shaped, and functionalized components, *a priori*, to design similar spherical hosts in a way similar to the self-assembly exhibited by **1**.

The hydrogen bond pattern that holds **1** together is complex. Although the hydroxyl hydrogen atoms were not locateable, it is possible to deduce the pattern knowing the positions and numbers of hydrogen bond donors and acceptors. In principle, 64 hydrogen bond donors and 112 hydrogen bond acceptor sites are available to form **1**. There are, however, two restrictions that arise from the positions of these sites that require the number of hydrogen bonds that define **1** to equal 60 which, as a result,

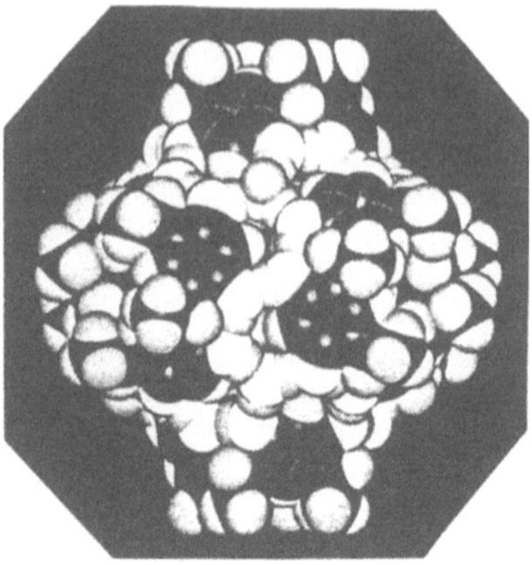

Figure 4. The structure of **1** viewed along the crystallographic 2-fold rotation axis.

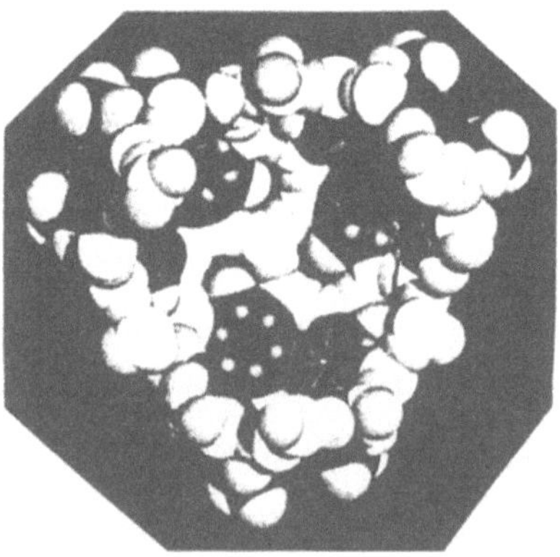

Figure 5. Cut-away view of the structure of **1** viewed along the crystallographic 3-fold rotation axis.

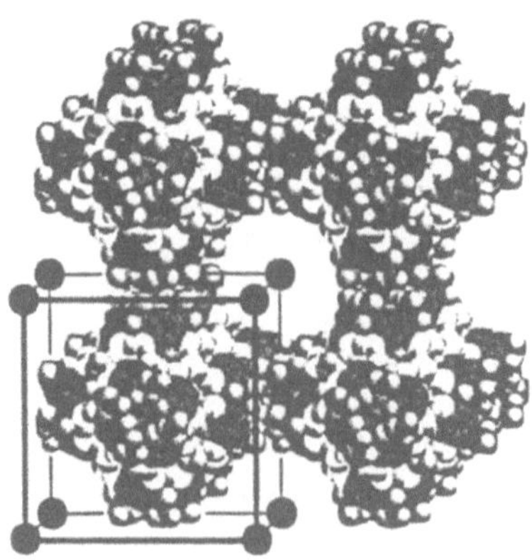

Figure 6. Solid state packing of **1**, where the solid grey spheres also represent **1**.

cause four hydroxyl hydrogen atoms to point away from its surface. The first restriction lies in the positions of the water molecules. Their location on the three fold axes limits the number of hydrogen bonds in which the water molecules can participate that contributes to **1** to 24; in other words, each water molecule is only capable of participating in three hydrogen bonds along the surface of **1**. The second restriction arises from the presence of the four intramolecular hydrogen bonds at the corners of the calixarenes. In particular, these interactions force four hydroxyl groups to point their hydrogen atoms above each macrocycle which, as a consequence, makes each calixarene a quadruple hydrogen bond donor.[15] Since each calixarene uses two hydroxyl hydrogen atoms to form two hydrogen bonds to two calixarenes, this leaves two hydroxyl hydrogen atoms per calixarene which are used to form 12 hydrogen bonds to the water molecules. Consequently, in order to form the 12 hydrogen bonds, the water molecules, owing to their positions on the three-fold axes, must position four hydrogen atoms, each from a different water molecule, away from the surface of **1**. To complete the hydrogen bond array, the water molecules form 12 hydrogen bonds to 12 hydrogen bond acceptor sites located on the calixarenes.

A closer inspection of the hydrogen bond pattern of **1** reveals that each edge of the water cuboid consists of a polar chain of five cooperative hydrogen bonds (Fig. 8). As a result, each edge is capable of changing the sense of its direction *via* interconversion of two water molecules from a hydrogen bond donor to a hydrogen bond acceptor and vice versa. Furthermore, since four water molecules point four hydrogen atoms away from the surface of **1**, the edges of the cuboid are capable of undergoing such proton rearrangements simultaneously, which, in effect, makes **1** a spherical proton pump able to channel the four dangling protons to the eight corners of the water cuboid.

Although each of the dangling protons may be placed at any one of the eight corners of the water cuboid, it is most convenient, for symmetry reasons, to place them at the vertices of a tetrahedron such that the cuboid, ignoring the calixarenes, possesses D_{2d} symmetry (Fig. 9).[16] This arrangement furnishes the cuboid with four chiral faces and two achiral faces which, upon completing the hydrogen bonds along each edge, gives rise to three distinct calixarenes, two of which are chiral (C_4) and one of which is achiral (C_{2v}) (Fig. 10). Notably, the chiral calixarenes are associated with the chiral faces such that their handedness matches while the achiral calixarenes are associated with the achiral faces. As a consequence of this arrangement, **1** possesses D_2 symmetry.[11]

Evidence supporting **2b** to maintain the structure of **1** in solution has been obtained *via* 1D and 2D ^{1}H NMR measurements. Indeed, the spectrum of **2b** in benzene-d_6, at increasing concentrations, displays resonances attributed to the chiral and achiral calixarenes as well as the eight water molecules. Indeed, the NMR observations, coupled with a molecular mass determination in benzene (7066 g mol^{-1}),[16] provide convincing evidence supporting ***2b** to maintain the structure of the spheroid in solution.*

Calix[4]resorcinarene **2c**, C-phenylethylcalix[4]resorcinarene, can form a spherical entity as for **1**, but in the presence of solvent molecules such as propan-2-ol, which can act simultaneously as hydrogen bond donors and acceptors, **2c** can also self-assemble to form a dimeric carcerand-like complex. Utilizing an intricate array of hydrogen bonds between solvent and calix[4]resorcinarene, pairs of concave molecules associate indirectly in a rim-to-rim fashion to form a 230 Å^3 supramolecular

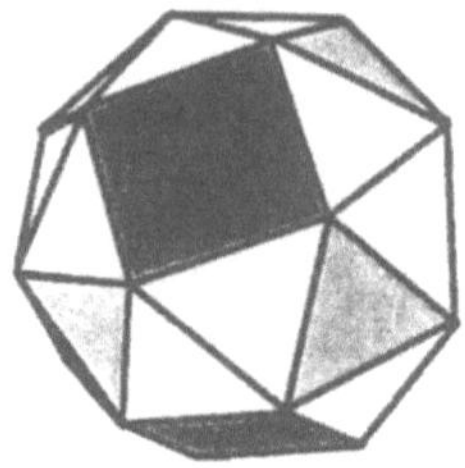

Figure 7. The snub cube, one of the 13 Archimedean solids. The square faces correspond to the calixarenes; the eight shaded triangels that adjoin three squares correspond to the water molecules of **1**.

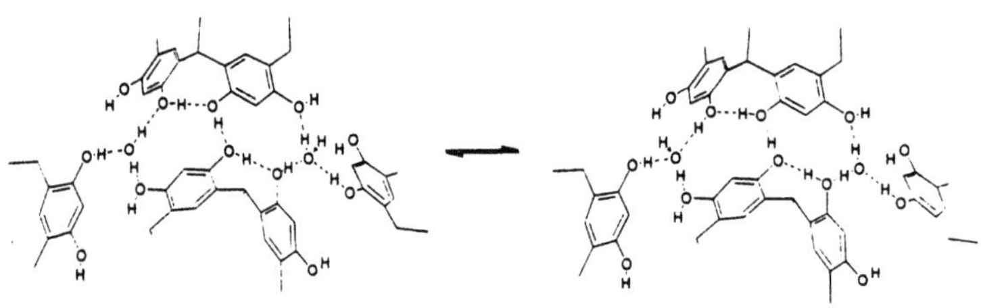

Figure 8. The five hydrogen bonds that comprise each edge of the water cuboid.

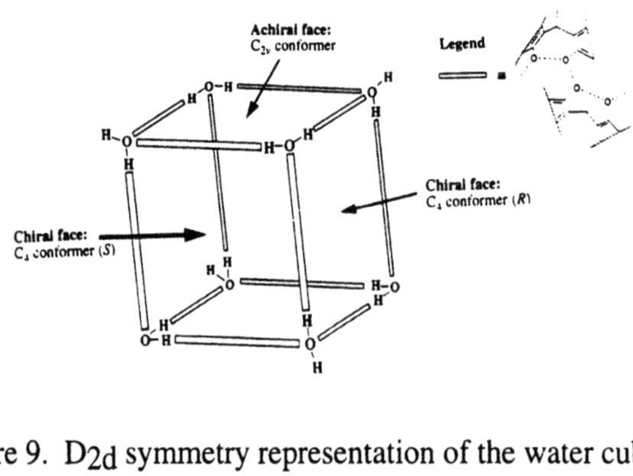

Figure 9. D_{2d} symmetry representation of the water cuboid.

cavity.[12] It has been subsequently demonstrated that water can function, as expected, in the role of propan-2-ol.[18]

We are keenly interested in the supramolecular complexation of fullerenes,[19] and have been investigating the possibility of encapsulating C_{60} within a calix[4]resorcinarene hexamer. Since a solitary C_{60} molecule is too small to fill the void efficiently, the complex would also require encapsulation of a significant amount of solvent in order to stabilize the structure. During the course of our attempts at assembling such a system, we have instead produced a dimeric carcerand-like complex in which two concave calix[4]resorcinarene molecules are linked indirectly by hydrogen bond bridges involving eight propan-2-ol solvent molecules. Crystallographic characterization of a dimeric system of this nature is unprecedented and is an important extension of the work initiated by Rebek.[8,9]

Dark red crystals suitable for X-ray diffraction analysis were grown by slow diffusion of propan-2-ol into a solution of **2c** and C_{60} (5:1 molar ratio) in *o*-dichlorobenzene.[20] The most striking feature of the structure is that pairs of concave calix[4]resorcinarene molecules are arranged in a rim-to-rim fashion to form dimers as shown in Figure 11. The eight hydroxyl atoms of one calix[4]resorcinarene form hydrogen bonds with eight oxygen atoms belonging to propan-2-ol solvent molecules. The latter, in turn, form hydrogen bonds to a second calix[4]resorcinarene molecule, thus completing the dimer. No hydroxyl hydrogen atoms were located and hydrogen bonds are inferred from short O···O contacts. The proximate hydroxyl O···O contacts within each calixresorcinarene are 2.771(7) Å, implying that these oxygen atoms are also hydrogen bonded to one another. The unique calix[4]resorcinarene to solvent O···O distances are 2.660(9) and 2.743(9) Å. The two calix[4]resorcinarene molecules each have C_{4v} symmetry and are related to one another by a mirror plane at *x*,*y*,0 which passes through all of the propan-2-ol oxygen atoms. The assembly of the dimer is thus facilitated by the formation of sixteen intermolecular hydrogen bonds while a further eight intramolecular hydrogen bonds impart structural rigidity of the calix[4]resorcinarene molecules. The symmetry relationship between the constituents of the dimer requires all of the hydrogen atoms involved in hydrogen bonding to be disordered.

The effective van der Waals volume of the cavity was calculated[14] to be 230 $Å^3$ and its cross section, measured at *x*,*y*,0, is approximately equal in dimensions to a molecule of *o*-dichlorobenzene. In order to maintain structural stability, it is probable that the cavity contains several solvent molecules, perhaps of both *o*-dichlorobenzene and propan-2-ol. Relatively large peaks of electron density within the cavity appear to indicate the presence of chlorine atoms, but crystallographically imposed symmetry precludes the use of a cogent model for included solvent molecules. As shown in Figure 12, the calix[4]resorcinarene dimers are stacked in columns at 0,0,*z* and 0.5,0.5,*z* while the C_{60} molecules are similarly arranged along 0,0.5,*z* and 0.5,0,*z*. The calix[4]resorcinarene columns involve alternating head-to-head and tail-to-tail associations of the molecules. The latter result in the formation of lattice voids bounded by phenethyl moieties belonging to the host and also appear to contain disordered *o*-dichlorobenzene solvent.

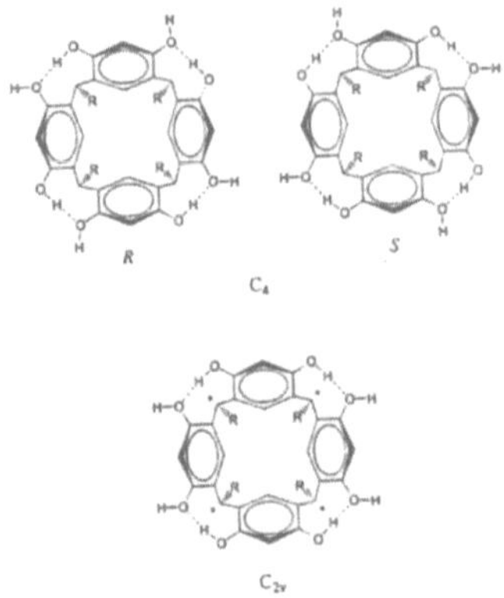

Figure 10. The three calixarene conformers that are associated with the faces of the water cuboid.

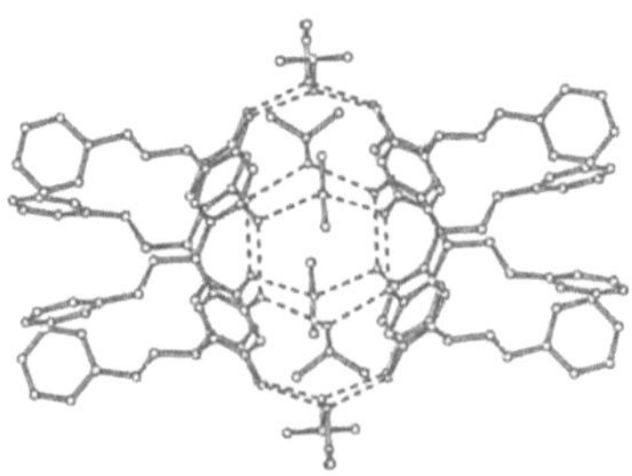

Figure 11. Projection showing the calix[4]resorcinarene dimer. Hydrogen atoms are omitted for clarity and hydrogen bonded interactions are shown as broken bars.

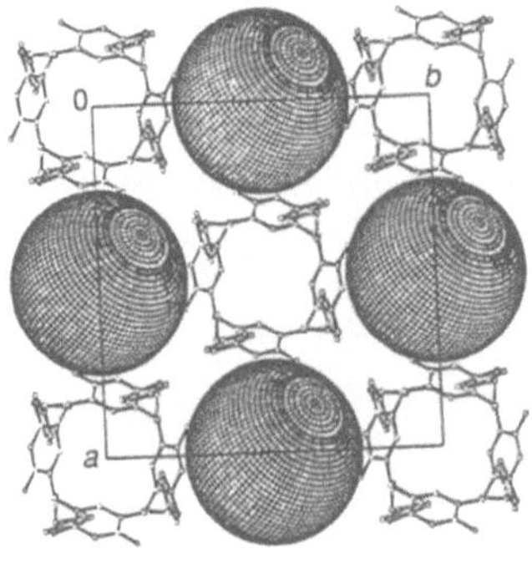

Figure 12. Packing diagram (only calix[4]resorcinarene and C_{60} molecules are shown) viewed along [001], parallel to the molecular columns.

We have shown that the solvent-assisted hexameric assembly of calix[4]resorcinarenes to form carcerand-like complexes has a dimeric analogue. Characterization of such systems is a significant step towards understanding the principles involved in the design of supramolecular cavities of diverse shapes and sizes and assemblies of this nature are important because, like their biological counterparts, the contents of large molecular voids are often considered[21] to represent an additional phase of matter.

ACKNOWLEDGEMENTS. We are grateful for funding from the National Science Foundation (NSF) and the Natural Sciences and Engineering Research Council of Canada (NSERC) for a research fellowship (L. R. M.).

References

1. Caspar, D., and Klug, A. (1962) *Cold Spring Harb. Symp. Quant. Biol.* **27**, 1.

2. Kroto, H. W., Heath, J. R., O'Brien, S. C., Curl, R. F., and Smalley, R. E. (1985) *Nature* **318**, 162.

3. Casjens, S. Nucleic acid packaging by viruses, in *Virus Structure and* 1995).

4. Schwarz, H., Weiske, T., Böhme, D. K., Hrusák, J. Exo- and endohedral fullerene complexes in the gas phase, in *Buckminsterfullerenes* (eds. Billups, W. E. & Ciufolini, M. A.) 257-283 (VCH, New York, 1993).

5. Sherman, J. C., and Cram, D. J. (1989) *J. Am. Chem. Soc.* **111**, 4527.

6. Garel, L., Dutasta, J.-P., and Collet, A. (1993) *Angew. Chem. Int. Ed .Engl.* **32**, 1169.

7. Timmerman, P., Verboom, W., van Veggel, F. C., van Duynhoven, J. P., and Reinhoudt, D. N. (1994) *Angew. Chem. Int. Ed .Engl.* **33**, 2345.

8. Kang, J., and Rebek, J. Jr. (1997) *Nature* **385**, 50.

9. Shimizu, K. D., and Rebek, J. Jr. (1995) *Proc. Natl. Acad. Sci. USA* **92**, 12403.

10. MacGillivray, L. R., and Atwood, J. L. (1997) *Nature* **389**, 469.

11. Wenninger, M. J. *Polyhedron Models* (Cambridge University Press, New York, 1971).

12. Rose, K. N., Barbour, L. J., Orr, G. W., and Atwood, J. L. (1998) *Chem. Comm.*, 407.

13. MacGillivray, L. R., and Atwood, J. L. (1997) *J. Am. Chem. Soc.* **119**, 6931.

14. Barbour, L. J., *CAVITY*, (unpublished), A program to compute the volume available to a sphere of given radius within a molecular cavity. coordinates of all the atoms surrounding the cavity. A cube of length *j* the atoms. If the cube can reside within any sphere of radius *r*, which were used. van der Waals radii were obtained from Bondi, A. (1964) *J. Phys. Chem.* **68**, 441, University of Missouri-Columbia, USA (1997).

15. Wales, D. J., and Ohmine, I. (1993) *J. Chem. Phys.* **98**, 7257.

16. Mathiowitz, E., Jacob, J. S., Jong, Y. S., Carino, G. P., Chickering, D. E., Chaturvedi, P., Santos, C. A., Vijayaraghavan, K., Montgomery, S., Bassett, M., and Morrell, C. (1997) *Nature* **386**, 410.

17. Aoyama, Y., Tanaka, Y., Sugahara, S. (1989) *J. Am. Chem. Soc.* **111**, 5397.

18. Murayama, K., and Aoki, K. (1998) *Chem. Comm.*, 607.

19. Atwood, J. L., Koutsantonis, G. A., and Raston, C. L. (1994) *Nature* **368**, 229; Atwood, J. L., Barnes, M. J., Gardiner, M. G., and Raston, C. L., (1996) *Chem. Comm.* 1449; Raston, C. L., Atwood, J. L., Nichols, P. J., and Sudria, I. B. N., (1996) *Chem. Comm.*, 2615; Barbour, L. J., Orr, G. W., and Atwood, J. L. (1997) *Chem. Comm.*, 1439.

20. Direct methods yielded all unique calix[4]resorcinarene non-hydrogen atoms and the structure was expanded by difference Fourier synthesis. The structure consists of alternating columns of hydrogen bonded calixarene dimers and C_{60} molecules. The latter are situated on positions of relatively high symmetry, i.e. the center of each C_{60} molecule lies on the intersection of a 4_2 screw axis and an *n*-glide plane. Careful inspection of difference electron density maps revealed that the fullerene is orientationally disordered over two distinct positions. The disorder could be modeled accurately by introducing two half-occupancy, superimposed C_{60} molecules (coordinates for the idealized C_{60} molecule were retrieved from the Cambridge Crystallographic Database) into the model. Two unique peaks of significant electron density were located within the cavity of the dimer. These peaks are presumably due to the presence of a disordered *o*-dichlorobenzene molecule and were assigned, according to their relative intensities, as carbon and chlorine atoms and refined isotropically. A similar model was used to account for a pair of peaks located in the lattice, outside the cavity. Both disordered models for *o*-dichlorobenzene are situated on positions of 4/m site symmetry.

21. Jacopozzi, P. and Dalcanale, E. (1997) *Angew. Chem. Int. Ed. Engl.* **36**, 613.

Correspondence and requests for materials should be addressed to J.L.A. (e-mail: chemja@showme.missouri.edu).

SYNTHETIC SUPRAMOLECULAR CHEMISTRY

S.J. CANTRILL, M.C.T. FYFE, F.M. RAYMO, J.F. STODDART
Department of Chemistry and Biochemistry
University of California at Los Angeles
405 Hilgard Avenue, Los Angeles CA 90095-1569, USA

1. Organic Synthesis for the Next Century

For many decades, the construction of organic compounds in the laboratory has relied on the remarkable abilities of the 20th century 'alchemists' — namely, synthetic organic chemists — to make and break covalent bonds. Careful selection of functional groups and reaction conditions, in conjunction with protection/deprotection protocols, constitute the 'secrets' and 'tricks' of their 'art' which is commensurate with 'traditional' organic synthesis [1,2]. Indeed, relying on multistep reaction sequences, the total syntheses of structurally intricate molecular compounds which are constructed entirely using covalent bonds — *e.g.*, brevetoxin B [3], palytoxin [4], and the calichearubicins [5] — have been realized in recent times. These very elegant and successful syntheses have required enormous intellectual and hands-on effort by large teams of chemists over rather long periods of time — very often, several years. Moreover, these extremely complex, and often particularly beautiful, examples represent close to state-of-the-art as far as 'traditional' organic synthesis is concerned. Alas, they also highlight the difficulties and limitations associated with classical organic syntheses — specifically, that the multistep aspect of such syntheses can be extremely laborious and time-consuming. With the possible exception of some dendritic structures [6], it is becoming apparent that the construction of nanoscopic structures, of the same complexities as those found in biological systems, using these classical methods is out of the reach of even the most talented and optimistic of the 20th century chemists!

Fortunately, the advent of supramolecular chemistry [7] — another form of 'magic art', but on an even larger scale — will significantly broaden the number of 'secrets' and 'tricks' available to the 'alchemists' of the 21st century. Noncovalent bonding interactions can be utilized to drive the construction of supermolecules [8] that can associate with one another to form species of an even higher level of complexity, *viz.*, supramolecular arrays and macroscopic conglomerates, with amazing degreees of control and with extremely high efficiencies. In the example illustrated in Figure 1, a π-electron

G. Tsoucaris (ed.), Current Challenges on Large Supramolecular Assemblies, 17–35.

deficient cyclophane, incorporating bipyridinium recognition sites, binds [9] a π-electron rich acyclic guest containing a hydroquinone recognition site, as a result of π–π stacking, [C–H···O] hydrogen bonding and [C–H···π] interactions, to form a pseudorotaxane supermolecule. This supermolecule polymerizes noncovalently to form a supramolecular array by virtue of strong hydrogen bonding interactions involving the dimeric matching of its carboxylic acid termini. Furthermore, π–π stacking interactions are established between adjacent supramolecular arrays, ultimately helping to afford the most robust of macroscopic conglomerates — a crystal. Thus, the process evolves spontaneously from the simple host and guest to the final three-dimensional aggregate without requiring any external intervention — all today's 'alchemist' has to do, is to mix the two components in solution and wait for the crystals to appear!

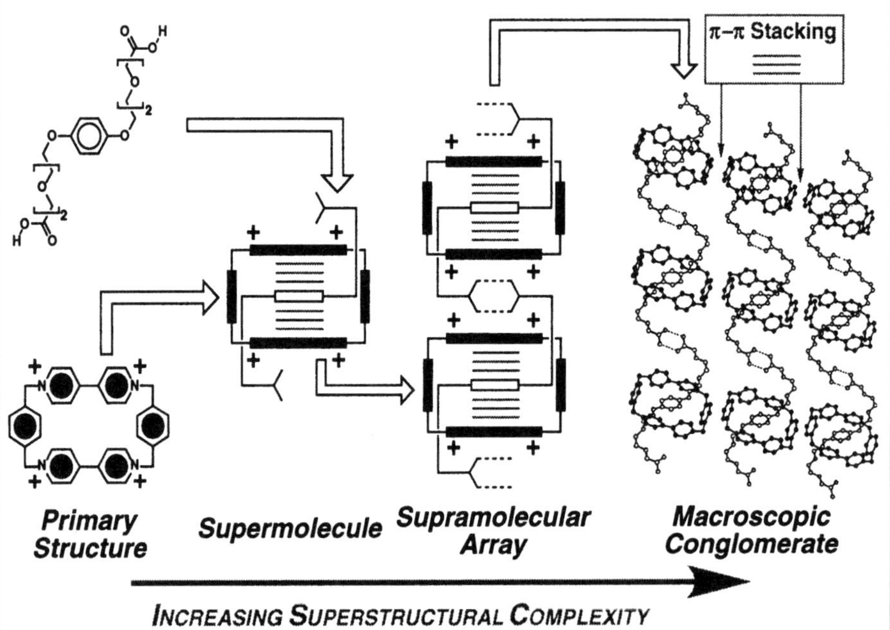

Figure 1. Supramolecular synthesis in action.

The introduction of appropiate recognition sites into simple so-called *tectons* [10] has to rely upon well-established classical synthetic procedures. However, once synthesized, such 'magic' building blocks possess the requisite information needed to self-assemble spontaneously into a supermolecule, which can then, by virtue of its preprogrammed stereoelectronic information, grow further into a supramolecular array and, ultimately, even into a macroscopic conglomerate — *e.g.*, a crystal. Although, the supramolecular synthesis of such polymolecular systems in solution is still something of a challenge, numerous examples of well-defined, self-assembling systems in the solid state are now known [11,12] and the once 'black art' of *crystal engineering* [13] is beginning to become more and more of a respectable and acceptable science.

Two areas of contemporary chemical synthesis have been inspired (Figure 2) by supramolecular chemistry: these are (1) *supramolecular synthesis* and (2) *supramolecular assistance to molecular synthesis*. The impetus for the development of both of these aspects of *synthetic supramolecular chemistry* [14] has been *self-assembly* [15,16]. In supramolecular synthesis, noncovalent bonding interactions are employed to generate multicomponent *supermolecules* and/or *supramolecular arrays* — *i.e.*, the equivalent of small molecules and polymers at the supramolecular level. Supramolecular assistance to molecular synthesis is achieved, again, by employing noncovalent bonding interactions which guide the formation of covalent and/or mechanical bonds to yield discrete *molecular species*.

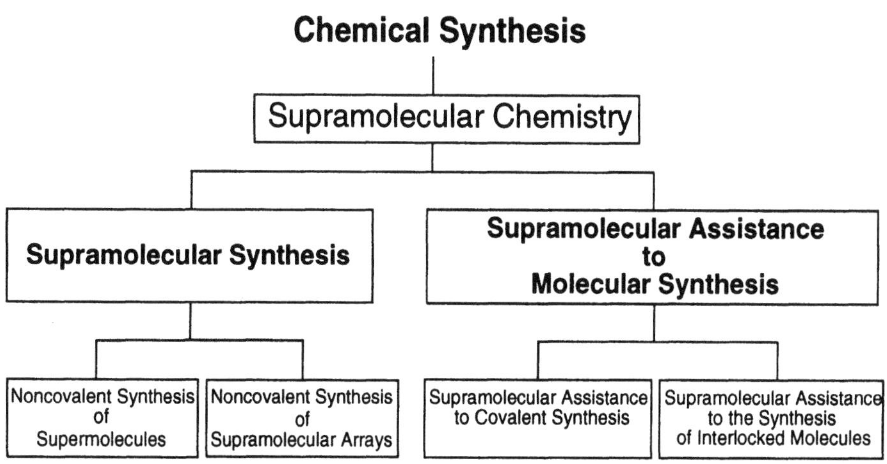

Figure 2. Flow chart illustrating the role of supramolecular chemistry in chemical synthesis.

2. Noncovalent Synthesis of Supermolecules

The macrocyclic polyethers **1–4** (Figure 3) are capable of binding secondary dialkylammonium salts, such as **5**$^{+}$– **8**$^{4+}$, with pseudorotaxane geometries both in solution and in the solid state [17]. In this case, the noncovalent bonding interactions responsible for the molecular recognition event are mainly [N^+–H···O] and [C–H···O] hydrogen bonds, involving the NH_2^+ and the adjacent CH_2 groups as the hydrogen bond donors and the polyether oxygen atoms as the acceptors. In some instances, these hydrogen bonds are supplemented by π–π interactions between the aromatic units of the host and those of the guest.

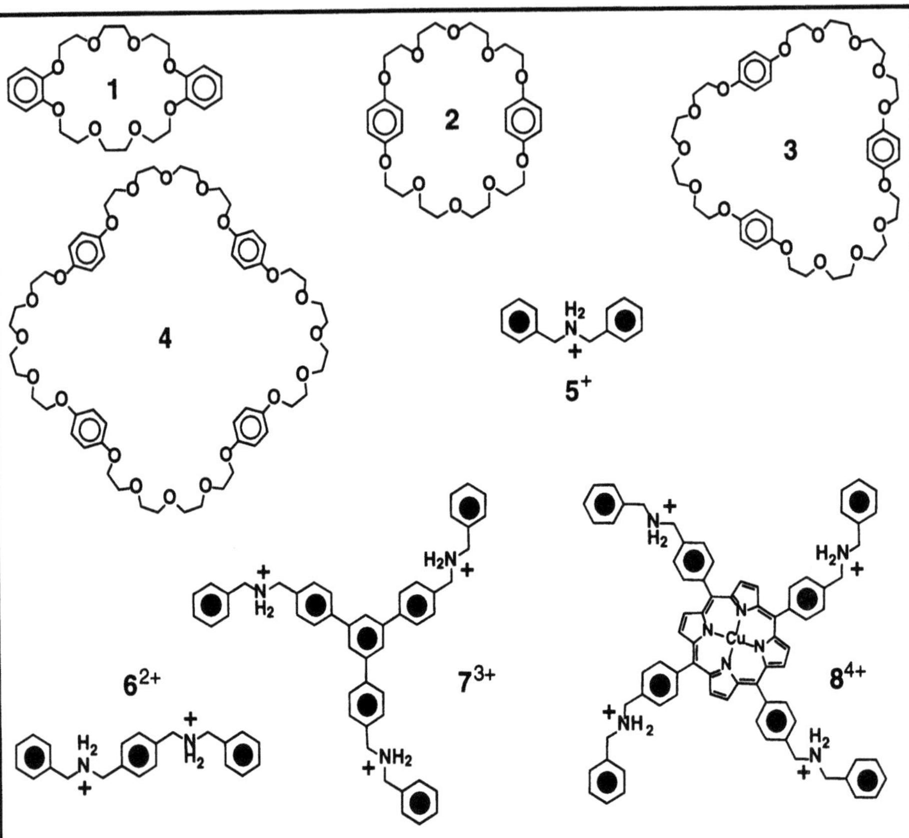

Figure 3. The building blocks — tectons.

The ^1H NMR spectrum of a CD_3CN solution containing the host **1** and the guest **5**·PF_6 in a 1:1 ratio showed (Figure 4) sharp and well-resolved signals for three different species — the uncomplexed host and guest plus a 1:1 complex [18]. The slow kinetics of complexation and decomplexation, together with NOE experiments performed in $CDCl_3$, suggest that the acyclic guest is inserted through the cavity of the macrocyclic host with a rotaxane-like geometry in solution. In particular, by irradiating the α–, β– and γ–OCH_2 protons of the host, NOE enhancement was observed for the resonances associated with the N^+CH_2 protons of the guest and *vice versa*. The association constant (K_a) of this 1:1 complex was determined in a range of solvents by ^1H NMR spectroscopy. In all cases, single point determination of the K_a value was possible as a result of the slow kinetics of complexation and decomplexation on the ^1H NMR timescale. The K_a values range from 360 to 27000 M^{-1} on going from $(CD_3)_2CO$ to $CDCl_3$ at 25 °C and are correlated to the Gutmann donor number of the solvent. This trend reflects the fact that hydrogen bonding interactions are the main driving force for complexation: indeed, when a strongly coordinating solvent, such as $(CD_3)_2SO$, is employed, no complexation is observed at all.

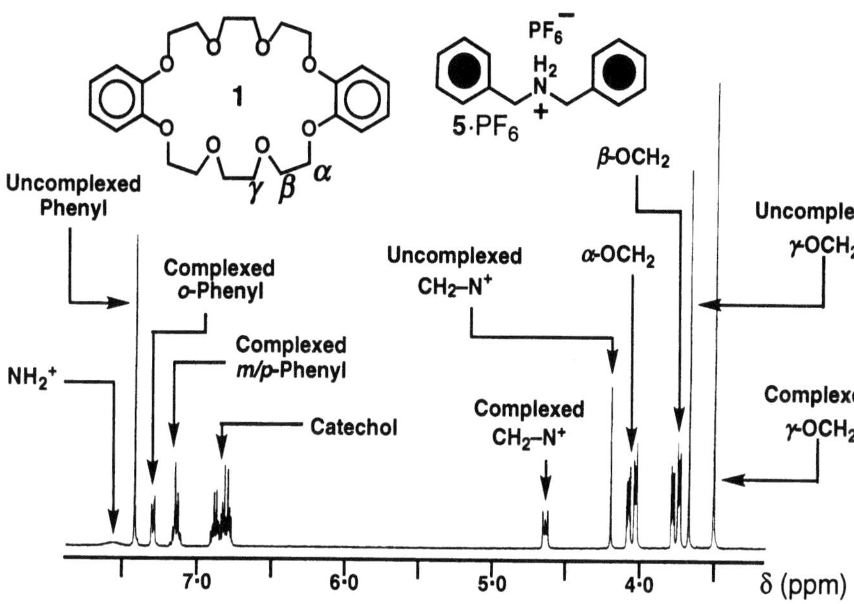

Figure 4. Partial ^{1}H NMR spectrum of a 1:1 mixture of **1** and **5**·PF_6 in CD_3CN at 25 °C.

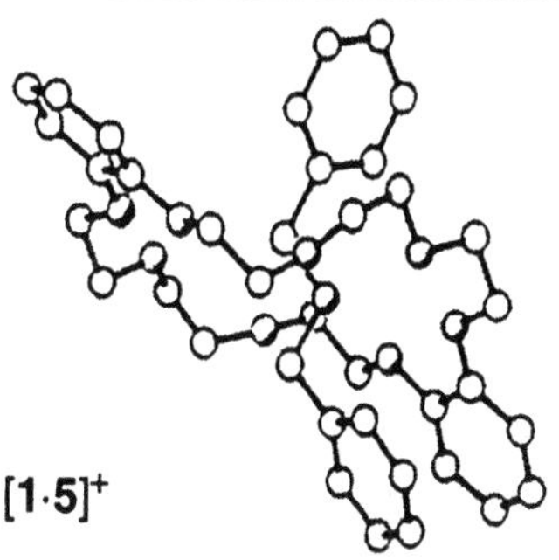

$[\mathbf{1}\cdot\mathbf{5}]^+$

Figure 5. One of the two crystallographically independent 1:1 complexes $[\mathbf{1}\cdot\mathbf{5}]^+$ observed in the unit cell.

Single crystal X-ray analysis of the complex $[\mathbf{1}\cdot\mathbf{5}]^+$ revealed two crystallographically-independent 1:1 complexes in the unit cell: one of them is shown in Figure 5. In both instances, pseudorotaxane geometries with short distances separating three of the polyether oxygen atoms of the host and the NH_2^+ and CH_2 protons of the guest are observed, in agreement with the solution studies.

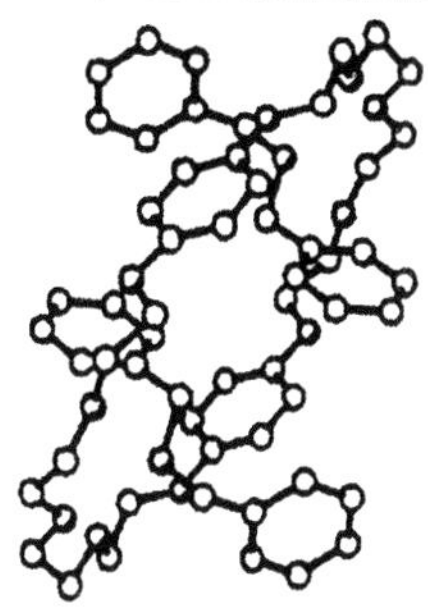

$[\mathbf{2}\cdot(\mathbf{5})_2]^{2+}$

Figure 6. The 1:2 complex $[\mathbf{2}\cdot(\mathbf{5})_2]^{2+}$ formed in the solid state.

The macrocyclic polyether **2** possesses a cavity large enough to accommodate two acyclic guests $\mathbf{5}^+$. Indeed, a single

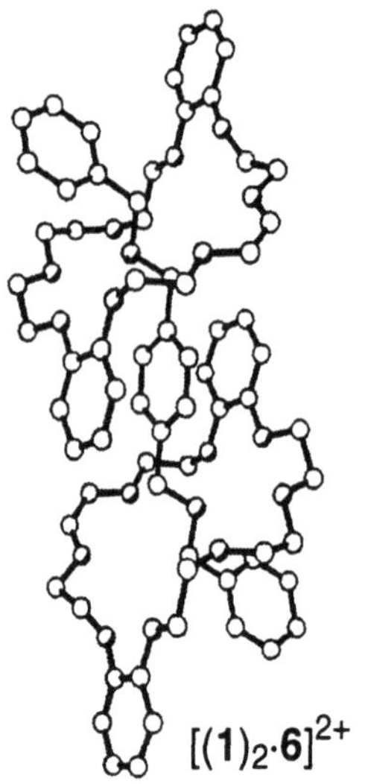

$[(\mathbf{1})_2{\cdot}\mathbf{6}]^{2+}$

Figure 7. Geometry adopted by the 1:2 complex $[(\mathbf{1})_2{\cdot}\mathbf{6}]^{2+}$ in the solid state.

crystal X-ray analysis has revealed (Figure 6) the formation of a 1:2 complex in the solid state [19]. This double-stranded pseudorotaxane is again stabilized by $[N^+\text{–}H\cdots O]$ hydrogen bonds between the NH_2^+ groups of the two guests and the oxygen atoms of the polyether chains. However, in the solid state, no $[C\text{–}H\cdots O]$ hydrogen bonding interactions are observed. ^1H NMR Spectroscopy revealed that the 1:2 stoichiometry is maintained in solution, although, in this case, the complexation and decomplexation processes are fast on the ^1H NMR timescale as a result of the relatively large cavity of the host.

The acyclic guest $\mathbf{6}^{2+}$ possesses two NH_2^+ groups and forms (Figure 7) a 2:1 complex with the macrocyclic polyether **1** in the solid state [19]. In this complex, the thread-like guest is encircled by two macrocyclic hosts as a result of the 'usual' $[N^+\text{–}H\cdots O]$ and $[C\text{–}H\cdots O]$ hydrogen bonds. Additionally, the *p*-xylyl spacer of the guest is 'sandwiched' between two catechol rings — one from each host — as a result of $\pi\text{–}\pi$ stacking interactions. In a CD_3CN solution of an equimolar mixture of **1** and $\mathbf{6}{\cdot}2PF_6$, both 1:1 and 2:1 complexes were detected by ^1H NMR spectroscopy, in addition to the 'free' host and guest, as a result of the slow rates of complexation and decomplexation on the ^1H NMR timescale.

The single crystal X-ray analysis of the complex formed between the macrocycle **2** and the acyclic guest $\mathbf{6}^{2+}$ shows (Figure 8) the generation of a double-stranded, doubly-encircled supermolecule [19]. This 2:2 complex consists of two acyclic guests co-threaded through the cavities of two adjacent macrocyclic hosts and is held together by the 'usual' $[N^+\text{–}H\cdots O]$ and $[C\text{–}H\cdots O]$ hydrogen bonding interactions. In addition, its formation is assisted by the two PF_6^- counterions which are positioned respectively above and below the cavities of the two macrocyclic components in the solid state. In solution, ^1H NMR spectroscopy indicates the existence of only a 1:1 complex.

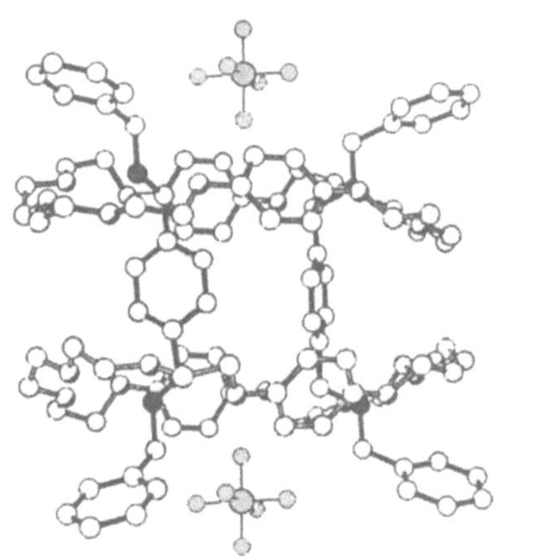

$[(\mathbf{2})_2{\cdot}(\mathbf{6})_2]^{4+}$

Figure 8. The 2:2 complex $[(\mathbf{2})_2{\cdot}(\mathbf{6})_2]^{4+}$ formed in the solid state, showing the relative locations of the two PF_6^- counterions

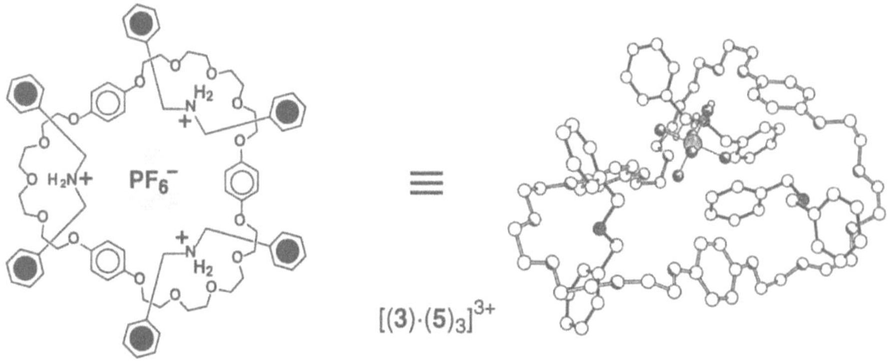

Figure 9. Geometry adopted by the 1:3 complex $[\mathbf{3}\cdot(\mathbf{5})_3]^{3+}$, showing the partially encapsulated PF_6^- anion in the solid state.

The macrocyclic polyether **3** possesses a cavity larger than that of **2** and is able to bind three acyclic guests $\mathbf{5}^+$, as revealed (Figure 9) by single crystal X-ray analysis [20]. Each guest is involved in [N⁺–H···O] and [C–H···O] hydrogen bonding interactions with the polyether chains of the host. In addition, the host and guests sustain [C–H···F] hydrogen bonds [21] with a PF_6^- anion located in the cleft generated by the saddle-like geometry of the 1:3 complex.

Moreover, the total encapsulation of a PF_6^- anion is observed (Figure 10) in the 1:4 complex formed between the macrocyclic polyether **4** and the acyclic guest $\mathbf{5}^+$ in the solid state [20]. In the two crystallographically-independent complexes observed in the unit cell, four thread-like guests are inserted through the cavity of the host as a result of a combination of [N⁺–H···O] and [C–H···O] hydrogen bonds. In addition, in both cases, a PF_6^- anion is fully encapsulated in the center of the 1:4 complex as a consequence of a series of [C–H···F] hydrogen bonds involving the cooperative action of the host **4** and the four guests $\mathbf{5}^+$.

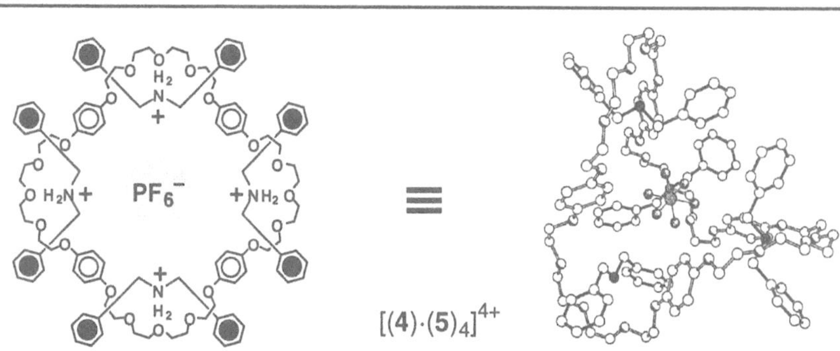

Figure 10. One of the two crystallographically-independent 1:4 complexes $[\mathbf{4}\cdot(\mathbf{5})_4]^{4+}$ observed in the unit cell, illustrating the fully encapsulated PF_6^- anion.

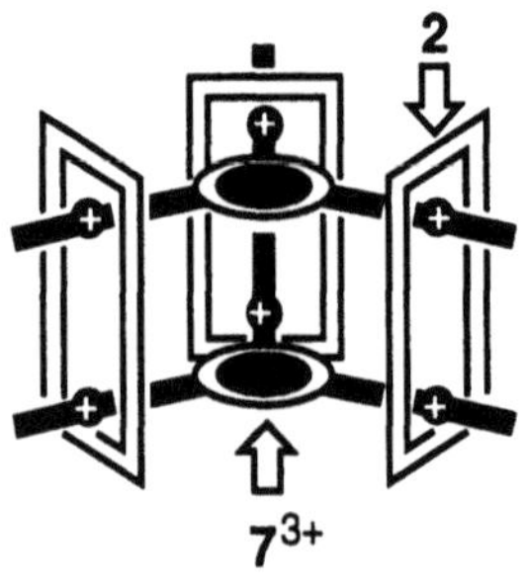

Figure 11. Cartoon representation of the 3:2 complex $[(\mathbf{2})_3 \cdot (\mathbf{7})_2]^{6+}$.

The geometry of the 1:2 complex $[\mathbf{2} \cdot (\mathbf{5})_2]^+$ (Figure 6) suggested the design of branched tectons for the construction of more complex multicomponent supermolecules. The branched building block $\mathbf{7}^{3+}$ possesses one NH_2^+ recognition site in each of its three arms, each of which is attached to a central benzenoid core. Single crystals grown from a $MeNO_2$ solution containing **2** — a host which can bind (*vide supra*) two linear guests simultaneously — and $\mathbf{7}^{3+}$ in a 3:2 ratio were subjected to X-ray analysis which revealed [22] the formation of the 3:2 complex $[(\mathbf{2})_3 \cdot (\mathbf{7})_2]^{6+}$. This supermolecule is schematically depicted in Figure 11 — it is composed of two parallel branched components $\mathbf{7}^{3+}$ and three macrocyclic hosts **2**, each encircling two NH_2^+-containing arms.

The self-assembly of the cage-like supermolecule $[(\mathbf{2})_3 \cdot (\mathbf{7})_2]^{6+}$ suggested the design of a tetrafurcated tecton incorporating one NH_2^+ recognition site in each of its four arms. The tetracation $\mathbf{8}^{4+}$ possesses four NH_2^+ containing arms attached to a central porphyrin core and is a suitable candidate for the construction of a 4:2 supermolecule in conjunction with the host **2**. Indeed, a six component complex is observed (Figure 12) in the solid state [23]. The two porphyrin-based components $\mathbf{8}^{4+}$ are held together by the four macrocyclic hosts **2**, each encircling two NH_2^+-containing arms. In addition to the 'usual' [N⁺–H···O] and [C–H···O] hydrogen bonds, the complex is stabilized by π–π stacking interactions between the two porphyrin cores. Interestingly, EPR spectroscopic investigations, performed in $MeCN/CH_2Cl_2$ (4:6), show that the 4:2 complex $[(\mathbf{2})_4 \cdot (\mathbf{8})_2]^{8+}$ is also formed in solution.

In order to exploit the recognition motif characteristic of the complexes illustrated in Figures 5–12 to generate supramolecular arrays, we have designed a self-complementary tecton possessing a macrocyclic polyether 'head' and an NH_2^+-containing 'tail'. This molecule is represented schematically in Figure 13 and — at least in principle — should be capable of forming an infinite supramolecular array. However, the formation of discrete supermolecules is entropically favored and the single crystal X-ray analysis of $\mathbf{9}^+$ reveals (Figure 14) the formation of a dimeric complex [24]. Interestingly, the achiral form of the dimeric complex is not observed in the crystal, which contains both enantiomers of the chiral form.

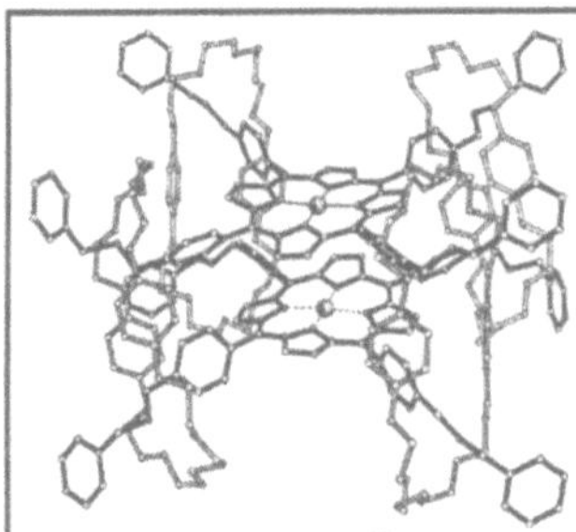

$[(\mathbf{2})_4 \cdot (\mathbf{8})_2]^{8+}$

Figure 12. The 4:2 complex $[(\mathbf{2})_4 \cdot (\mathbf{8})_2]^{8+}$ formed in the solid state.

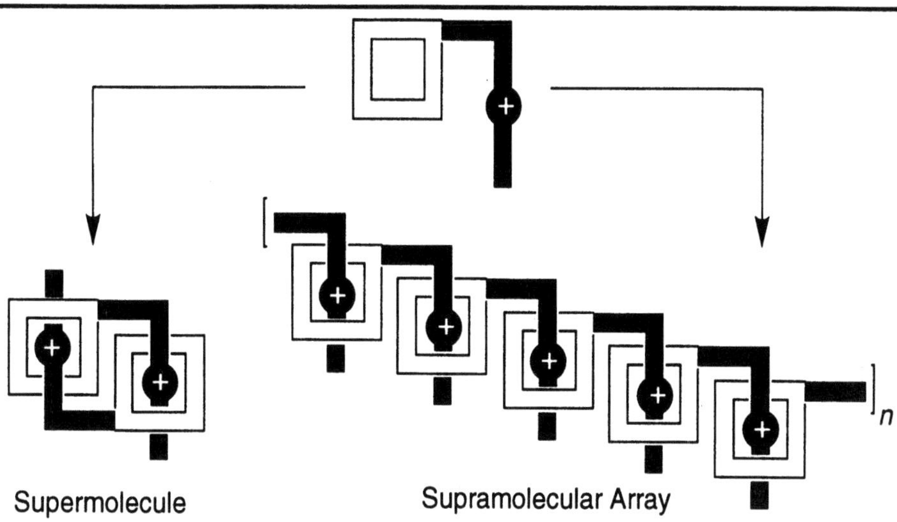

Figure 13. Schematic representation of a supermolecule and of an infinite supramolecular array formed by a self-complementary tecton.

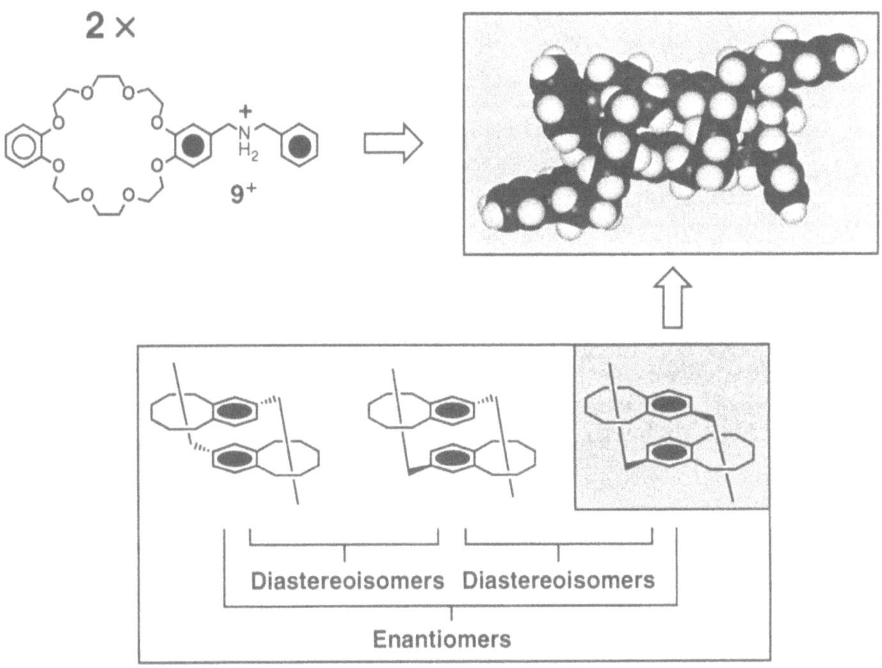

Figure 14. The three possible diastereoisomeric forms of the dimer complex $[(\mathbf{9})_2]^{2+}$ and a space-filling representation of its solid state geometry.

3. Noncovalent Synthesis of Supramolecular Arrays

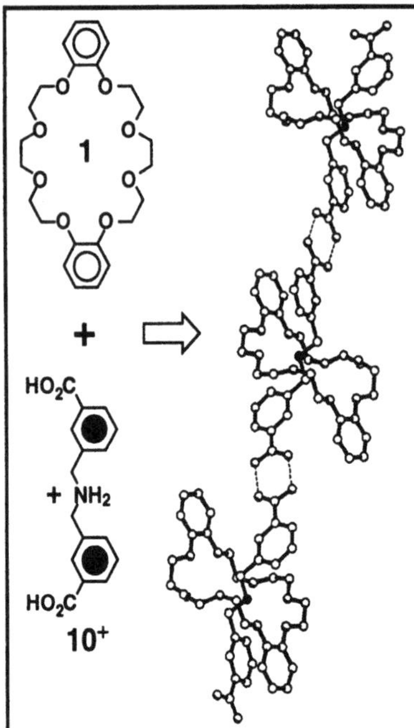

Figure 15. The hydrogen bonded supramolecular polymer formed by **1** and **10**$^+$ in the solid state.

Under appropriate conditions, carboxylic acids form hydrogen bonded dimers as a result of their self-complementarity. This recognition motif has been widely employed to construct well-defined supermolecules and supramolecular arrays (*vide* Figure 1), both in solution and in the solid state [25,26]. One of the most striking examples is the interpenetrating hydrogen bonded network formed by trimesic acid in the solid state, which was reported [27] for the first time as early as 1969, although at the time it was not recognized as an example of crystal engineering. Inspired by these beautiful supramolecular arrays, we have introduced [28,29] carboxylic acid groups into some of our tectons in order to encourage the formation of supramolecular arrays which are held together by a network of hydrogen bonds. The cationic tecton **10**$^+$ incorporates one NH_2^+ recognition site and two carboxylic acid groups. The ^{1}H NMR spectrum of an equimolar CD_3CN solution of **10**$^+$ and the macrocyclic host **1** revealed sharp and well-resolved signals for three species — the uncomplexed host and guest and a 1:1 complex — as in the case of the parent guest **5**$^+$. The slow kinetics of complexation and decomplexation in CD_3CN at 25°C on the ^{1}H NMR timescale enabled a single-point determination of the association constant (880 M^{-1}). Interestingly, the presence of electron withdrawing groups on both aryl rings of **10**$^+$ reinforces the binding event, presumably by enhancing the acidity of the NH_2^+ and of the benzylic CH_2 protons involved in the [N^+–H···O] and [C–H···O] hydrogen bonds. Single crystal X-ray analysis of the 1:1 complex reveals (Figure 15) the formation of a main-chain hydrogen bonded pseudopolyrotaxane — adjacent 1:1 complexes are held together by the

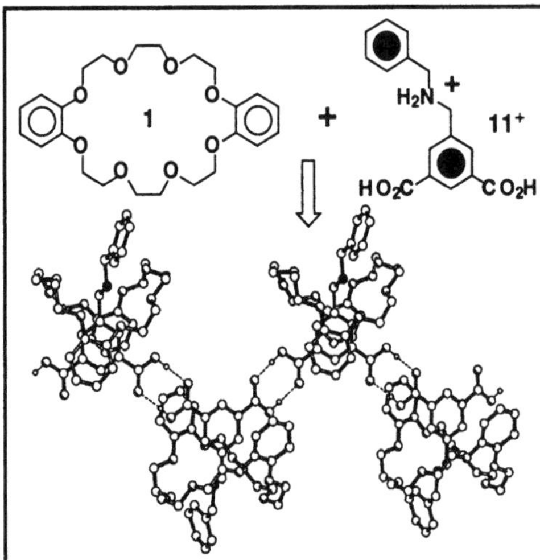

Figure 16. The hydrogen bonded supramolecular polymer formed by **1** and **11**$^+$ in the solid state.

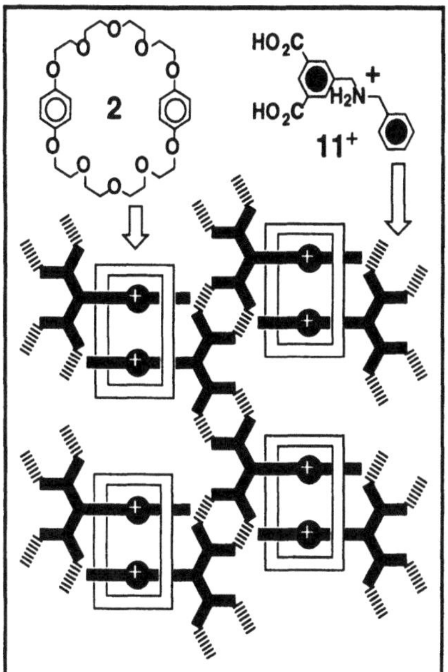

Figure 17. Schematic representation of the hydrogen bonded network formed by **2** and 11^+ in the solid state.

expected hydrogen bonds between the carboxylic acids, leading to a supramolecular polymer in the solid state.

The cationic tecton $\mathbf{11}^+$ incorporates two carboxylic acid groups on the same aryl ring. This compound is bound in solution by the macrocyclic polyether **1** with a 1:1 stoichiometry and a K_a value of 1500 M^{-1} in CD_3CN at 25°C. However, in the solid state (Figure 16), a side-chain pseudopolyrotaxane, consisting of a ribbon-like backbone, held together by carboxylic acid dimers and bearing macrocycle-appended tails, is formed.

The macrocyclic host **2** binds two thread-like guests $\mathbf{11}^+$ in its cavity as a result of [N^+–H···O] and [C–H···O] hydrogen bonds. The formation of a supramolecular array, such as the one schematically represented in Figure 17, is expected as a result of hydrogen bonding interactions between the self-complementary carboxylic acid groups. Indeed, the single crystal X-ray analysis of this 1:2 complex reveals (Figure 18) an interwoven supramolecular cross-linked polymer — a two-dimensional pseudopolyrotaxane — where the cross-linking of the isophthalic acid tapes is induced by the macrocyclic polyether **2**.

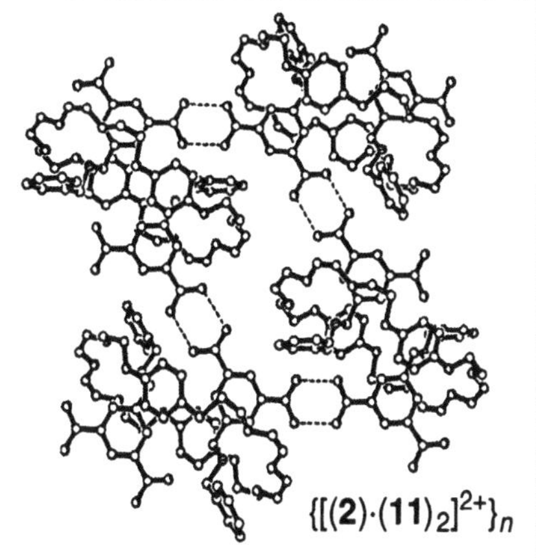

Figure 18. The two-dimensional pseudopolyrotaxane $\{[\mathbf{2}\cdot(\mathbf{11})_2]^{2+}\}_n$ formed in the solid state.

4. Supramolecular Assistance to Covalent Synthesis

We have developed [30] a number of template-directed approaches for synthesizing mechanically interlocked molecules and their 'free' components. These procedures rely on the supramolecular assistance provided by appropriate templates and are schematically illustrated in Figure 19. The template-directed synthesis of a macrocycle can be realized by reacting acyclic precursors in the presence of a linear template. As a result of noncovalent bonding interactions, the acyclic precursors are preorganized by the template so that the subsequent macrocyclization reaction is favored with respect to competing polymerizations. After the formation of covalent bonds, a *supramolecular* complex — namely, a [2]pseudorotaxane which dissociates to regenerate the template in addition to the 'free' macrocycle — is formed. By contrast, when a macrocyclic template is employed, a *molecular* compound — namely, a [2]catenane — is obtained after the macrocyclization reaction. This *molecular* species cannot dissociate into its constituent components as a result of the mechanical bond holding them together, *i.e.*, the template remains irreversibly 'trapped' within the mechanically-interlocked molecule. Similarly, when a [2]rotaxane is constructed, after the formation of two covalent bonds between an acyclic guest species and two bulky stoppers, no dissociation can occur and, once again, the template constitutes an integral part of the final assembly.

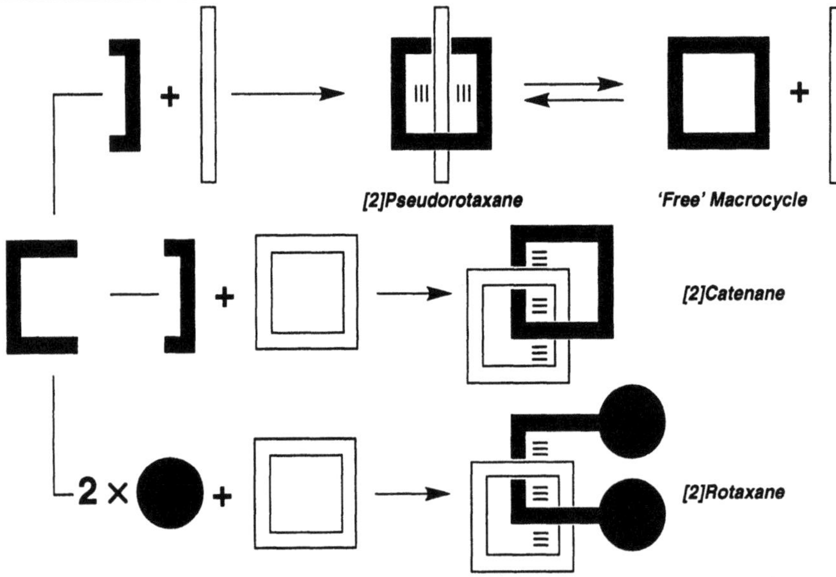

Figure 19. Schematic representation of template-directed synthetic approaches to mechanically-interlocked molecules and their 'free' components.

On reacting (Figure 20) the acyclic bis(hexafluorophosphate) salt **13**·$2PF_6$ with 1,1′-dibromomethyl-4,4′-biphenyl in the presence of the ferrocene-based template **14**, a supramolecular complex is formed [31]. This [2]pseudorotaxane is stabilized by π–π stacking interactions between the π-electron deficient bipyridinium units and the π-electron rich cyclopentadienyl rings. In addition, the complex is reinforced by [C–H···O] hydrogen bonds between the acidic protons in the α-positions (with respect to the nitrogen atoms on the bipyridinium units) and the polyether oxygen atoms. Dissociation of the complex into its host and guest components gave back the recyclable template and the tetracationic cyclophane **15**·$4PF_6$ in an isolated yield of 32%.

Interestingly, when the reaction was performed in the absence of any template, the tetracationic cyclophane **15**·4PF$_6$ was obtained in the very poor yield of 2%.

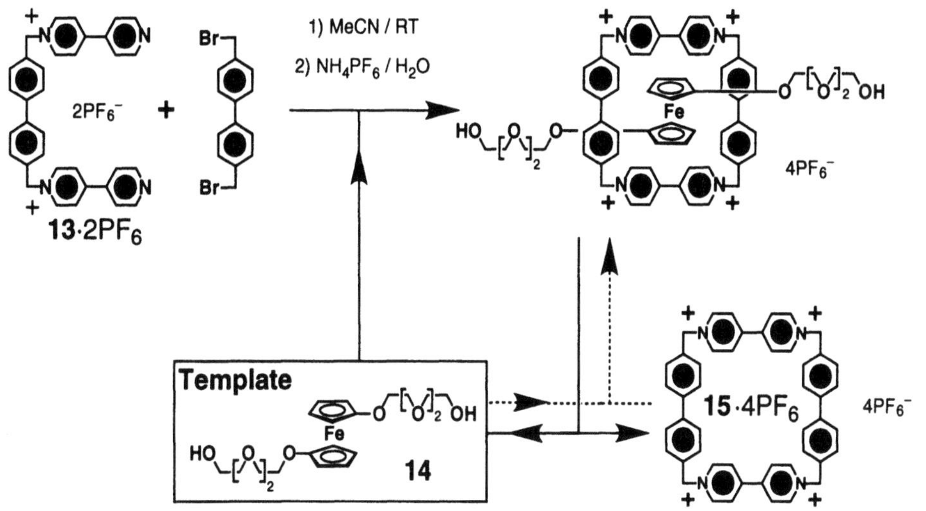

Figure 20. Template-directed synthesis of the tetracationic cyclophane **15**·4PF$_6$.

Reaction (Figure 21) of the bis(hexafluorophosphate) salt **16**·2PF$_6$ with 1,4-bis(bromomethyl)benzene affords a tricationic intermediate, which is bound by the macrocyclic polyether **2**. The formation of this supramolecular intermediate is a result of (i) π–π stacking interactions between the π-electron deficient bipyridinium units and the π-electron rich hydroquinone rings, (ii) [C–H···O] hydrogen bonds between the protons in the α-positions (with respect to the nitrogen atoms on the bipyridinium units) and the polyether oxygen atoms, and (iii) [C–H···π] interactions between the hydroquinone ring protons and the *p*-xylyl spacers. The macrocyclization of the π-electron deficient component is highly favored within this *supramolecular* complex and the formation of a second covalent bond affords [32] the [2]catenane **17**·4PF$_6$ — a *molecular* compound — in a yield of 70%. The template-directed synthesis of the [2]rotaxane **19**·4PF$_6$ follows a very similar mechanism. Reaction (Figure 21) of the bis(hexafluorophosphate) salt **16**·2PF$_6$ with the benzylic chloride **18** yields a tricationic intermediate which is bound by the macrocyclic polyether **2**. The covalent attachment of a second bulky stopper transforms [33] this *supramolecular* complex into a *molecular* compound — namely, the [2]rotaxane **19**·4PF$_6$ — in an isolated yield of 18%.

An alternative synthetic approach to self-assembling a [2]rotaxane is illustrated in Figure 22. By heating a solution containing preformed macrocyclic and dumbbell-shaped components at an appropriate temperature, the free energy barrier associated with the formation of the [2]rotaxane can be overcome, allowing the macrocycle to slip on

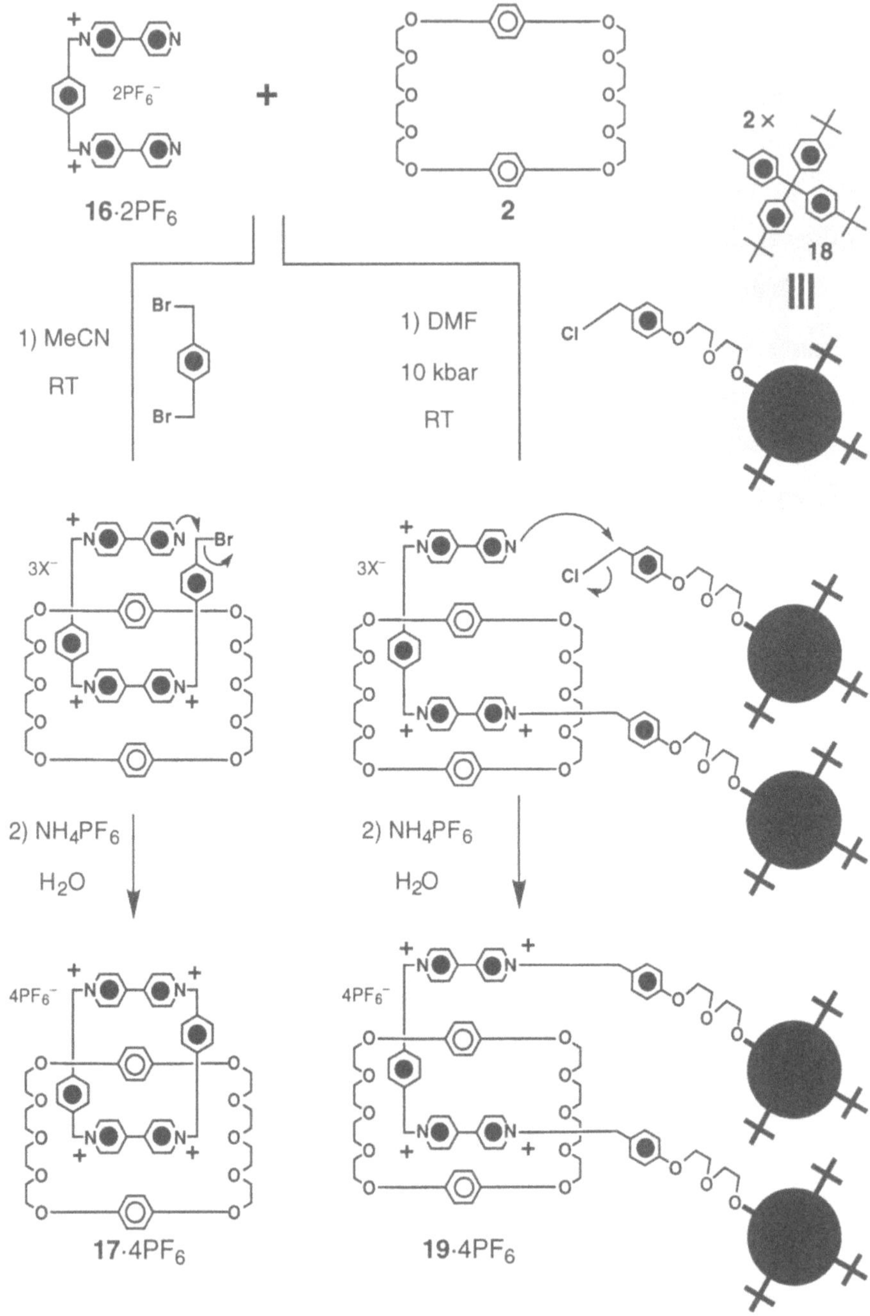

Figure 21. Template-directed syntheses of the [2]catenane **17**·4PF_6 and of the [2]rotaxane **19**·4PF_6.

over the stoppers of the dumbbell-shaped component. However, the opposing slipping-off process, involving the dismembering of the [2]rotaxane, can also occur at this temperature. After a certain period of time, an equilibrium between the species present in solution — one that is governed by the noncovalent bonding interactions existing within the [2]rotaxane — is achieved. By cooling down the solution to room temperature, the [2]rotaxane becomes kinetically-stable — *i.e.*, the system does not possess enough energy to overcome the energy barriers associated with the slipping-on and slipping-off processes — and can be isolated from its precursors by column chromatography.

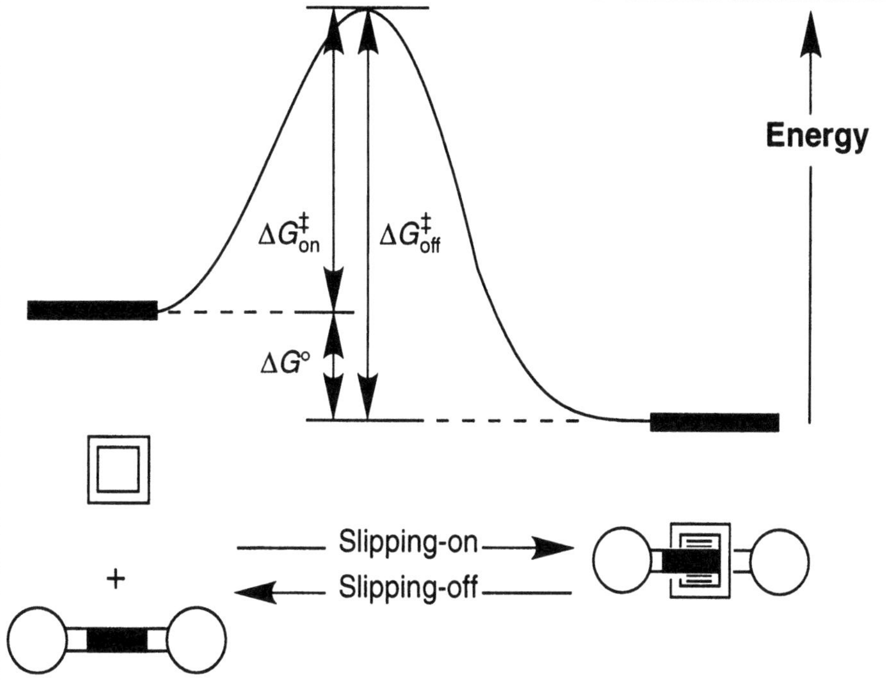

Figure 22. Schematic representation of the slipping approach to self-assembling a [2]rotaxane.

The major requirement for the slipping approach to self-assembling rotaxanes is the size complementarity between the stoppers and the macrocycle. Only when this complementarity is achieved, is the formation of a rotaxane feasible. The dumbbell-shaped compounds **20**$^{2+}$– **23**$^{2+}$ (Figure 23) possess a bipyridinium recognition site, in addition to tetraarylmethane-based stoppers that differ in the nature of their R substituents. By heating a MeCN solution containing one of the dumbbell-shaped compounds and the macrocyclic polyether **2**, a [2]rotaxane is formed [33] when R is equal to H, Me and Et. On the other hand, no [2]rotaxane is obtained when R is equal to *i*-Pr. Thus, the stoppers incorporating *i*-Pr groups as their R substituents are too bulky for the passage of the macrocycle over them.

The slippage procedure was also employed to self-assemble rotaxane-like species incorporating NH_2^+ recognition sites. Heating a CH_2Cl_2 solution of the macrocyclic polyether **1** and the NH_2^+-containing guest **24**$^+$ at 40 °C affords (Figure 24) the rotaxane-like compound **25**$^+$ which can be isolated and was characterized in solution as well as in the solid state [34]. However, dissociation of **25**$^+$ into its separate components — namely, **1** and **24**$^+$ — occurs in $(CD_3)_2SO$ — a solvent possessing a strong hydrogen bond acceptor group. Thus, **25**$^+$ behaves as a rotaxane in nonpolar solvents but as a pseudorotaxane in solvents possessing a high Gutmann donor number. We believe that this example shows (Figure 25) that the transition between rotaxanes and pseudorotaxanes is a gradual one in which pseudorotaxanes acquire progressively more rotaxane-like character until they are completely interlocked.

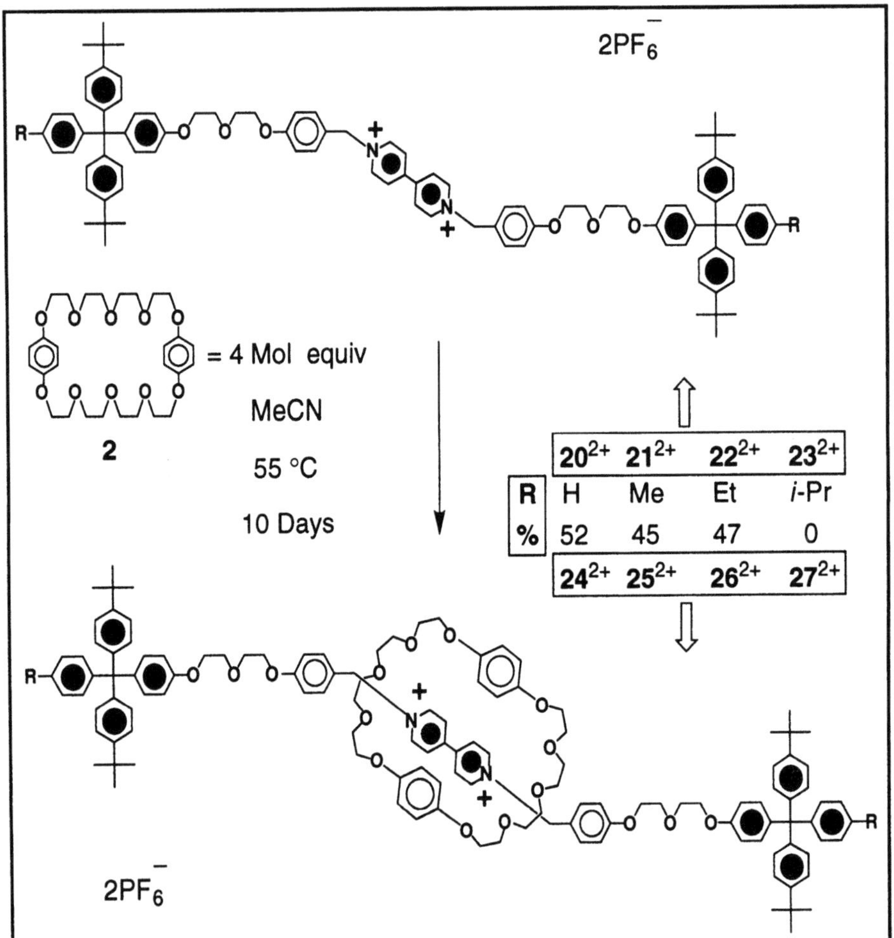

Figure 23. Self-assembly of the [2]rotaxanes **24**$^{2+}$– **26**$^{2+}$ by slippage.

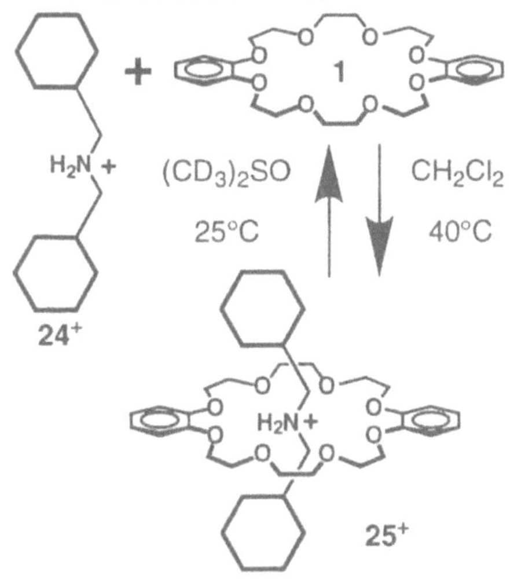

Figure 24. Self-assembly of the rotaxane-like species 25^+ by slippage and its dissociation in $(CD_3)_2SO$.

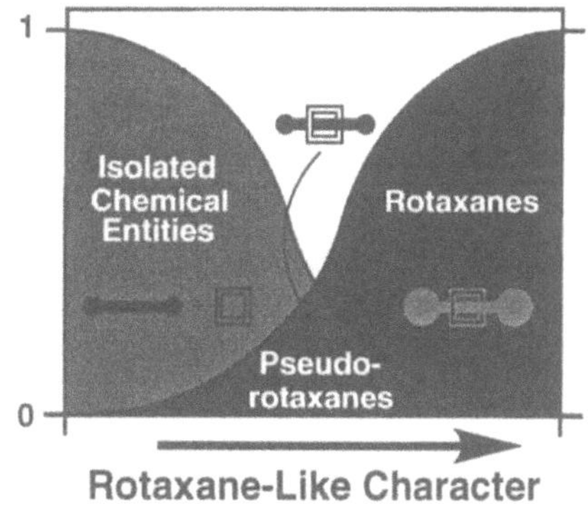

Figure 25. Pseudorotaxanes lie at the intersection of the domains characterizing isolated chemical entities and rotaxanes. The numbers 1 and 0 have been assigned arbitrarily to the respective species belonging either totally or not at all to one of the two domains.

5. Conclusions

- *Supramolecular synthesis* often equates with *strict self-assembly*, whereas *supramolecular assistance to molecular synthesis* often equates with *self-assembly with covalent modification.*

- When complementary recognition sites (**A/B**) are present in the same molecule, *supermolecules* will often be formed in preference to *supramolecular arrays*, particularly if the **A–X–B** molecules are flexible. The entropic cost of forming $[\mathbf{A–X–B}]_n$ *supramolecular arrays* rather than $n[\mathbf{A–X–B}]$ *supermolecules* may be reduced by introducing more rigidity into the **A–X–B** molecules.

- The use of *orthogonal recognition sites* in *noncovalent synthesis* allows good control of supramolecular and molecular structures to ensue. This is comparable to employing numerous different covalent bond-manipulating reactions, in a synthetic sequence, for the total synthesis of a complex natural product.

- As in *covalent synthesis*, it is possible to manipulate systems undergoing *noncovalent synthesis* in either *thermodynamic or kinetic regimes*, or indeed in a mixture of both, *e.g.*, *slippage* employs both kinetic and thermodynamic regimes, operating in tandem, for the synthesis of rotaxane-like structures. The dividing line between these two regimes is not clear cut.

6. Reflections

We have demonstrated how a number of different recognition motifs can be exploited for the *noncovalent synthesis* of supermolecules and supramolecular arrays. Furthermore, employing concurrently such orthogonal recognition algorithms diversifies dramatically the range of self-assembled systems that are accessible to the contemporary supramolecular chemist. They permit the construction of supermolecules and supramolecular arrays that would have been inconceivable not all that long ago. In addition, we have also demonstrated how noncovalent bonding interactions — in some instances, aided and abetted by mechanical bonds — can be employed to assist in the *covalent synthesis* of molecular structures which cannot be obtained easily while employing only the conventional methods of traditional organic synthesis. Synthetic supramolecular chemistry — the chemical synthesis for the next century — is already upon us!

7. References and Notes

1. a) Eds. Barton, D. and Ollis, W.D. (1979) Comprehensive Organic Chemistry, Pergamon, Oxford. b) Corey, E.J. and Cheng, X.–M. (1989) The Logic of Chemical Synthesis, Wiley, New York. c) Nicolaou, K.C. and Sorensen E.J. (1995) Classics in Total Synthesis, VCH, Weinheim.
2. a) Whitesides, G.M. (1990) *Angew. Chem., Int. Ed. Engl.*, **29**, 1209–1218. b) Seebach, D. (1990) *Angew. Chem., Int. Ed. Engl.*, **29**, 1320–1367.
3. Nicolaou, K.C. (1996) *Angew. Chem., Int. Ed. Engl.*, **35**, 589–607.
4. Suh, E.M. and Kishi, Y. (1994) *J. Am. Chem. Soc.*, **116**, 11205–11206.
5. Depew, K.M., Zeman, S.M., Boyer, S.H., Denhart, D.J., Ikemoto, N., Danishefsky, S.J., and Crothers D.M. (1996) *Angew. Chem., Int. Ed. Engl.*, **35**, 2797–2801.
6. a) Tomalia, D.A. and Durst, H.D. (1993) *Top. Curr. Chem.*, **165**, 193–313. b) Fréchet, J.M.J. (1994) *Science*, **263**, 1710–1715. c) Newkome, G.R., Moorefield, C.N., and Vögtle, F. (1996) Dendritic Macromolecules, VCH, Weinheim. d) Zeng, F. and Zimmerman, S.C. (1997) *Chem. Rev.*, **97**, 1681–1712. e) Moore, J.S. (1997) *Acc. Chem. Res.*, **30**, 402–413.
7. a) Vögtle, F. (1991) Supramolecular Chemistry, Wiley, New York. b) Lehn, J.–M. (1995) Supramolecular Chemistry, VCH, Weinheim. c) Eds. Atwood, J.L., Davies, J.E.D., MacNicol, D.D., and Vögtle, F. (1996) Comprehensive Supramolecular Chemistry, Pergamon, Oxford.
8. Whitesides, G.M., Simanek, E.E., Mathias, J.P., Seto, C.T., Chin, D.N., Mammen, M., and Gordon, D.M. (1995) *Acc. Chem. Res.*, **28**, 37–44.
9. Asakawa, M., Ashton, P.R., Brown, G.R., Hayes, W., Menzer, S., Stoddart, J.F., White, A.J.P., and Williams, D.J. (1996) *Adv. Mater.*, **8**, 37–41.
10. Simard, M., Su, D., and Wuest, J.D. (1991) *J. Am. Chem. Soc.*, **113**, 4696–4698.
11. a) Etter, M.C. (1990) *Acc. Chem. Res.*, **23**, 120–126. b) Etter, M.C., MacDonald, J.C., and Bernstein, J. (1990) *Acta Crystallogr., Sect. B*, **46**, 256–262. c) Etter, M.C. (1991) *J. Phys. Chem.*, **95**, 4601–4610. d) Desiraju, G.R. (1991) *Acc. Chem. Res.*, **24**, 290–296. e) Desiraju, G.R. (1995) *Angew. Chem., Int. Ed. Engl.*, **34**, 2311–2327. f) Desiraju, G.R. (1996) *Acc. Chem. Res.*, **29**, 441–449. g) Desiraju, G.R. (1997) *Chem. Commun.*, 1475–1482.
12. a) Aakeröy, C.B. and Seddon, K.R. (1993) *Chem. Soc. Rev.*, **22**, 397–407. b) MacDonald, J.C. and Whitesides, G.M. (1994) *Chem. Rev.*, **94**, 2383–2420. c) Zaworotko, M.J. (1994) *Chem. Soc. Rev.*, **23**, 283–288. d) Bernstein, J., Davis, R.E., Shimoni, L., and Chang, N.L. (1995) *Angew. Chem., Int. Ed. Engl.*, **34**, 1555–1573. e) Burrows, A.D., Chan, C.W., Chowdhry, M.M., McGrady, J.E., and Mingos, D.M.P. (1995) *Chem. Soc. Rev.*, **24**, 329–339.
13. Desiraju, G.R. (1989) Crystal Engineering, Elsevier, Amsterdam.
14. Fyfe, M.C.T. and Stoddart, J.F. (1997) *Acc. Chem. Res.*, **30**, 393–401.
15. a) Lehn, J.–M. (1990) *Angew. Chem., Int. Ed. Engl.*, **29**, 1304–1319. b) Lindsey, J.S. (1991) *New J. Chem.*, **15**, 153–180. c) Whitesides, G.M., Mathias, J.P., and Seto, C.T. (1991) *Science*, **254**, 1312–1319. d) Lawrence, D.S., Jiang, T., and Levett, M. (1995) *Chem. Rev.*, **95**, 2229–2260. e) Conn, M.M. and Rebek, J., Jr. (1997) *Chem. Rev.*, **97**, 1647–1668. f) Linton, B. and Hamilton, A.D. (1997) *Chem. Rev.*,

97, 1669–1680. g) Stupp, S.I., LeBonheur, V., Walker, K., Li, L.S., Huggins, K.E., Keser, M., and Amstutz, A. (1997) *Science*, **276**, 384–389.

16. a) Philp, D. and Stoddart, J.F. (1996) *Angew. Chem., Int. Ed. Engl.*, **35**, 1154–1196. b) Raymo, F.M. and Stoddart, J.F. (1996) *Curr. Op. Coll. Interf. Sci.*, **1**, 116–126. c) Gillard, R.E., Raymo, F.M., and Stoddart, J.F. (1997) *Chem. Eur. J.*, **3**, In press. d) Raymo, F.M. and Stoddart, J.F. (1998) *Chemtracts*, **11**, In press.
17. Glink, P.T., Schiavo, C., Stoddart, J.F., and Williams, D.J. (1996) *Chem. Commun.*, 1483–1490.
18. Ashton, P.R., Campbell, P.J., Chrystal, E.J.T., Glink, P.T., Menzer, S., Philp, D., Spencer, N., Stoddart, J.F., Tasker, P.A., and Williams, D.J. (1995) *Angew. Chem., Int. Ed. Engl.*, **34**, 1865–1869.
19. Ashton, P.R., Chrystal, E.J.T., Glink, P.T., Menzer, S., Schiavo, C., Stoddart, J.F., Tasker, P.A., and Williams, D.J. (1995) *Angew. Chem., Int. Ed. Engl.*, **34**, 1869–1871.
20. Fyfe, M.C.T., Glink, P.T., Menzer, S., Stoddart, J.F., White, A.J.P., and Williams, D.J. (1997) *Angew. Chem., Int. Ed. Engl.*, **36**, 2068–2070.
21. a) Grepioni, F., Cojazzi, G., Draper, S.M., Scully, N., and Braga, D. (1997) *Organometallics*, **16**, In press. b) Braga, D., Grepioni, F. (1998) Current Challenges on Large Supramolecular Assemblies, Ed. Tsoucaris, G., Kluwer Academic Publishers, Dordrecht, In press.
22. Ashton, P.R., Collins, A.N., Fyfe, M.C.T., Glink, P.T., Menzer, S., Stoddart, J.F., and Williams, D.J. (1997) *Angew. Chem., Int. Ed. Eng.*, **36**, 59–62.
23. Feiters, M.C., Fyfe, M.C.T., Martínez–Díaz, M.–V., Menzer, S., Nolte, R.J.M., Stoddart, J.F., van Kan, P.J.M., and Williams, D.J. (1997) *J. Am. Chem. Soc.*, **119**, 8119–8120.
24. Ashton, P.R., Cantrill, S.J., Fyfe, M.C.T., Glink, P.T., Stoddart, J.F., and Williams, D.J. (1998) *Angew. Chem., Int. Ed. Engl.*, **37**, Submitted.
25. Ribbon–like supramolecular arrays are formed in the solid state by simple molecules such as terephthalic acid and isophthalic acid, see: a) Bailey, M. and Brown, C.J. (1967) *Acta Crystallogr., Sect. B*, **22**, 387–391. b) Alcala, R. and Martinez–Carrera, S. (1972) *Acta Crystallogr., Sect. B*, **28**, 1671–1677.
26. For two striking examples of self–assembling supermolecules held together by carboxylic acid dimers, see: a) Yang, J., Marendez, J.–L., Geib, S.J., and Hamilton, A.D. (1994) *Tetrahedron Lett.*, **35**, 3665–3668. b) Zimmerman, S.C., Zeng, F., Reichert, D.E.C., and Kolotuchin, S.V. (1996) *Science*, **271**, 1095–1098.
27. Duchamp, D.J. and Marsh, R.E. (1969) *Acta Crystallogr. Sect. B*, **25**, 5– 19. For later discussions and analyses of the interpenetrating network formed by trimesic acid in the crystal, see: a) Herbstein, F.H., Kapon, M., and Reisner, G.M. (1981), *Proc. R. Soc. Lond.*, **376**, 301–318. b) Herbstein, F.H., Kapon, M., and Reisner, G.M., (1985) *Acta Crystallogr., Sect. B*, **41**, 348–354. c) Herbstein, F.H., Kapon, M., and Reisner, G.M. (1987) *J. Inclusion Phenom.*, **5**, 211–214.
28. Ashton, P.R., Collins, A.N., Fyfe, M.C.T., Menzer, S., Stoddart, J.F., Williams, D.J. (1997) *Angew. Chem., Int. Ed. Engl.*, **36**, 735–739.
29. Ashton, P.R., Fyfe, M.C.T., Hickingbottom, S.K., Menzer, S., Stoddart, J.F., White, A.J.P., and Williams, D.J. (1998) *Chem. Eur. J.*, **4**, In press.
30. a) Philp, D. and Stoddart, J. F. (1991) *Synlett*, 445–458. b) Amabilino, D.B. and Stoddart, J.F. (1995) *Chem. Rev.*, **95**, 2725–2828. c) Pasini, D., Raymo, F.M., and Stoddart, J.F. (1995) *Gazz. Chim. Ital.*, **125**, 431–435. d) Belohradsky, M., Raymo, F. M., and Stoddart J. F. (1996) *Collect. Czech. Chem. Commun.*, **61**, 1–43. e) Amabilino, D.B., Raymo, F.M., and Stoddart, J.F. (1996) Comprehensive Supramolecular Chemistry, Vol. 9, Eds. Hosseini, M.W. and Sauvage, J.–P., pp. 85–130, Pergamon Press, Oxford. f) Raymo, F.M. and Stoddart, J.F. (1996) *Pure Appl. Chem.*, **68**, 313–322. g) Belohradsky, M., Raymo, F.M., and Stoddart, J.F. (1997) *Collect. Czech. Chem. Commun.*, **62**, 527–557. h) Raymo, F. M. and Stoddart, J.F. (1997) *Pure Appl. Chem.*, **69**, 1987–1997.
31. Asakawa, M., Ashton, P.R., Menzer, S., Raymo, F.M., Stoddart, J.F., White, A.J.P., and Williams, D.J. (1996) *Chem. Eur. J.*, **2**, 877–893.
32. a) Anelli, P.L., Ashton, P.R., Ballardini, R., Balzani, V., Delgado, M., Gandolfi, M.T., Goodnow, T.T., Kaifer, A.E., Philp, D., Pietraszkiewicz, M., Prodi, L., Reddington, M.V., Slawin, A.M.Z., Spencer, N., Stoddart, J.F., Vicent, C., and Williams, D.J. (1992) *J. Am. Chem. Soc.*, **114**, 193–214. b) Capobianchi, S., Doddi, G., Ercolani, G., Keyes, J.W., and Mencarelli, P. (1997) *J. Org. Chem.*, **62**, 7015-7017.
33. Ashton, P.R., Ballardini, R., Balzani, V., Belohradsky, M., Gandolfi, M.T., Philp, D., Prodi, L., Raymo, F.M., Reddington, M.V., Spencer, N., Stoddart, J.F., Venturi, M., and Williams, D.J. (1996) *J. Am. Chem. Soc.*, **118**, 4931–4951.
34. Ashton, P.R., Baxter, I., Fyfe, M.C.T., Raymo, F.M., Spencer, N., Stoddart, J.F., White, A.J.P., Williams, D.J. (1998) *J. Am. Chem. Soc.*, **120**, Submitted.

BEYOND SUPRAMOLECULAR CHEMISTRY: SUPRAMOLECULAR AND ÜBERSUPRAMOLECULAR ASPECTS OF CHEMICAL STRUCTURE AND REACTIVITY: STATIC AND OSCILLATING ELECTRIC AND MAGNETIC FIELDS, COHERENCE, AND COOPERATIVITY

NICHOLAS J. TURRO
Departments of Chemistry and Chemical Engineering, Columbia University
New York, NY 10027 (USA)

1. Paradigms. An Overview.

An accepted scientific theory may be viewed as an intellectual structure universally accepted by a community of scientists and termed the community's **paradigm.**[1] A mature science, such as Chemistry, is usually governed by a set of paradigms setting the standards for legitimate work within the science it governs. The paradigms, being the set of beliefs driving the intellectual activities, strategies, and methodologies (what can be done, what is worth doing, and how to do it) of the practitioners, coordinate and direct the puzzle-solving activity of normal everyday science. The existence of a paradigm capable of supporting a normal science tradition has been put forth as the most important characteristic that distinguishes a mature science from a developing science or a non-science. The nature of a paradigm is to possess intellectual and physical components and connections that are accepted as necessary and sufficient to describe all observable phenomena.

Thus, the paradigm dictates which entities may exist and may not exist in the Universe and, in doing so, determines the legitimate puzzles that practitioners may attempt to solve; it assures the practitioners of the existence of solutions to these puzzles and it guides the construction and development of instrumentation and instrumental techniques that test the theoretical aspects of the paradigm. The paradigm also provides the means of certification of practitioners. A research puzzle in Chemistry is proposed by a mentor to an eager young apprentice. Since the paradigm guarantees a solution, it is the skill of the young apprentice that is being tested in the ensuing research, not the paradigm! When students demonstrate that they can solve chemical

G. Tsoucaris (ed.), Current Challenges on Large Supramolecular Assemblies, 37–49.

puzzles with proficency, they obtain a "card" admitting them to the profession. Normal scientists presuppose that a paradigm provides the technical means of solving puzzles posed within it, and its structure provides the discipline that must be exerted in solving the puzzle. The excitement of the hunt and the detective work accompanying the solution of the puzzle provide the mentor with a passion for his profession. The student must provide the hard work and the energy! Failure to solve a puzzle is seen as a failure of the scientist, rather than an inadequacy of the paradigm. To blame the paradigm for any failure to solve a puzzle, is to play a carpenter who blames his tools for poor work.

The paradigm of molecules and the covalent bond between atoms has been a powerful and unifying paradigm for chemistry and photochemistry for the last half of the millenium. A "paradigm shift" has been occurring during the past two decades, namely the emergence of the paradigm of supramolecular chemistry.[2] Although its full acceptance may still be argued, we do not do so here. It is the purpose of this article to attempt to learn from the paradigm shifts and intellectual revolutions of the past and to attempt to extrapolate to the future, the next millenium!

2. From Atoms to Molecules to Supermolecules. From Molecular Photochemistry to Supramolecular Photochemistry.

Let us review the powerful paradigms that have developed in Chemistry over the last 5 or so decades and search for common components that might provide clues for the prediction of the new emerging paradigms. Chemistry is an ever-expanding "universe" at the microscopic level requiring mastery of the invisible at ever-increasing levels of compexity. Starting with atoms, assemblies of a single nucleus and orbiting electrons, the chemist has learned to understand atomic structure through mastery of the principles of electronic configurations behaving as a unit in conjunction with a nucleus. Proceeding to molecules, assemblies of two or more atoms, the chemist then learned to understand molecular structure through mastery of the principles of the covalent bond and a collection of atoms behaving as a unit. Most recently, chemists have attacked the next level of complexity, supermolecules or supermolecular assemblies consisting of two or more molecules behaving as a unit, whose structure will be understood through mastery of the principles of the intermolecular bond.[2]

Let us now consider the emerging paradigm of supramolecular chemistry, a term which has been widely accepted, to embrance Chemistry "beyond the covalent bond". Two or more molecules which behave as a unit as the result of intermolecular bonding may be considered a "supermolecule" or "supramolecular assembly". The investigation of the structure, dynamics and chemistry of supermolecules defines the emerging discipline of "supramolecular chemistry". There is still a healthy debate over terminology and the universe of what the discipline of supramolecular chemistry will encompass. Until recently, this situation was similarly acute for "supramolecular photochemistry", a subdiscipline for which the practitioners have tended to focus on molecular features and to be generally uninvolved in the development of supramolecular chemistry as a field. However, beginning in the 1990s an awareness of the potential of supramolecular ideas to invigorate the field has developed, and photochemistry is truly being transformed, as supramolecular photochemistry burgeons as an important and exciting subdiscipline.[3] Furthermore, as the paradigm of molecular photochemistry developed, it naturally integrated the disciplines of spectroscopy, chemical physics and organic and inorganic chemistry. As a result, photochemists have been required to accept a range of terminologies and conceptual viewpoints in order to deal effectively with photochemical systems.

For example, since photochemistry involved the interaction of light and matter, photochemists must consider the nature and behavior of photons and light waves (chemical physics and spectroscopy) in addition to the nature and behavior of molecules (organic and inorganic chemistry). As a result of collaborations and "intellectual mixing", ideas which are peculiar and appear farfetched to an organic chemist may be considered routine to a spectroscopist or chemical physicist and *vice-versa*. This fact is important, because the knowledge that certain phenomena exist in one field of science provides an "existence theorem proof" of the possibility of extending the phenomena to another field of science. We shall come back to this point as a basis for making our predictions about the future!

2.1 THE SUPRAMOLECULAR PARADIGM

Structure is at the heart of systematic scientific thinking; the paradigm of molecular structure is at the heart of virtually all chemical thinking. Any molecular structure may be decomposed into the following forms or levels: (1) composition; (2) constitution; (3) configuration; and (4) conformation. At the level of composition only the number and kinds of "elements" or atoms of the structure are of chemical interest. At the level of constitution, not only is the composition of chemical interest, but also the number of possible ways in which the atoms are connected to one another. A unique *molecular structure* at this level may be identified as one for which all of the atoms of the composition possess a unique connectivity relationship. At the level of configuration, in addition to the composition and constitution, the number of possible dispositions in three dimensional space of "neighboring atoms" about a central atom is of chemical interest. A unique structure at this level may be identified as one for which all of the neighboring atoms surrounding a central atom possess a distribution in three dimensional space that is unique. Finally, at the level of conformation, the possible shapes of a given composition, constitution and specific set of configurations is of chemical interest. A unique structure at this level may be identified as one which differs from others in terms of shapes resulting from rotations about single bonds.

Chemists have grown so familiar with the paradigm of molecular structure that its validity and centrality to all of chemistry are no longer seriously debated. However, chemists have recently developed a paradigm at a level beyond that of molecular structure, namely the **supramolecular level**. Supramolecular structures are related to molecules in the same way that molecular structures are related to atoms. Although atoms play a critical role in the paradigm of molecular structure, it is the molecular structure which is generally at the center of the chemistry of molecules. The "intellectual glue" of molecular structure is the concept of the covalent bond and the stereochemistry associated with the consequences of the covalent bonds. The paradigm of the covalent chemical bond provides the rules that govern the structures, dynamics, characteristics and transformations of molecules. A paradigm of non-covalent bonds provides the rules governing the structures, dynamics, characteristics and transformations of supramolecular assemblies.

Just as the level of atomic structure is inadequate for understanding aspects of chemistry where molecular aspects dominate, the level of molecular structure is inadequate for understanding aspects of chemistry where supramolecular aspects dominate. The emerging paradigm of supramolecular systems emphasizes the chemistry of the **intermolecular bond which holds molecules together, rather than the covalent bond, which holds atoms together.** Non-covalent intermolecular bonds are more varied and complex than covalent intramolecular bonds. In general, intermolecular, non-covalent bonds are held together by weaker forces than intramolecular, covalent bonds. For example, the important forces holding molecules together consist of dispersion forces, hydrogen bonds, hydrophobic bonds, etc. Although these forces, **per**

atom contact, are small, in many supramolecular systems many such weak bonds are formed in a cooperative manner, thereby causing the supramolecular structure to be energetically stable; however, because each "bonding contact" is relatively weak, the bonds may be broken relatively easily so that the structure may have a certain amount of flexibility. To a photochemist, the cooperative behavior of molecules in a supermolecule has an analogy in the coherence of photons in a laser, i.e., laser light is "superphotonic."

2.1.1 *Beyond the Supramolecular Paradigm. From Atomic Chemistry to Superdupermolecular Chemistry*

With the above ideas in mind, let us now attempt to extrapolate them and try to imagine the paradigm that will be used by chemists and photochemists in the 21st century. Prediction is difficult, especially for the future, but it is entertaining to try. First let us review the structural basis of current atomic, molecular and supramolecular concepts from the organic , chemical physics, and spectroscopic points of view and then attempt to see what is missing and what we can learn from the intellectual jumps from the atomic to the molecular to the supramolecular to the "superdupermolecular" paradigm of the next millenium!

In Chemical Physics the paradigm of the "Hamiltonian" determines the way practitioners think about chemical problems. Chemical physicists employ an abstract mathematical Hamiltonian, based on physical models, and solve wave equations to come to conclusions concerning observables. The organic chemist uses molecular structure and chemical equations based on these structures instead of Hamiltonians to solve chemical problems. Both are useful but different representations of the **same** chemical systems. Therefore, we shall use a hybrid language in an attempt to extract some of the more useful features of both representations.

We take as a given the idea that representations of molecular structure are the bedrock of organic thinking and so powerful intellectually, that they must be employed. However, we take as an advantage of the Hamiltonian approach its universality in including all forms of interactions and energetic considerations in a systematic and potentially quantitative manner. In the language of the organic chemist. a proper chemical structure contains all the information required to solve any chemical problem of interest. In the language of the chemical physicist, a complete Hamiltonian contains all of the information required to solve any chemical problem of interest. The problem is that usually neither the organic chemist nor the chemical physicist have a "proper" structure or a "complete" Hamiltonian. Worse, they often do not realize that this is the case and are frustrated by their inability to solve the problem at hand! They don't understand why the paradigm isn't working!

For example, the organic chemist may be considering a chemical problem at the molecular level, whereas the proper structure is supramolecular. The chemical physicist may be considering a chemical problem with a Hamitonian that is missing an interaction or is not starting from the correct physical assumptions. Let us see how the chemical physicist handles the problem and then extend it to suggest how the organic chemist can handle the problem.

A key concept in setting up a Hamiltonian is the selection of a physical model that determines the overall energetic characteristics of the system of interest. This selection is based on experience and judgement in the same way that the organic chemist selects a starting structure to attack a chemical problem. The starting Hamitonian is called the "Zeroth" order and then direct significant interactions are added as "First Order" and indirect weaker interactions are added as "Second Order". Although second order effects may be tiny in energy, they can be important if they serve as the key to the structural changes that determine the rate of a dynamic process. We shall see below that this is the case for intersystem crossing in geminate radical pairs. Since only singlet geminate pairs are reactive, a triplet geminate pair is inert toward radical pair reactions until some tiny perturbation causes a spin change from triplet to singlet.

In the language of topology, atomic chemisty, molecular chemistry and supramolecular chemistry may be viewed as topologically equivalent, i.e., in going from atomic chemistry to molecular chemistry, the structure of atoms (electrons and nuclei) are suppressed and collapsed to a single point, whose internal structure may be ignored to Zero Order. In going from molecular chemistry to supramolecular chemistry, the structure of supramolecular systems (molecules) are suppressed and collapsed to a single point, whose internal structure may be ignored to Zero Order.

We consider in this article situations for which the Zero Order approximation for atomic, molecular and supramolecular chemistry is not sufficient to explain experimental observations. We term these situations as manifestations of Überatomic, Übermolecular and Übersupramolecular chemistry. For example, the properties of matter in the presence of an oscillating electromagnetic field or in the presence of a laboratory magnetic field often cannot be understood without explicit consideration of the interaction of the fields with the structure representing the matter. Often, features explicitly absent from the atomic, molecular or supramolecular structure traditionally employed to represent matter, are dominant in determining the observation. This is particularly common in the case of electron and nuclear spins, which are not explicitly represented.

The purpose of this article is to explore situations for which übermolecular effects, i.e., weak energetic interactions which are generally not considered to be important by organic chemists will be important for supramolecular system and provide examples which may provide a scaffolding for extension to other systems. Consideration of these übermolecular features will form the basis of a superdupermolecular photochemistry.

3. From the Paradigm of Molecular Photochemistry to the Paradigm of Supramolecular Photochemistry.

Molecular photochemistry[4] has a mature paradigm providing the practitioner with a conceptual basis for generating a "map and a clock" for examining the mechanisms of photochemical reactions. The "map" is the connectivity of the chemical entities whose structures must be characterized and the reaction steps and/or structural changes that interconvert the chemical structure, whose dynamics are of greatest importance in understanding the details of the mechanism of a photochemical reaction. In addition, the paradigms of modern molecular photochemistry provide the practitioner with (1) a blueprint whose theoretical parameters are of greatest import, (2) the way to connect the theoretical parameters with experimental parameters, and (3) techniques that will provide the appropriate data to be transformed into experimental parameters. Finally, the paradigm provides the information needed to formulate a comprehensive theoretical basis for quantitative computation to simulate experimental data.

All of the features of the paradigm of modern molecular photochemistry were readily incorporated into the developing paradigm of supramolecular photochemistry. However, in addition, the cross fertilization with chemical physicists and spectroscopists has provided a viewpoint on imagining what lies beyond supramolecular photochemistry.

Let us now present a specific example of a supramolecular photochemical system extremely sensitive to "übersupramolecular" effects on reactivity, i.e. the geminate radical pair. Geminate radical pairs are produced by photochemical reactions of ketones. Let us consider the triplet radical pairs commonly produced by the α-cleavage reaction of ketones. The intersystem crossing step from the triplet geminate pair to a singlet geminate pair determines the reactivity of the pair. In general, this step is very inefficient in ordinary solvents (molecular photochemistry), because the diffusional separation of the pair is much faster than intersystem crossing. However, it has been shown that the "complexing" of the radical pair as a substrate with a micelle or zeolite to form a supramolecular system can enhance the probability of geminate recombination tremendously.[5]

This result in itself is not particularly exciting, since it can be viewed as a simple "constraint" or inhibition of diffusion provided by the host micelle or zeolite. However, the BIG surprise comes about when it is found that gigantic magnetic isotope and magnetic field effects occur when gemimate radical pairs are produced in supramolecular systems, but not in molecular (ordinary non-viscous solvent) systems.

In the case of geminate radical pairs, the most important supramolecular correlations are the geminate atomic composition of the pair and the spin correlation of the pair. Both are intermolecular features that distinguish radical pairs and become, therefore, truly supramolecular when they influence chemical reactivity.

Consider a system for which photolysis of a supramolecular system consisting of a (guest) ketone associated with a (host) micelle produces a new supramolecular system consisting of two subunits, a triplet geminate radical pair. The micelle host has been termed a **supercage** and **the geminate pair guest in the micelle may be considered to be a stoichiometric entity or a supermolecule.** The term supercage is intended to describe a supramolecular system possessing some of the characteristics of a homogeneous solvent "cage", but in addition, possesses certain special supramolecular characteristics of a guest host complex. For example, the boundaries of the geminate pair-micelle supermolecule constrain separation of the pair, thereby discouraging long trajectories and encouraging or forcing reencounters on the pair. The supercage structure enhances the lifetime of the triplet geminate pair by orders of magnitude over that of its lifetime in a solvent cage (of a non-viscous homogeneous liquid). Depending on the details of the supramolecular structure (hydrophobic characteristics of the subunit radicals in the pair and the hydrophobicity and size of the micelle), the rate of escape of one or both radicals of the geminate from the supercage may be competitive with geminate reactions within the supercage. The radical subunits of the biradical undergo trajectories in space that are similar to those of the geminate pair in a micelle supercage, but instead of being random motions in a restricted supercage, the biradical subunits undergo motions in space that are correlated by the flexible connector of covalent bonds holding the subunits together. An important and very intriguing supramolecular issue is whether the flexible connector of the biradical supermolecule provides an intrinsic interaction between the subunits that is absent in the geminate pair in the supercage.

Consider the photolysis[5] of dibenzyl ketone, DBK, adsorbed in the inside of a micelle or in a zeolite cage. Let us assume that the usual situation for solution with respect to the photophysics of the ketone are appropriate for the complexed ketone, i.e., the intersystem crossing from the singlet excited state produced by light absorption crosses quantitatively and rapidly to the triplet excited state which then cleaves quantitatively to a spin and composition correlated triplet geminate radical pair, $^3[C_6H_5CH_2CO\cdot\ \cdot CH_2C_6H_5]_{GP}$. In this case the composition is correlated to that of the starting ketone, i.e., each atom in the radical pair is correlated with each atom of the parent ketone. The electron spin of the geminate radical pair is correlated to that of the starting parent ketone, i.e., the orbitally uncoupled electron spins are parallel and correspond to a triplet state. The geminate radical pair also possesses a constitution correlation identical to that of the starting material in all respects except for the disconnection of the bond that has undergone α-cleavage. As time goes on, the three correlations of spin, composition and constitution will be lost as the triplet relaxes to singlets or to uncorrelated doublets and as the geminate composition becomes lost as the result of diffusional separation and randomization of the radicals in the global space of the solution. Finally, the constitution correlation becomes lost as the $C_6H_5C_2CO$ radical loses CO *via* bond cleavage. The time scales for each correlation loss are different and, as suggested by the $t = R^2/D$ relationship, they depend on several factors that are functions of the size (R) of the available space. In this regard, the different topological regions of a given system may have different effects on the evolution of the triplet excited state and the subsequent radical intermediates.[5,6] Importantly this simple equation ties together the "space" of the supermolecular host (through R) with the time of some resonance process.

The loss of spin correlation of the geminate radical pair mediates the probability of competition between primary geminate recombination and decarbonylation. The loss of spin correlation is a strong function of the intramolecular magnetic fields due to nuclear hyperfine coupling and to spin orbit coupling of the odd electrons in the radical pair, and to the separation of the radical centers in space. The strength of the hyperfine coupling in the radical pair is distance independent, but its ability to induce intersystem crossing is strongly distance dependent. The strength of the spin orbit coupling is strongly distance dependent, being related to the magnitude of orbital overlap between the two radical centers, and falling off roughly exponentially as a function of the radical pair separation. The magnitude of the singlet-triplet energy gap depends on distance in a manner similar to that of the spin-orbit coupling. Although the situation is obviously complex, the essential points are readily comprehended and the qualitative effects are easy to identify. For example, if the size of the restricted space is very small relative to the size of the radical pair, the singlet-triplet gap will be large (enforced orbital overlap) and the intersystem crossing will be slow. Decarbonylation should occur preferentially over primary geminate coupling. On the other hand, if the size of the restricted space is large relative to the size of the radical pair, the triplet-singlet separation will tend to zero and either hyperfine or spin orbit induced intersystem crossing will be effective. If the size of the topological region is adequate, the probability of reencounter of the singlet radical pair will be correlated with the rate of intersystem crossing and the formation of geminate products will compete with decarbonylation. However, if the size of the space is too large, separated radical pairs will tend to lose their compositional correlation as they spread throughout the space and encounter radicals generated from other parent ketones. Finally, the constitutional composition will be lost if the primary geminate pair is separated long enough for loss of carbon monoxide to occur. Several situations covering these possibilities have been observed experimentally in several supramolecular systems which we discuss below. Interestingly, these effects are modulated by J, the exchange interaction, perhaps the most subtle of intermolecular interactions in radical pairs.

4. The Role of the Environment on the Photochemistry of Dibenzyl Ketone.

Various reaction media can impose different constraints on the molecular freedom of dibenzyl ketone. It is not surprising to observe completely different photochemical results in different reaction media considering that the space available to the reactant is of great importance in modulating the competition between the alternative deactivation and decay processes. These may all be characterized as supramolecular photochemical systems. Prospective reaction environments may range from the relatively tight and tailor-made medium of the crystalline solid state to the somewhat larger environments of inclusion complexes, zeolites and micelles. According to the topological regions where the reaction takes place, restricted environments and organized media can be classified as topologically closed or topologically open environments. Closed environments are those systems in which the entire reaction sequence occurs under the influence exerted by a unique topological space. Reactions occurring solely at the surface of a stretched polymer, at the external surface of a solid support, or within the boundaries imposed by the host of an inclusion complex can all be considered under this category. Topologically open environments, on the other hand, are those in which the substrates, excited states or reaction intermediates are able to partition between more than one of the topological regions available in the system.

4.1 THE EXTRAPOLATION TO SUPERDUPER MOLECULAR PHOTOCHEMISTRY

With this background we can consider the question, "What comes after the supermolecule and supramolecular assemblies"? An obvious, straightforward and important response is "super"supermolecules or "super" supramolecular assemblies. However, in this manuscript we propose to consider a concept that allows the consideration of terms that may be missing in the Hamiltonians of systems at any level of complexity and, therefore, take the system to a level of complexity that is "sideways" rather than "vertical". We shall use the terms "ultraelectronic", "ultraatomic", "ultramolecular" and "ultrasupramolecular" to describe systems for which static or oscillating electric and magnetic fields are included in the Hamiltonian. The static and oscillating fields may be applied externally by the experimentalist as electromagnetic radiation or laboratory electrostatic or magnetic fields. In addition, electric and magnetic fields resulting from the components of the system at the atomic and molecular and supermolecular level also occur.

A number of these effects have actually been demonstrated to occur for geminate radical pairs.[7] For example, the application of static external magnetic fields can "hold" the electron spin orientation of the triplet state and allow the correlation of the triplet pair to be preserved. To be effective, the pair must remain "interacting" for a relatively long period of time. This is where the supramolecular nature of the pair comes in. If the geminate character and interaction of the pair is preserved, oscillating and static magnetic fields can influence the triplet to singlet interconversion and thereby effect the reactivity of the pair. One of the electron spins can be "locked" by a certain frequency of microwave electromagnetic radiation, and intersystem crossing may be induced by using radiofrequency radiation to cause simultaneous nuclear and electronic spin flips. These ideas suggest that the superdupermolecular organic photochemistry will involve the integration of the concepts of supramolecular chemistry, spectroscopy and chemical physics. The notion of magnetic fields as an überchemistry can also be extended to the concept of electric fields, both static and oscillating.

5. Conclusion

Photochemistry has long possessed the richness of multidimensionality in structure, dynamics, time, space and energy. Photochemists must keep track of time scales, energy surfaces, magnetic effects, etc. in examining the mechanisms of photoreactions. Perhaps the most important consequence of "thinking supramolecularly" is the ability to "see" apparently disparate chemical systems as "the same thing" in the sense that techniques and strategies for mechanistic investigation can be mapped from a well established supramolecular system onto a new one, once the topological supramolecular connections have been established. In this respect the move from molecular chemistry to supramolecular chemistry has the flavor of moving from "static molecular structures" to conformationally flexible structures. The next level of expanded thinking could be to consider the special features of übersupramolecular systems allowing for the enhancement of magnetic field and electric field effects that are absent or very weak in molecluar systems.

Acknowledgements

The author thanks the KAST (Kanagawa Academy of Science and Technology) and the National Science Foundation for their generous support of this research.

References

1. Kuhn, T. S. (1970) The Structure of Scientific Revolutions, University of Chicago Press, Chicago.

2. Lehn, J. M. (1995) Supramolecular Chemistry, VCH, New York.

3. Balzani, V. (ed) (1987) Supramolecular Photochemistry, Reidel, Dordrecht; Balzani, V. and Scandola, F. (1991) Supramolecular Photochemistry, Ellis Horwood, New York; H. J. Schneider and H. Dürr (eds) (1991) Frontiers in Supramolecular Organic Chemistry and Photochemistry, VCH, Weinheim.

4. Turro, N. J. (1991) Modern Molecular Photochemistry, University Science Books, Mill Valley, CA .

5. See for reviews: Turro, N.J. (1995) *Pure & Appl. Chem.*, **67,** 199; Turro, N.J. Buchachenko, A.L.and Tarasov, V.F. (1995) *Acc. Chem. Res.*, **28,** 69 .

6. Tarasov, V.F., Ghatlia, N.D., Buchachenko, A.L. and Turro, N.J. (1992) J. Am. Chem. Soc., **114**, 9517.

7. Buchachenko, A. L. (1995) *Chem. Rev.*, **95,** 2507.

NON-COVALENT SYNTHESIS AND CHARACTERIZATION OF LARGE SUPRAMOLECULAR ASSEMBLIES

D.N. REINHOUDT*, P. TIMMERMAN, F.C.J.M. VAN VEGGEL
Laboratory of Supramolecular Chemistry and Technology,
University of Twente, P.O. Box 217, 7500 AE Enschede,
The Netherlands

ABSTRACT: This review describes the synthesis and characterization of two different types of supramolecular assemblies. As non-covalent interactions hydrogen bonding and coordination chemistry have been used. Various techniques were studied, like NMR spectroscopy, X-Ray crystallography, ESMS and MALDI TOF mass spectrometry, and AFM to characterize these assemblies both in the solid state and in solution. Assembly of the calix[4]arene boxes via hydrogen bonding gives particles that are 3-5 nm in size, whereas self-assembly of the metallo-pincer ligands leads to spherical assemblies with diameters in the range of 100 to 400 nm, with standard deviations in the order of 10-15 %. The controlled assembly approach allows a precise control of size and assemblies with molecular weights up to 10,000 Dalton have been realized. The combination of hydrogen bonding and metal-coordination has been achieved in the synthesis of metallodendrimers with a hydrogen-bonded core.

1. Introduction

Nature constructs receptors by the combination of a limited number of building blocks such as amino acids, nucleotides and saccharides. Systematic variation of the monomer sequence gives access to an infinite number of bioreceptors. The price Nature pays for this strategy is high molecular weight which may even be larger when the biologically active species is not monomeric but composed of several subunits connected by non-covalent bonds. Over the past decade we have more or less adopted this strategy for the synthesis of artificial receptors. We use medium-sized, rigid modules to which functional groups can be attached and which we combine to larger receptor molecules.
Calix[4]arenes and resorcin[4]arenes are representatives of such modules but also metallosalenes, porphyrins, terphenyls and cyclodextrins have been used to achieve specific shapes and functions [1]. In terms of size this strategy reaches its limitation with molecules like the holand (**1**), which was synthesized in 17 steps in order to combine two calix[4]arenes with two resorcin[4]arenes in a very rigid, preorganized network.

G. Tsoucaris (ed.), Current Challenges on Large Supramolecular Assemblies, 51–66.

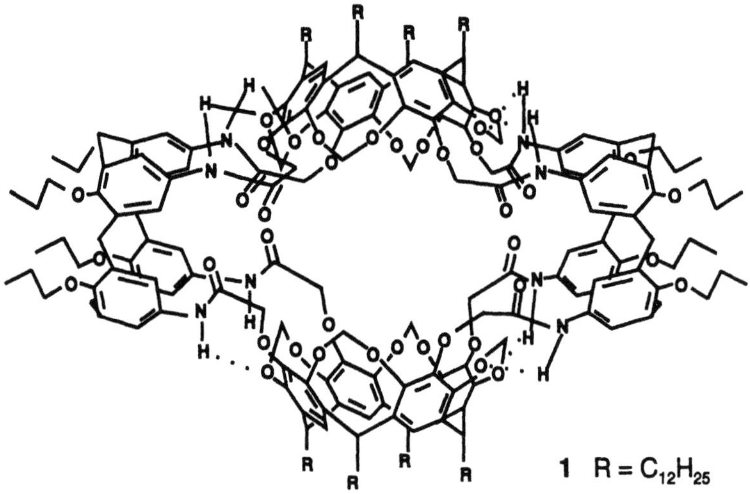

1 $R = C_{12}H_{25}$

Therefore we are currently investigating an alternative way to combine such modules *viz* via non-covalent bonds. In order to arrive at the desired final structure, the modules should contain all chemical information to direct the assembly process in a unique way. We investigate two approaches to assemble large systems (4-100 kD) *viz* via hydrogen bonding of functionalized calix[4]arenes and via coordination chemistry of functionalized porphyrins and metallopincer ligands (vide infra).

In the area of hydrogen-bonded aggregates one of the fundamental problems is that the individual bonding interactions are weak but that multiple hydrogen bond acceptors and donors in one molecule render these extremely insoluble. We try to solve this problem by the assembly of nine individual components in *box-like* structures. This has the advantage that the individual components with only three or six hydrogen-bond donors or acceptors are still sufficiently soluble in organic solvents, but that the ultimate assembly is held together by 36 hydrogen bonds [2].

In the alternative approach we use coordinative interactions between organic ligands and metal centers (Pt, Pd) incorporated in so-called pincer structures. These building blocks can (self-)assemble upon removal of a protecting, kinetically stable, ligand at the coordinated metal center. Assembly processes can be either convergent or divergent to give metallodendrimer structures.

Finally we have combined both strategies in a hybrid structure of dendritic wedges held together in a cyclic rosette motif.

2. Hydrogen-Bonded Assemblies

2.1. SELF-ASSEMBLY OF HYDROGEN-BONDED MOLECULAR BOXES

Calix[4]arene **1**, diametrically substituted at the upper rim with two melamine units, spontaneously forms a well-defined box-like assembly in the presence of two equivalents of 5,5-diethylbarbiturate (**DEB**) (see Figure 1) [3]. The self-assembled calix[4]arene box $(\mathbf{1})_3{\bullet}(\mathbf{DEB})_6$, consisting of nine different components, is held together by a total of 36 hydrogen bonds. It is stable in apolar solvents like chloroform and toluene even at 10^{-4} M, but the stability progressively decreases when significant amounts (10-20%) of highly polar solvents (like DMSO and methanol) are added.

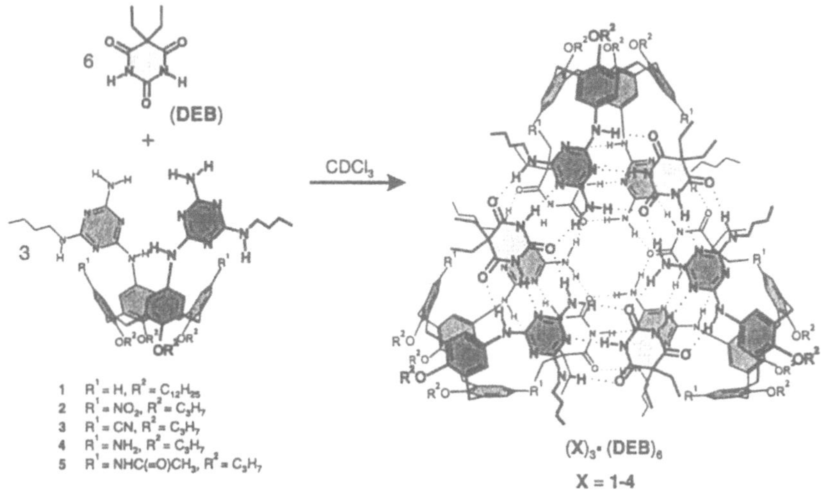

Figure 1. Formation of calix[4]arene box-like assemblies $(X)_3{\bullet}(DEB)_6$ (X = 1-4).

2.2. CHARACTERIZATION

2.2.1. ^1H NMR Spectroscopy

Titration of dimelamine **1** with **DEB** in $CDCl_3$ shows several characteristic features (Figure 2). Already at low concentrations of **DEB** two signals become evident at very low field. These resonances, which remain at δ = 14.10 and 13.32 regardless of the **1**:**DEB** ratio, are assigned to the hydrogen-bonded NH protons of **DEB** in the complex. These protons are observed at different chemical shifts as a result of the unsymmetrical substitution of the melamine units, which gives both protons a chemically different environment in the complex. When the amount of **DEB** is increased, the signal corresponding to the NH_2 protons of free **1** decreases in intensity but remains at the same position. At the same time, two signals are observed around δ = 6.9 (e) and 6.7 (f) for both the NH_2 protons. Two additional signals appear at δ = 8.37 (c) and 7.43 (d), which correspond to the two secondary amine protons of **1** in the hydrogen-bonded complex. The aromatic protons of the melamine substituted aromatic ring of calix[4]arene **1** give rise to signals at δ = 7.15 (g) and 6.03 (h). In free **1** these protons display broad signals at δ = 6.65-6.05. The resonance at δ = 6.03 is in accordance with a *pinched cone* structure in which the two melamines approach each other [3].

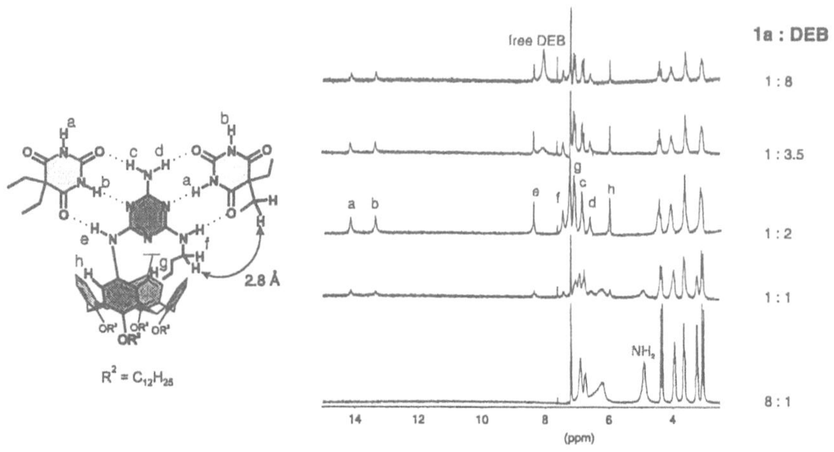

Figure 2. ¹H NMR titration (400 MHz) of 1 with DEB in $CDCl_3$ + intramolecular NOE distances.

At a ratio of **1**:**DEB** of 1:2 the spectrum is sharp, which indicates the absence of free **1**. The 1:2 ratio is consistent with the box-like assembly that is represented in Figure 1. When more than two equivalents of **DEB** are added, the signal for the NH protons of free **DEB** is observed alongside the two signals for the hydrogen-bonded **DEB**. This indicates that exchange between hydrogen-bonded and free **DEB** is slow on the NMR time scale.

2.2.2. *Single Crystal X-Ray Diffraction Studies*

Upon recrystallization from toluene the assembly $(\mathbf{2})_3\bullet(\mathbf{DEB})_6$ (R^1 = NO_2) forms large single crystals. The X-ray structure, which is shown in Figure 3, provides the first crystallographic information for this type of molecular boxes [4]. The structure shows that the calix[4]arene units are fixed in a *pinched cone* conformation, which is the only conformation that allows simultaneous participation of the calix[4]arene units in both rosette motifs. The two rosettes tightly stack on top of each other with an interatomic separation of 3.5 Å at the edges to 3.2 Å in the centre of the box, which leaves little space for guest molecules. Interestingly, the structure reveals that the two rosette motifs are oriented in an *anti-parallel* fashion (see also Figure 1), which means that the assembly is *stereogenic.*

2.2.3. *2D NOE Spectroscopy*

Evidence for the fact that the solid state structure resembles the structure in solution was obtained by measuring the distances between the relevant protons in the box-like assembly with 2D NOESY experiments using the initial rate approximation. The distances are slightly larger than those observed in the crystal structure, but still provide sufficient evidence for the proposed assembly structure. A particularly strong NOE connectivity was observed between the NCH_2 protons of **2** (R^1 = NO_2) and one of the ethyl-CH_2 groups of **DEB** (see Figure 2). The interatomic distance of these protons *within one single* rosette is too large (*ca.* 6.4 Å) to cause the observed NOE connectivity. Therefore, it must arise from a proximity effect between the two rosette motifs, which consequently puts a limit to their mutual orientation. In the *anti-parallel* orientation (also observed in the solid state) the two protons are 2.8 Å apart, which is in perfect agreement with the strong NOE observed. The presence of this NOE connectivity provides strong evidence for the fact that the structure in solution closely resembles the solid state structure as determined by X-ray crystallography [4].

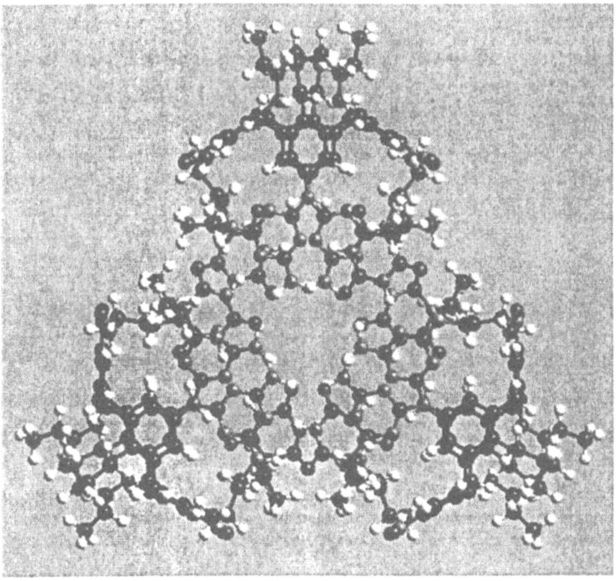

Figure 3. X-Ray crystal structure of assembly $(2)_3$•$(DEB)_6$ (crystals obtained from toluene).

2.2.4. Ag^+ Assisted MALDI TOF Mass Spectroscopy

Additional evidence for the self-assembly of the boxes was obtained using Ag^+ assisted MALDI TOF mass spectroscopy [5]. This novel technique is based on the remarkably high affinity of silver cations for cyano groups and it provides for a *non-destructive* way to generate positively charged hydrogen bonded assemblies that can be easily detected by MALDI-TOF mass spectrometry. A sample prepared by stirring assembly $(\mathbf{3})_3$•$(\mathbf{DEB})_6$ (R^1 = CN) with 2.0 equiv. of silver trifluoroacetate gives rise to an intense signal at *m/z* 4220.0 (calc. for $^{13}C_2{}^{12}C_{214}H_{282}N_{54}O_{30}$•$^{107}Ag^+$-complex: 4221.9) in the MALDI-TOF mass spectrum (see Figure 4). Signals corresponding to partially formed aggregates or higher oligomers were not observed in the mass spectrum, which illustrates the unprecedented mildness of this technique. Formation of the silver complex of assembly $(\mathbf{3})_3$•$(\mathbf{DEB})_6$ (R^1 = CN) clearly results from coordination of a silver cation to one of the cyano groups, since neither of the assemblies $(\mathbf{1})_3$•$(\mathbf{DEB})_6$ (R^1 = H) or $(\mathbf{2})_3$•$(\mathbf{DEB})_6$ (R^1 = NO_2) shows any significant signal between *m/z* 1500 and 8000 in the corresponding MALDI-TOF mass spectrum [5].

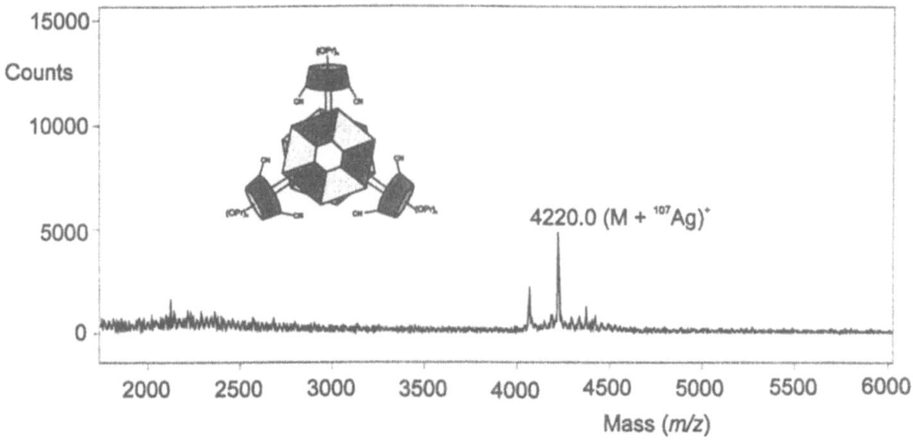

Figure 4. MALDI TOF mass spectrum of assembly $(3)_3$•$(DEB)_6$ *in the presence of 2.0 equiv. of* $AgCF_3COO$.

2.3. FUNCTIONALIZED HYDROGEN-BONDED ASSEMBLIES.

In the hydrogen-bonded assembly *six* functionalizable sites R (see Figure 1) are gathered around the box-like cavity. In view of our objective to generate functional group diversity in or around non-covalently assembled cavities, we studied how the nature of the functionality at positions R affects the stability of such calix[4]arene box-like assemblies in solution [4].

The introduction of polar substituents at positions R, such as nitro or cyano, does not influence the stability of the cyclic hydrogen bonded assembly in solution to any significant extent. Compounds **2** (R^1 = NO_2) and **3** (R^1 = CN) readily assemble in chloroform in the presence of **DEB**, the ^{1}H NMR spectra of which are virtually identical to that of $(\mathbf{1})_3$•$(\mathbf{DEB})_6$ (R^1 = H). Titration of **3** (R^1 = CN) with **DEB** clearly proved the 3:6 stoichiometry of the box-like assembly formed.

Surprisingly, the stability of assembly $(\mathbf{4})_3$•$(\mathbf{DEB})_6$ (R^1 = NH_2) is highly solvent dependent. The ^{1}H NMR spectrum in toluene-d_8 is very well-defined and shows two singlets at 14.55 and 13.85 ppm for the magnetically nonequivalent barbiturate NH protons. In $CDCl_3$, however, the spectrum is very broad and does not show any resonance in the region between 13 and 15 ppm, which indicates a preference for non-specific oligo- or polymer formation in this solvent. The decreased stability of $(\mathbf{4})_3$•$(\mathbf{DEB})_6$ in the more polar solvent $CDCl_3$ is most probably related to the hydrogen bond donating ability of the two NH_2 groups.

Compound **5** [R^1 = NHC(O)CH_3] does not form the box-like assembly in either $CDCl_3$ or toluene-d_8. Both ^{1}H NMR spectra exhibit broad resonances indicating non-specific aggregation. Also in this case the formation of intramolecular hydrogen bonds, *i.e.* between the amide carbonyl and the melamine NH proton, seems to be primarily responsible for paralyzing the assembling ability of **5**.

3. Metallodendrimers

3.1. SELF-ASSEMBLY APPROACH

Our first approach is schematically depicted in Figure 5. Two metal centers are bound kinetically inert to a "pincer" type ligand bridged via a spacer that contains a ligand that **cannot** intramolecularly bind to the metal centers. The fourth coordination position of the metal is occupied by a weakly bound ligand (e.g. acetonitrile). After removal of this weakly bound ligand, the ligand in the dinuclear building block will occupy the fourth position intermolecularly. This will create two reaction sites to which two building blocks can be added, etc.

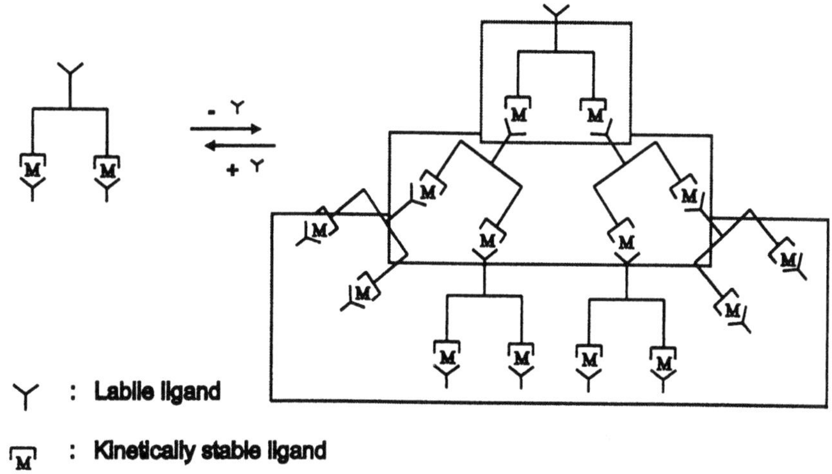

Figure 5: Schematic self-assembly strategy

The building block that satisfies all demands is shown in Figure 6 (BB_{Cl}) [6]. It is made as the bis-chloride adduct, which is an overall neutral compound that can easily purified by column chromatography. The chlorides can be quantitatively be removed by reaction with two equivalents of $AgBF_4$ in acetonitrile.

After removal of the two coordinated acetonitriles by gentle heating in nitromethane (polar but non-coordinative) the formation of large assemblies is induced. The ^{1}H NMR spectrum of this assembly shows broad peaks, which sharpened after the addition of small amounts of acetonitrile. Removal of the acetonitrile restored the original broad spectrum, showing that the assembly and disassembly is a completely reversible process. The growth could also easily be followed by IR spectroscopy. The uncoordinated cyano group exhibits an absorption at 2252 cm^{-1}, whereas the coordinated cyano group has an absorption at 2290 cm^{-1}. The size as determined by QELS (quasi-elastic light scattering) in nitromethane solution was 180 nm. The AFM picture after spreading the assembly on a glass or gold support is given in Figure 7. The mean size is 205 nm with a standard deviation in the order of 25-30 nm. The absorption at 2290 cm^{-1} was also observed by grazing-angle FT-IR spectroscopy, in agreement with the bulk spectrum.

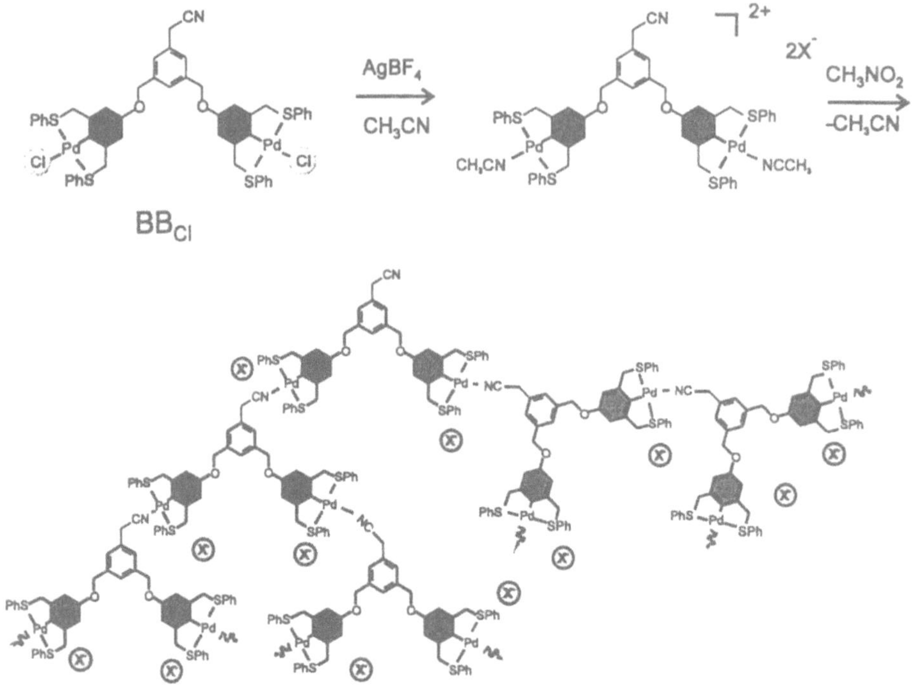

Figure 6: Self-assembly of dendritic metallodendrimer

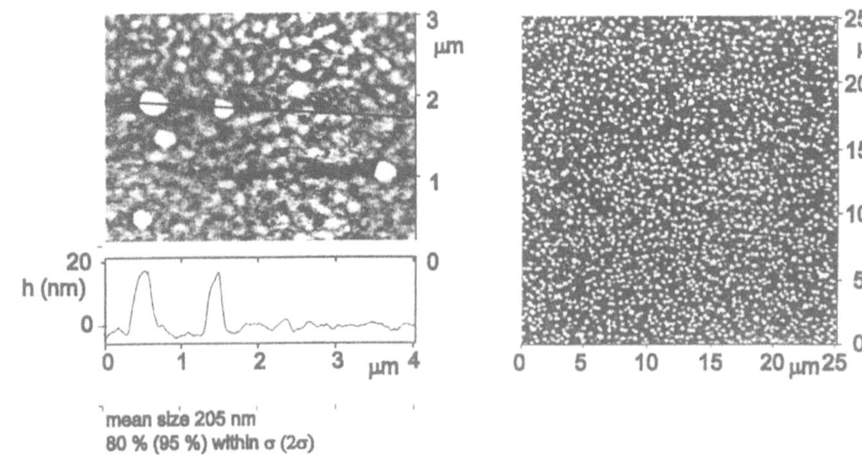

Figure 7: AMF picture of flattened spheres on gold.

Finally, we have characterized the assemblies by TEM (Figure 8). The contrast was easily obtained without the need of shadowing, as a result of the presence of à large number of palladium ions. This is a direct proof of the concept shown in Figure 4.

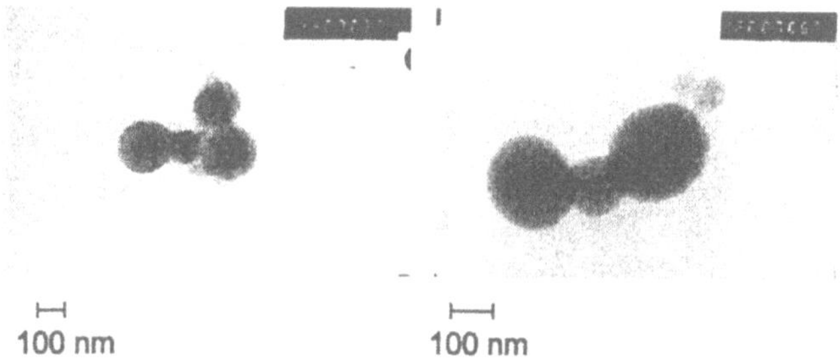

Figure 8: TEM images

In order to control the size of the metallodendrimers a structural variation was carried out. The substituent on the sulfurs have been varied from ethyl, *tert*-butyl, phenyl, to naphthyl, as well as the counter ions (BF_4^-, ClO_4^-, triflate, tosylate, and BPh_4^-) [7]. This simple structural variation allows the formation of assemblies in the range of 50 to 200 nm (radius), with a relatively narrow size distribution (standard deviation in the order of 10 to 15 %). In general the larger the substituents and/or the larger the counter ions the smaller the assemblies (Figure 9). We do not have an explanation for the anomalous behavior of BF_4^-. From AFM data sizes of 160 (± 20) nm for R = naphthyl and X^- = BF_4^- and 110 (± 20) nm for R = naphthyl and X^- = OTf^-. This is in good agreement with the QELS data. In case of R = phenyl and X^- = Tosylate and PF_6^- extensive clustering made size determination unreliable. The phosphorous analogue of BB_{Cl} and G_0 have been made allowing the incorporation of Ni^{2+}, Pt^{2+}, and Pd^{2+}, but the assemblies proved very sensitive to oxidation [8].

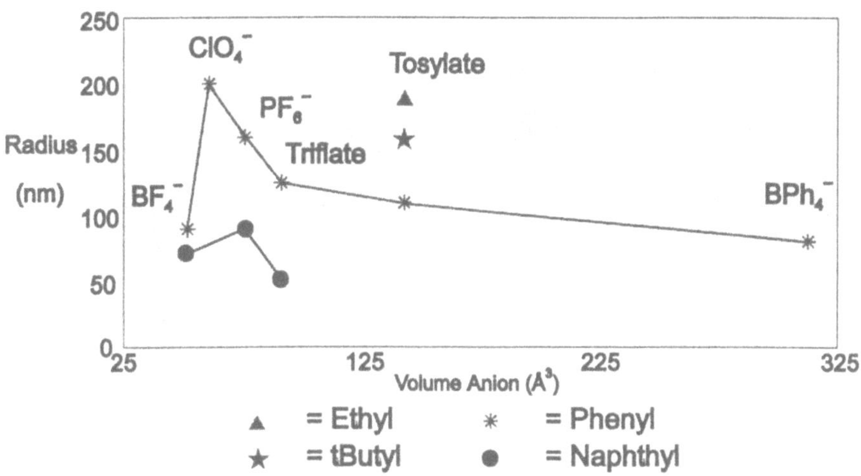

Figure 9: Sphere size vs. anion size (Connoly Surface).

3.2. CONTROLLED ASSEMBLY

In order to synthesize monodisperse assemblies of nanosize dimensions we have changed the strategy a little. In the divergent growth we start from the trifunctional core G_0, which is activated for further growth by removal of the three chloride ligands with $AgBF_4$. To this activated core three equivalents of the building block BB_{Cl} are added giving the first generation (G_1) (Figure 10). This two-step sequence has been used up to G_5. The infrared spectra show the characteristic absorption around 2290 cm^{-1} for the coordinated cyano group and the ^1H NMR spectra are informative up to G_3 with respect to the ratio of the signals due to the CH_2CN and CH_2O moieties. Generations G_1 to G_4 could be characterized by ES-MS (Figure 11), showing the molecular ion minus the loss of counter ions and chloride ligands (G_1). The fragmentation observed in the spectrum of G_2 is believed to have taken place in the mass spectrometer [9].

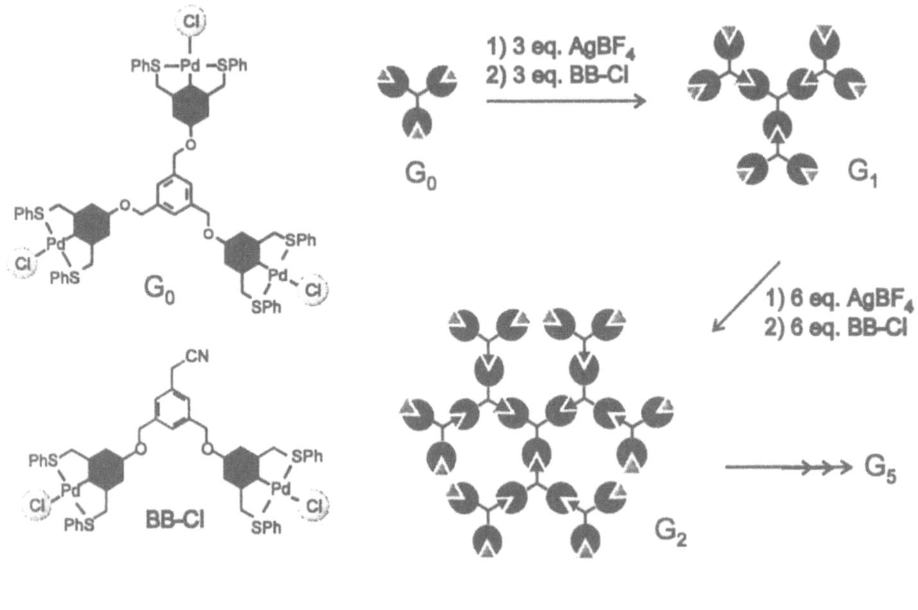

Figure 10: Strategy of controlled assembly.

Generation five could not be studied by ES-MS due to the too low resolution of the analyzer, but TM-AFM was successful (Figure 12). Figure 12 shows flattened spherical aggregates with an approximate diameter of 15-20 nm and a height of 4.2 nm [10].

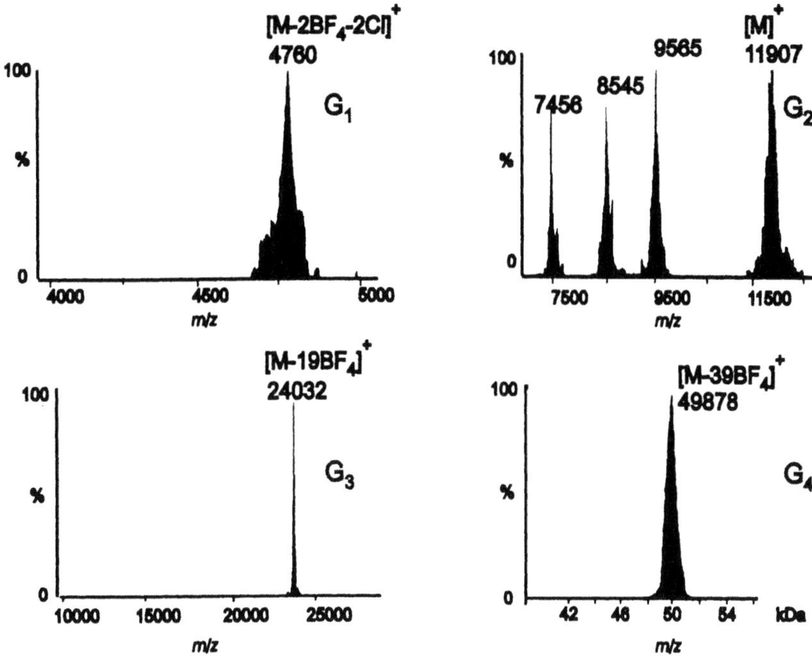

Figure 11: Deconvoluted electrospray mass spectra of G_1-G_4.

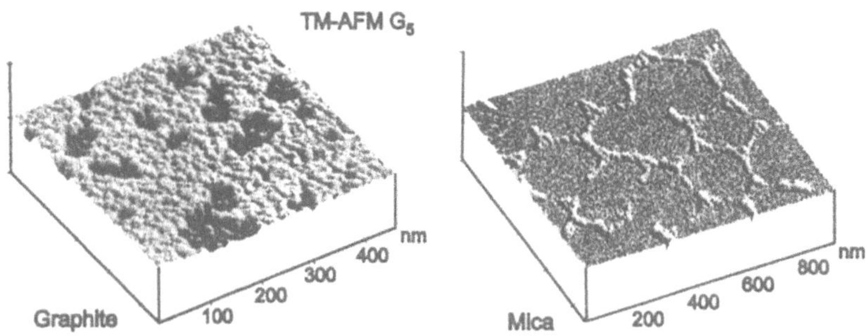

Figure 12: TM-AFM picture of G_5.

Sofar, no true functionality is present in the assemblies. Porphyrins have a.o. interesting optical properties and therefore a porphyrin core and a porphyrin building block have been synthesized (Figure 13) [11]. With the controlled assembly, assemblies with 3 (G_0-Por_3, not shown), 6, and 12 porphyrins, respectively, at the outer generation have been synthesized (Figure 14). G_0-Por_3 could a.o. be chararcterized by field-desorption mass spectrometry. Upon assembling the mono-pyridyl-porphyrin to the palladium center, the ^{1}H NMR spectra show a down field shift of the CH_2S groups. IR and NMR spectroscopy gave the expected absorption of the cyano groups and ratio of the various methylene groups, respectively. Similarly, 4 or 8 porphyrins have been assembled around the tetrafunctional porphyrin core (Figure 14). The MALDI-TOF spectra of these assemblies show the molecular ion at 4961 and 11547 Dalton, respectively (Figure 15).

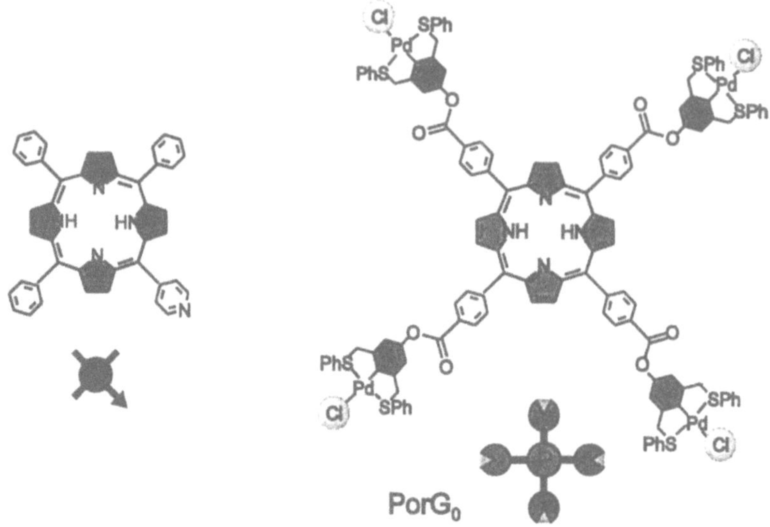

Figure 13: Porphyrin-containing building blocks.

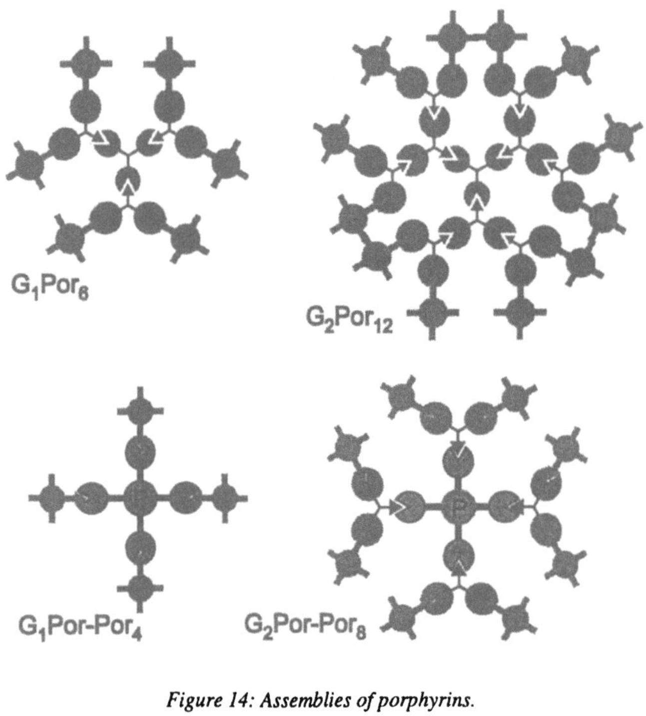

Figure 14: Assemblies of porphyrins.

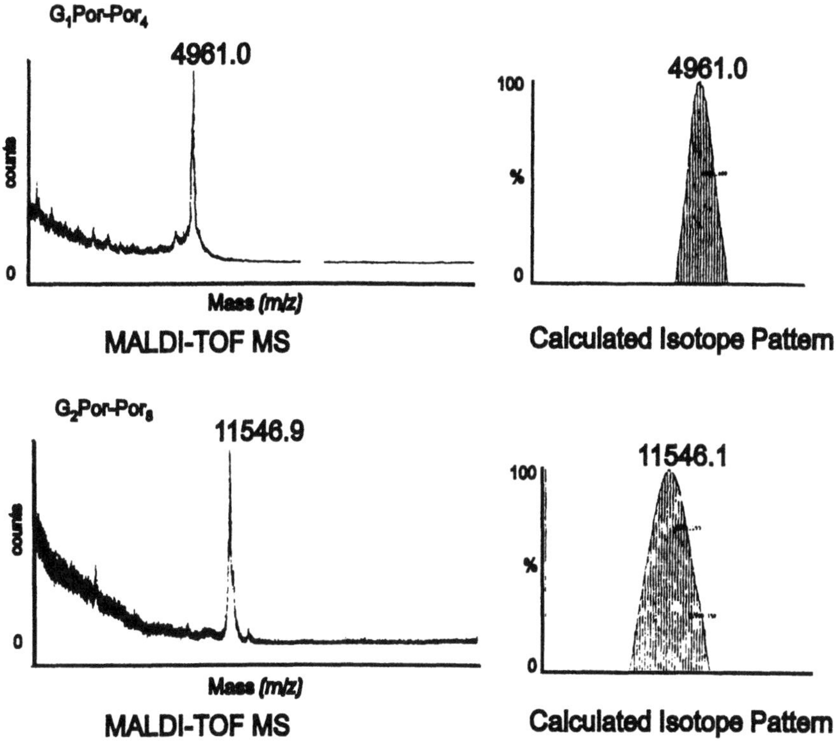

Figure 15: MALDI-TOF spectra of $G_1Por\text{-}Por_4$ and $G_2Por\text{-}Por_8$.

4. Combining Hydrogen Bonding and Metallodendrimers

In order to enlarge the scope of the assembly process we have combined the coordination chemistry with hydrogen bonding. If melamines with large substituents on two of the amino groups are used the formation of a [3+3] hydrogen bonded *rosette* (RG_n), is favored over non-cyclic species. We have synthesized a barbiturate with a pincer-palladium complex attached. Metallodendrimers with up to three generations were added in the controlled assembly approach [12]. Addition of a sterically demanding melamine derivative to a solution in CD_2Cl_2 of the barbiturate gave spontaneously the formation of rosettes (Figure 16).

The low temperature NMR spectra clearly show the hydrogen-bonded protons at low field, which could all be assigned by 2D NOESY, COSY, and TOCSY experiments (Figure 17). The hydrogen bonding did not interfere with the coordination chemistry, making it an "orthogonal" tool in the development of new assembly strategies. The rotation correlation times τ_c for RG_n (n = 1-3) were determined as 1.32, 2.75, and 3.55 x 10^{-9} s, respectively, in agreement with increasing size (M_{calcd} = 7266, 14177 and 27972 Da for RG_1, RG_2, RG_3, respectively) of the assemblies.

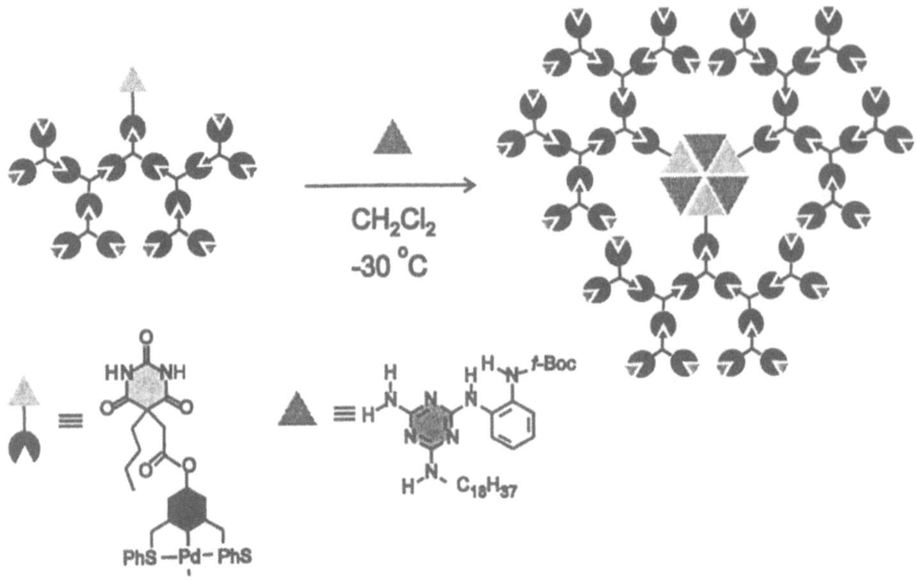

Figure 16: Formation of a rosette RG_3 with metallo-wedges.

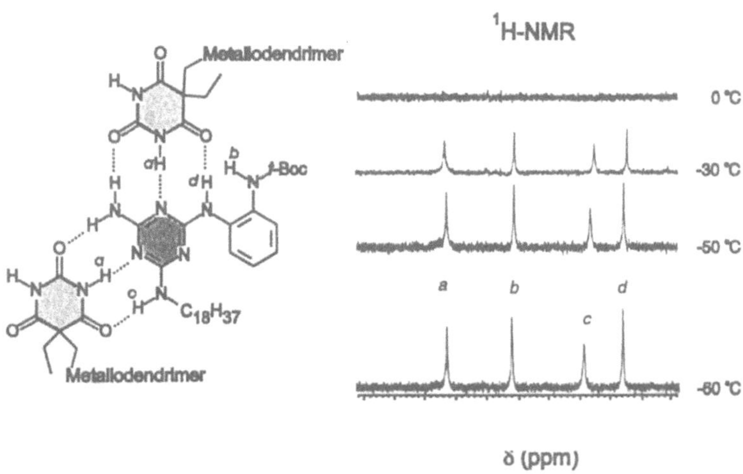

Figure 17: Low temperature spectra of rosette with metallo-wedges.

5. Conclusions

In this chapter we have presented the complete characterization of hydrogen bonded assemblies of calix[4]arenes by ^{1}H NMR spectroscopy, single crystal X-ray diffraction and MALDI TOF mass spectrometry. The assemblies, which are highly stable in apolar solvents, are *stereogenic* as a result of an anti-parallel orientation of the two rosette motifs.

Furthermore, three strategies are discussed to synthesize nanosize structures using coordination chemistry. The self-assembly approach allows the fast formation of spherical assemblies with diameters of 100 to 400 nm, with a relatively narrow size distribution. Monodisperse assemblies can be made with the controlled assembly approach, which comprises a simple two-step procedure. The third method exploits both coordination chemistry and hydrogen bonding, thus greatly enlarging the scope of these non-covalent strategies.

6. Acknowledgement

The work described in this paper is the result of the efforts of more than 10 coworkers whose names appear in the references. We thank R. Fokkens and N. Nibbering (University of Amsterdam) for the FD-MS, ES-MS, and MALDI-TOF MS measurements. We are grateful for financial contributions of the NATO (research grant 930740) and the EC (Marie Curie research training grant ERBFMBICT 961445).

7. References

1. a) Timmerman, P., Verboom, W., Van Veggel, F.C.J.M., Van Hoorn, W.P., and Reinhoudt, D.N. An organic molecule with a rigid cavity of nanosize dimensions, *Angew. Chem. Int. Ed. Engl.* **33**, 1292-1295; b) Higler, I., Timmerman, P., Verboom, W., and Reinhoudt, D.N. (1996) Combination of calix[4]arenes and resorcinarenes for the complexation of steroids, *J. Org. Chem.* **61**, 5920-5931; c) Timmerman, P., Verboom, W., Van Veggel, F.C.J.M., Van Duynhoven, J.P.M., and Reinhoudt, D.N. (1994) A novel type of stereoisomerism in calix[4]arene-based carceplexes, *Angew. Chem. Int. Ed. Engl.* **33**, 2345-2348; d) Van Dienst, E., Snellink, B.H.M., Von Pieckartz, I., Engbersen, J.F.J., and Reinhoudt, D.N. (1995) Novel water-soluble cyclodextrin-calix[4]arene host molecules with strongly enhanced binding properties, *J. Chem. Soc., Chem. Commun.* 1151-1152; e) Rudkevich, D.M., Verboom, W., and Reinhoudt, D.N. (1995) Capped biscalix[4]arene-Zn-porphyrin: metalloreceptor with a rigid cavity, *J. Org. Chem.* **60**, 6585-6587.
2. Whitesides, G.M., Simanek, E.E., Mathias, J.P., Seto, C.T., Chin, D.N., Mammen, M., and Gordon, D.M. (1995) Noncovalent synthesis: using physical-organic chemistry to make aggregates, *Acc. Chem. Res.* **28**, 37-44 and references cited therein.
3. Vreekamp, R. H., Hubert, M., Van Duynhoven, J.P.M., Verboom, W., and Reinhoudt, D. N. (1996) Molecular Boxes based on calix[4]arene double rosettes, *Angew. Chem., Int. Ed. Engl.* **35**, 1215-1218.
4. Timmerman, P., Vreekamp, R. H., Hulst, R., Verboom, W., Reinhoudt, D. N., Rissanen, K., Udachin, K. A., and Ripmeester J. (1997) Noncovalent assembly of functional groups on Calix[4]arene Molecular Boxes, *Chem. Eur. J.* **3**, 1823-1832.
5. Jolliffe, K.A., Crego Calama, M., Fokkens, R., Nibbering, N.M.M., Timmerman, P., and Reinhoudt, D. N. Characterization of hydrogen bonded supramolecular assemblies by Ag^+ assisted MALDI TOF mass spectrometry, submitted for publication.
6. a) Huck, W.T.S., Van Veggel, F.C.J.M., Kropman, B.L., Blank, D.H.A., Keim, E.G., Smithers, M.M.A., and Reinhoudt, D.N. (1995) Large self-assembled organopalladium spheres, *J. Am. Chem. Soc.* **117**, 8293-8294; b) Huck, W.T.S., Van Veggel, F.C.J.M., and Reinhoudt, D.N. (1997) Self-assembly of hyperbranched spheres, *J. Mater. Res.* **7**, 1213-1219.
7. Huck, W.T.S., Snellink-Ruël, B.H.M., Lichtenbelt, J.W.Th., Van Veggel, F.C.J.M., and Reinhoudt, D.N. (1997) Self-assembly of hyperbranched spheres, *Chem. Commun.* 9-10.

8. Huck, W.T.S., Snellink-Ruël, B.H.M., Van Veggel, F.C.J.M., and Reinhoudt, D.N. (1997) New building blocks for the noncovalent assembly of *homo-* and *hetero*-multinuclear metallodendrimers, *Organometallics*, **16**, 4287-4291.
9. Huck, W.T.S., Van Veggel, F.C.J.M., and Reinhoudt, D.N. (1996) Controlled assembly of nanosized metallodendrimers, *Angew. Chem., Int. Ed. Engl.* **35**, 1213-1215.
10. Huck, W.T.S., Van Veggel, F.C.J.M., Sheiko, S., Möller, M., and Reinhoudt, D.N. Molecular thin films of metallodendrimers, submitted.
11. Huck, W.T.S., Rohrer, A., Anilkumar, A.T., Fokkens, R., Nibbering, N.M.M., Van Veggel, F.C.J.M., and Reinhoudt, D.N. Noncovalent synthesis of multi-porphyrin systems, *New J. Chem.*, accepted.
12. Huck, W.T.S., Hulst, R., Timmerman, P., Van Veggel, F.C.J.M., and Reinhoudt, D.N. (1997) Noncovalent synthesis of nanostructures: combining coordination chemistry and hydrogen bonding, *Angew. Chem., Int. Ed. Engl.* **36**, 1006-1008.

SELF-ASSEMBLING STRUCTURES BASED ON HYDROGEN BONDING SUBUNITS

ABDULLAH ZAFAR§ AND ANDREW D. HAMILTON§*
§*Department of Chemistry, University of Pittsburgh, Pittsburgh, PA, 15260*
*and *Department of Chemistry, Yale University, New Haven, CT, 06511.*

1. Introduction

Over the past quarter of the century organic chemistry has been widely used as a tool to investigate and mimic biological processes[1]. The closing gap between physical and biological disciplines has lead to the evolution of new areas such as bioorganic chemistry and molecular recognition. Conventionally, the area of molecular recognition emerged as scientists investigated the interaction of a larger molecule "the receptor" with a smaller molecule called "the substrate"[2]. The earlier investigations in this area were focused on developing synthetic receptors for small organic molecules containing important functionalities such as urea, carboxylic acids, amides, etc[3]. The following section will cover briefly, some classical receptor substrate complexes that will lay the ground for a later discussion on the nature of self-assembling structures.

2. Molecular Recognition

One of the earliest examples of host-guest chemistry was the discovery by Pedersen of crown ethers that bind alkali metals with high selectivity[4]. Figure 1 shows the binding of K^+ by dibenzo-18-Crown-6. The binding is extremely strong and is widely used to dissolve potassium salts in organic media. This work reflects the importance of shape and size in receptor design, as 18-Crown-6 does not bind the smaller Li^+ ion with similar efficiency.

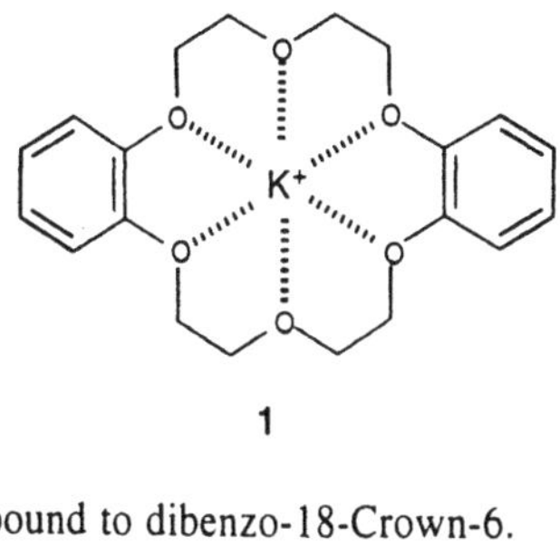

1

Figure 1. Potassium ion bound to dibenzo-18-Crown-6.

Lehn and coworkers extended the concept of crown-ethers by developing aza-crown-ethers and cryptates[5]. They also introduced the concept of anion binding and

G. Tsoucaris (ed.), Current Challenges on Large Supramolecular Assemblies, 67–85.

developed polyguanidinium and amidinium salts that bind anions by virtue of salt bridge formation and hydrogen bonding[6]. Some of the receptors that were developed by Lehn and coworkers are shown in Figure 2.

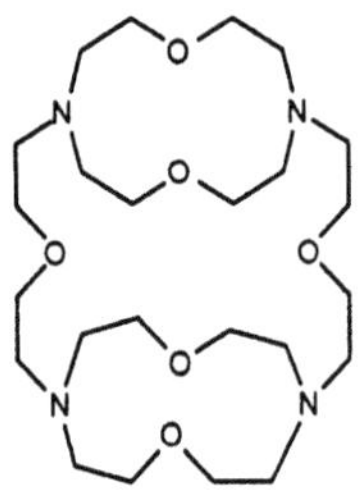

2

a

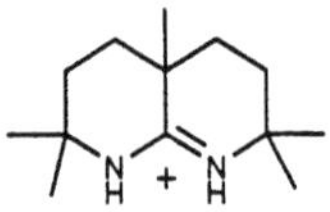

3

b

Figure 2. a) Lehn's Cryptate. b) An example of anion recognition motif.

The formation of an enzyme-substrate complex is an essential step in all enzyme catalyzed transformations[7]. Enzymes have cleft like cavities[8] that bind the substrates by making a number of favorable contacts through several different types of interactions i.e. hydrogen bonding, electrostatic, hydrophobic and π-stacking[9]. A number of investigators have designed both macrocyclic and cleft like receptors to bind small organic molecules with the goal that they may lead towards the design of artificial enzymes.

Hamilton and coworkers have reported a macrocyclic receptor **4** for barbiturates that binds barbital with an association constant of 6×10^5 in dichloromethane (Figure 3)[10]. The stability of the complex decreases by at least three orders of magnitude when oxypyridine macrocycle **5** was used ($K_a = 3.1 \times 10^2$). The difference in binding constants between the two reflects the replacement of one hydrogen bond donor amide by an ether oxygen which generates unfavorable interaction with barbital.

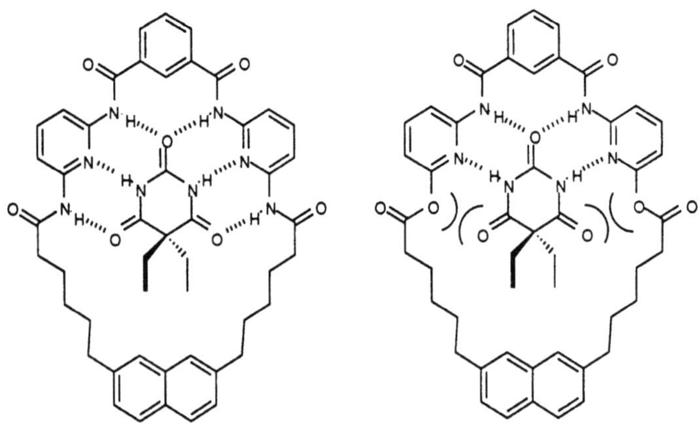

4

5

Figure 3. Macrocyclic receptors **4** and **5** bound to barbital.

Calix-4-arenes have been a very attractive template for receptor design because of their hydrophobic core and cup shaped cavity. Proper functionalization at the periphery allows the formation of a cleft that may bind the substrate with some selectivity. Aoyama's cyclic tetramer (**6**) of resorcinol (Figure 4) binds sugars and dihydroxy compounds selectively. It can be used to extract aldopentoses and aldohexoses into carbon tetrachloride[11]. The selectivity is governed by the stereochemistry of the hydroxyl groups on the C-3 and C-4 atoms of the sugar (cis >> trans). In this way, a mixture of cis and trans 1-4 cyclohexanediol can be extracted stereoselectively.

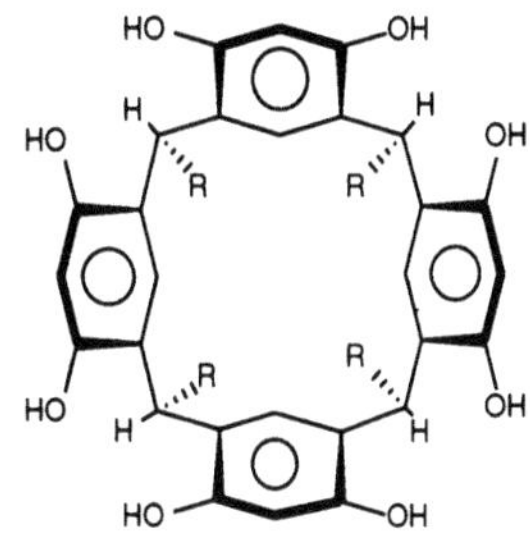

R = $(CH_2)_{10}CH_3$

6

Figure 4. Ayoyama's macrocyclic receptor showing the hydrophobic core and hydrophilic edges.

Recognition of individual amino acids is necessary for proper biological function[12]. In a synthetic system Clark Still has reported that the macrocyclic receptor **7** binds peptides stereoselectively and discriminates between the D and L forms of a dipeptide (Figure 5)[13]. The receptor has been successfully used to separate the D and L isomers of a dipeptide with an enantiomeric excess of >90%.

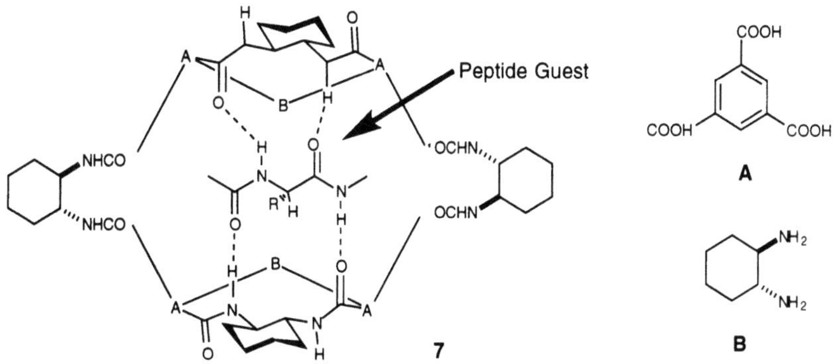

Figure 5. Still's macrocyclic receptor binding a dipeptide.

Rebek and coworkers have focused on the development and identification of templates that can be used in receptor design. Receptors based on Kemp's triacid bind biologically important substrates such as adenines, diketopiperazines and barbiturates[14]. The chiral receptor binds only one enantiomer of diketopiperazines as shown in Figure 6. The other enantiomer can not be bound so strongly because it will encounter severe steric repulsion between the R groups and the naphthalene ring. Another interesting development from Rebek and coworkers has been receptor design based on the xanthene template[15]. They

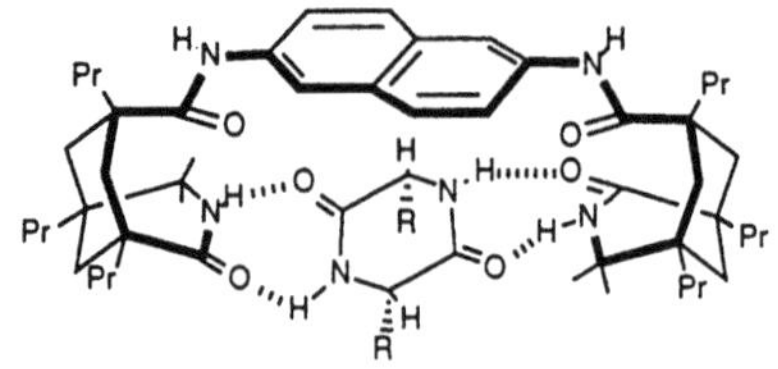

8

Figure 6. Rebek's Kemp's triacid receptor bound to diketopiperazine.

synthesized a variety of derivatives by connecting two xanthene templates using a rigid spacer[16]. Substitution of various chiral auxiliaries on one carboxylic acid generates a chiral environment that may be used to carry out stereoselective transformations such as resolution of a racemic mixture of chiral amines using the stable anhydride **9** (Figure 7)[17].

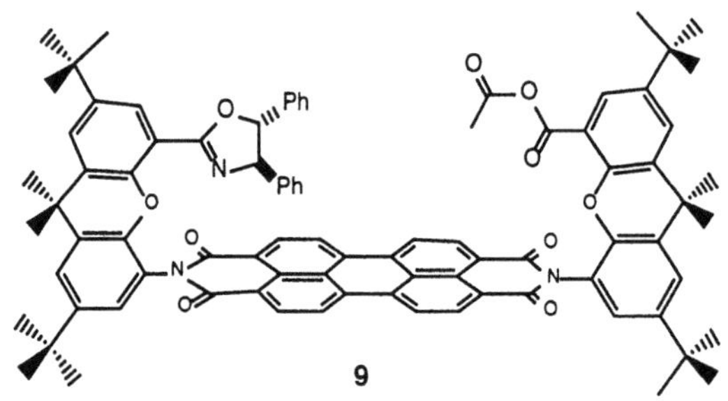

9

Figure 7. Rebek's chiral anhydride.

3. Hydrogen Bonding in Molecular Recognition

Hydrogen bonding in non covalent association has been extensively studied both in synthetic and biological systems. Two intriguing properties of hydrogen bonding are directionality and selectivity. A very large number of crystal structures show that the average angle between a donor N-H and the acceptor C=O in a hydrogen bond of the type N-H....O=C is 160° to 165°[18]. The directionality of hydrogen bonding provides potential control over the structure of the receptor-substrate complex and is an integral part of the secondary and tertiary structure of proteins[19]. Hydrogen bonding plays a key role in the formation of double helical DNA. The formation of a few hydrogen bonds between three or four base pairs triggers a highly cooperative process that results in the formation of a DNA double helix[20].

Due to this important role, hydrogen bonding has been extensively studied and its important factors analyzed. The development of a variety of quantum mechanical methods for small molecules enabled Umeyama and Morokuma to explore the exact nature of hydrogen bonding[21]. They proposed the following equation to describe the energetics of the hydrogen bonding.

$$\Delta E = ES + PL + EX + CT + MIX$$

The components have the following physical meaning.

ES is the electrostatic interaction of undistorted electron distribution. This includes the interaction of all the permanent charges and multipoles. It can be either attractive or repulsive.

PL is the effect of distortion of electron distribution on X-H by acceptor A. This is always an attractive interaction.

EX is the interaction caused by exchange of electrons between the donor and acceptor. It is a short range interaction and is repulsive in nature.

CT is the charge transfer or electron delocalization interaction and is covalent in nature.

MIX is the coupling term i.e. the difference between the total SCF (self consistent field) interaction energy and the sum of all the above four terms.

In order to estimate the strength of a hydrogen bond in chloroform, Schneider plotted the free energy of complexation for some small receptor-substrate complexes containing amide functionality of the type R-CONHR'. According to their study ΔG for one hydrogen bond is approximately -1.2 kcal mol^{-1}. The interaction becomes stronger in CCl_4 and is estimated to be -2.2 kcals mol^{-1}[22a]. However, if one of the partners is charged such as CH_3COO^-...HOH, the energy may be more than -3 kcals/mol.

4. Hydrophobic and π-Stacking Interactions

Hydrophobic interactions are relatively weak but represent essential stabilizing forces in biological systems. Hydrophobic interactions play a key role in protein folding which protects the hydrocarbon chains from exposure to water and exposes the polar side chains to the surface[22b]. It is estimated that for each square angstrom that is removed from exposure to water there is a gain of approximately 0.05 kcal/mol in the binding energy.

Recently Wilcox and coworkers have designed a "molecular torsion balance" and estimated the free energy of edge-to-face π-stacking interaction[23]. They used the well-known Troger's base as a scaffold to design the molecular torsion balance (Figure 8). The rotation about the phenyl-phenyl bond is slow on the NMR time scale (E_a = 16 kcal mol^{-1}). They have estimated the energies of edge-to-face interaction to be between 0.2-0.7 kcal mol^{-1} for different esters in a chloroform solution.

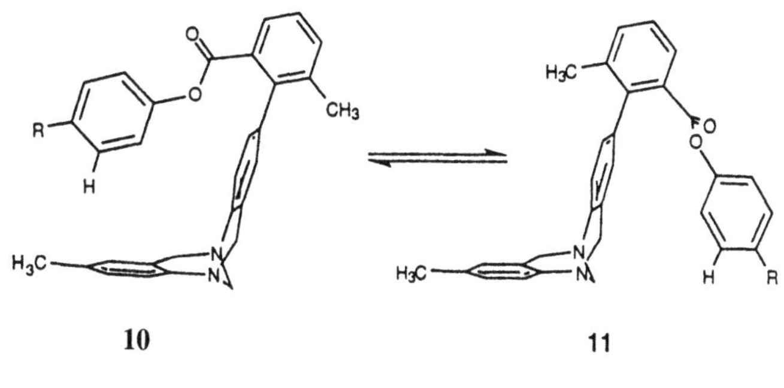

10 **11**

Figure 8. Conformational states for aryl esters (Molecular torsion balance).

5. Molecular Self-Assembly

Molecular Self-Assembly is one of the most important process involved in the synthesis of biological macromolecules. The spontaneous association of thousands of small subunits form macromolecular structures in the size range of 1 to 10,000 nm. The process of self-assembly therefore is highly convergent[24]. The common feature found in most biological assemblies is the utilization of weak, noncovalent binding interactions that allow the formation of thermodynamically stable products and also check and remove errors in the final product[25].

Recently, there has been growing interest in developing synthetic self-assembling systems to mimic and understand the biological self-assembly. The development of nanoscale electronic devices that can function as switches and communication devices (input/output) has also fascinated the scientific community for more than a quarter of a century. Structures on the nanometer size scale can not be made by using conventional lithographic techniques due to the limitation of insulation barriers (0.2-1 μm), electronic cross talk between devices, and quantum mechanical problems[26]. Self-assembly has also been shown to have great potential in the areas of catalysis and the development of ion and molecular transport channels. It has been widely proposed as a viable strategy to develop molecular systems that are beyond the capability of conventional organic synthesis[27].

6. Self-Assembly in Biology

Two complementary strands of oligonucleotides spontaneously self-assemble under physiological conditions to form the DNA double helix (Figure 9)[28]. The nucleation i.e. the formation of first 3 base pairs has a positive free energy. After the recognition of three to four base pairs any subsequent base pairing rapidly decreases the free energy and therefore the propagation of the double helix is highly cooperative. This process reveals a very important principle of self-assembly, that the formation of complex structures using a large number of favorable interactions is relatively easier than forming a small structure using a few favorable interactions.

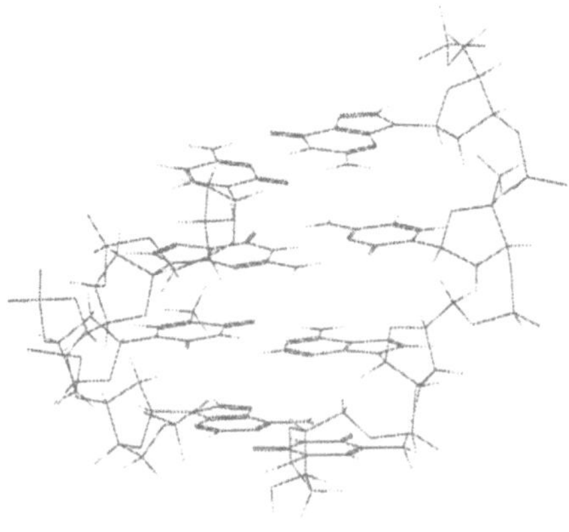

Figure 9. A wire and frame representation of two complementary strands of DNA, generated by MacroModel.

The self-assembly of tobacco mosaic virus (TMV) has also been studied in great detail. It has a helical structure composed of a single strand of RNA surrounded by helical discs of 2130 identical proteins each comprising 158 amino acids[29]. The virus can disassemble into its component parts in vitro and can be reassembled accurately in vitro to give the fully functional virus. The mechanism of self-assembly involves the aggregation of proteins to from helical discs that are about two and a half turns long. These helical discs then slide over the loop of RNA until the virus is complete. It has been estimated that only 53 out of 135 genes contain information directly related to the assembly of this complex structure.

7. Synthetic Self-Assembling Systems

During the last ten years a variety of self-assembling systems have been reported[30]. Due to the involvement of a wide variety of disciplines in the area, it is not possible to cover all the different types of self-assembling systems. This section will cover the self-assembling systems that are stabilized by hydrogen bonding and hydrophobic interactions.

7.1 MELAMINE-CYANURIC ACID SYSTEMS

A very popular approach to control the size and shape of the aggregate is to incorporate complementary binding groups in such a way that the process of aggregation may terminate with cyclization. The best binding motifs for such cyclic assemblies came from the Cambridge Structural Database. For instance, Whitesides and coworkers have exploited the lattice of melamine and cyanuric acid cocrystal (M:CA). Both melamine and cyanuric acid offer nine hydrogen bonding sites and form an infinite lattice composed of hexagons (Figure 10). Each hexagon has three melamines, and just as many cyanuric acid groups, hydrogen bonded to each other.

CA

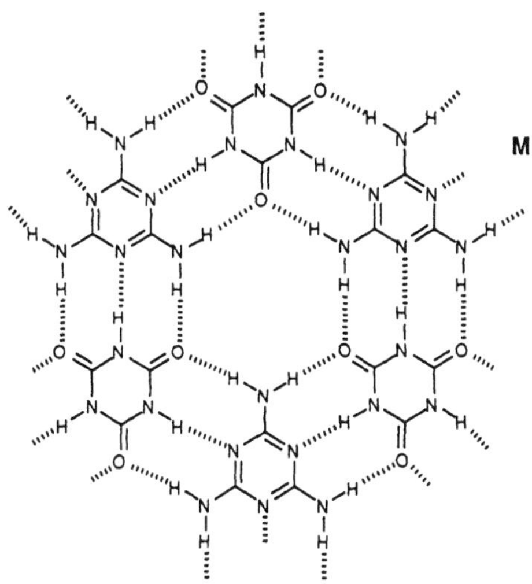

M

12

Figure 10. A schematic representation of M:CA lattice.

Various issues in self-assembly were addressed using the M:CA binding motif. Whitesides et al. synthesized a family of melamines and cyanuric acids connecting a varying number of these molecules with the help of appropriate spacers and studied their aggregation properties in solution using various analytical techniques. They have also estimated the relative stability of different aggregates in solution by computational and solution methods.

Melamine derivative hubM3 (**13**) and cyanuric acid derivative benzCA2 (**14**) (Figure 11) form a 2+3 aggregate (**15**) as shown in Figure 12. The aggregate was characterized by the appearance of new peaks in the ^1H NMR spectrum in $CDCl_3$. The increased solubility of up to three equivalents benzCA2 in a chloroform solution of hubM3 provided evidence for the stoichiometery of the complex. The complex was stable enough to survive gel permeation chromatography (GPC) and the elution time gave an estimate of the molecular weight of the complex. The molecular weight was further confirmed by vapor pressure osmometery (VPO)[31].

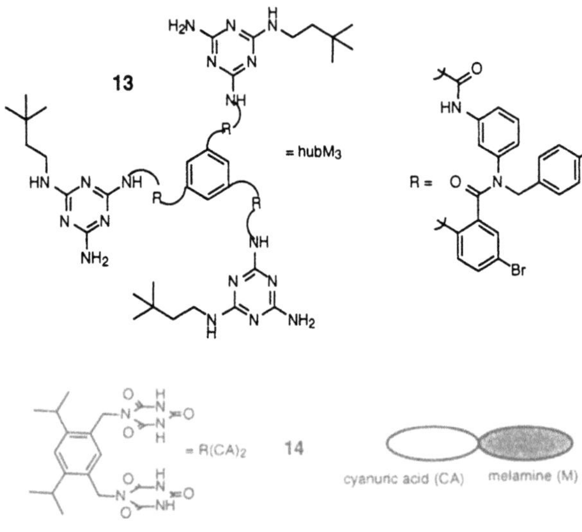

Figure 11. HubM3 and benz(CA)2.

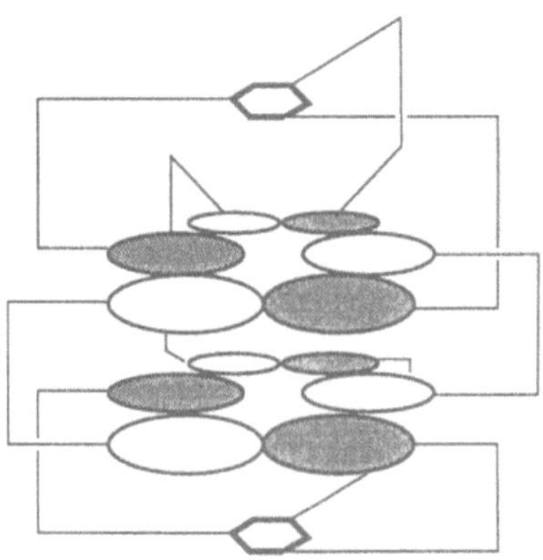

15

Figure 12. A schematic representation of the proposed supramolecular complex formed by hubM3 and benz(CA)2.

Another remarkable self-assembling system based on the M:CA lattice forms a 9+1 aggregate[32]. Hub(MMM)3 and neohex(CA) forms a 9+1 aggregate (**16**) as shown in Figure 13. Though it is entropically unfavorable to bring ten particles together, the preorganization and increased area of favorable contact is presumably the driving force for the aggregation. The aggregate is held together by 54 hydrogen bonds and shows slow exchange by NMR between different conformers.

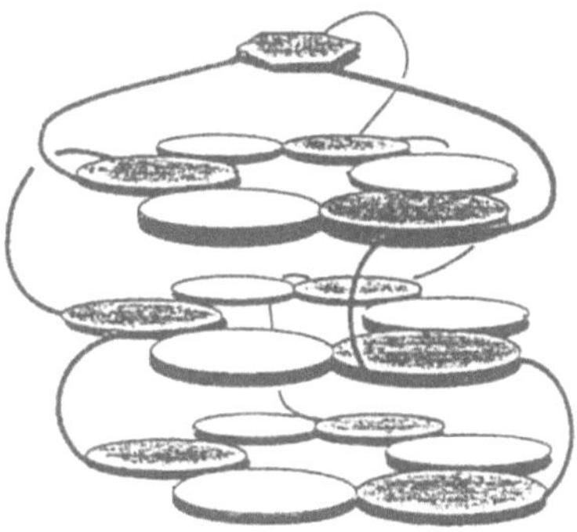

16

Figure 13. A schematic representation of the proposed 9+1 supramolecular complex formed by hub(MMM)3 and neohex(CA)(Reproduced with permission)

The strategy of using the substituents to favor formation of cyclic arrays over linear infinite arrays has also been used successfully by Whitesides et. al. They have reported a variety of structural motifs formed by barbituric acid/cyanuric acid and melamine[33]. They proposed that an aggregate of melamine and barbituric acid is dependent upon steric interactions. For example, a cyclic hexamer is formed in the case where the substituent R is a 4-tert-butylphenyl group. Substituents smaller than tert-butyl lead to the formation of two dimensional infinite arrays (Figure 14).

a

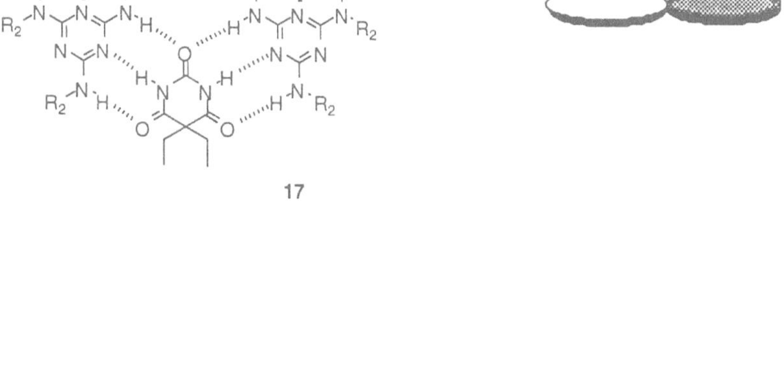

17

b

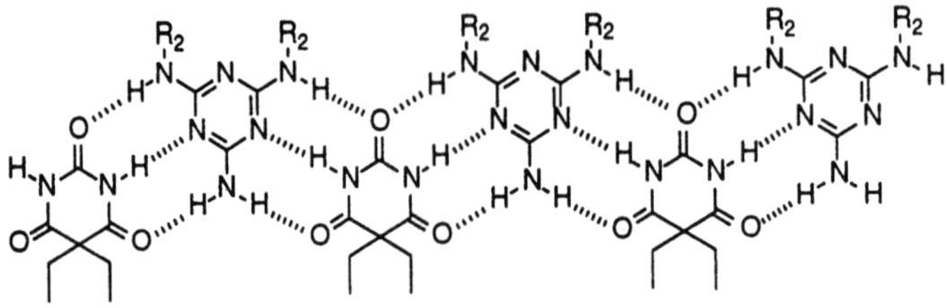

18

c

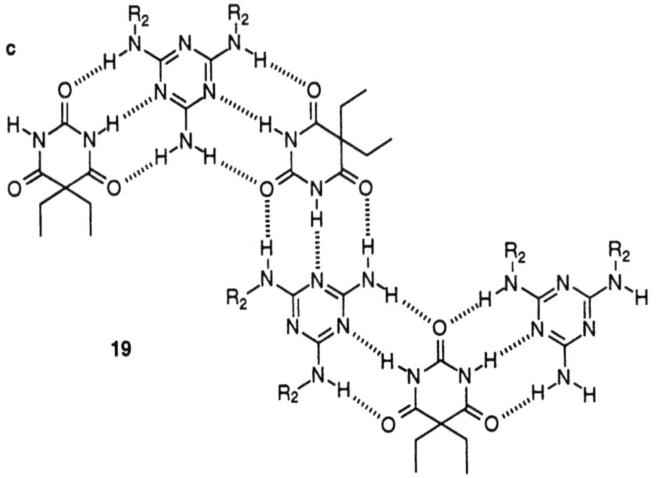

19

Figure 14. a) 3+3 cyclic hexamer (R = 4-tert-butylphenyl) b) linear tape (R = para-toluyl). c) crinkled tape (R = 4-methylbenzoate).

7.2 ISOPHTHALIC ACID SYSTEMS

The dimerization of carboxylic acids is a very well studied and widely observed phenomenon. It has long been known that trimesic acid forms cyclic hexameric arrays in the solid state (Figure 15a)[34]. These cyclic hexamers are interlinked and intercalated. One would expect that isophthalic acid would also have the capability to form a similar cyclic hexamer in the solid state. However, isophthalic acid forms a linear infinite array in the solid state. Hamilton and coworkers observed that if 5-hydroxyisophthalic acid is substituted with an alkyl chain of appropriate length (C_7 to C_{10}) then it forms a cyclic hexamer (Figure 15b) similar to that observed in the case of trimesic acid[35]. They proposed that the alkyloxy groups at C-5 position serves the purpose of interlinking the cyclic hexamers by hydrophobic interactions. Isophthalic acid is incapable of forming the cyclic hexamer because of the absence of such interaction.

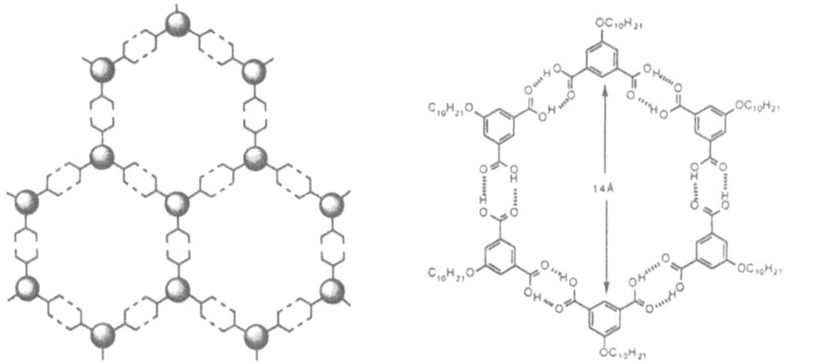

a 20 b 21

Figure 15. a. Schematic representation of trimesic acid lattice. b. X-ray crystal structure of 5-decyloxyisophthalic acid.

7.3 PYRIDONE SYSTEMS

2-pyridones are isostructural with carboxylic acids. The crystal structure of 2-pyridone shows a hydrogen bonded dimer analogous to the carboxylic acid dimer. They are also known to exist as dimers in solution[36]. Wuest and coworkers have synthesized dipyridones **22** and **23** using acetylene as a rigid spacer[37]. The X-ray crystal structure reveals that the asymmetric dipyridone **22** exists as a dimer in the solid state. NMR and VPO studies confirmed the presence of dimers in chloroform even at concentrations as low as 3.6 x 10^{-4}M. In contrast, symmetrical dipyridone **23** forms linear ribbons in the solid state. Aggregation of both systems was found to be sensitive to the solvent polarity and little aggregation was observed in methanol. These results prove the importance of self-complementarity in the design of self-assembling systems.

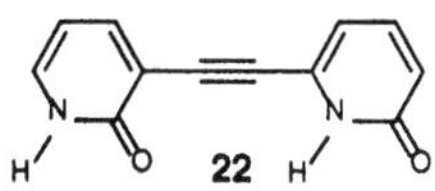

22

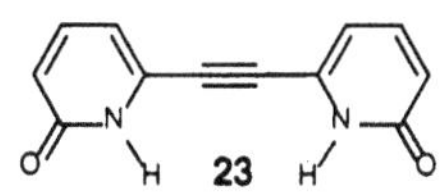

23

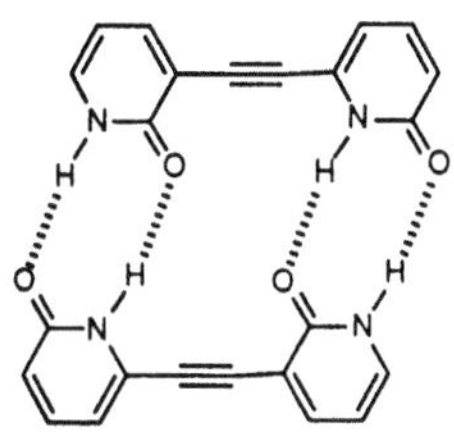

24

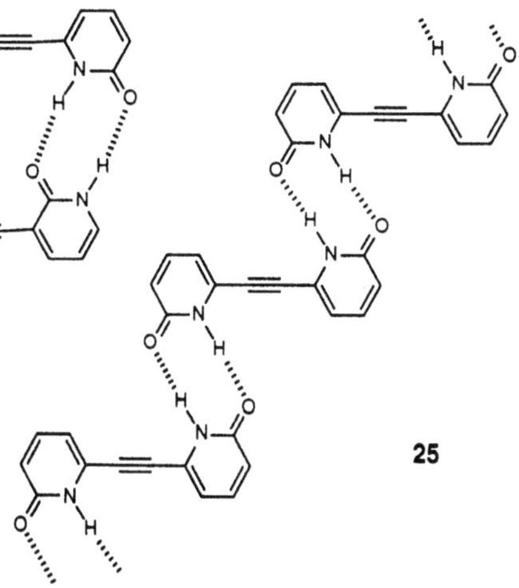

25

Zimmerman and coworkers made pyridoquinones that can form both the cyclic trimer or the linear oligomer (Figure 16)[38]. They proposed that for smaller aggregates, cyclic arrays are favored over the linear arrays as the former has a greater enthalpic stabilization due to the formation of 2 hydrogen bonds per molecule in the cyclic trimer as opposed to only (2n+2)/(n+2) in the similar linear aggregate. They have shown by NMR dilution studies and VPO that the cyclic trimer is the predominant species in a chloroform solution.

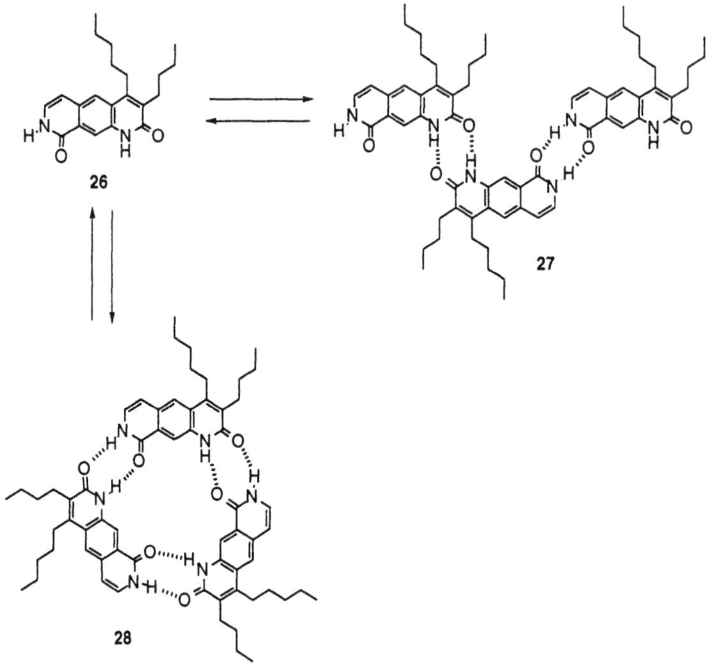

Figure 16. Zimmerman's pyridoquinone in equilibrium with the two possible aggregates.

7.4 OTHER CYCLIC SYSTEMS

Lehn and his colleagues have described the assembly of a bis-porphyrin containing supramolecular cage[39]. The porphyrin subunit contains two 5-alkyluracil groups. The

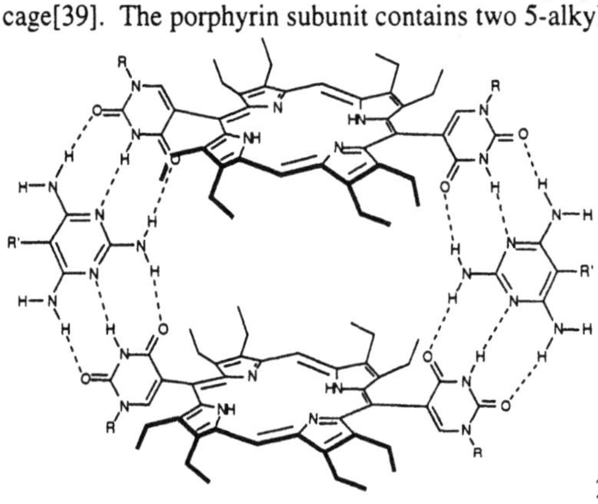

29

solution of bis-porphyrin in CD_2Cl_2 shows a mixture of syn and anti isomers. When 2,4,6-triaminopyrimidine is added to the solution the anti isomer decreased to about 20% (3 days) due to the formation of a 2+2 cyclic bis-porphyrin complex **29**. The formation of the complex was supported by electrospray mass spectrometry, VPO and NMR spectroscopy.

Tetramerization of guanosine monophosphate in vivo in the presence of potassium to form G-quartets is a very well studied phenomenon. The planar quartets stack over each other to form cylinders. The structure has been confirmed by various techniques including X-Ray diffraction. Ciuchi and coworkers used folic acid **30** that is isostructural to guanosine to demonstrate the tetramerization[40].

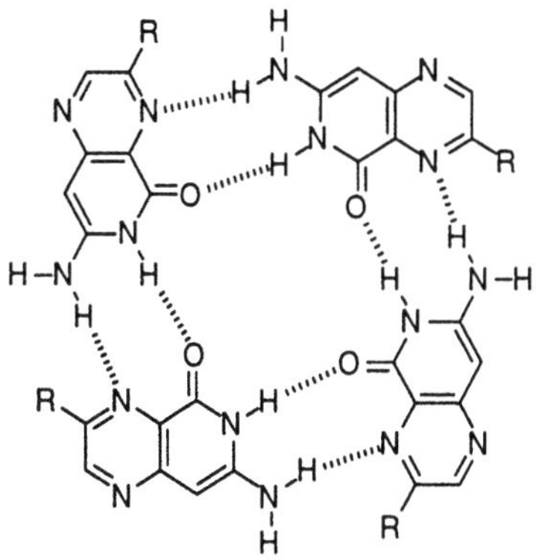

30

Davis and his colleagues have recently demonstrated that isoguanosine forms tetramers in organic solvents[41]. They also observed that the tetramers are predominant in the presence of potassium (Figure 17) while monomer is predominant in the presence of sodium and lithium. This work demonstrates the remarkable size selectivity of the isoguanosine quartet.

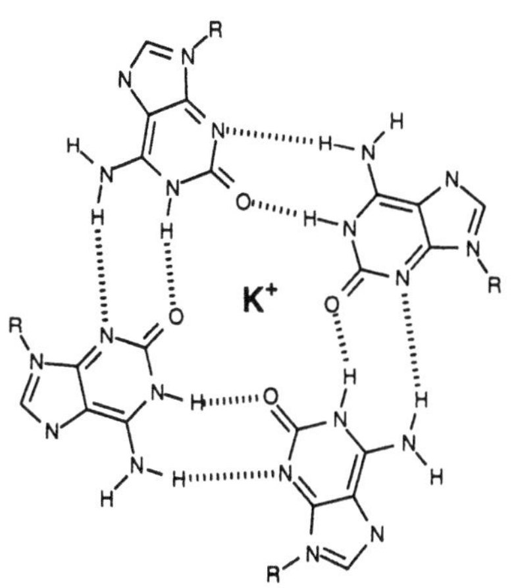

Figure 17. Isoguanosine quartet

7.5 LINEAR INFINITE ARRAYS

The relationship between functional group orientation in the monomer subunit controls the nature of the multicomponent complex. This has been demonstrated by Hamilton and coworkers by using a simple bis-acylaminopyridine. In the protonated form this compound was predicted to assume conformation B (Figure 18a). Indeed when bis-(4-nitrophenyl) hydrogen phosphate was added in a 1:1 ratio to a solution of **31**, a supramolecular complex was formed (Figure 18b). The resulting 1:1 complex is composed of alternating anionic and cationic subunits, while **31** is just a simple receptor[42].

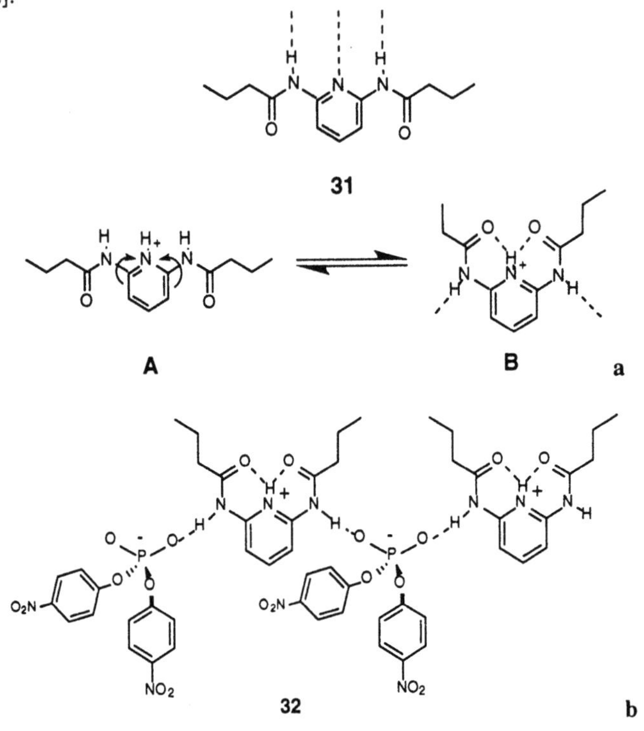

Figure 18. a) The equilibrium in protonated bis-acylaminopyridine. b) The supramolecular infinte array formed by bis-acylaminopyridine with phosphate.

Carboxylic acids are known to bind acylaminopyridines. The binding between a carboxylic acid and acylaminopyridine is significantly stronger than the self-association between each individual subunit[43]. Also, small changes in the distance of the functional groups on a subunit have a profound effect on the outcome of the final complex. Hamilton et al. synthesized a series of bis-acylaminopyridines with a variety of spacers and complexed them with bis-carboxylic acids of different length spacers[44] In all the cases they observed a 1:1 complex between aminopyridine and carboxylic acid. In addition, it is important to note that both bis-aminopyridine and the bis-carboxylic acid may be present in two different conformations, cis or trans. If both of them are in a cis conformation the result is the receptor substrate complex **33** (Figure 19a), while 1,12-dodecanedicarboxylic acid produces an infinite array (**34**) of alternating diacid/diamine subunits (Figure 19b).

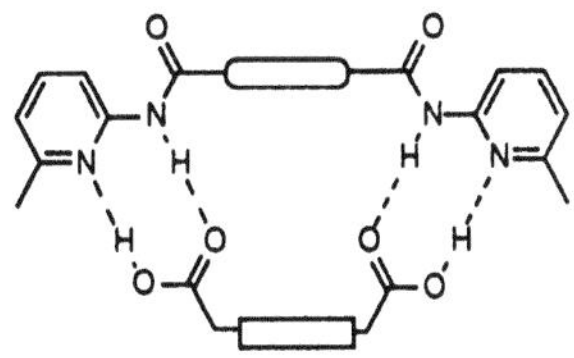

a

33

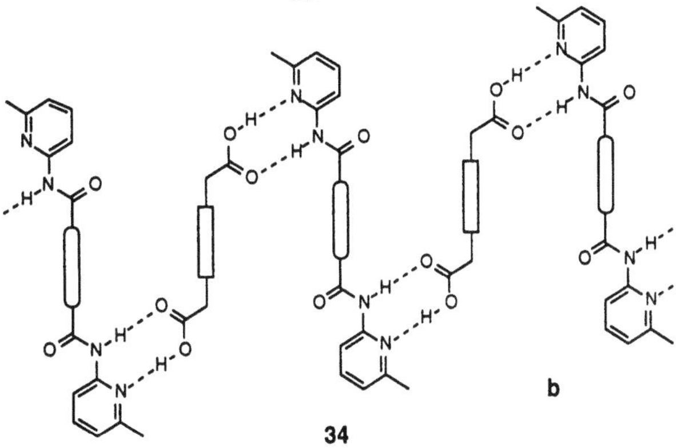

b

34

Figure 19. a) The receptor substrate complex. b) The linear infinte array.

7.6 HELICAL ARRAYS

Lehn and his colleagues have used self-assembling helix formation to conduct a chiral resolution[45] They prepared tartaric acid based bis-acylaminopyridine **35** and a complementary bis-uracil **36** (Figure 20a). Equimolar mixtures of L-tartaric acid based species form a right-handed helix while D-tartaric acid based compounds produce a left-handed helix. A mixture of D-aminopyridine and L-uracil generate right-handed helices while L-aminopyridine and D-uracil form left-handed helices (Figure 20b). In contrast, no helicity is observed with compounds containing meso-tartaric acid. A mixture of all four D and L isomers forms both right-handed (L,L) and left handed (D,D) superhelices indicating that only configurationally compatible subunits are added to the growing chiral helices.

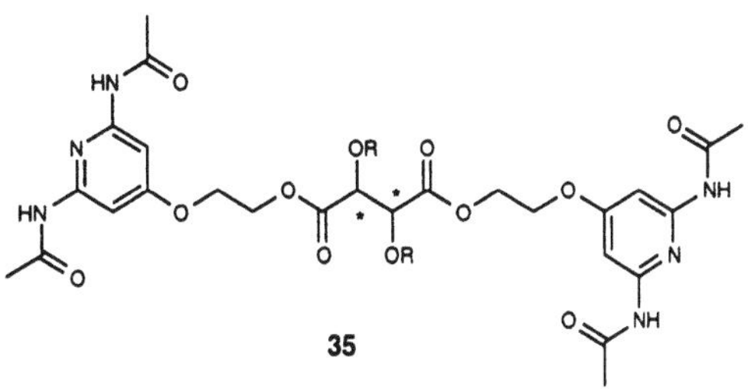

35

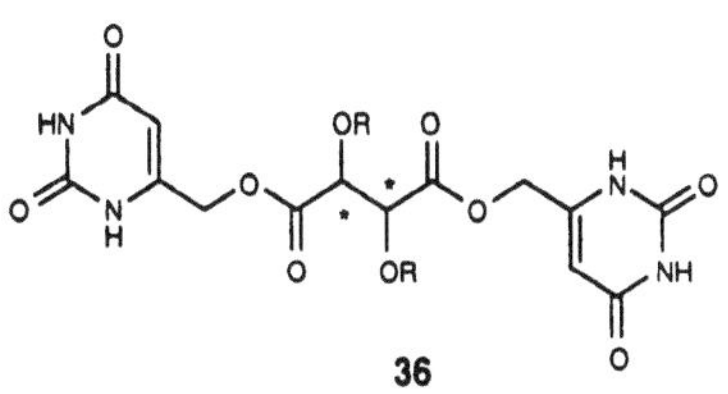

36

Figure 20a. Lehn's bis-aminopyridine **35** and bis-uracil **36**.

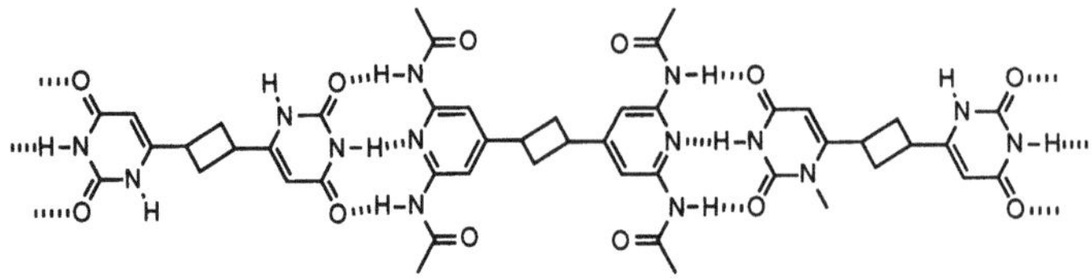

37

Figure 20b. Schematic representation of the proposed helical array formed by bis-acylaminopyridine and bis-uracil.

Hamilton and coworkers have made a bis-acylaminopyridine **38** with a 1,3-disubstituted phenyl spacer. The aminopyridine groups in this derivative do not lie in the same plane. One lies 20.5° above the isophthalate ring and the other lies 9.4° below. A 1:1 complex is formed with heptanedioic acid and crystal structure analysis reveals the formation of a double helical entity (Figure 21).

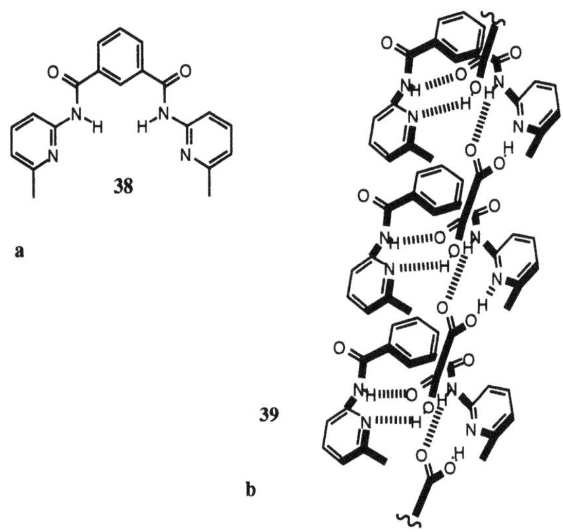

38

a

39

b

Figure 21. a) Isophthaloyl-bis-aminopyridine. b) The helical supramolecular complex formed by **38**.

The stability of helix depends upon the length of the diacid in the solid state. A helix does not form when the bis-acylaminopyridine crystallized with glutaric acid. Instead the aminopyridines take up a trans conformation and the aggregate forms a planar ribbon structure[46].

7.7 CYLINDRICAL ARRAYS

The design of artificial peptides to create receptors is becoming an active area of research. Ghadiri and coworkers have shown that cyclic peptides consisting of alternating D- and L-aminoacid sequences (**40**)[47] take a flat donut-like conformation (Figure 22). They self-assemble via hydrogen bonding between the amide N-H and the carbonyl groups to form nanotubules (Figure 23). The formation of the nanotubes may be better understood by considering them as "cyclic stacked ß-sheets". The authors have also reported the formation of a dimer when alternating amide N-H groups are substituted with methyls and formation of a dimer when alternating amide N-H groups are substituted with methyls and one face of the ring is blocked. The dimer has recently been stabilized by covalent bond formation using an alkene metathesis reaction[48]. These cyclic peptides have been shown to work as ion channels across the membrane with comparable efficiency to natural ion channels such as gramicidin[49]. The size of the internal cavity can be altered to allow the passage of larger molecules such as glucose through bilayer membranes[50].

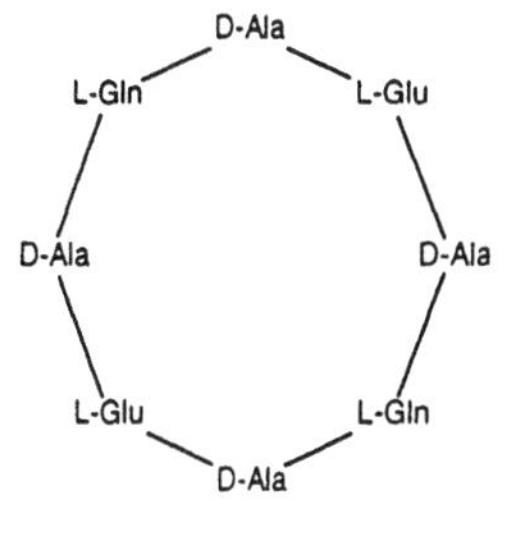

40

Figure 22. The schematic diagram of Ghadiri's cyclic peptide.

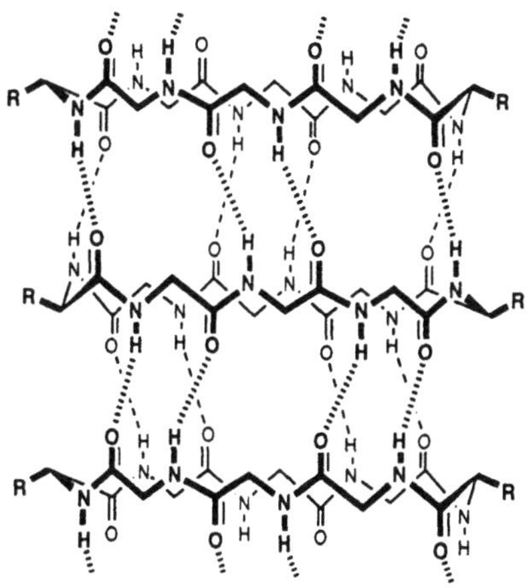

41

Figure 23. The proposed structure of the nanotubes formed by the cyclic peptide **40**.

7.8 SELF-ASSEMBLING MOLECULAR ENCAPSULATERS

Recently, Rebek and coworkers have designed a bis-cyclic urea that self-assembles into a dimer in a way that is similar to the formation of a hollow sphere by the two halves of a tennis ball[51]. They synthesized a family of glycourils (**42**) with different spacers (Figure 24a)[52]. These molecules are self-complementary and dimerize in chloroform (Figure 24b). The resulting semispherical capsule (**43**) is capable of encapsulating small organic molecules such as methane. They have also reported the formation of larger capsules that not only bind larger organic molecules such as benzene and adamantane dicarboxylic acid but also catalyze pericyclic reactions.

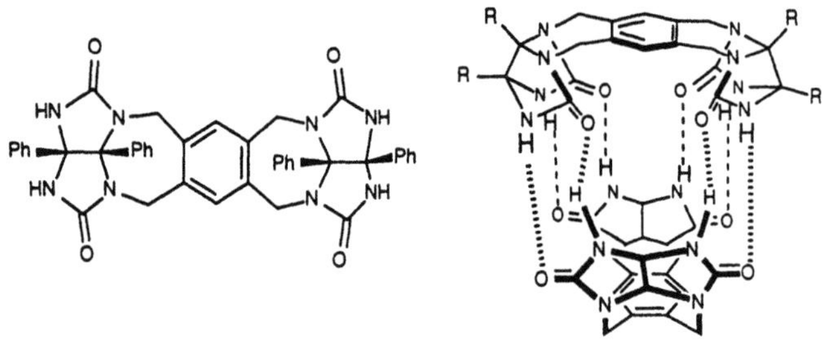

a 42 43 b

Figure 24: a. Rebek's bis-cyclic urea (**42**). b. Proposed structure of the dimer formed by **42**.

Acknowledgment

We thank the National Science Foundation (CHE 9213937) for support of this work.

8. References

1. Wilson, E. K. *Chem. Eng. News* **1996**, *74*, 39-44.
2. Lehn, J.-M. *Angew. Chem. Int. Ed. Engl.* **1988**, *27*, 89-112.
3. Julius Rebek, J. *Angew. Chem. Int. Ed. Engl.* **1990**, *29*, 245-255.
4. Pedersen, C. J. *J. Am. Chem. Soc.* **1967**, *89*, 7017-7036.
5. Lehn, J. M.; Stubbs, M. E. *J. Am. Chem. Soc.* **1974**, *96*, 4011-4012.
6. Dietrich, B.; Fyles, D. L.; Fyles, T. M.; Lehn, J.-M. *Helv. Chim. Acta* **1979**, *62*, 2763-2787.
7. Stryer, L. In *Biochemistry*; 3rd ed.; L. Stryer, Ed.; W. H. Freeman and Company: New York, **1988**, pp 177-200.
8. Koshland, D. E. *Sci. Amer.* **1973**, *229*, 52-64.
9. Fersht, A. *Structure and Mechanism;* 2nd ed.; W. Freeman: **1983**.
10. Chang, S.-K.; Engen, D. V.; Fan, E.; Hamilton, A. D. *J. Am. Chem. Soc.* **1991**,*113*, 7640-7645.
11. Aoyama, Y.; Tanaka, Y.; Sugahara, S. *J. Am. Chem. Soc.* **1989**, *111*, 5397-5401.
12. Clark, B. F. C.; Marcker, K. A. *Sci. Amer.* **1968**, *218*, 36-42.
13. Yoon, S. S.; Still, W. C. *J. Am. Chem. Soc.* **1993**, *115*, 823-824.
14. Jeong, K. S.; Tjivikua, T.; Muehldorf, A.; Deslongchamps, G.; Famulok, M.; Rebek, J., Jr. *J. Am. Chem. Soc.* **1991**, *113*, 201-209.
15. Nowick, J. S.; Ballester, P.; Ebmeyer, F.; Rebek J., Jr. *J. Am. Chem. Soc.* **1990**, *112*, 8902-8906.
16. Shimizu, K. D.; Dewey, T. M.; Rebek J., Jr. *J. Am. Chem. Soc.* **1994**, *116*, 5145-5149.
17. Rojas, C. M.; Rebek J., Jr. *Bioorg. Med. Chem. Lett.* **1996**, *6*, 3013-3016.
18. Murray-Rust, P.; Glusker, J. P. *J. Am. Chem. Soc.* **1984**, *106*, 1018-1025.
19. Creighton, T. E. *Proteins: Structures and Molecular Principles*; W. H. Freemen: New York, **1983**.
20. Stryer, L. *Biochemistry;* 3rd ed.; W. H. Freeman and Company: New York, **1988**, pp 177-200.
21. Umeyama, H.; Morokuma, K. *J. Am. Chem. Soc.* **1977**, *99*, 1316-1332.
22. a) Schneider, H. J. *Angew. Chemie. Int. Ed. Engl.* **1991**, *30*, 1417. b) Symons, M. C. R. In *Principles of Molecular Recognition*; First ed.; A. D.

Buckingham, A. C. Legon and S. M. Roberts, Eds.; Blakie Academic & Professional: New York, **1993**, pp 43-78.

23. Paliwal, S.; Geib, S.; Wilcox, C. S. *J. Am. Chem. Soc.* **1994**, *116*, 4497-4498.
24. Lindsey, J. S. *New J. Chem.* **1991**, *15*, 153-180.
25. Stoddart, J. F. In *Host Guest Molecular Interactions: From Chemistry to Biology*; Wiley-Interscience: **1991**, pp 5-22.
26. Philp, D.; Stoddart, J. F. *Angew. Chem. Int. Ed. Engl.* **1996**, *35*, 1154-1196.
27. Whitesides, G. M.; Mathias, J. P.; Seto, C. T. *Science* **1991**, *254*, 1312-1319.
28. Felsenfeld, G. *Sci. Amer.* **1985**, *253*, 58-67.
29. Namba, K.; Stubbs, G. *Science* **1986**, *231*, 1401-1406.
30. Lawrence, D. S.; Jiang, T.; Levett, M. *Chem. Rev.* **1995**, *95*, 2229-2260.
31. Seto, C. T.; Whitesides, G. M. *J. Am. Chem. Soc.* **1991**, *113*, 712-713.
32. Mathias, J. P.; Simanek, E. E.; Seto, C. T.; Whitesides, G. M. *Macromolecular Symposia* **1994**, *77*, 157-166.
33. Zerkowski, J. A.; Seto, C. T.; Whitesides, G. M. *J. Am. Chem. Soc.* **1992**, *114*. 5473-5475.
34. Duchamp, D. J.; Marsh, R. E. *Acta Cryst.* **1969**, *B-25*, 5-19.
35. Yang, J.; Marendez, J. L.; Geib, S. J.; Hamilton, A. D. *Tetrahedron Lett.* **1994**, 3665-3668.
36. Ducharme, J.; Wuest, J. D. *J. Org. Chem.* **1988**, *53*, 5787-5789.
37. Gallant, M.; Viet, M. T. P.; Wuest, J. D. *J. Am. Chem. Soc.* **1991**, *113*, 721-723.
38. Zimmerman, S. C.; Duerr, B. F. *J. Org. Chem.* **1992**, *57*, 2215-2217.
39. Drain, C. M.; Fischer, R.; Nolen, E. G.; Lehn, J. M. *J. Chem. Soc. Chem. Commun.* **1993**, 243-245.
40. Ceuchi, F.; Nicola, G. D.; Franz, H.; Gottarelli, G.; Mariani, P.; Bossi, M. G. P.; Spada, G. P. *J. Am. Chem. Soc.* **1994**, *116*, 7064-7071.
41. Davis, J. T.; Tirumala, S.; Jenssen, J. R.; Radler, E.; Fabris, D. *J. Org. Chem.* **1995**, *60*, 4167-4176.
42. Geib, S. J.; Hirst, S. C.; Vicent, C.; Hamilton, A. D. *J. Chem. Soc. Chem. Commun.* **1991**, 1283-1284.
43. Garcia-Tellado, F.; Goswami, S.; Chang, S. K.; Geib, S. J.; Hamilton, A. D. *J. Am. Chem. Soc.* **1990**, *112*, 7393-7394.
44. Garcia-Tellado, F.; Geib, S. J.; Goswami, S.; Hamilton, A. D. *J. Am. Chem. Soc.* **1991**, *113*, 9265-9269.
45. Gulick-Krzywicki, T.; Fouquey, C.; Lehn, J. M. *Proc. Natl. Acad. Sci.* **1993**. *90*.
46. Fan, E.; Vicent, C.; Geib, S. J.; Hamilton, A. D. *Chem. Mat.* **1993**, *6*, 1113-1117.
47. Ghadiri, M. R.; Granja, J. R.; Milligan, R. A.; McRee, D. E.; Khazanowich, N. *Nature* **1993**, *366*,
48. Clark, T. D.; Ghadiri, M. R. *J. Am. Chem. Soc.* **1995**, *117*, 12364-12365.
49. Wallace, B. A.; Ravikumar, K. *Science* **1988**, *241*, 182-187.
50. Granja, J. R.; Ghadiri, M. R. *J. Am. Chem. Soc.* **1994**, *116*, 10785-10786.
51. Wyler, R.; Mandoza, J. d.; Rebek, J., Jr. *Angew. Chem. Int. Ed.* **1993**, *32*, 1699-1701.
52. Valdes, C.; Spitz, U. P.; Toledo, L. M.; Kubik, S. W.; Rebek J., Jr. *J. Am. Chem. Soc.* **1995**, *117*, 12733-12745.

HYDROGEN BONDS IN ORGANISED MONOLAYERS

D. MÖBIUS[1], O. FELIX[2], and M.W. HOSSEINI[2]
[1]Max-Planck-Institut für biophysikalische Chemie, D-37070 Göttingen, Germany
[2]Laboratoire de Chimie de Coordination Organique, Institut Le Bel, Université Louis Pasteur, 4 rue Blaise Pascal, 67000 Strasbourg, France

Non-amphiphilic molecules in the adjacent aqueous solution may bind to the head groups of insoluble monolayers by hydrogen bond interactions as verified by monolayer techniques and spectrosopic methods after transfer to solid substrates. Multiple site interactions of a solute with different amphiphiles involving both electrostatic attraction and hydrogen bonds can cause lateral organisation of the amphiphiles with formation of systems that show molecular recognition. The hydrogen bonds in these layers are directed towards the aqueous subphase. Large aggregates with hydrogen bonds in the monolayer plane are formed by using ionic and hydrogen bond interactions.

1. Introduction

Among the various intermolecular interactions, hydrogen bonds play an important role in arranging different molecules to large supramolecular assemblies. Prominent examples are found in biopolymers where hydrogen bonds contribute strongly to determining the tertiary structure of proteins and control base pairing of nucleic acids. A large field with a major contribution of hydrogen bonds is molecular recognition in biosystems like antibody-antigen interactions. Studies of molecular recognition in organised molecular assemblies like monolayers, vesicles and micelles have recently been reviewed [1]. The main difference between H bond interactions and electrostatic interactions between charged molecules aside of the magnitude is the directionality of H bonds. Therefore, molecules can form H bonds in particular geometries only. This feature opens possibilities to organise different molecules in a well defined way to hetero-dimers or extended aggregates. Organised monomolecular layers at the air-water interface and as components of monolayer systems on solids are excellent model structures for mechanistic studies as well as for the construction of artificial devices capable of molecular recognition using hydrogen bonds. An important phenomenon here is the competition in hydrogen bond interactions with the recognition site between the target molecule and water. A large variety of investigations has been performed using different hydrogen bond donors and acceptors to mimick the interactions observed in biosystems. A small selection of examples is given in the following to illustrate the strategies developed and the detection techniques used along with the results of some recent investigations.

G. Tsoucaris (ed.), Current Challenges on Large Supramolecular Assemblies, 87–96.

2. Base Pairing Interactions

Monolayers of an amphiphilic diaminotriazine are capable of specifically binding barbituric ancid and thiobarbituric acid due to hydrogend bond interactions between the diaminothriazine head group and the imide moiety of barbiturates [2]. The surface pressure-area isotherm of the amphiphilic diaminotriazine derivative 3 is modified in the presence of various molecules in the aqueous subphase as shown in Figure 1.

3 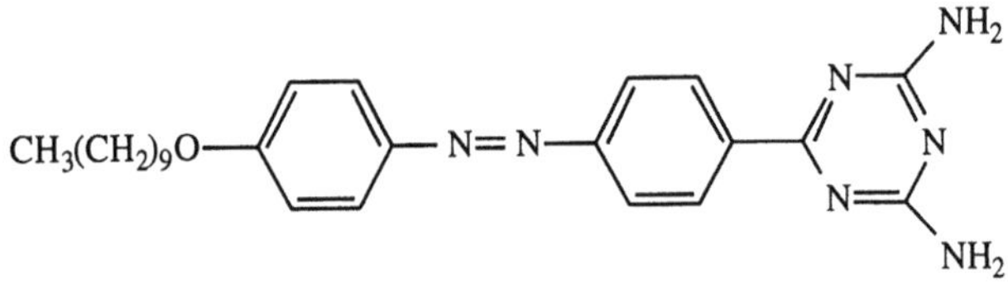

5 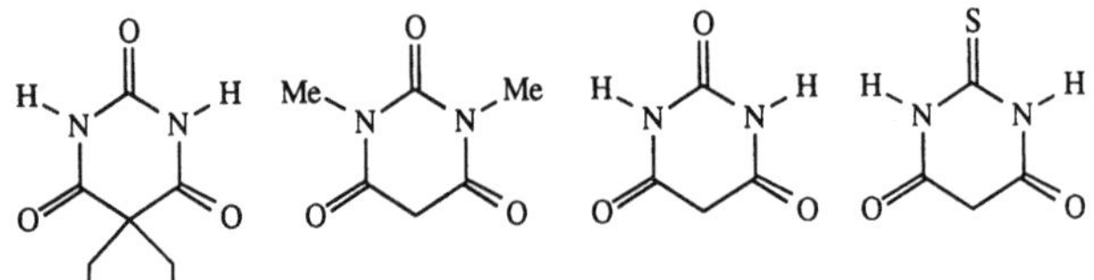

6 7 8

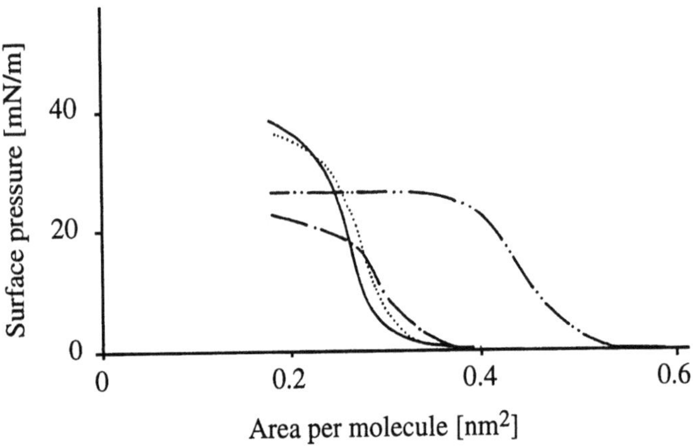

Figure 1. Surface pressure-area isotherms of the amphiphilic diaminotriazine derivative 3 on pure water (solid line) and on 0.01 M solutions of the compounds 5 (dotted line), 7 (dotted-dashed line) and 8 (double dot-dashed line); the interactions of 5, 7, and 8 modify the isotherm of 3, the dimethyl barbituric acid 6 did not alter the isotherm of 3 (modified from [2]).

The diaminotriazine derivative 3 also showed selective binding of nucleosides as well as nucleic acid bases dissolved in the aqueous subphase [3]. Evidence for binding was obtained from FTIR reflection spectra of multilayer systems (LB-layers) assembled by monolayer transfer onto glass slides covered with a 100 nm thick vapor-deposited gold film. Additional quantitative binding data from X-ray photoelectron spectroscopy confirmed these observations. Clearly, a selectivity was found with binding of thymidine, uridine and thymine in preference of adenosin and adenine, respectively. This seems to be due to the fact that the former group of bases and nucleic acids form 3 H bonds per base with the amphiphilic diaminotriazine, whereas adenine and adenosin form 2 H bonds per base only.

The interactions between equal and complementary nucleobases have been studied directly by measuring surface forces [4]. The bases were attached to the hydrophobic ends of long chain ammonium salts. The monolayers were transferred from the air-water interface to mica anchoring the layers with the ammonium groups on the substrate. The force-distance profiles indicate attractive forces between a surface of the adenine derivative and a surface of the thymin derivative at all pH values investigated. On the other hand, thymine-thymine contacts showed attractive behaviour at low pH (5.6, pure water) and repulsive behaviour at high pH (ionization). Adenine-adenine pairs show the opposite behaviour. Therefore, the interactions between non-complementarty pairs are regulated by surface charge.

A variety of amphiphiles with nucleobases and nucleosides have been synthesized and the interaction of monolayers of these materials with monomeric as well as polymeric nucleobases have been studied [5]. Complementary nucleobases dissolved in the subphase expand the monolayer more than non-complementary bases. The effect of monomeric bases on the surface pressure-area (π-A) isotherms is much smaller than that of the polymeric bases, demonstrating again the necessity of a minimum number of H bonds per base to be formed or cooperative interactions for binding by a monolayer at the air-water interface.

3. Multisite Recognition

A molecule like flavine adenine dinucleotide (FAD) provides different sites for formation of H bonds and ionic interactions. In an elegant work, this has been used to achieve molecular patterning and multi-site recognition, respectively, in a mixed monolayer of two [6] or three different amphiphiles [7]. The binding of FAD has been characterized by the changes of the surface pressure-area isotherm in the presence of FAD in the aqueous subphase as well as by IR and XPS spectroscopy of transferred multilayer systems on solid support. Most striking is the difference between the scanning force microscopic images of the two-component mixed monolayers on the aqueous subphase in absence and in presence of FAD, see Figure 2 [6]. The monolayer components, octadecyl-guanidinium G, and di-octadecyl-orotate O, are laterally rearranged due to specific recognition by FAD. The methyl end group of G is at a height different from that of methyl end groups of O when the system is deposited on a solid surface (mica). However, on water the different hydrophilic head groups are totally immersed in the water, thus exposing the methyl end groups at approximately the same height.

4. Large Aggregate Formation

Hydrogen bonds play an inportant role in protein interactions. The efficient molecular recognition in biosystems has been adapted for use in artificial structures such as mono-layers. An example is the strong interaction between biotin and streptavidin with biotinylated lipids in the monolayer and streptavidin in the aqueous subphase [8]. Based on the fact, that streptavidin has four binding sites, this specific interaction has recently been used to form two-dimensional aggregates in monolayers of a tetrabiotinylated porphyrin via binding of streptavidin from the subphase [9].

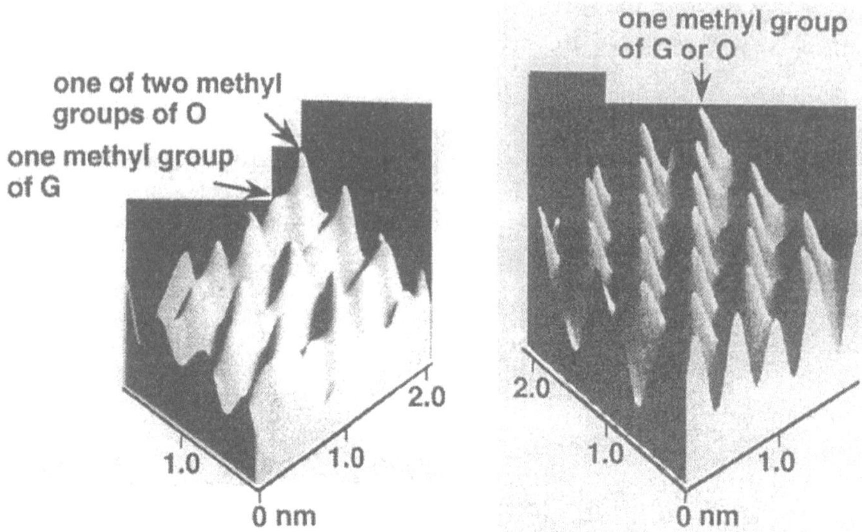

Figure 2: SFM image of a mixed monolayer of G and O, molar ratio 1:1, prepared on an aqueous subphase with FAD and transferred to mica (left); mixed monolayer of G:O = 1:1 on pure water (right); reproduced from [6].

Well defined extended linear aggregates have been formed in organic solution by hydrogen bond interactions of two different molecules, like barbituric acid and 2,4,6-triaminopyrimidine (TAP) [10,11]. This system with numerous H bonds per TAP molecule forming extended molecular strands is particularly suited for studies in monolayers at the air-water interface.

Amphiphilic derivatives of barbituric acid have been used for a spectroscopic investigation of the hydrolysis of this derivative in organized monolayers due to binding of TAP [12]. The amphiphilic barbituric acid derivative ABA forms associates as concluded from the reflection spectrum of the monolayer. These associates disappear upon binding of TAP from the aqueous subphase at bulk pH = 3. At this pH no hydrolysis occurs, however at a bulk pH = 6.5 the disappearance of the main band in the reflection spectrum at 460 nm indicates hydrolytic cleavage of the C=C bond of the chromophore (retro-Knoevenagl reaction) with formation of barbituric acid and the corresponding amphiphilic aldehyde as characterized by the new band with maximum at 350 nm (see Figure 3).

In the absence of TAP, a very slow hydrolysis is observed at bulk pH = 10 with formation of the aldehyde. These observations clearly show that TAP binds to the head group region of the monolayer of the amphiphilic derivative of barbituric acid. This is possible by formation of hydrogen bonds only. The C=C double bond is activated by the H bond to TAP, and therefore, TAP catalyzes the hydrolysis.

The ratio of the amphiphilic barbituric acid ABA and TAP may be strictly defined by co-spreading of both at the air-water interface. Although TAP is quite soluble in water it remains in the monolayer due to its interactions with ABA. For a molar ratio of ABA:TAP = 3:1, each TAP molecule catalyses the hydrolysis of two ABA molecules. Since the barbituric acid formed and the TAP are no longer anchored in the monolayer, they dissolve after the reaction in the aqueous subphase. Consequently, the reaction stops after hydrolysis of 2/3 of the ABA in the monolayer [12].

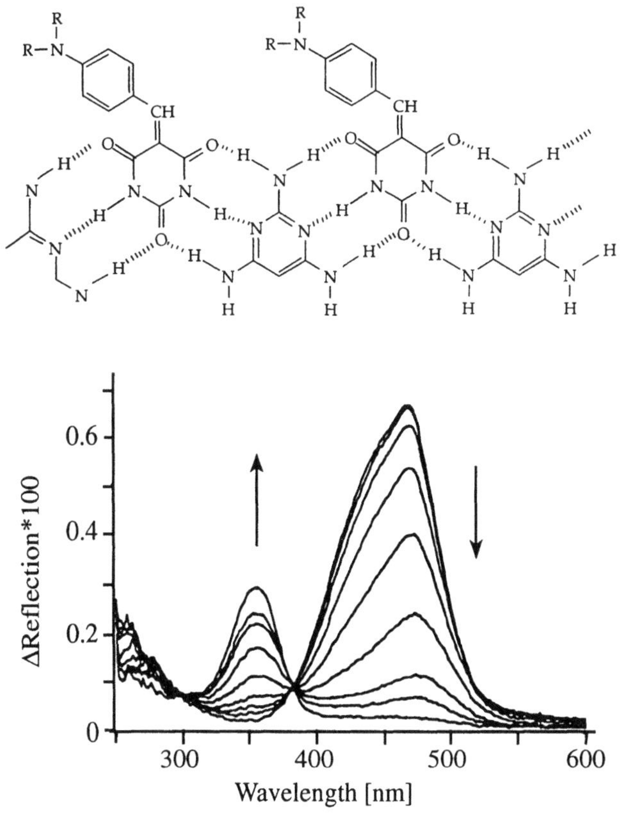

Figure 3. Structure of molecular strands formed in monolayers of the amphiphilic derivative of barbituric acid, ABA, by incorporation of 2,4,6-triamino-pyrimidine, TAP, from the aqueous subphase (top); reflection spectra (bottom) of a monolayer of ABA on 10^{-4} M TAP, pH = 6.5; the band at 460 nm decreases and that at 360 nm increases with increasing time: hydrolysis of ABA (λ_{max} = 460 nm) and formation of the corresponding aldehyde (λ_{max} = 350 nm); reproduced from [12].

Aggregates of TAP and barbituric acid form also in aqueous solution. Cleavage of ABA at the air-water interface is observed on a subphase containing a 1:1 aggregate of TAP and barbituric acid (10^{-4} M, pH = 6.5), although much slower than in the case of a sub-phase containing only TAP at same concentration and pH [13]. Monolayers of an amphiphilic derivative of 2,4,6-triaminotriazin (melamine), bind barbituric acid from the aqueous subphase at a ratio of (1.1 ± 0.1):1. After transfer on mica, such complex monolayers show a very regular molecular packing in the scanning force microscope in reasonable agreement with packing of alkyl chains [14].

5. Combination of Hydrogen Bonds with Ionic Interactions

Particularly strong interactions can be expected in the case of a combination of directionally controlled hydrogen bonding with attractive electrostatic forces, i.e. ion pairing. Evidence for the formation of extended strands by interaction of alkylsulfonate monolayers with guanidinium cations in the aqueous subphase was obtained from measurements of surface pressure-area isotherms [15]. Formation of infinite molecular tapes has been observed even in aqueous solution by self-assembly of complementary dianionic and dicationic hydrogen bond donors and acceptors [16]. Evidence for such interactions in monolayers at the air-water interface is obtained from π-A isotherms of mixtures of the amphiphilic diamidine THA (structure see Figure 4) and terephthalic acid TPS as well as nitro-isophthalic acid, NIP.

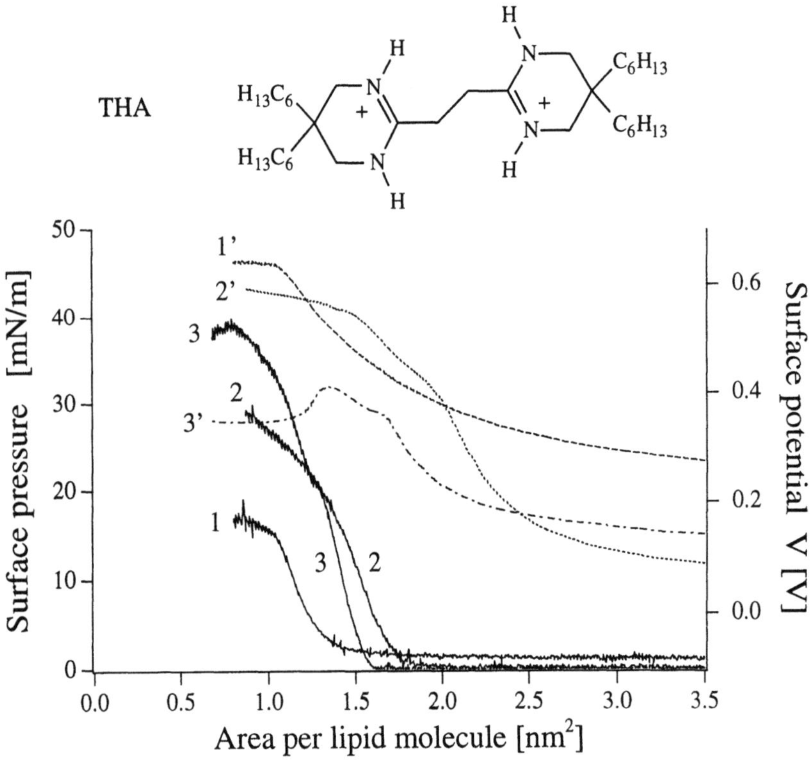

Figure 4. Surface pressure (π, full lines) and surface potential (ΔV, dotted lines) plotted vs. area per THA molecule, A, of monolayers of THA (1, 1') and of mixtures THA:TPS = 1:1 (2, 2') and THA:NIP = 1:1 (3, 3'); subphase water, 20°C (bottom); structure of THA (top).

The monolayers were formed by co-spreading of the components. Figure 4 shows π-A (full lines) and surface potential-area (ΔV-A, dotted lines) isotherms of monolayers of the pure THA (1,1'), and of mixtures THA:TPS = 1:1 (2,2') and THA:NIP = 1:1 (3, 3'). A is the area per THA molecule spread. The presence of terephthalic acid, TPS, as well as nitro-isophthalic acid, NIP, clearly modifies the π-A and ΔV-A isotherms of THA. The collapse pressure is strongly increased, and the surface potential is considerably diminished in the mixed monolayers. In case of the mixed monolayer THA:TPS the area per THA molecule at π = 10 mN/m is 1.53 nm^2 (curve 2) which is in excellent agreement with an area of 1.5 nm^2 estimated for the pair THA/TPS from X-ray crystallographic data [16] assuming dense packing of parallel flat lying strands of THA:TPS = 1:1. The reduction of the surface potential at 10 mN/m comparing THA:TPS with = 1:1 THA is $\Delta\Delta V$ = -0.067 V. Such a reduction may be attributed to salt formation in the monolayer plane (no contribution to the surface potential) in the mixed monolayer strongly diminishing the contribution of the diffuse double layer present in the case of the pure THA monolayer to the surface potential (measuring the normal component of dipole moments only).

The behaviour of the mixed monolayer THA:NIP = 1:1 is similar to that of THA:TPS. The area per THA is A = 1.42 nm^2. The surface potential is more strongly reduced in the presence of NIP, i.e. $\Delta\Delta V$ = -0.202 V, indicating an additional influence of the nitro group as compared to the case of THA:TPS.

The modification of both π-A and ΔV-A isotherms of THA due to the presence of terephthalic acid and nitro-isophthalic acid, respectively, in the spreading solution clearly demonstrates that these components are present in the monolayer at the air-water interface. This may be confirmed by spectroscopic measurements of the monolayers.

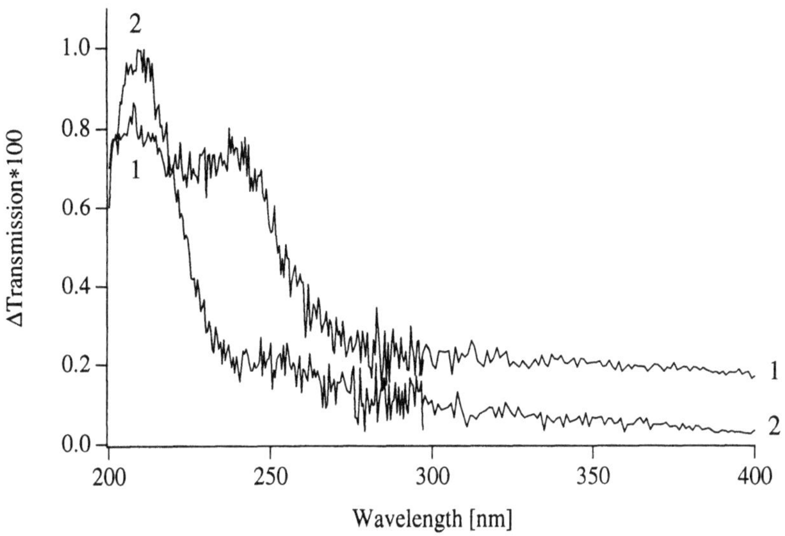

Figure 5: Absorption spectra of single monolayers transferred on quartz plates; (1) THA:TPS = 1:1; (2) THA:TPS = 1:1; the absorption here is the difference in transmission ΔT between a reference section of the slide without the mixed monolayer and the sample section with monolayer.

The mixed monolayers have been transferred from the air-water interface directly onto quartz plates at a surface pressure of 10 mN/m. Figure 5 shows the absorption spectra (measured as difference ΔT in transmission of a reference section of the slide without the mixed monolayer and the sample section with the monolayer) of single monolayers of THA:TPS = 1:1 (curve 1, shifted up for clarity) and of THA:NIP = 1:1 (curve 2).

The monolayer of THA:TPS = 1:1 shows a maximum around 209 nm and another band with maximum at 240 nm. In solution, the main absorption band of TPS is at 233 nm (another small band at 285 nm) and that of THA is at 222 nm. Therefore, we attribute the band at 209 nm mainly to THA and the band at 240 nm to TPS. The monolayer of THA:NIP = 1:1 shows a strong band with maximum at 210 nm and a weak band at 250 nm. In solution, NIP has a band with maximum at 225 nm and a weaker band at 255 nm. Consequently, the band at 210 nm is due to both THA and NIP, whereas the weak band at 250 nm is attributed to NIP. The shifts observed in the monolayers may be caused by interactions between the components as indicated by slight deviations of the spectra of the mixtures in solution from additivity (not shown). The absorption spectra of single monolayers on quartz reconfirm the presence of terephthalic acid and of nitro-isophthalic acid in these monolayers based on H bonds and electrostatic interactions with THA.

From the absorption ΔT at the maximum, the surface density σ of the absorbing species can be calculated in a first approximation from the extinction coefficient ε in solution assuming that this is unchanged [17,18] according to

$$\sigma/(\mathrm{nm}^{-2}) = [6.023 \cdot 10^{23} \cdot \Delta T]/[f_o \cdot 2.303 \cdot \varepsilon/(\mathrm{Lmol}^{-1}\mathrm{cm}^{-1}) \cdot 1000 \cdot 10^{14}] \qquad (1)$$

where f_o is a factor taking the different average orientations of the transition moment in the monolayer as compared to the solution into account. For an orientation of the transition moments in the layer plane, f_o = 1.5. With the extinction coefficients measured in the mixed solutions the value σ = 0.37 nm^{-2} is obtained for TPS as compared to σ = 0.65 nm^{-2} calculated from the π-A isotherm. In the case of NIP the values are σ = 0.59 nm^{-2} (absorption spectroscopy) and σ = 0.70 nm^{-2}. This calculation reconfirms the conclusion based on the isotherm measurements that both TPS and NIP are present in the mixed monolayers.

A possible reason for the difference between the surface density of TPS on the quartz slide (as calculated form ΔT) and the surface density at the air-water interface is incomplete transfer of TPS. In contrast to monolayers of long chain ammonium salts those of fatty acids are not transferring easily onto the hydrophilic quartz surface. Therefore, the mixed monolayer of THA:TPS = 1:1 was transferred onto a hydrophobised quartz plate (coated with a monolayer of eicosylamine). The absorption spectrum of this layer system is shown in Figure 6 along with the schematic system structure. The absorption ΔT(230 nm) = 0.0087, and with the same orientational factor of 1.5, the surface density σ = 0.59 nm^{-2} is calculated with Eq. 1 in good agreement with the value of σ = 0.65 nm^{-2} calculated from the π-A isotherm.

Although these results clearly indicate binding of both TPS and NIP to monolayers of THA formed by co-spreading of the components from a mixed solution, the interaction by hydrogen bonds should be unambiguously verified by FTIR of transferred monolayers. Preliminary studies of THA:TPS monolayers by Brewster angle microscopy did not show the expected optically anisotropic domains associated with infinite strands. This, however, could be due to the limited lateral resolution of the instument.

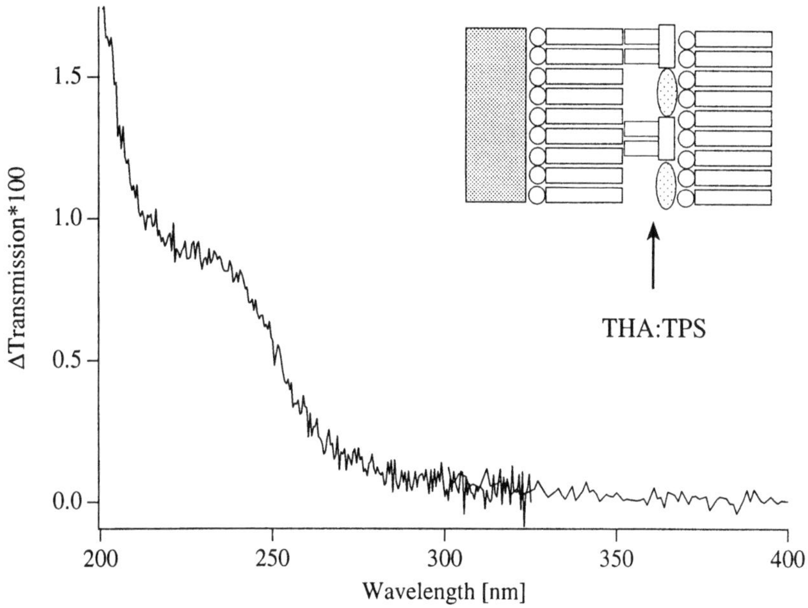

Figure 6. Absorption spectrum of a monolayer of THA:TPA = 1:1 transferred on a hydrophobic quartz plate (system structure shown schematically).

6. Conclusion

Hydrogen bond interactions in monolayers, sometimes in combination with electrostatic interactions, give rise to molecular patterning, molecular recognition, binding of a catalyst and the formation of large ordered aggregates. Truely supramolecular structures are spontaneously formed in monolayers at the air-water interface due to the directionality of the hydrogen bond interactions.

7. References

1. Paleos, C.M. and Tsiourvas, D. (1997) Molecular recognition of organized assemblies via hydrogen bonding in aqueous media, *Adv. Mater.* **9**, 695-710.
2. Honda, Y., Kurihara, K., and Kunitake, T. (1991) Molecular recognition of barbiturates by diaminotriazine functionalized monolayers, *Chem. Lett.* **1991**, 681-684.
3. Kurihara, K., Ohto, K., Honda, Y., and Kunitake, T. (1991) Efficient, complementary binding of nucleic acid bases to diaminotriazine-functionalized monolayers on water, *J. Am. Chem. Soc.* **113**, 5077-5079.
4. Kurihara, K., Abe, T., and Nakashima, N. (1996) Direct measurement of attraction for a complementary pair of apposed nucleic acid base monolayers, *Langmuir* **12**, 4053-4056.

5. Ahlers, M., Ringsdorf, H., Rosemeyer, H., and Seela, F. (1990) Orientation, recognition and photoreaction of nucleolipids in model membranes, *Colloid Polym Sci* **268**, 132-142.
6. Oishi, Y., Torii, Y., Kuramori, M., Suehiro, K., Ariga, K., Taguchi, K., Kamino, A., and Kunitake, t. (1996) Two-dimensional molecular patterning through molecular recognition, *Chem. Lett.* **1996**, 411-412.
7. Taguchi, K., Ariga, K., and Kunitake, T. (1995) Multi-site recognition of flavin adenine dinucleotide by mixed monolayer on water, *Chem. Lett.* **1995**, 701-702.
8. Ahlers, M., Blankenberg, R., Haas, H., Möbius, D., Möhwald, H., Müller, W., Ringsdorf, H., and Siegmund, H.-U. (1991) Protein interactions with ordered lipid films: Specific and unspecific binding, *Adv. Mater.* **3**, 39-46.
9. Fukushima, H., Taylor, D.M., Morgan, H., Ringsdorf, H., and Rump, E. (1995) The specific adsorption of streptavidin to a tetrabiotinylated porphyrin monolayer at the air-water interface, *Thin Solid Films* **266**, 289-291.
10. Lehn, J.-M., Mascal, M., DeCian, A., and Fischer, J. (1990) Molecular recognition directed self-assembly of ordered supramolecular strands by cocrystallization of complementary molecular components, *J. Chem. Soc., Chem. Commun.* **1990**, 479-481.
11. Zerkowsky, J.A., Seto, C.T., Wierda, D.A., and Whitesides, G.M. (1990) Design of organic structures in the solid state: Hydrogen-bonded molecular "tapes", *J. Am. Chem. Soc.* **112**, 9025-9026.
12. Ahuja, R.C., Caruso, P.-L., Möbius, D., Paulus, W., Ringsdorf, H., and Wildburg, G. (1993) Formation of molecular strands by hydrogen bonds at the gas-water interface: molecular recognition and quantitative hydrolysis of barbituric acid lipids, *Angew. Chem. Int. Ed. Engl.* **32**, 1033-1036.
13. Bohanon, T.M., Denzinger, S., Fink, R., Paulus, W., Ringsdorf, H., and Weck, M. (1995) Barbituric acid/2,4,6-triaminopyrimidine aggregates in water and their competitive interaction with a monolayer of barbituric acid lipids at the gas-water interface, *Angew. Chem. Int. Ed. Engl.* **34**, 58-60.
14. Koyano, H., Yoshihara, K., Ariga, K., Kunitake, T., Oishi, Y., Kawano, O., Kuramori, M., and Suehiro, K. (1996) Atomic force microscopic observation of a dialkylmelamine monolayer on barbituric acid, *Chem. Commun.* **1996**, 1769-1770.
15. Frostman, L.M. and Ward, M.D. (1997) Nucleation of molecular crystals beneath guanidinium alkanesulfonate monolayers, *Langmuir* **13**, 330-337.
16. Hosseini, M.W., Ruppert, R., Schaeffer, P., De Cian, A., Kyritsakas, N., and Fischer, J. (1994) A molecular approach to solid-state synthesis: Prediction and synthesis of self-assembled infinite rods, *J. Chem. Soc., Chem. Commun.* **1994**, 2135-2136.
17. Grüniger, H., Möbius, D., and Meyer, H. (1983) Enhanced light reflection by dye monolayers at the air-water interface, *J. Chem. Phys.* **79**, 3701-3710.
18. Orrit, M., Möbius, D., Lehmann, U., and Meyer, H. (1986) Reflection and transmission of light by dye monolayers, *J. Chem. Phys.* **85**, 4966-4979.

Abbreviations

ABA amphiphilic barbituric acid derivative
FAD flavin-adenine-dinucleotide
NIP nitro-isophthalic acid
TAP trisamino-pyrimidine
THA tetrahexyl-diamidinium
TPS terephthalic acid

SUPRAMOLECULAR ARCHITECTURES ON LIQUID SURFACES. FROM 2-D ASSEMBLIES TO 3-D CRYSTALS

I. WEISSBUCH, R. POPOVITZ-BIRO, I. KUZMENKO,
H. RAPPAPORT, M. BERFELD, S. GUO,
L. LEISEROWITZ and M. LAHAV
Department of Materials and Interfaces,
The Weizmann Institute of Science
76100 Rehovot, Israel

1. Introduction

Self-assembly of molecules into structured clusters is an important intermediate state *en route* to the growth of three-dimensional crystals or the formation of supramolecular architectures. We directed our efforts towards the preparation of functional clusters at the air-aqueous solution interface for several reasons: most important, thin crystalline films can be easily prepared at the water surface; furthermore, it is possible, at the air-liquid interface, to combine water-soluble and water-insoluble components to form multi-component films; and thirdly, the flat nature of the water surface is ideal for performing surface X-ray diffraction measurements (GIXD) using synchrotron radiation for structure elucidation of the films, an analytical method that has emerged over the last decade. The thin-film clusters generated at the air-solution interface can be deposited on solid surfaces and then analyzed by other methods such as scanning force microscopy (SFM) and cryo-electron microscopy (cryo-TEM).

G. Tsoucaris (ed.), Current Challenges on Large Supramolecular Assemblies, 97–114.

For some years we had applied these methods for the study of the early steps of crystal nucleation; more recently we focused our attention on the preparation of complex architectures related to various problems of current interest in science. These include the early stages of the crystalline aggregation of cholesterol at the water surface in relation to gallstone precipitation, the spontaneous separation of racemates into enantiomorphous 2-D domains, which may have bearing on the generation of homochirality on earth, the formation of interdigitated crystalline thin films at the air-water interface and the organization of channel-forming molecules and of natural ionophores within a hydrophobic membrane-like environment. The structure elucidation of latter systems is related to processes of ion transfer through membranes. Finally, we studied the self-organization of organometallic supramolecular architectures and their conversion into organic-inorganic composite materials by topotactic gas-solid reactions. In these composites the inorganic particles are of quantum size and are arranged in the form of a one-dimensional superlattice within the organic matrix. Such architectures will hopefully display magnetic, conducting or optoelectronic properties.

2. Self-assembled multilayers at the air-solution interface

During systematic studies on ice nucleation induced by long-chain monoalcohols[1], $C_nH_{2n+1}OH$ (n=13-31), spread on drops of water we observed that in the series of α,ω-diols[2], HO-$(CH_2)_n$-OH n=16-30, molecules with n=even are efficient promoters of ice formation, inducing freezing of super-cooled water at about -5C°, temperature higher by a few degrees than those obtained from the monoalcohols of corresponding length. This observation implied that the diols are arranged in a manner akin to that of the long-chain monoalcohols so that the long axes of their hydrocarbon chains are aligned almost perpendicular to the water surface and their hydroxy groups in contact with the water form a two-dimensional net similar to that of hexagonal ice. This structure is at variance with a suggested notion that an α,ω-diol alkyl chain adopts a hair-pin conformation to permit both OH groups to dip into the water. Grazing incidence X-ray diffraction (GIXD) studies of several diols, for example $HO(CH_2)_{22}OH$, demonstrated the formation of crystalline bilayers on the water surface.[3] Transfer of these films onto freshly cleaved mica yielded crystallites with an average thickness of four layers according to the height analysis of the SFM images.[4] The discrepancy between the two methods may be resolved by assuming either a multilayer buildup on transfer to mica or an interlayer crystalline misregistry of the films on water so that the number of layers, as detected by the Bragg-rod analysis, appears to be less than what they actually are.

The formation of crystalline films, several layers thick, amenable to characterization provides a means to probe more directly the role played by tailor-made additives on the early steps of crystal nucleation. By applying the above mentioned analytical methods it could be demonstrated that the addition of small quantities of additives that structurally ressemble the molecules of the multilayers such as monoalchools or monoacids bearing longer hydrocarbon chains transform the multilayers into monolayers.[5]

Other bolaamphiphiles, $X\text{-}(CH_2)_n\text{-}X$, such as α,ω-alkane dicarboxylic acids, X=COOH, also form well characterized clusters at the interface. In the system $HOOC(CH_2)_{22}COOH$ a coexistence of dimorphic aggregates composed of molecules whose long molecular axes are aligned either vertically or making a tilt of about 35° from the perpendicular to the water surface.[6]

The formation of crystalline multilayers of the bolaamphiphiles on water led to the realization that unfunctionalized long-chain alkanes should also self-organize on the water surface to form multilayer crystallites, as was demonstrated by GIXD and specular X-ray reflectivity studies on the systems C_nH_{2n+2} n=23-30.[7] The structures of these clusters was determined, to near atomic resolution, making use of the known packing arrangements of the corresponding 3-D crystals. For C_nH_{2n+2} $n > 31$ monolayers were formed and as the chain length decreased from $C_{29}H_{61}$ to $C_{23}H_{48}$ the number of layers increased from 2 to about 20. (Fig. 1). With the exception of $C_{24}H_{50}$, for which all the hydrocarbon chains are tilted by about 16° from the normal to the layer plane and related by translation only, the chains in the other crystalline films are aligned normal to the layer plane and adopt the herringbone motif as in the corresponding 3-D crystal structures. For very long chains, such as $C_{50}H_{102}$, the molecules form a crystalline

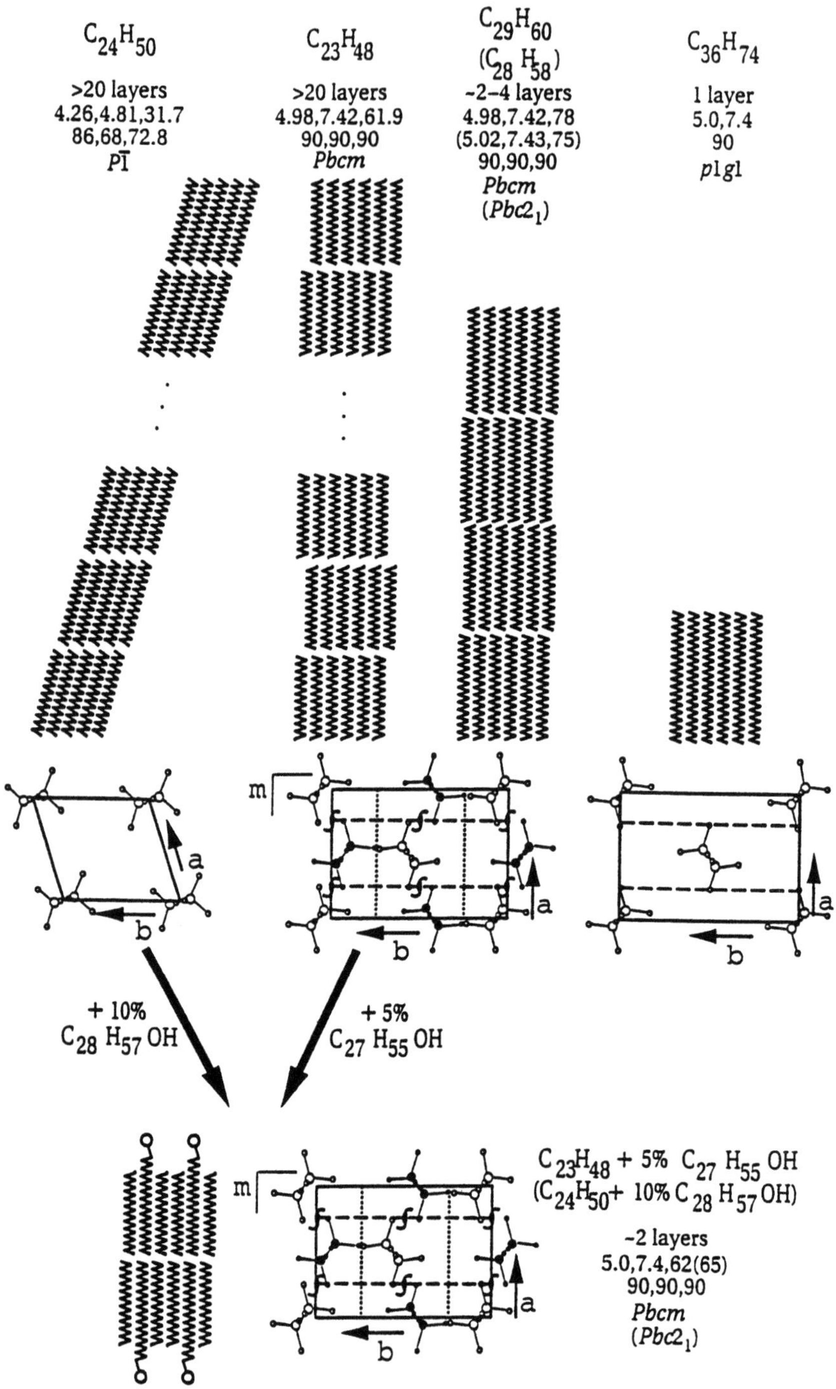

Figure 1 Schematic representation of the layer-by-layer building of C_nH_{2n+2} alkane crystallites on a water subphase, as determined by GIXD, shown as side-view of the molecuyles together with the packing arrangements along the molecular axis.

monolayer on water[8], but the Bragg rods are skewed suggesting that the monolayer films are not flat, perhaps assuming a concave-like shape, like that of the bottom surface of an oil drop on the water surface.

Aromatic hydrocarbons such as oligothiophenes, H-(C_4H_2S)$_n$-H, n=4-6, were also found to self-organize on the water surface.[9] Multilayers were formed by the shorter homologues n= 4, 5, whereas hexathiophene (n=6) yielded primarily a monolayer film. According to GIXD measurements the latter system is dimorphic, as was also observed for the films of HOOC-(CH_2)$_{22}$-COOH on water. In one dimorph of hexathiophene the long molecular axis is aligned perpendicular to the layer plane, in the other dimorph the chain axis is tilted by an angle of 23.5°, as in the three-dimensional crystal structure. The coexistence of different structured clusters in the early stages of their formation has bearing on the kinetics of crystallization of polymorphic crystalline forms.

3. Self-aggregates of cholesterol

Another amphiphilic system that was recently investigated in some details, with regards to the formation of gallstones, is cholesterol. Upon deposition of cholesterol on water it forms spontaneously clusters of a monomolecular thickness where the molecules exihibit herring-bone arrangement in a hexagonal type structure, probably of the rotator phase, with a pronounced librational motion about the long molecular axis (Fig. 2).[10]

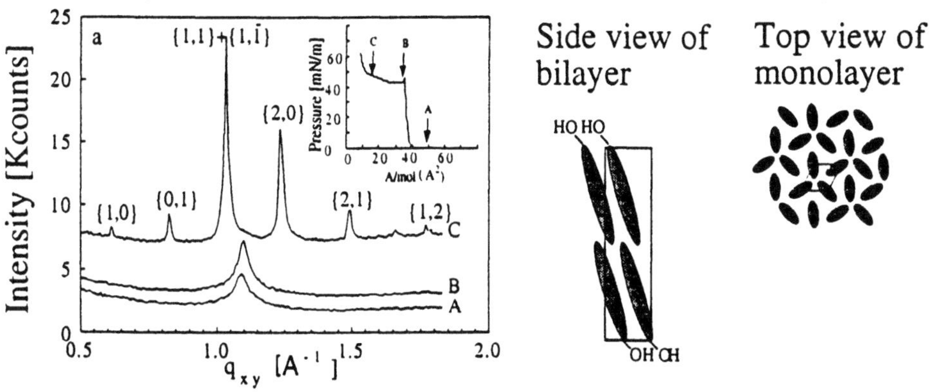

Figure 2 *left* GIXD diffraction pattern of cholesterol film at the air-water interface, inset showing the surface pressure-area isotherm with the arrows indicating the points at which measurements were performed. *right* Packing arrangements of the bilayer (side view) and of the monolayer (top view).

Upon compression, a crystalline bilayer is formed that assumes a structure simmilar albeit different from that of the cholesterol monohydrate 3-D crystal, as determined from the GIXD data.

Three or more layers should perhaps yield the macroscopic structure. The morphology and the thickness of these films are strongly affected by the presence of other steroids that inhibit the cluster formation. The ability to monitor the structures of these early clusters directly provides new means to design stereospecific nucleation inhibitors of cholesterol crystal growth.

4. Spontaneous separation of racemic systems into 2-D enantiomorphous crystals

The spontaneous separation of enantiomers from a racemic mixture has intrigued scientists since this phenomenon was discovered by Pasteur a century and half ago. Unlike 3-D crystals, the detection of spontaneous separation in 2-D is not straightforward. The method of comparing surface pressure-area isotherms of racemic and chiral Langmuir monolayers[11] is not fully reliable for the following reasons: It is based on the concept that the shape of the isotherm reflects a phase transformation from a gaseous to a liquid to a homogeneous crystalline monolayer. However, according to grazing incidence X-ray diffraction studies, amphiphiles spontaneously form crystalline monolayer islands at very low surface pressures thus the shape of the surface pressure-area isotherm depends not only on molecular packing within the islands, but also upon other factors such as size and shape of the islands and the percentage of nonordered molecules that lie on island peripheries. In principle, the outline of 2-D single crystals delineated by straight boundary lines at the air-liquid interface should reflect the plane group symmetry of the crystal and thus may display chiral hemihedral shapes in the event of enantiomorphous crystals. The latter can be detected by Brewster angle and epifluorescence microscopy. However, observed 2-D domains were never observed to be single crystals but rather assemblies of crystallites, often assuming dendridic morphologies; thus optical methods do not suffice to establish spontaneous separation of enantiomers in monolayer crystals.

Experiments involving detection of chiral segregation in 2-D crystals only began to bear fruit with the use of techniques that probe the structure of thin crystalline films at the nanometer scale. For example, a monolayer, deposited on mica, of a tetracyclic alcohol[12] displaying mirror-image domains and therefore of opposite chirality has been observed by scanning force microscopy and separation of enantiomers of a liquid crystalline biphenyl derivative[13] on graphite by scanning tunneling microscopy. The spontaneous chiral segregation in Langmuir films of a racemic mixture of myristoyl-alanine has been recently inferred from grazing incidence X-ray diffraction measurements.[14]

The possibility that chiral information may have been transferred across an interface as in important step in the transformation from a racemic chemistry to chiral biology prompted us to investigate the separation of a racemic mixture, at the air-water interface, into left- and right-handed 2-D arrays of hydrophobic α-amino acids. In order to induce racemic mixtures of amphiphiles to form, by design, either *homo*- or *hetero*-chiral monolayer clusters on the water surface, it is necessary to first review the fundamental differences in the packing of molecules in three- and two-dimensional crystals. The most common symmetry elements in 3-D crystals are center of inversion, glide plane, twofold screw axis and translation. In 2-D crystals, but at a liquid surface for amphiphilic molecules that are specifically oriented *vis-a-vis* the water surface, the situation is simpler. The center of inversion, the glide with its plane parallel to the water surface and the corresponding twofold screw symmetry are precluded. The only remaining symmetry elements are translation and a glide whose plane is perpendicular to the water surface. Thus, in principle, it should be easier to bring about spontaneous segregation of enantiomeric territories at the solution surface if the glide symmetry and possible formation of solid solutions between the two enantiomers is prevented. There is, however, one difficulty in such an approach: the hydrophobic moiety of amphiphilic molecules is generally composed of a long aliphatic chains which tend to pack in the herring-bone motif generated by glide symmetry. Such a molecule may require a functional group within the chain that, together with the hydrophilic group, will promote only translational packing.

Reasoning on the above lines was applied for the design of racemic mixtures of α-amino acid amphiphiles, X-HC(NH_3^+)CO_2^-, to form either of two hydrogen-bonded monolayer motifs on water.[15] In one motif, the molecules are related by glide symmetry as in the racemic crystal structure of norleucine. In the other, the molecules are related by translation as in the α-form of glycine. For the first motif, X was chosen to be a long chain alkyl C_nH_{2n+1}-, yielding the amphiphile labelled C_n-gly, to induce the herring-bone motif and thus glide symmetry. In the second motif, X was chosen to be $C_nH_{2n+1}CONH(CH_2)_4$-, yielding an amphiphile that is a derivative of lysine, labelled C_n-lys, that contains a hydrogen-bonding amide group to promote translation along a 5Å axis as does glycine itself.

The crystalline structure of monolayers of these two types of amphiphiles on water were elucidated by GIXD.[15] The molecules of racemic (R,S)-C_n-gly,n= 12,16 spontaneously assemble on water into *heterochiral* crystalline domains in a rectangular unit cell containing two molecules related by glide symmetry in a herring-bone arrangement of the chains, as anticipated (Fig. 3*right*).

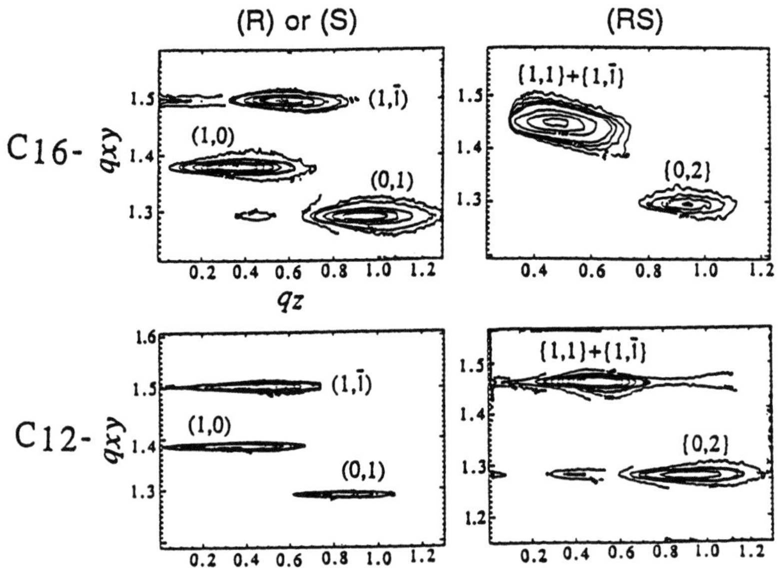

Figure 3. GIXD patterns shown as two-dimensional intensity contour plots I(q_{xy}, q_z) measured from monolayers of enentiomerically pure and racemic C_n-gly spread on water.

The enantiomerically pure (S)-C_n-gly, n = 12,16 self-assemble into crystalline domains with molecules related by translation in an oblique unit cell (Fig. 3*left*).

On the other hand, the crystalline monolayers of C_n-lys, n = 17,21, both in racemic (R,S) and in chiral (S) form, are essentially isostructural, (Fig. 4), the molecules being related by translation in an oblique unit cell in which the glycine head-groups, $^{+}H_3NCHCO_2^{-}$, are interlinked by a two-dimensional hydrogen-bonded motif as in the 3-D crystal of α-glycine. Thus the introduction of an amide group into the chain induces the racemic mixture to segregate into domains of opposite chirality.

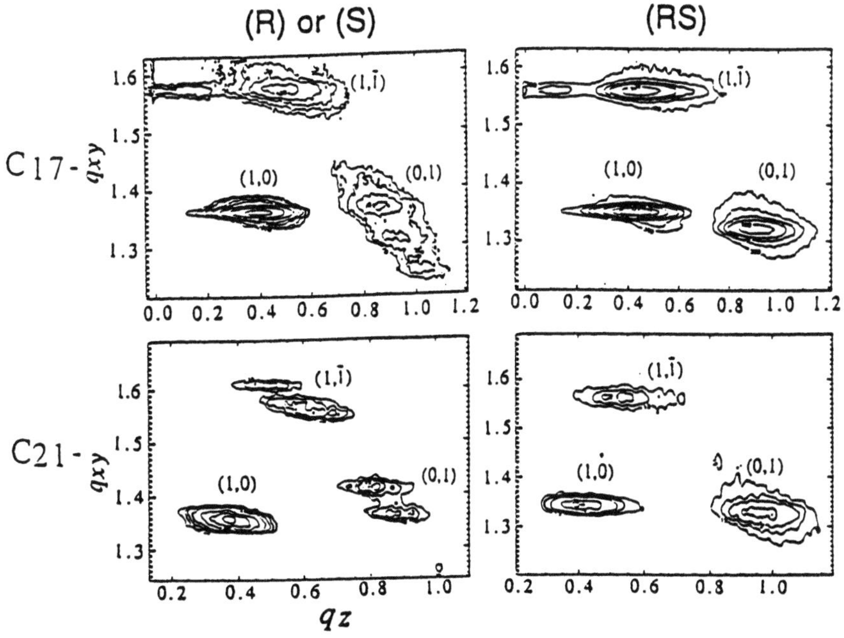

Figure 4. GIXD patterns shown as two-dimensional intensity contour plots $I(q_{xy}, q_z)$ measured from monolayers of enentiomerically pure and racemic C_n-lys n=17,21 spread on water.

Spontaneous separation of enantiomers of more complex monolayer systems has been achieved making use of acid-base interactions in mixtures, taking advantage of the fact that the water-soluble 'homochiral' salt $C_6H_5CH(CH_3)NH_3^{+}$:$C_6H_5CH(OH)COO^{-}$, is more stable than its diastereomeric counterpart. 1:1 mixtures of (R) or (S) *p*-tetradecyl-phenylethylamine, $C_{14}H_{29}$-C_6H_4-$CH(CH_3)NH_2$, and (R') or (S') *p*-pentadecyl-mandelic acid ,C_6H_5-CH(OH)COOH, form on water stable crystalline monolayers independent of molecular chirality. The unit cells of the (R,R') and (R,S') crystalline films each contain one diastereomeric pair of molecules. Their GIXD patterns are distinctly different. An equimolar mixture of the four components gave rise to a diffraction pattern almost identical to that of the (R,R') mixture, signifying a separation of (R,R') and (S,S') territories. More detailed studies, however, demonstrated that a substantial degree of mixing of the enantiomers takes place within these phases.[16]

5. Diastereomeric complexes of copper / α-amino acids at the air-solution interface

The driving force for the separation of the enantiomers of the amphiphilic α-amino acids, discussed above, is provided by the hydrogen-bonds of the glycine head-group and the amide group in the hydrocarbon chain. Addition, into the subphase, of copper acetate whose Cu^{2+} ions bind to the glycine moieties, affects the hydrogen-bonding properties of the amphiphile so as to preclude spontaneous separation of a racemic mixture, according to GIXD measurements. The diffraction experiments on copper acetate solution also demonstrated superlattice formation of the monolayer-bound copper ions. Injection of each enantiomer of the water-soluble (*R'*) and (*S'*) α-amino acids serine, alanine or valine into the aqueous copper acetate subphase under the monolayer of enantiomerically pure (*S*) C_{15}-lys led to different surface pressure-area isotherms; the isotherms being more expanded for the amphiphile and solute of the same absolute configuration (Fig. 5)[17].

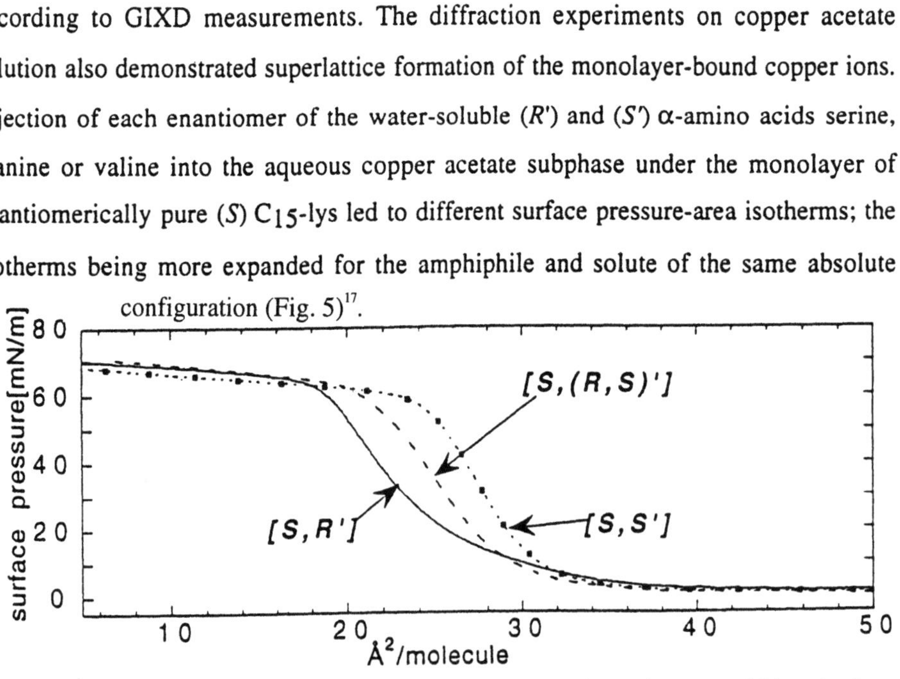

Figure 5. Surface pressure-area isotherms showing the effect of injection of the water-soluble serine into the copper acetate aqueous solution beneath the monolayer of (*S*)-C_{15}-lys. *[S,R']*, *[S,(R,S)']*, and *[S,S']* refer to the injection of (R'), (R,S)' and (S') serine beneath the (*S*)-C_{15}-lys monolayer.

Changes in the monolayer GIXD pattern were observed after injection of serine. These patterns were different for the (*R*)' and of (*S'*) serine, implying not only reaction of the Cu^{2+} bound polar head groups of (*S*) C_{15}-lys with serine but different packing arrangements for the two (*S*-Cu-*S'*) and (*S*-Cu-*R'*) heterocomplexes. The configuration of the ligands about the Cu^{2+} ion in these two systems proved to be the same as in the 3-D crystal structures of the copper complexes of (*S*)-serine and (*RS*)-serine which display syn and anti arrangements of their ligands, respectively. This conclusion was based on comparative X-ray photoelectron spectroscopy (XPS) studies of the deposited films of the two heterocomplexes and the 3-D crystals. We thus explain the more expanded isotherms of the amphiphile and solute of the same absolute configuration by assuming that the water-soluble complexes (*S'*-Cu-*S'*) with a syn configuration can bind to the periphery of monolayer islands of the *S*-Cu-*S'* system, but not to the monolayer of *S*-Cu-*R'* which has an anti configuration. This proposed adsorption is akin to the enantioselective adsorption of chiral tailor-made additives to the surfaces of growing 3-D crystals.

6. Interdigitated architectures at the air-solution interface

The self-assembly, at the air-water interface, of chain-like molecules into thin crystalline films was extended to the generation of crystalline films made up of two different and complementary molecules, one water-insoluble and the other water-soluble. The latter is intercalated between the polar head groups of the Langmuir monolayer (Scheme 1). In a search for appropriated bimolecular systems that satisfy the above criteria we made use of the diastereomeric phenethylamine mandelate salts. The 3-D crystal structure, composed of phenethylamine, $C_6H_5CHCH_3NH_2$ (PEA) and mandelic acid $C_6H_5CH(OH)CO_2H$ (MA) molecules of the same handedness, is characterized by hydrogen-bonded bilayers. The phenyl rings are positioned and oriented in such a way that is compatible with the formation of an interdigitated arrangement when either the acid or the amine would be modified by attaching a long hydrocarbon chain in the para position of its phenyl ring. The 3-D crystal structure of phenethylamine mandelate salt composed of molecules of opposite handedness also contains hydrogen-bonded bilayers, but whose phenyl rings are not aligned in a way suitable for an interdigitated arrangement.

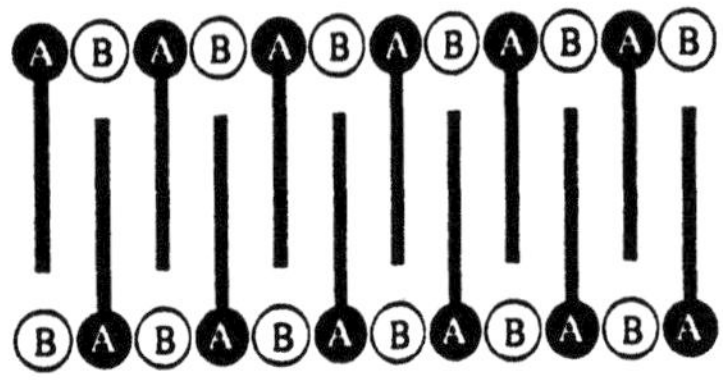

Scheme 1

The surface pressure-area isotherms of (R) pentadecyl-mandelic acid p-$C_{15}H_{31}$-$C_6H_4CH(OH)CO_2H$ (C15-MA) on aqueous solutions of (R')- and of (S')-PEA have similar shapes (Fig. 6). Both isotherms are expanded and reach a plateau at a kink with a molecular area $A \approx 40A^2$ and $\Pi \approx 42$ mN/m. GIXD patterns were observed only for the diastereomeric system (R,R') and only upon compression beyond the kink point.

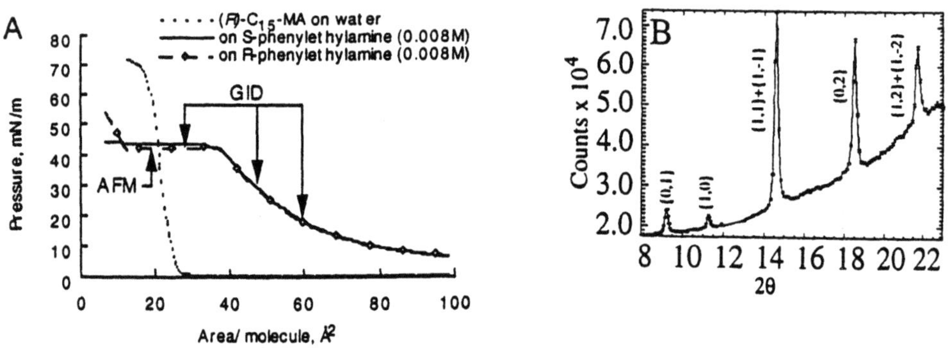

Figure 6. Surface pressure-area isotherm sof C_{15}-MA spread on water and aqueous solution of PEA.

The GIXD data was interpretated in terms of an interdigitated crystalline trilayer about 40Å thick, whose detailed packing arrangement (Fig. 7) was determined by X-ray structure factor analysis of the Bragg rod data, making use of atomic coordinates obtained from the interdigitated complex as it appears in the 3-D crystal.[18] The film at the air-solution interface consists of a crystalline interdigitated bilayer and a monolayer whose alkyl chains exposed to the air are disordered because of the gaps created by the interleaved PEA units. This model was also supported by SFM studies. The corresponding (R,S') salt appears to form a bilayer, but which does not diffract.

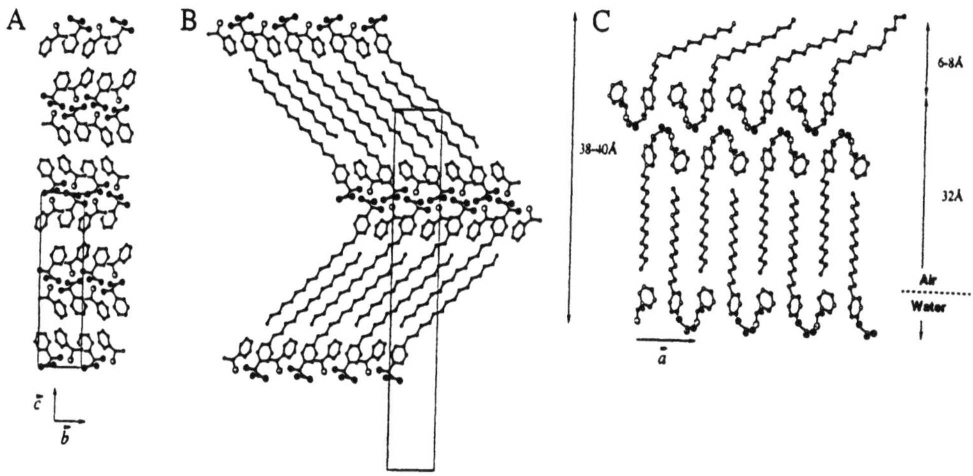

Figure 7. (A,B) Molecular packing arrangements of the 3-D crystals of (R-MA,R-PEA) and (R-C_{15}-MA, R-PEA) viewed along the *a* axis. (C) Packing arrangement of the interdigitated trilayer (R-C_{15}-MA, R-PEA) viewed along the *b* axis.

More recently, in collaboration with G. von Kiedrowsky and M. Kindermann we have demonstrated the presence of interdigitation for other systems including amphiphilic acids with short chain amphiphilic amidiniums, studies performed also with relevance to the origin of self-replicating systems.

7. Oriented Multilayers of organometallic films

The formation of interdigitated systems in the previous examples demonstrated that water-insoluble and water-soluble components can interact to form oriented crystalline architectures at the air-solution interface. We have discussed earlier that the long-chain α,ω-dicarboxylic acids, when deposited on water, form crystallites in which the molecules are aligned with their long axes making an angle ~35° with the normal to the water surface. A very different orientation of these molecules was observed by GIXD on aqueous subphases containing divalent Cd^{2+} and Pb^{2+} ions. The dicarboxylic acids react with the ions and form the corresponding di-salt. The salt molecules self-assemble with their hydrocarbon chains parallel to the solution surface forming multilayer crystalline domains, about 50Å thick (Scheme 2), with the metal ions arranged in layers separated by a distance dependent on the length of the hydrocarbon chain, as determined by GIXD.[6]

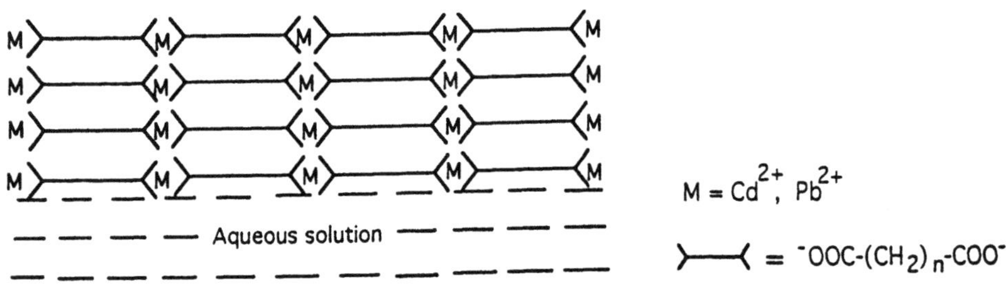

Scheme 2

The film thickness could be controlled, but within limits, by varying the quantity of deposited acid on the surface of the aqueous solution. The crystalline multilayers preserve their orientation during the transfer from the water surface onto a mica support, as visualised by SFM. The films were transferred onto glass slides and reacted with H_2S gas. The progress of the reaction was monitored by X-ray photoelectron spectroscopy (XPS) measurements. Cadmium sulphide and lead sulphide quantum dots with a moderate homodispersity and of diameter ranging from 2 to 4 nm were formed.[19]

In order to obtain such nanoparticles arranged in a one-dimensional superlattice the distance between the neighboring layers of metal ions has to be much longer than that found in the dicarboxylic acids used. For this purpose, we made use of the salts of fatty acids, $[CH_3\text{-}(CH_2)_n\text{-}COO\text{-}]_2M$, n=16-28, M=Cd, Pb, which have a tendency to crystallize in the form of bilayers, where the distance between neighboring ion layers ranges between 4nm and 8nm. The crystalline salt powders were reacted with H_2S to form quantum size PbS disc-like particles arranged in the form of a one-dimensional superlattice.[20] The corresponding cadmium salts did not yield a superlattice of CdS nanoparticles on reaction with H_2S, despite being isostructural with the lead salts. Mechanistic studies are underway in order to understand the differences between the two systems and in order to design organic matrices for the preparation of superlattices of the CdS quantum particles. Recently, this goal was accomplished by the use of thioacid derivatives.

In the previous systems we transformed ion layers into inorganic quantum particles with only a moderate degree of monodispersity. In order to generate particles of a single size, it is imperative to prepare films and single crystals where the inorganic component is composed of a finite number of ions. Supramolecular architectures such as grids, cages, ladders etc., prepared by J.-M Lehn[21] and his school might be excellent candidates for such studies. In cooperation with the Lehn group we investigated the formation of oriented thin films of silver and the cobalt grids on the air-solution interface by depositing the corresponding free ligand on aqueous subphases containing silver or cobalt ions. The structures of some of these grids were assigned by GIXD, XPS and other spectroscopic measurments.

8. Orientation and structure of ionophores in a membrane-like environment

The ability to probe the structures of the supramolecular architectures at the water surfaces allow us to provide some information regarding the organization of ionophores at the water surface in particular within hydrophobic and hydrophylic environments. In this respect we analysed the structures of several ionophors including valinomycin and nonactin on pure water and on aqueous solutions on a variety of salts. GIXD studies demonstrated that valinomycin, a natural ionophore, forms a crystalline monolayer when deposited on aqueous solutions of potassium, (Figs. 8,9) amonium or rubidium ions.[22] On the other hand, in the presence of hydrophobic monolayers of phospholipids or stearic acids that provide a membrane-like environment the same molecules organize in the form of multilayers forming channel-like architectures. Similar studies applying the same methods on the cyclic peptides that form channels structures showed that when deposited on water the molecules reside parallel to the water surface. On the other hand when these molecules are deposited in the presence of phospholipids the channels are emmbedded in the phospholipid film and reside perpendicular to the water surface.

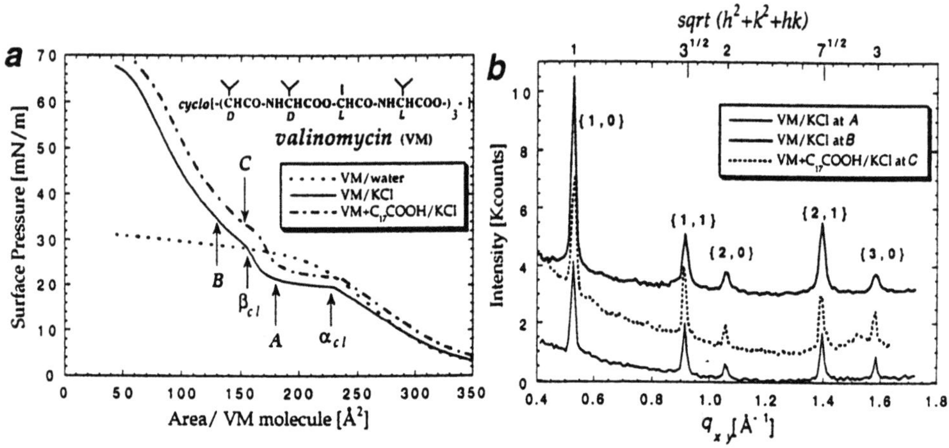

Figure 7. (a) Surface pressure-area isotherms of valinomycin (VM) on water, on 1 M KCl solution and in an equimolar mixture with stearic acid on 1 M KCl. (b) The GIXD patterns of the VM-KCl system, without and with atearic acid.

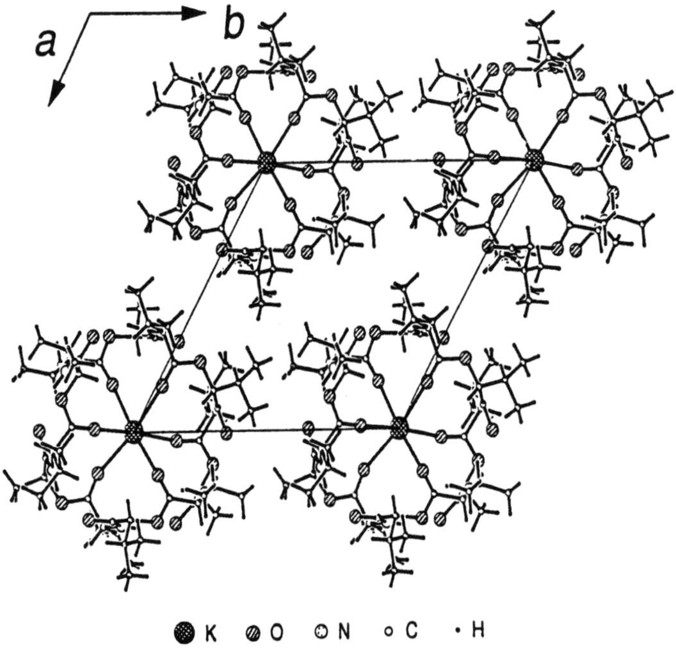

Figure 8. The molecular packing in the 2-D hexagonal unit cell of the VM-potassium cation complex, assuming the bracelet0like shape as in the $C222_1$ polymorph.

9. Outlook

In the present article we describe our structural studies on architectures that reside at the air-solution interface, by applying modern surface-sensitive methods. The organization of these molecules at the interface could be investigated by X-ray diffraction studies using synchrotron radiation and thus obtain structural information that usually one can obtain for probing the bulk structure of such systems. The demonstration that surface X-ray diffraction techniques are not confined to amphyphilic molecules alone but might comprise a variety of hydrophobic water-insoluble molecules opens new ways to investigate self-organization in two-dimensions *en-route* to crystal formation. Hopefully these methods will be applied in the future for the investigation of systems of higher complexity. In particular one should attempt to extend and apply these methods to less smooth surfaces such as the solid-liquid inteface.

Acknowledgements. We thank the G.M.J. Schmidt Minerva Centre on supramolecular architecture and the ISF for financial support.

10. References

(1) Popovitz-Biro, R.; Wang, J. L.; Majewski, J.; Shavit, E.; Leiserowitz, L.; Lahav, M., *J. Am.Chem.Soc.*, (1994), **116**, 1179.

(2) Popovitz-Biro, R.; Edgar, R.; Majewski, J.; Cohen, S.; Margulis, L.; Kjaer, K.; Als-Nielsen, J.; Leiserowitz, L.; Lahav, M., *Croatica Chemica Acta*, (1996), **69**, 689-708.

(3) Majewski, J.; Edgar, R.; Popovitz-Biro, R.; Bouwman, W. G.; Als-Nielsen, J.; Lahav, M.; Leiserowitz, L., *Angew.Chem.Int.Ed.Engl.*, (1995), **34**, 649.

(4) Popovitz-Biro, R.; Majewski, J.; Margulis, L.; Cohen, S.; Leiserowitz, L.; Lahav, M., *J.Phys.Chem.*, (1994), **98**, 4970.

(5) Popovitz-Biro, R.; Majewski, J.; Margulis, L.; Cohen, S.; Leiserowitz, L.; Lahav, M., *Adv.Mater.*, (1994), **6**, 956.

(6) Weissbuch, I.; Guo, S.; Cohen, S.; Edgar, R.; Howes, P.; Kjaer, K.; Als-Nielsen, J.; Lahav, M.; Leiserowitz, L., *Adv. Mat.*, (1998), in press.

(7) Weinbach, S. P.; Weissbuch, I.; Kjaer, K.; Bouwman, W. G.; Als-Nielsen, J.; Lahav, M.; Leiserowitz, L., *Adv. Mater.*, (1995), **7**, 857.

(8) Weissbuch, I.; Popovitz-Biro, R.; Lahav, M.; Leiserowitz, L.; Kjaer, K.; Als-Nielsen, J., in: Advances in Chemical Physics, Eds. S. Rice and I. Prigogine, Publ.. J. Wiley: New-York, 1997; Vol. 102, pp 39-121.

(9) Isz, S.; Weissbuch, I.; Palacin, S.; Ruaudell-Texier, A.; Bouwman, W. G.; Kjaer, K.; Als-Nielsen, J.; Leiserowitz, L.; Lahav, M., *Chemistry A Eur. J.*, (1997), **3**, 930.

(10) Lafont, S.; Rapaport, H.; G.J. Somjen; A. Renault; Kjaer, K.; J. Als-Nielsen; Leiserowitz, L.; Lahav, M., *J.Phys.Chem.*, (1998), in press.

(11) Rose, P. L.; N.G.Harvey; E.M.Arnett In *Advances in Physical Organic Chemistry,*Academic Press Limited: 1993; Vol. 28; pp 45.

(12) Eckardt, C. J.; Peachey, N. M.; Swanson, D. R.; Takacs, J. M.; Khan, M. A.; Gong, X.; Kim, J.-H.; Wang, J.; Uphaus, R. A., *Nature*, (1993), **362**, 614.

(13) Stevens, F.; Dyer, D. J.; Walba, D. M., *Angew. Chem. Int. Ed. Engl.*, (1996), **35**, 900-901.

(14) Nassoy, P.; Goldmann, M.; Bouloussa, O.; Rondelez, F., *Phys.Rev.Letters*, (1995), **75**, 457.

(15) Weissbuch, I.; Berfeld, M.; Bouwman, W.; Kjaer, K.; Als-Nielsen, J.; Lahav, M.; Leiserowitz, L., *J. Am. Chem. Soc.*, (1997), **119**, 933.

(16) Kuzmenko, I.; Kjaer, K.; Als-Nielsen, J.; Lahav, M.; Leiserowitz, L., (1998), submitted.

(17) Berfeld, M., *Ph.D thesis*, (1997), Feinberg Graduate School, Rehovot, Israel..

(18) Kuzmenko, I.; Buller, R.; Bouwman, W. G.; Kjaer, K.; Als-Nielsen, J.; Lahav, M.; Leiserowitz, L., *Science*, (1996), **274**, 2046.

(19) Guo, S.; Popovitz-Biro, R.; Weissbuch, I.; Cohen, H.; Hodes, G.; Lahav, M., *Adv. Mat.*, (1998), in press.

(20) Guo, S.; Popovitz-Biro, R.; Arad, T.; Hodes, G.; Leiserowitz, L.; Lahav, M., (1998), submitted.

(21) Lehn, J.-M., 1995, Supramolecular Chemistry Concepts and Perspectives, VCH, Weinheim, Germany.

(22) Rappaport, H.; Kuzmenko, I.; Bouwman, W.; Kjaer, K.; Als-Nielsen, J.; Leiserowitz, L.; Lahav, M., *J. Am. Chem. Soc.*, (1997), **119**, 11211.

SUPRAMOLECULAR SYNTHESIS OF ORGANIC AND METAL-ORGANIC LAMINATES WITH AFFINITY FOR AROMATIC GUESTS: HYDROPHOBIC CLAY MIMICS

KUMAR BIRADHA, DOROTHY DENNIS, VERNA A. MACKINNON, COREY SEWARD & MICHAEL J. ZAWOROTKO*
Department of Chemistry, Saint Mary's University, Halifax, Nova Scotia, B3H 3C3, Canada

ABSTRACT New laminated structures based upon purely organic or metal-organic materials are reported. These structures can be regarded as hydrophobic clay mimics because they intercalate a wide range of aromatic molecules and these guest molecules can be readily desorbed and adsorbed or exchanged for other guest molecules. These compounds might therefore be regarded as being general purpose adsorbents for aromatic molecules.

1. Introduction

Crystal engineering offers the intriguing promise of facile development of a new generation of functional solids which have been designed from first principles. Therefore, a degree of control over bulk properties that is not inherently present in naturally occurring compounds is potentially offered. The relatively recent rise in interest in the concept of crystal engineering[1,2] owes much to advancements in supramolecular chemistry and recognition, understanding and exploitation of *supramolecular synthons*.[3] A property of solids that has attracted considerable attention from chemists and crystal engineers is the ability of a solid to adsorb and/or desorb guest molecules. Clays, via intercalation between 2D layers, and zeolites, which contain channels and cavities because of rigid 3D frameworks, are natural prototypes and are of considerable commercial interest because of their widespread applications in separations and catalysis.[4] There are now examples of metal-organic[5,6] and organic[7,8] zeolite mimics which incorporate organic guests and there are also "organic clay mimics" which are able to exchange metal cations.[9] However, there are not to our knowledge examples of hydrophobic clay mimics. In this contribution we report the first examples of this new class of compound, simple organic salts and metal-organic coordination polymers that self-assemble into laminated structures which are structurally related to clays but are inherently hydrophobic because of their chemical nature.

G. Tsoucaris (ed.), Current Challenges on Large Supramolecular Assemblies, 115–132.

2. Strategy

The concept of *crystal engineering*, originally introduced in 1971 in the context of stereochemical control of photochemical reactions,[10] encompasses all aspects of design in the solid state and therefore has no boundaries in terms of the chemical type of the moieties involved or which intermolecular/ionic forces can be exploited. However, although even the weakest of noncovalent interactions are directional enough to be exploitable, coordination polymerization involving transition metals would seem to offer a particularly strong opportunity for chemists to directly impact materials science. The "golden rule" of crystal engineering is that the driving force for formation of crystals is molecular recognition and self-organization. This idea is especially appropriate for organic salts or coordination polymers. The former are particularly relevant as organic salts can form very strong charge assisted hydrogen bonds with predictable motifs. The latter can generate 1D, 2D or 3D architectures with a reasonable degree of control and predictability because metal coordination geometries are so well documented and understood. Both classes of compound are inherently modular and they are discussed separately below.

2.1 ORGANIC CLAYS

Trimesic acid, H_3TMA, and trimellitic acid, H_3TML, are two inexpensive and thermally and chemically robust carboxylic acids with exodentate functionality in two dimensions. They are therefore ideal templates for generation of 2D hydrogen bonded arrays.[11-18] $[NR_2H_2]_2[HTMA]$ and $[NR_2H_2]_2[HTML]$, which are afforded in simple one-pot processes by the appropriate acid-base reaction in MeOH or EtOH, exhibit such structures because of the complementary nature of the strong hydrogen bond donors of the cations and the strong hydrogen bond acceptors of the anions. For R = propyl, hexyl, octyl or decyl interdigitated laminates with poor ability to incorporate guest molecules are formed.[18] However, if R = benzyl, interdigitation is eschewed in favour of incorporation of solvent or aromatic guest molecules. X-ray crystallographic characterization of 23 host-guest complexes based upon $[N(benzyl)_2H_2]_2[HTMA]$ or $[N(benzyl)_2H_2]_2[HTML]$ hosts has revealed that aromatics as diverse in size and electronic character as nitrobenzene, anisole, veratrole, 1,4-dimethoxybenzene, 1,3,5-trimethoxybenzene, m-xylene, mesitylene, tetramethylbenzene, pentamethylbenzene, hexamethylbenzene, dibenzylamine, naphthalene, 1-methylnaphthalene, pyrene and ferrocene can be incorporated as guests by intercalation. For the HTMA salts the connectivity of the hydrogen bonding network that occurs within the laminates does not change if the guest is different, Fig. 1. The hydrogen bond network within the HTML salts (Fig. 2) is also invariable within the set of HTML salts studied. Furthermore, the networks in both HTMA and HTML salts are adaptable enough to permit at least three crystal packing modes: corrugated sheets; flat sheets with identical adjacent layers and

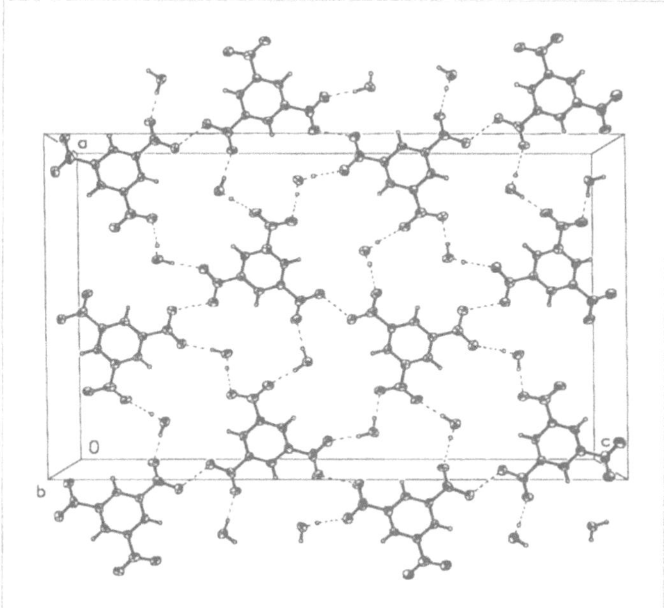

Figure 1. The 2D hydrogen bond network in $[NR_2H_2][HTMA]$ compounds (R = benzyl, alkyl). R substituents are omitted for the sake of clarity.

guest environment; flat sheets with alternating layers and guest environment. These structures are exemplified by the illustrations presented in Figs. 3-5.

For 23 of 24 compounds the proportion of guest varies from 16.6% to 26.3% by mass and is based upon the relative size of the guest molecule and the packing mode adapted by the laminates. The unit cell lengths in these 23 compounds are based upon multiples of ca. 12 x 17 x 21Å. The 12Å distance represents the approximate interlayer separation whereas 17 x 21Å represents the dimensions of the repeat unit within the laminate (Fig. 1). An analogy might reasonably be drawn between the new compounds and lipid bilayers since the former are effectively an infinite stack of hydrophobic bilayers. As might be expected,[19] interactions between guest molecules and the host frameworks are based upon a plethora of edge-to-face and face-to-face interactions between various aromatic moieties. In the absence of guest molecules or the presence of a very small number of guest or solvent molecules interdigitation of benzyl groups occurs and interplanar separations are reduced to 8-9Å.[20]

Our observations with monoalkylammonium salts are consistent with those described above. For example, for R_2 = H, benzyl or R_2 = H, phenylethyl, structures based upon alternating hydrophobic and hydrophilic layers are obtained. An example of one such structure based upon HTMA and phenylethylammonium is presented in Figure 6. This structure is quite remarkable because the asymmetric unit contains 24 organic residues and 10 water molecules and crystallizes in space group P1. The formula is $C_{200}H_{244}O_{58}N_{16}$ and the hydrogen bonding in the laminate also generates a repeat unit of ca. 17 x 21Å. Such structures might reasonably be called *supramolecular bilayers.*

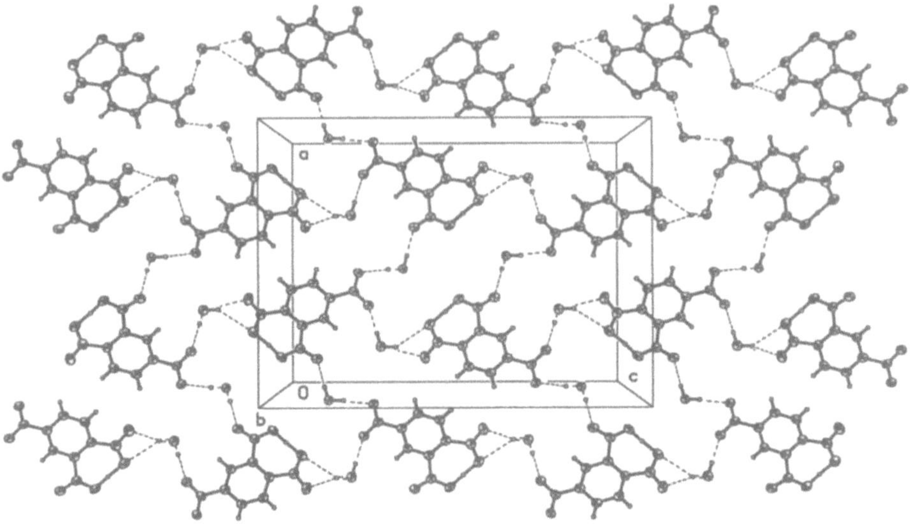

Figure 2. The 2D hydrogen bond network in $[NR_2H_2][HTML]$ compounds (R = benzyl, alkyl). R substituents are omitted for the sake of clarity.

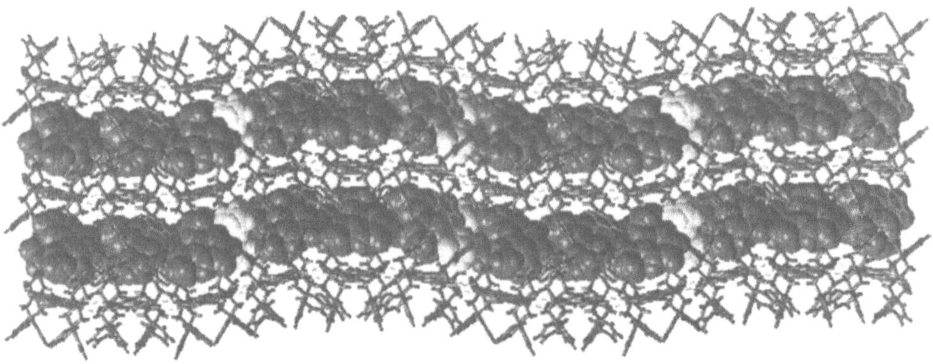

Figure 3. An illustration of the crystal packing in the supramolecular laminate based upon $[N(benzyl)_2H_2]_2[HTMA]$• veratrole•1/3EtOH, Guest molecules are in space-filling mode and the structure can be described as consisting of corrugated sheets. The packing of guest molecules is not in phase with that of the host framework and causes the b-axis to be tripled. Crystal data: monoclinic, space group $P2_1/c$, a = 11.5765(6), b = 49.905(3), c = 21.505(1)Å, β = 90.929(1)°, V = 12423(1)Å^3, Z = 12, ρ = 1.22 Mgm^{-3}, R = 0.076, wR_2 = 0.139 for 11030 out of 21508 reflections with $I>2\sigma(I)$. Data were collected on a Siemens SMART/CCD diffractometer at 193K and the structure was solved and refined using SHELX/TL.

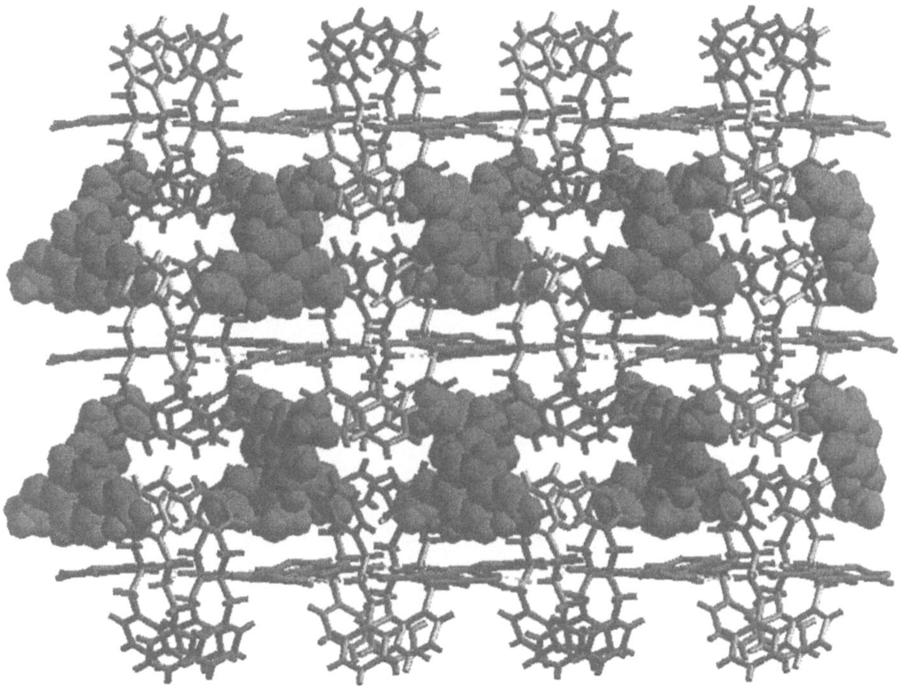

Figure 4. An illustration of a second mode of crystal packing seen in the supramolecular laminates based upon $[N(benzyl)_2H_2]_2[HTMA]$ and $[N(benzyl)_2H_2]_2[HTML]$. Guest molecules are in space-filling mode. The structure can be described as a flat structure in which adjacent layers and guests are identical. Interlayer separations are 1.20 nm. Crystal data: $[N(benzyl)_2H_2]_2[HTML]$•1.5 *p*-dimethoxybenzene, orthorhombic, space group Aba2, a = 21.414(1), b = 17.161(1), c = 24.051(1)Å, V = 8840.4(9)Å^3, Z = 8, ρ = 1.22 Mgm^{-3}, R = 0.086, wR_2 = 0.113 for 4670 out of 9675 reflections with $I>2\sigma(I)$. Data were collected on a Siemens SMART/CCD diffractometer at 193K and the structure was solved and refined using SHELX/TL.

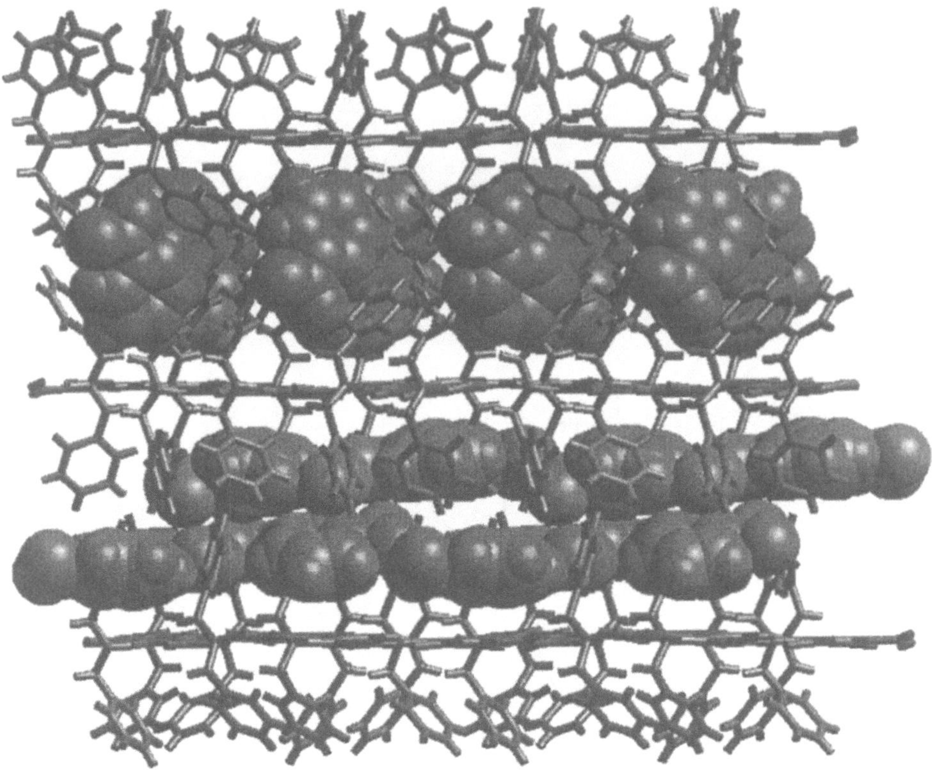

Figure 5. An illustration of the third mode of crystal packing exhibited by supramolecular laminates based upon $[N(benzyl)_2H_2]_2[HTMA]$ and $[N(benzyl)_2H_2]_2[HTML]$. Guest molecules are in space-filling mode. The packing can be described as a flat sheet in which guest molecules have alternating packing modes. Interlayer separations are 1.14 and 1.21nm. Crystal data: $[N(benzyl)_2H_2]_2[HTML]$ •1.75nitrobenzene, triclinic, space group P-1, a = 17.089(1), b = 21.417(1), c = 24.602(2)Å, α = 106.568(1), β = 95.664(1), γ = 90.405(1)°, V = 8582.4(9)Å^3, Z = 8, ρ = 1.27 Mgm^{-3}, R = 0.084, wR_2 = 0.2039 for 15223 out of 36581 reflections with $I>2\sigma(I)$. Data were collected on a Siemens SMART/CCD diffractometer at 193K and structures were solved and refined using SHELX/TL.

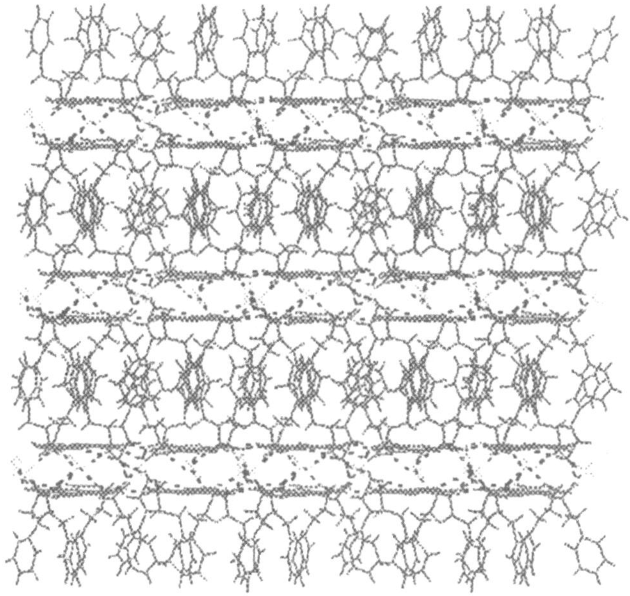

Figure 6. Crystal Structure of 8[HTMA]•16[HPhenethylamine]$_2$•10 H_2O. Crystal data: [N(phenylethyl)H_3]$_2$[HTMA] •1.25 H_2O, triclinic, space group P1, a = 13.9519(7), b = 16.9398(8), c = 21.6289(11)Å, α = 89.628(1), β = 74.455(1), γ = 89.867(1)°, V = 4924.7Å^3, Z = 8, ρ = 1.23 Mgm^{-3}, R = 0.081, wR_2 = 0.219 for 12193 out of 23666 reflections with I>2σ(I). Data were collected on a Siemens SMART/CCD diffractometer at 193K and structures were solved and refined using SHELX/TL.

2.2 METAL-ORGANIC CLAYS

Our investigations have also focused upon transition metal sustained laminated structures. As revealed in the schematic below, it is reasonable to envisage a generic class of laminated or 2D compounds which is based upon a square planar or octahedral

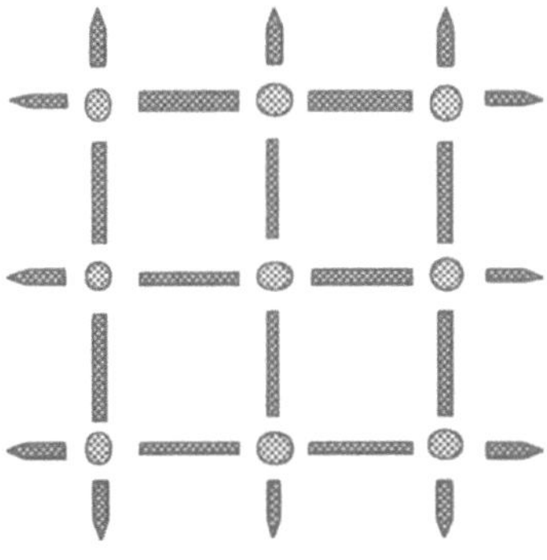

metal and linear "spacer" ligands. The first open framework examples of these compounds were based upon Cd(II) and 4,4'-bipyridine and were reported by Fujita quite recently.[21] Fig. 7 reveals how we have extended this series of compounds so as to include larger open framework coordination polymers which have cavities with an effective cross-sectional area of 11 x 11 Å. The cavities in these compounds are therefore large enough to enclathrate more than one aromatic guest molecule at a time. Fig. 7 reveals the network structure of $[Ni(dipy\text{-}Et)_2(NO_3)_2]_n$ • 2 veratrole and Fig. 8 focuses upon guest-guest and host-guest interactions inside each square box. Close examination of intermolecular contact distances reveals a plethora of stacking (both edge-to-face and face-to-face) and C-H···O hydrogen bonds. It should therefore be clear that the veratrole (1,2-dimethyoxybenzene) molecules are compatible with the inner features of the cavity. We have formed a number of related compounds, including one that contains triple-decker stacks of 1-methylnaphthalene molecules within the cavity and one that contains two different guest molecules, p-dichlorobenzene and anisole.

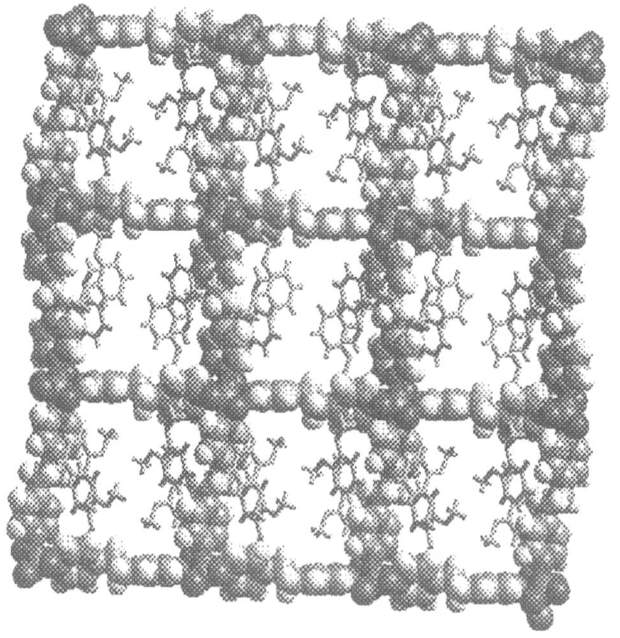

Figure 7. An overhead view of the open framework square grid coordination polymer $[Ni(dipy\text{-}Et)_2(NO_3)_2]_n$ •2 veratrole. Crystal data: 0.10 x 0.20 x 0.20 mm, pale blue rectangular crystal, triclinic, P-1 with a = 12.4022(9), b = 13.5639(10), c = 13.5930(9) Å, α = 61.5050(10), β = 87.1850(10), γ = 89.535(2)°, Z = 2, V = 2006.9(2) Å^3, ρ_{calc} = 1.369 Mgm^{-3}, μ = 0.55 mm^{-1}, 3.28>2θ>55.88°, T = 290K, R = 0.0612, wR_2 = 0.1604, 4810 out of 8821 reflections with I > 2.0 σ(I) and 518 parameters. Data were collected on a Siemens SMART/CCD diffractometer.

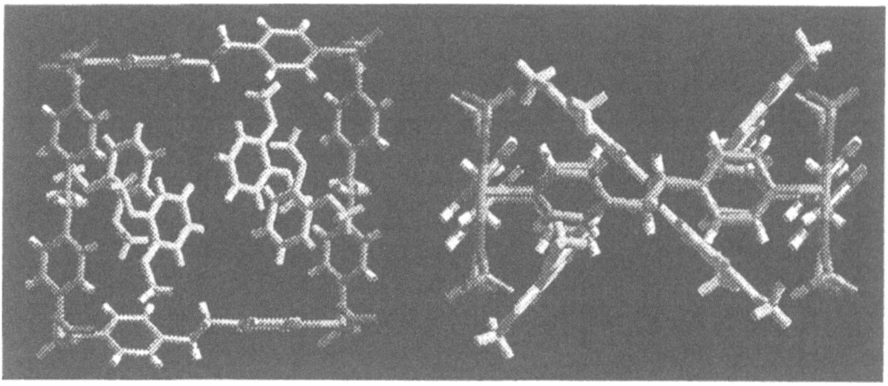

Figure 8. Overhead and side views of a single square box in the open framework square grid coordination polymer $[Ni(dipy\text{-}Et)_2(NO_3)_2]_n$ • 2 veratrole. Numerous intermolecular contacts of ca. 3.8Å (plane-plane, CH···arene) and ca. 2.7Å (C-H···O) occur between guest molecules and the host framework.

The smaller square grids $[M(4,4'\text{-bipyridine})_2(NO_3)_2]$, M = Co, **1**; Ni, **2** generate clathrates with aromatic hydrocarbon molecules as diverse in size as benzene and pyrene and with both electron rich (napthalene, anisole, veratrole) and electron poor (nitrobenzene) guests. Tables 1 and 2 reveal salient features of these 19 compounds.

In terms of guest:host stoichiometry and crystal packing of the square grids there are three types of compound. Type A compounds are isostructural with one another: they crystallize in C2/c with similar cell parameters; they have 2:1 guest:host stoichiometry; their interplanar separations are ca. 6Å. In these compounds the guest molecules form edge-to-face and face-to-face interactions with the hydrocarbon portion of the 4,4'-bipy moieties and between themselves. These interactions are presumably a primary driving force for the clathration of the guests and a major mitigating factor against interpenetration. Indeed, it is interesting to note that none of the 19 compounds reported herein display interpenetration and that stacking interactions are seen in all 19 compounds. An overhead view of the crystal structure of $[Co(4,4'\text{-bipyridine})_2(NO_3)_2]$•2 benzene which illustrates how the aromatic guest molecules interact with the square grids is presented in Figure 9.

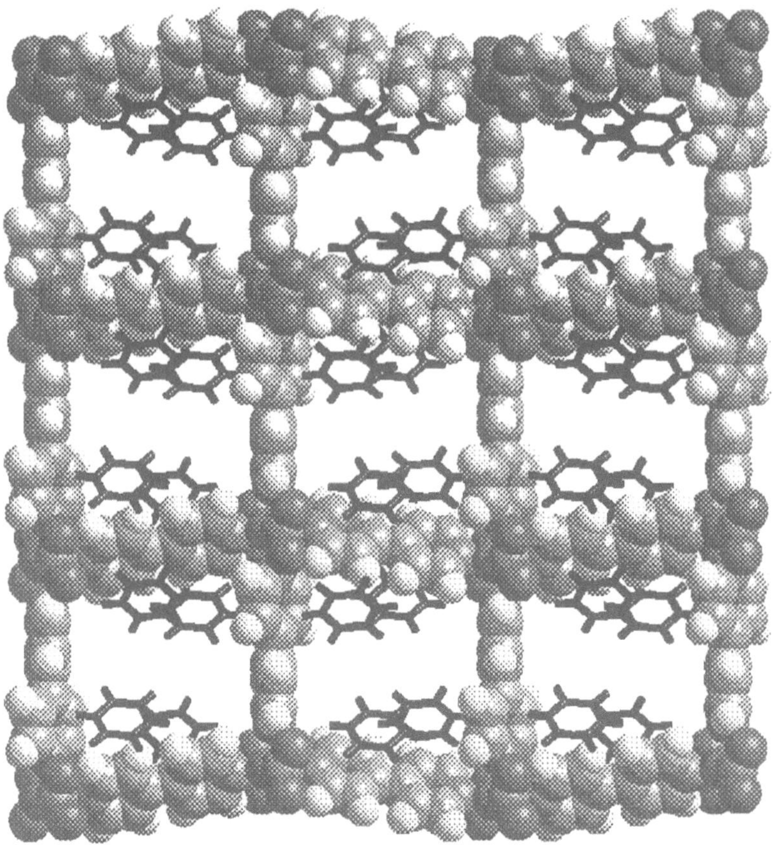

Figure 9. An overhead view of the crystal packing in $[Co(4,4'\text{-bipyridine})_2(NO_3)_2]$•2 benzene.

The crystal packing of the square grids in type A compounds appears to be influenced by C-H•••O hydrogen bond interactions between 4,4'-bipy ligands of one square grid and nitrate anions of the adjacent square grid. These interactions occur in all of the compounds that display the type A crystal packing. It should be noted that pairs of nitrate groups adapt an orientation consistent with 2-fold or inversion symmetry and can therefore be regarded as being *trans-* to one another. As would be expected, the interplanar separations do not vary significantly within the type A compounds (Table 1). The square grids do not align with a unit cell face and adjacent grids are slipped in one direction by ca. 20%, i.e. every sixth layer repeats.

In the case of type B compounds, the interlayer separation is somewhat larger but there is still C-H•••O hydrogen bonding between 4,4'-bipy ligands and nitrate ligands of adjacent grids. However, the positioning of the grids is quite different and this facilitates inclusion of one guest molecule in the centre of each grid and a larger interlayer separation (Table 1). The other guest molecule(s) lies between the grids and engages in stacking interactions with the 4,4'-bipy ligands and themselves. This is exemplified by compounds **1i** and **2i**. It is perhaps counterintuitive that in **1i** and **2i** a larger amount of a larger guest is held than in the type A compounds. Indeed, for compounds **1i** and **2i**, the proportion of the crystal that is occupied by guest molecule is 44% by weight and an even larger percentage by volume. However, the third naphthalene guest molecule sits comfortably inside the square grid. Naphthalene is roughly rectangular but the cavity is able to adapt because one pair of bipy molecules lies flat whereas the other orients vertically. The remaining naphthalene molecules form an infinite stack via edge-to-face interactions similar to those seen in pure naphthalene.[22] In other words, it is reasonable to question whether the interactions between the naphthalene molecules determine the cavity shape and crystal packing of the square grid polymers rather than *vice-versa*. There are very few examples of naphthalene clathrates for comparison, the most closely related system being $[Ni(SCN)_2(4\text{-methylpyridine})_4] \bullet 2\text{naphthalene}$.[23]

Type B compounds differ from type A compounds in the orientation of the nitrate ligands and the way in which adjacent layers stack. Nitrate ligands on adjacent layers form O•••H-C hydrogen bonds with 4,4'-bipy ligands on adjacent layers in such a manner that larger interplanar separations are facilitated. The nitrate ligands are oriented on the same side of the metal in a cis- type arrangement. There is less uniformity in the crystal packing in type B compounds. Four of the type B compounds have 3:1 stoichiometry, are isostructural and, as for the type A compounds, they crystallize in C2/c or Cc (**2b**, **1c**, **1i**, **2i**). **2b** is somewhat anomalous because it crystallizes in Cc and there is almost a tripling of the a-axis and the cell volume. Furthermore, **2b** is not isostructural with **1b**, which is a type A compound. The only other Co/Ni analogues which are not isostructural are **1c** and **2c** and **1g** and **2g**. **2c** crystallizes in $P2_1/n$ with 2.5 guest molecules per metal centre and is isostructural with **2g**. **1g** has 3:1 stoichiometry but crystallizes in $Pna2_1$ rather than C2/c. The other two type B compounds are **1h** and **2h** and they also crystallize in $P2_1/n$ but with different cell parameters.

The crystal packing in type C compounds is also controlled by weak interactions between adjacent layers and is a hybrid of the type A and type B compounds. Intralayer structure is similar in all 19 compounds with M-N and M-O distances being within expected ranges and C-H•••O interactions between nitrato groups and 4,4'-bipy ligands.

It should be clear that compounds with similar structural features to those described above should exist for a large range of metals, ligands and ions. What is particularly intriguing is that we are not aware of any natural analogues of these compounds, i.e. to our knowledge there are no naturally occurring square grid compounds which contain hydrophobic cavities within the grid.[24]

Table 1. A comparison of salient structural features of compounds **1a-j** and **2a-j**.

Type A	Guest	Cell Vol. (Å^3)	Interlayer Separation (Å)	Dimensions of Square (Å)	Angle (°)
1a	C_6H_6	3045.64 (7)	5.9	11.431 X 11.530	86.0, 94.0
2a	C_6H_6	2999.5(4)	5.9	11.413 X 11.324	93.4, 86.6
1d	$C_6H_5(CH_3)$	3143.1(3)	6.0	11.454 X 11.518	86.1, 93.9
1f	$C_6H_5NO_2$	3131.8 (3)	6.1	11.449 X 11.496	94.3, 85.7
2f	$C_6H_5NO_2$	3100.6 (3)	6.0	11.355 X 11.400	93.9, 86.1
1g	C_6H_5Cl	3159.3 (3)	6.1	11.528 X 11.479	94.1, 85.9
2g	C_6H_5Cl	3107.5 (7)	6.2	11.411 X 11.146	86.4, 93.6
1h	$C_6H_4Cl_2$	3253.0 (3)	6.2	11.489 X 11.519	93.7, 86.3
2h	$C_6H_4Cl_2$	3179.7 (4)	6.2	11.358 X 11.406	86.4, 93.6
Type B					
2e	C_8H_{10}	3752.99(14)	7.4	11.312 X 11.282	89.0, 89.3, 90.4, 91.3
1i	$C_6H_5OCH_3$	4022.6 (3)	7.7	11.406 X 11.421	92.7, 89.5, 89.7, 88.1
2i	$C_6H_5OCH_3$ MeOH	3778.3(5)	7.4	11.302 X 11.335	91.2, 89.1, 88.1, 91.6
1j	$C_6H_5(OCH_3)_2$	8399.4(9)	7.9	11.424 X 11.593	95.4, 86.1, 90.7, 89.5
2j	$C_6H_5(OCH_3)_2$	8109.9(7)	8.0	11.302 X 11.382	95.7, 90.3, 85.8, 89.3
Type C					
1b	$C_{10}H_8$	4217.4 (10)	8.0	11.494 X 11.434	90.2, 89.8
2b	$C_{10}H_8$	4116.1 (4)	8.0	11.331 X 11.381	90.4, 89.6
2c	$C_{16}H_{10}$	4092.4(4)	7.9	11.368 X 11.388	87.6, 91.7, 91.6, 89.5
2d	C_7H_8	11671.27(14)	7.6	11.279 X 11.329	91.0, 91.1, 88.8, 89.0
1e	C_8H_{10}	4177.2 (4)	7.9	11.482 X 11.454	90.7, 89.3

Table 2. A comparison of salient crystallographic details for compounds **1a-j** and **2a-j**.

Type A	FW (g/mol)	Space Group	a (Å)	B (Å)	c (Å)	α(°)	ß(°)	γ(°)	Wt. % guest
1a	651.54	C 2/c	21.3256(3)	11.5305(1)	12.6079(2)	90	100.767(1)	90	24.0
2a	651.31	C 2/c	21.164(2)	11.4128(9)	12.6597(10)	90	101.206(1)	90	24.0
1d	679.27	C 2/c	21.7636(13)	11.5176(7)	12.8555(8)	90	102.7406(11)	90	27.1
1f	741.54	C 2/c	21.7003(11)	11.4961(6)	12.8598(6)	90	102.526(1)	90	33.2
2f	741.32	C 2/c	21.5968(13)	11.4000(7)	12.9303(8)	90	103.104(1)	90	33.2
1g	720.43	C 2/c	21.8140(12)	11.5281(20)	12.8767(7)	90	102.674(1)	90	31.3
2g	720.20	C 2/c	21.678(3)	11.4111(10)	12.9139(16)	90	103.401(11)	90	31.3
1h	789.32	C 2/c	21.6458(13)	11.5193(7)	13.3717(8)	90	102.668(1)	90	37.3
2h	789.09	C 2/c	21.5155(17)	11.4059(9)	13.3137(10)	90	103.294(1)	90	37.3
Type B									
2e	760.50	P 2_1/n	15.9478(4)	14.7158(3)	15.9917(3)	90	90.1532(10)	90	34.9
1i	819.72	P na2_1	22.8114(11)	15.4407(7)	11.4207(9)	90	90	90	39.6
2i	634.27	P2_1/n	16.0055(12)	14.7585(11)	15.9953(11)	90	90.166(2)	90	21.9
1j	734.58	P 2_1/n	19.7897(12)	11.6414(7)	37.5619(23)	90	101.928(1)	90	32.6
2j	734.36	P 2_1/n	19.6175(10)	11.3822(6)	37.026(2)	90	101.208(1)	90	32.6
Type C									
1b	879.83	C 2/c	16.1658(20)	11.4940(16)	22.855(4)	90	96.73(3)	90	43.7
2b	879.60	C 2/c	16.1014(8)	11.3812(6)	22.6453(11)	90	97.306(1)	90	43.7
2c	899.59	Pn	11.3602(6)	22.7707(13)	15.8580(9)	90	93.356(9)	90	45.0
2d	771.50	Cc	47.0450(3)	11.2791(1)	22.6565(1)	90	103.8757(2)	90	35.8
1e	813.82	C 2/c	16.1344(9)	11.4822(6)	22.9072(12)	90	100.162(1)	90	39.1

2.3 SUPRAMOLECULAR ISOMERISM

The issue of supramolecular isomerism in coordination polymers was recently raised by us.[23] Indeed, several of the compounds described herein also display supramolecular isomerism. We have seen that square grids based upon [M(dipy-Et)$_2$(NO$_3$)$_2$], M = Co, Ni or Cu can have two basic supramolecular isomers. In the case of M = Cu and benzene as a guest, a 3–D frame related to that of NbO was observed.[25] The schematic below illustrates the nature of this frame. Thus far, it is the only example of this type of structure that we have observed. This compound might be called a *structural supramolecular isomer* of the square grid structures described earlier.

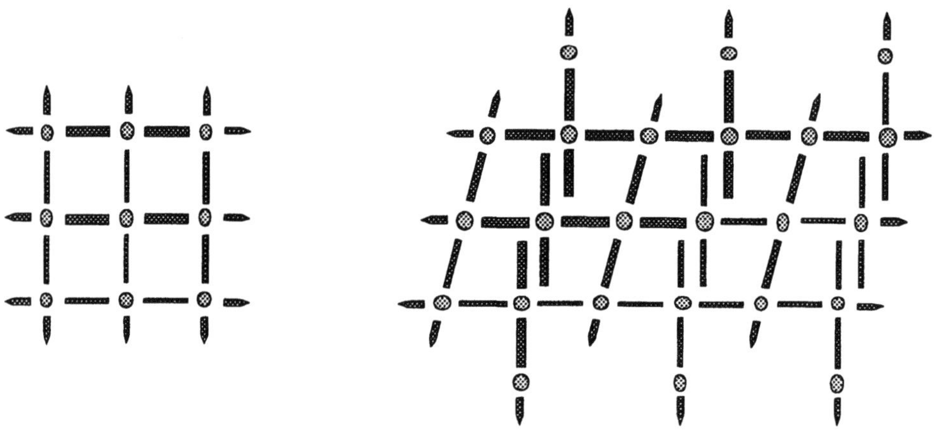

We have also observed several examples of catenated or interpenetrated square grid structures. Such compounds still have relatively large hydrophobic channels despite the presence of interpenetration and they might be called *catenane supramolecular isomers*. These compounds are represented schematically below.

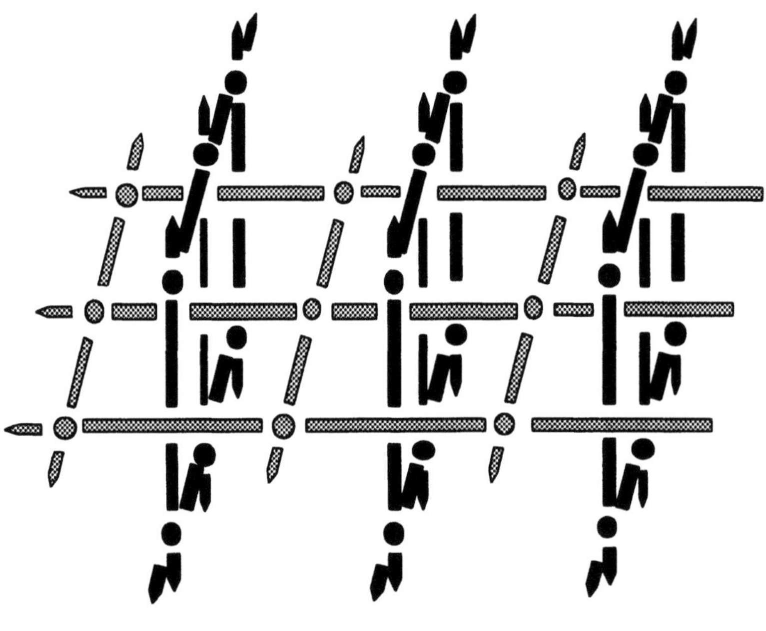

2.4 GUEST REMOVAL AND/OR EXCHANGE

We have conducted preliminary experiments on the ability of the host compounds described herein to desorb and/or exchange guest molecules. TGA analysis of a number of the coordination polymers reveals that irreversible decomposition of the host framework does not occur until after guest molecules are removed. A typical TGA plot is revealed in Figure 10. Guest molecules can also be removed if the host-guest compounds are exposed to vacuum for several minutes to several hours, depending upon the volatility of the guest. Exchange of guests is also possible. For example, both the metal-organic and organic laminates will exchange guests if they are placed under a methanol solution of a new guest. In all cases loss of crystallinity accompanies loss or

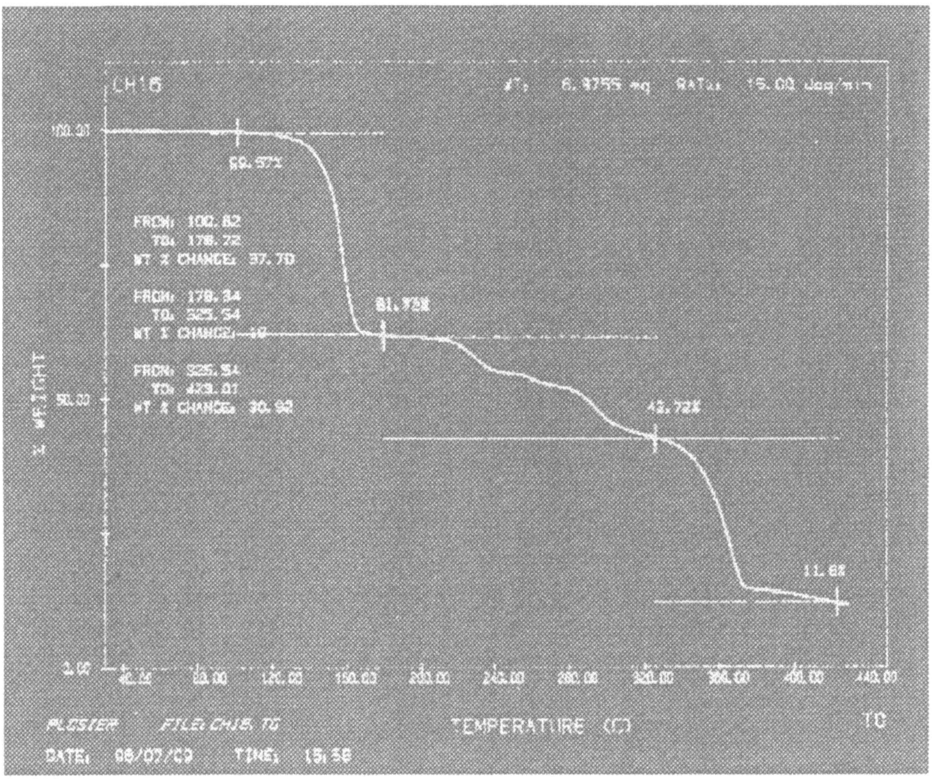

Figure 10. TGA plot obtained from crystals of [Co(4,4'-bipyridine)$_2$(NO$_3$)$_2$]•2 o-dichlorobenzene

exchange of guest but this does not appear to destroy the laminate structure itself since guests can still be adsorbed or exchanged.

2.5 CCD X-RAY DIFFRACTOMETERS

It is already becoming clear that the advent of CCD X-ray diffractometers will have a profound influence upon our ability to study and understand the crystalline state. The implications are very broad since crystals can be regarded as *de facto* examples of molecular recognition and they are the prime source of information about *supramolecular synthons.*[3] In the case of the results reported herein the influence has been even more significant. Simply put, many of the structures presented herein could not have been resolved on a conventional X-ray diffractometer because of their weak scattering and large asymmetric units. For example, the structure presented in Figures 5 and 6 have to our knowledge more residues than any presently deposited in the CSD and it would have taken one month to collect data on a conventional diffractometer. Even then, successful solution of the structure would have been very unlikely. The same kind of crystallographic problems will exist for large discrete or 0-dimensional structures of the type that nanochemists and supramolecular chemists will be generating over the next few years. It will now be possible to study such compounds studied routinely if they can be crystallized (i.e. within hours of preparation).

3. Summary

There are several observations that we consider to be of particular relevance. In the case of the organic laminates, the inherent torsional flexibility of the benzyl groups and the ability of even strong hydrogen bonds to distort are manifested by generation of a number of cavity and/or channel geometries for a single host. The ability of the supramolecular laminates to form similar crystalline structures with such a wide range of guests can therefore be rationalized. In the case of both classes of compound, preliminary studies have indicated that guest molecules can be easily removed by heat and/or vacuum to afford a stable amorphous apohost or exchanged by contact with solvent that is rich in another guest molecule. When combined with their low cost, facile supramolecular synthesis, chemical stability and modular nature, this means that the new compounds have potential applications in the context of separations, sensors and as general purpose adsorbents.

The broader implications of recent developments in crystal engineering are potentially staggering. New classes of designed, fine-tunable solids might be readily at hand by using crystal engineering to effect supramolecular synthesis of new solids. Even from the results presented herein there are new compounds which are candidates for study in the context of heterogeneous catalysis, solid-state synthesis (especially photochemistry), separations media (including chiral porous solids), materials applications (e.g. NLO, magnetism, liquid crystallinity). Indeed, a supramolecular isomer of molecular ladders which can adsorb and desorb small gas molecules was recently reported.[27] In a broader sense it seems likely that crystal nucleation processes, supramolecular isomerism and polymorphism will become better understood over the next generation. There will also be implications for surface chemistry. Finally, preliminary examination of the metal-organic frameworks using Atomic Force

Microscopy reveals surfaces that are fully consistent with what would expect from the crystal structures, i.e. flat square grid arrays with cavities on the surface every 11.5Å or 14Å.[28] The implications of crystal engineering in the context of surface science are therefore also potentially high.

In summary, the compounds reported herein are facile to prepare laminated compounds which are sustained by noncovalent interactions, are inherently hydrophobic and, because of their 2D structures and flexibility, have a generic affinity toward intercalation of aromatic molecules. A key feature is that, unlike zeolite mimics, they are selective for chemical type more than size and shape. They therefore complement the stronger shape and size selectivity of 3D structures.

Acknowledgements

We thank the N.S.E.R.C. and E.S.T.A.C. of Canada for financial support of this research and Saint Mary's University for purchase of a Siemens SMART diffractometer. We also acknowledge the Centre Jacques-Cartier, France, for funding travel and collaboration with Dr. Tony Coleman, IBCP CNRS 412, Lyon, France.

4. References

1. Maddox, J. (1988) *Nature* **335**, 201.
2. Ball, P. (1996) *Nature* **381**, 648-650.
3. Desiraju, G.R. (1995) *Angew.Chem.Int.Ed.Engl.* **34**, 2328.
4. Liebau, F. (1985) *Structural Chemistry of Silicates*, Springer-Verlag, Berlin.
5. Yaghi, O.M, Li, G. and Li, H. (1995) *Nature* **378**, 703-706.
6. Venkataraman, D., Gardner, G.B., Lee, S. and Moore, J.S. (1995) *J. Am. Chem. Soc.* **117**,11600-11601.
7. Brunet, P., Simard, M. and Wuest, J.D. (1997) *J. Am. Chem. Soc.* **119**, 2737-2738.
8. Endo, K., Sawaki, T., Koyanagi, M., Kobayashi, K., Masuda, H. and Aoyama, Y. (1995) *J. Am. Chem. Soc.* **117**, 8341-8352.
9. Coleman, A.W., Bott, S.G., Morley, S.D., Means, C.M., Robinson, K.D., Zhang, H. and Atwood, J.L. (1988) *Angew. Chem. Int. Ed. Engl.* **27**, 1361-1362.
10. Schmidt, G.M.J. (1971) *Pure Appl. Chem.* **27**, 647.
11. Aakeröy, C.B. and Nieuwenhuyzen, M. (1994) *J. Am. Chem. Soc.* **116**, 10983-10991.
12. Coe, S., Kane, J.J., Nguyen, T.L., Toledo, L.M., Wininger, E., Fowler, F.W. and Lauher, J.W. (1997) *J. Am. Chem. Soc.* **119**, 86-93.
13. Pedireddi, V.R., Jones, W., Chorlton, A.P. and Docherty, R. (1996) Chem.Commun. 997-998.
14. Russell, V.A., Evans, C.C., Li W. and Ward, M.D. (1997) *Science* **276**, 575.
15. Melendez, R.E., Sharma, C.V.K., Zaworotko, M.J., Bauer, C. and Rogers, R.D. (1996) *Angew.Chem. Int. Ed. Engl.* **35**, 2213-2215.
16. Desiraju, G.R. (1989) *Crystal Engineering: The Design of Organic Solids*, Elsevier, Amsterdam.
17. Sharma, C.V.K. and Zaworotko, M.J. (1996) *Chem. Commun.*, 2655-2656.

18. Sharma, C.V.K., Bauer, C.B., Rogers, R.D. and Zaworotko, M.J. (1997) *Chem. Commun.*, 1559-1560.
19. Jorgensen, W.L. and Severance, D.L. (1990) *J. Am. Chem. Soc.* **112**, 4768.
20. Biradha, K., Dennis, D., MacKinnon, V. and Zaworotko, M.J., unpublished results.
21. Fujita, M., Kwon, Y.J., Washizu, S., Ogura, K. (1994) *J. Am. Chem. Soc.* **116**, 1151-1152.
22. Brock, C.P., Dunitz, J.D. (1982) *Acta Crystallogr.,Sect.B*, **38**, 2218.
23. Lipkowski, J. (1982), *Acta Crystallogr.,Sect.B*, **38**, 1745.
24. Hennigar, T.L., Losier, P., MacQuarrie, D.C., Zaworotko, M.J. and Rogers, R.D. (1997) *Angew. Chem., Int. Ed. Engl.*, **36**, 972-973.
25. Wells, A.F. (1984) *Structural Inorganic Chemistry, 5th ed.*, Oxford University Press, Oxford.
26. Power, K.N.; Hennigar, T.L.; Zaworotko, M.J. "The X-ray Crystal Structure of $\{Cu[1,2\text{-bis}(4\text{-pyridyl})ethane]_2(NO_3)_2\}_n$: The First Example of a Coordination Polymer that Exhibits the NbO 3D Network Architecture". *Chem. Commun.*, accepted for publication August 1997.
27. Kondo, M., Yoshitomi, T., Seki, K., Matsuzaka, H. and Kitagawa, S. (1997) *Angew. Chem., Int. Ed. Engl.*, **36**,
28. Coleman, A.W., Singfield, K. and Zaworotko M.J. unpublished results.

"CRYSTAL ENGINEERING" WITH TWO-DIMENSIONAL HYDROGEN BONDING NETWORKS

J. A. SWIFT, A. M. PIVOVAR, A. M. REYNOLDS, C. C. EVANS, V. A. RUSSELL, W. LI and M. D. WARD
Department of Chemical Engineering and Materials Science
University of Minnesota
Amundson Hall, 421 Washington Ave. SE
Minneapolis, MN 55455

ABSTRACT. One approach toward controlling the assembly of molecules into predictable crystalline architectures involves the use of higher dimensional supramolecular networks that are resilient toward changes in ancillary functional groups attached to molecular constituents of the network. The use of *n*-dimensional networks facilitates the design and synthesis of new materials by reducing crystal engineering to *3-n* dimensions. We describe herein the assembly of adjustable and highly porous host lattices based on robust two-dimensional guanidinium-sulfonate hydrogen bonded networks. The resilience of this network allows facile prediction of crystal structure, including architectural isomers that can be directed by judicious choice of guest molecules in solution. Two architectural isomers, a pillared bilayer form and a pillared brick form, have substantial porosity that is sustained by a diverse variety of guest molecules. Notably, the brick frameworks have packing fractions (without guest), nominally twice that of the bilayer materials. Molecular host lattices with nanometer scale porosity provide substantial opportunities in areas including crystallization-based chemical separations, synthesis of new functional materials, and topochemically-directed reactions.

1. Two-Dimensional Hydrogen-Bonded Networks

Organic molecular materials often possess interesting electronic properties, including nonlinear optical behavior, electrical conductivity, superconductivity, and ferromagnetism. Crucial to the design and synthesis of these materials is a thorough understanding, and ultimately control of, the assembly of constituent molecules into the supramolecular architecture that defines the solid state structure. The formation of ordered solid state networks with a desired arrangement and dimensionality relies on an appropriate "topological director," that is, a module having a well defined functional group that can recognize complementary functional groups on other like molecules (homomeric assembly) or different molecules (heteromeric assembly). A crucial property of directors is bonding which is strong and highly directional relative to competing interactions. Formation of extended networks also requires "polyvalent" modules, that is, molecules having more than one bonding functionality. These capabilities can be provided by molecules containing hydrogen bonding functionality.

G. Tsoucaris (ed.), Current Challenges on Large Supramolecular Assemblies, 133–144.

Several examples of ordered, extended hydrogen bonded networks have been reported that illustrate the important influence of this interaction on directing the organization of molecules in the crystalline solid state. These reports have demonstrated that the local supramolecular organization about each module can be predicted with reasonable confidence based on molecular topology. Flat molecules having one-dimensional (1-D) hydrogen-bonding topologies form "ribbon" or "tape networks,"[1,2,3,4,5] while tetrahedral-like topologies have afforded diamond-like networks [6,7,8]. However, control of three-dimensional (3-D) packing is often elusive, owing to the contribution of weak dispersive interactions which hold individual networks together in the crystal. These interactions are numerous, weak, and relatively isotropic, which makes control of their 3-D assembly difficult.

One strategy for overcoming this obstacle involves using robust two-dimensional (2-D) hydrogen bonding networks whose assembly in the third dimension is governed simply by ancillary groups extending from the networks. "Robust" is defined as the ability of the module to maintain its dimensionality and general structural features upon changes in ancillary groups or other molecular species in the lattice. The use of a 2-D network whose integrity is preserved during crystallization decreases the degrees of freedom during assembly, thereby reducing the 3-D design to a single dimension. In other words, n-dimensional modules can reduce the crystal engineering problem to 3-n dimensions. The ancillary groups can be designed to "glue" the 2-D networks together by strong, directional interactions such as hydrogen bonding, metal-ligand coordination, or covalent bonding. In this sense, the 2-D network becomes a fundamental "module" that is assembled in the remaining dimension.

2. Resilient 2-D Guanidinium-Sulfonate Hydrogen Bonding Networks

Recently, we reported layered molecular materials based on 2-D H-bonded networks composed of guanidinium cations (**G**) and the sulfonate anions (**S**) of alkane and arene substituted monosulfonates [9,10]. The topological equivalence of the guanidinium and sulfonate ions, and numerous strong (**G**)N-H···O(**S**) H-bonds, favored the formation of quasihexagonal 2-D **GS** networks in more than 40 different crystalline phases containing various sulfonate functionalities (Figure 1). All of the hydrogen bonding capacity is satisfied within this network, which is important in forming robust networks [11]. The assembly of the **GS** network can be viewed as the formation of a guanidinium-sulfonate hydrogen-bonded dimer which forms, in graph-set notation, an $R_2^2(8)$ ring [12,13]. These dimers then assemble into one-dimensional "ribbons" or "tapes" via identical $R_2^2(8)$ interactions. The remaining hydrogen bonding sites connect adjacent ribbons via the formation of $R_2^2(8)$ and $R_6^3(12)$ rings. Assembly in the third dimension occurs as densely packed bilayers or as continuous stacks of interdigitated single layers, depending on the van der Waals interactions between alkane or arene groups of the organosulfonate extending from the **GS** sheets (Figure 2A). The pervasiveness of the **GS** sheets was attributed to their ability to pucker about (**G**)N-H···O(**S**) H-bonding "hinges" joining adjacent 1-D ribbons. This puckering, which can be defined by interribbon dihedral angles of $\theta_{IR} < 180°$, enables the sheets to adapt to the steric demands of different ancillary alkyl or aryl groups. In some cases, these steric demands may be accommodated further by the formation of a shifted ribbon **GS** network, which retains five of the original six hydrogen bonds.

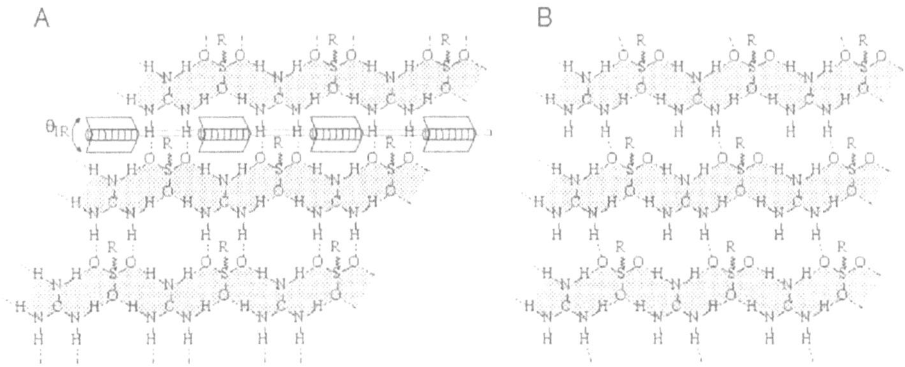

Figure 1. Schematic representations of the layered hydrogen-bonding motifs observed in guanidinium organosulfonates. Hydrogen-bonded ribbons are shaded, and the hydrogen-bond "hinge" position is indicated. (A) Quasihexagonal motif. (B) Shifted ribbon motif. R = alkyl or aryl group.

3. Rational Design of Porous Molecular Lattice

3.1. PRINCIPLES

The distinctive structural features of the **GS** networks prompted us to synthesize related materials in which the *mono*sulfonate components were replaced with *di*sulfonates [14,15]. We anticipated that this would afford porous pillared structures, reminiscent of pillared metal organodiphosphonates. This substitution would result in a twofold reduction of alkyl group density, leading to the formation of pores with sizes, heights, shapes, and chemical environments that could be manipulated by the choice of molecular pillar. The layer architecture would, in principle, depend upon the sizes of the pillar and guest molecules that may occupy these pores (Figure 2B). In the bilayer form, all pillars are projected from the same side of the **GS** sheet. The brick framework conceptually can be generated from the bilayer by shifting every other row of disulfonate pillars along a direction normal to the **GS** sheets so that all the sheets are connected continuously (pillars project from opposite sides of a given **GS** sheet). As the brick framework would have nominally twice the porosity of the bilayer, it seemed conceivable that large guest molecules incapable of fitting in the bilayer pores could template the formation of the brick architecture. Therefore, sterically undemanding pillars and small guests would favor a bilayer architecture, whereas large guests would favor a continuous brick architecture. The possibility also exists that host-guest interactions specific for the brick architecture can direct assembly toward this isomer.

Large pillars would also favor the brick architecture due to steric hindrance, including only small guests or possibly excluding guest molecules entirely. The 2-D stucture of the **GS** sheet should facilitate prediction of these architectural isomers from a simple geometric model that compares the area of a guest molecule, as projected onto the **GS** sheet, with the area available between the pillars in the different architectures.

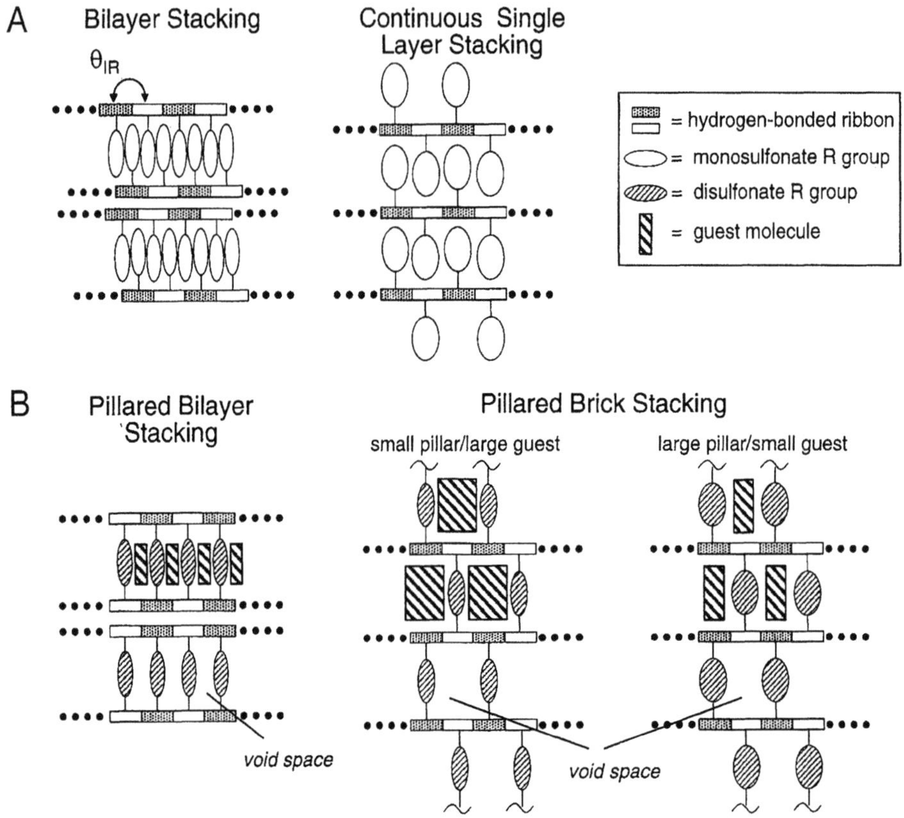

Figure 2. Schematic representations of layered materials synthesized from **G** cations and substituted monosulfonate and disulfonate anions (alkane, arene groups = R), as viewed along the H-bonded ribbon axis contained in the nominally planar **GS** networks. The white and shaded rectangles represent the narrow edges of the ribbons. (A) Two stacking arrangements of the **GS** networks observed for guanidinium monosulfonates. Bilayers are observed for R groups which are small enough to allow interdigitation in the nonpolar region separating the **GS** sheets. If the R groups are too large, they are oriented toward opposite sides of each sheet, which affords space for interdigitation and the continuous single layer stacking. (B) Possible layering architectures for guanidinium disulfonates. The twofold reduction in the number of R groups in the pillared region opens nanoscale pores capable of enclathrating molecular guests. The layering architecture is governed by the combined size of pillar and guest.

3.2. POROUS BILAYER FRAMEWORKS

We have implemented this design strategy for the disulfonate pillars **I** to **VI** (dithionate (**I**); 1,2-ethanedisulfonate (**II**); 1,4-butanedisulfonate (**III**); 2,6-naphthalenedisulfonate (**IV**); 4,4'-biphenyldisulfonate (**V**); 1,5-naphthalenedisulfonate (**VI**)). Crystalline forms of these materials can be obtained readily by standard crystallization methods. Pillars **I** to **III** form bilayer architectures in which planar **GS** sheets (θ_{IR} = 180°) are separated by the disulfonate pillars, which define 1-D pores. The 1-D bilayer pore width is established by the distance between sulfur atoms within a given **GS** sheet ($d_{SS} \approx 7.5$ Å), and the pore height is established by the length of the pillar. The bilayer thickness increases in the expected order **(G)$_2$I** < **(G)$_2$II** < **(G)$_2$III**. The pillared regions of **(G)$_2$I** (3.0Å) and **(G)$_2$II** (6.5Å) are devoid of guest or solvent molecules, but the greater void height in **(G)$_2$III** (8.3Å) accommodates two acetonitrile guest molecules per pillar. The butyl residues exhibit the less favorable gauche conformation, apparently to optimize packing of the acetonitrile guests. As expected from the axial length of **IV**, the pore height of **(G)$_2$IV** (9.5Å) is greater than that achieved with **I** to **III**. This enables inclusion of the larger benzonitrile guest molecules, which form 1-D arrays parallel to the pore direction.

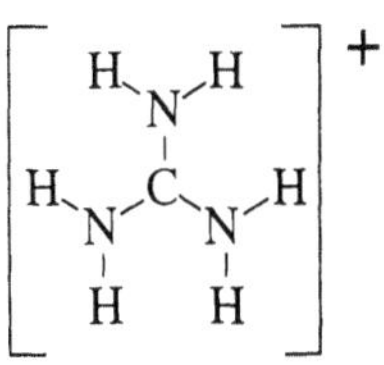

G

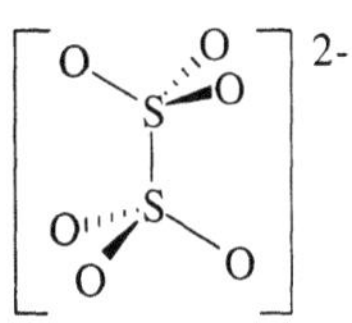

I

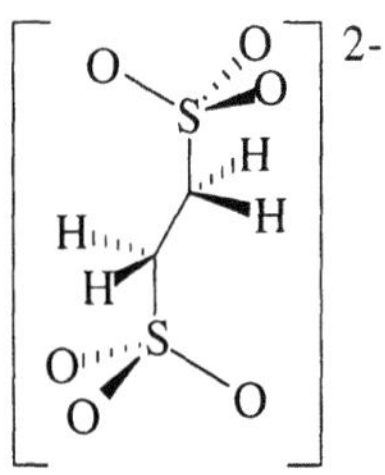

II

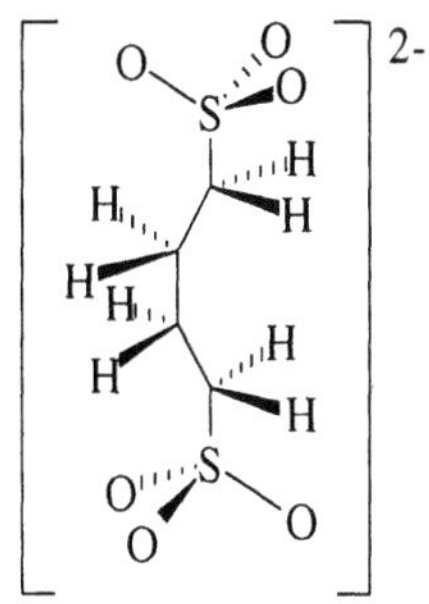

III

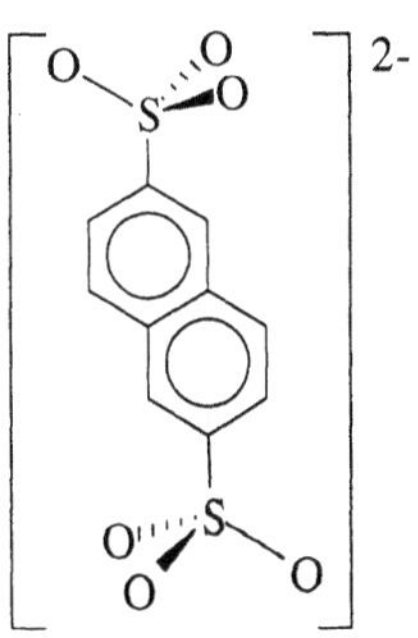

IV

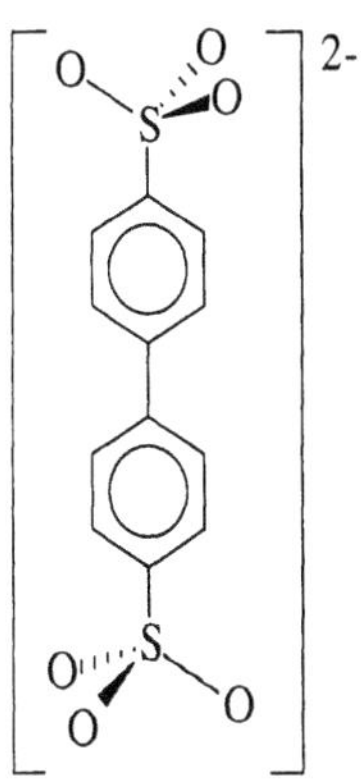

V

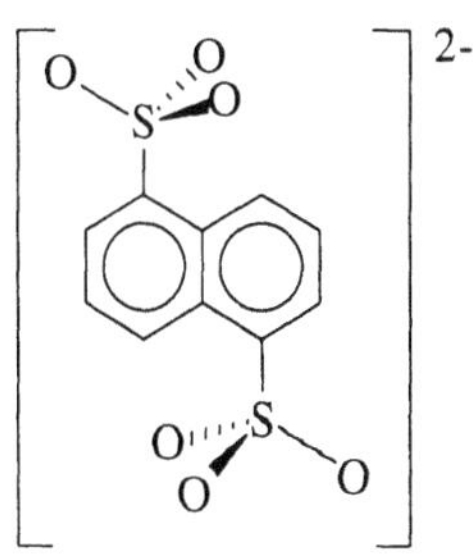

VI

The system most extensively investigated by our group has been that based on the 4,4'-biphenyldisulfonate pillar, **V**. Over 25 different mono- and disubstituted aromatic guests have been enclathrated in the $\mathbf{(G)_2V}$ bilayer architecture (Figure 3A). The 4,4'-biphenyldisulfonate pillars form densely packed walls which define guest-occupied 1-D pores. Small differences in the steric demands of these guests result in slightly different pillar tilt angles, which allows the 2-D networks to adjust to the vertical height of the guest so that void space can be minimized (i.e. the gallery height changes according to the steric demands of the guest) [16].

3.3. POROUS BRICK FRAMEWORKS

The brick architecture was first observed in our laboratory for the 1,5-naphthalenedisulfonate pillar, **VI** (Figure 3B). Crystalline $\mathbf{(G)_2VI}$ phases exhibit severely puckered quasihexagonal **GS** sheets in which the guanidinium-sulfonate ribbons are orthogonal to the pore direction. A consequence of this orthogonality, and the short span of the pillar, this phase exhibits 1-D pores with small cross-sections. The pore walls consist of naphthalene pillars whose molecular planes are parallel to the pore direction. Crystallization of $\mathbf{(G)_2VI}$ phases from MeOH solutions containing linear substituted alkanes such as 1-alkanenitriles, α,ω-alkanedinitriles, 1-alkanols, and triglyme afforded similar brick architectures in which 1-D pores are preferentially occupied by these linear guest molecules instead of MeOH. Although the topology of the $\mathbf{(G)_2VI}$ host lattices is independent of the identity of the guest, θ_{IR} ranges from 34° to 89°. Decreasing θ_{IR} values (increasing puckering about the interribbon (**G**)N-H···O(**S**) H-bonds) correspond to decreasing pore lattice constants, suggesting that the 1-D pores can adapt to different guest lengths, possibly driven by commensurism between the host lattice and guest molecules.

Our initial attempts to investigate template-directed architectural isomerism have employed the 4,4-biphenyldisulfonate pillar, **V** [17]. Simple models of the pore structure in the different $\mathbf{(G)_2V}$ architectures illustrate the dimensionality of the pores and the possible modes of guest occupation (Figure 4). The bilayer architecture can have 1-D pores, parallel or perpendicular to the **GS** ribbons, flanked by the aromatic planes of the 4,4'-biphenyldisulfonate pillars. The alternating orientation of the pillars in the brick architecture would result in wider 1-D pores if the aromatic planes of **V** were parallel to the **GS** ribbons, but a 2-D porous network would result if the pillar rotated about the sulfur-carbon bond so that the planes were orthogonal to the ribbons. The effective dimensions of the pores could be affected further by tilting of the pillars or conformational twisting about the central biphenyl C-C bond.

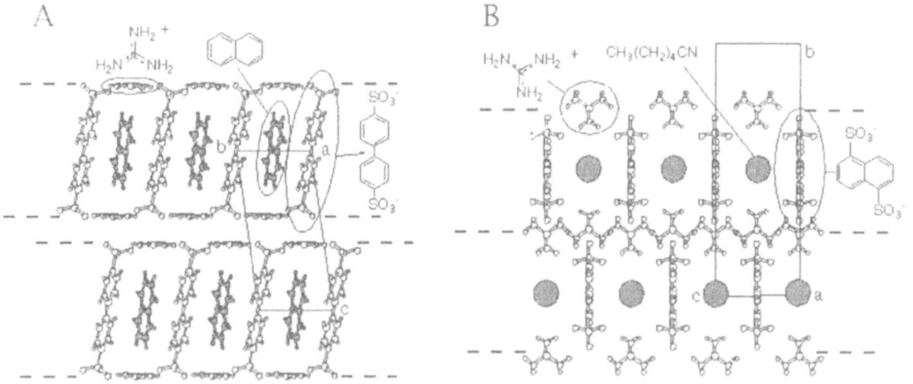

Figure 3. X-ray crystal packing diagrams of guanidinium organodisulfonate complexes. Guest molecules are shaded. (A) Bilayer architecture of **(G)$_2$V**•(naphthalene). (B) Brick architecture of **(G)$_2$VI**•(1-hexanenitrile).

Slow evaporation of supersaturated methanol solutions containing **G**, **V** and either 1,4-dibromobenzene, 1-nitronaphthalene, nitrobenzene, or 1,4-divinylbenzene afforded single crystals with the compositions **(G)$_2$(V)**•(1,4-dibromobenzene), **(G)$_2$(V)**•(1-nitronaphthalene), **(G)$_2$(V)**•2(nitrobenzene), and **(G)$_2$(V)**•1.5(1,4-divinylbenzene), respectively (Figures 5-7). Single crystal x-ray diffraction revealed that each of these compounds crystallized in the brick architecture with quasihexagonal **GS** sheets and guests included in pores between the sheets. Subtle differences in puckering angle between ribbons in the **GS** sheet (θ_{IR}), interlayer separations, pillar tilt, pillar conformation and guest occupancy can be attributed to the differing steric requirements of the guests or to host-guest interactions.

The frameworks of **(G)$_2$(V)**•(1,4-dibromobenzene) and **(G)$_2$(V)**•(1-nitronaphthalene) are essentially identical. 1-D pores are orthogonal to the **GS** ribbons and filled with either 1,4-dibromobenzene or 1-nitronaphthalene guest molecules. The pillars in these complexes are tilted substantially and the **GS** network is highly puckered, thereby closing the second pore direction that otherwise would be accessible to guests if the pillars were vertical. This results in a smaller void fraction than that of the ideal brick framework, with packing fractions of 0.50 and 0.49 calculated for guest-free **(G)$_2$(V)**•(1,4-dibromobenzene) and **(G)$_2$(V)**•(1-nitronaphthalene), respectively [18].

Parallel studies demonstrated that naphthalene, *p*-xylene and *p*-bromotoluene guests promoted the formation of bilayer architectures with guest-free packing fractions that were slightly greater than those of **(G)$_2$(V)**•(1,4-dibromobenzene) and **(G)$_2$(V)**•(1-nitronaphthalene) [19]. This indicates that the increased size of 1-nitronaphthalene guests probably is responsible for templating the assembly of the larger pores afforded by the brick architecture. However, molecular volumes of 1,4-dibromobenzene, *p*-xylene and *p*-bromotoluene are comparable, suggesting that specific host-guest interactions also may contribute to templating of the brick framework. The C-Br bonds of the 1,4-dibromobenzene guests are oriented vertically with the Br atoms directed away from the aromatic pillars and nestled in collapsed hexagonal pockets of the puckered **GS** sheet.

Notably, puckering of the **GS** sheet is not possible in the pillared bilayer architecture, thereby preventing this mode of host-guest interaction.

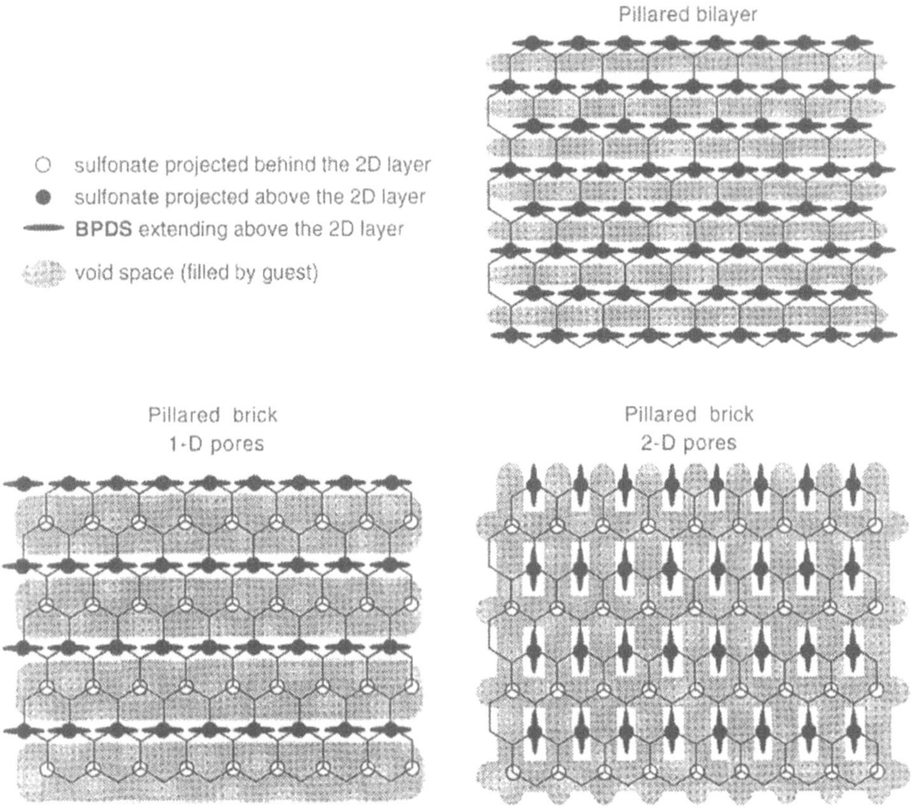

Figure 4. Schematic representations of the pores created in the pillared $(\mathbf{G})_2(\mathbf{V})$ bilayer and brick architectures as viewed normal to the **GS** network (the upper layer of **G** ions has been removed for clarity). If the aromatic planes of the pillars in the bilayer phase are either perpendicular (not shown) or parallel to the ribbon direction 1-D pores are formed. If the aromatic planes of the pillars are parallel to the ribbon direction in the brick phase, the width of the pores is nominally twice that of the bilayer because the pillars of adjacent **GS** ribbons project from different sides of the **GS** sheet. In contrast, if the planes of the pillars are orthogonal to the ribbon direction a 2-D porous network is created. BPDS = 4,4'-biphenyldisulfonate.

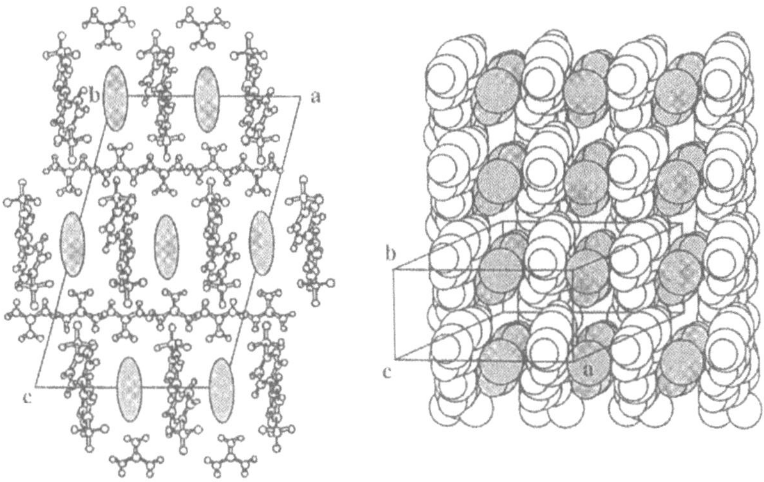

Figure 5. Molecular packing diagram illustrating the brick framework and guest occupancy of **(G)$_2$(V)**•(1,4-dibromobenzene). (left) View along 1-D pores. Shaded ovals represent 1,4-dibromobenzene molecules. (right) Space-filling diagram depicting the arrangement of guests in the pores. The **GS** ribbons are horizontal. The **G** ions and sulfonate oxygens of the top layer have been omitted so that the guests (shaded) can be viewed.

In contrast, the 4,4'-biphenyldisulfonate pillars in **(G)$_2$(V)**•2(nitrobenzene) are rotated such that their planes are parallel to the **GS** ribbon direction. This affords large 1-D pores (14.61 X 9.96Å), each filled with two π-stacks of nitrobenzene molecules in which the dipoles of neighboring nitrobenzenes within a π- stack are antiparallel [20]. This host framework has a very low guest-free packing fraction of 0.29. Although nitrobenzene molecules would fit edgewise in the narrower pores of the **(G)$_2$(V)** bilayer architecture, the small width of the bilayer pores would prevent the formation of π-stacks in which the cross-section of the molecule spans the pore width. This suggests that the formation of the brick architecture in **(G)$_2$(V)**•2(nitrobenzene) may be a consequence of templating by nitrobenzene π-stacks with the optimum antiparallel dipole-dipole arrangement. Close contacts existing between nitrobenzene guests and the **GS** sheet (3.08-3.34 Å) also may reflect favorable host-guest interactions that play a role in templating the brick framework.

The structure of **(G)$_2$(V)**•1.5(1,4-divinylbenzne) can be considered as intermediate between those described above. The planes of the pillars are rotated with respect to the ribbon direction by 60°. This affords two accessible orthogonal pores which are filled by a continuous 2-D array of guest molecules. The guest-free packing fraction of this framework is 0.29. The templating of the brick architecture by 1,4-divinylbenzene can be attributed to steric effects, as the smaller styrene molecule directs assembly into the bilayer structure. Interestingly, the 2-D 1,4-divinylbenzene network can be polymerized by heating the single crystals, suggesting the formation of molecularly-thick polymer films between the hydrogen-bonded layers. If these films

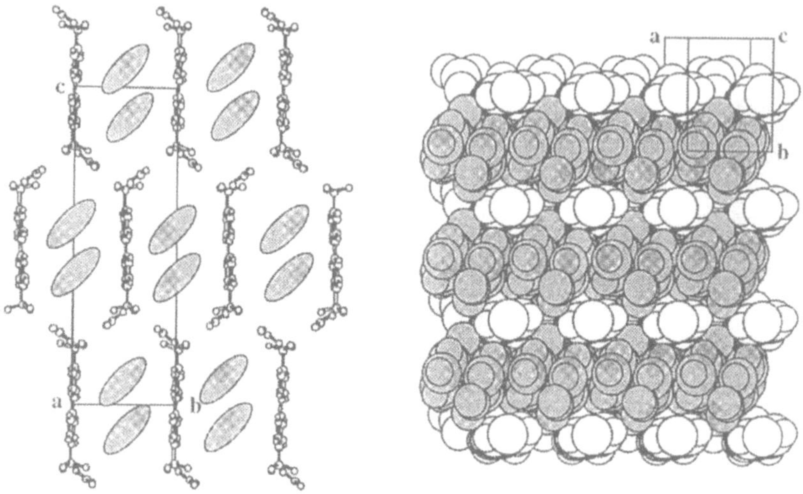

Figure 6. Molecular packing diagram illustrating the brick framework of $(\mathbf{G})_2(\mathbf{V})$•2(nitrobenzene). (left) View along 1-D pores. Shaded ovals represent nitrobenzene molecules. (right) Space-filling diagram depicting the arrangement of guests in the pores. The **GS** ribbons are horizontal. The **G** ions and sulfonate oxygens of the top layer have been omitted so that the guests (shaded) can be viewed.

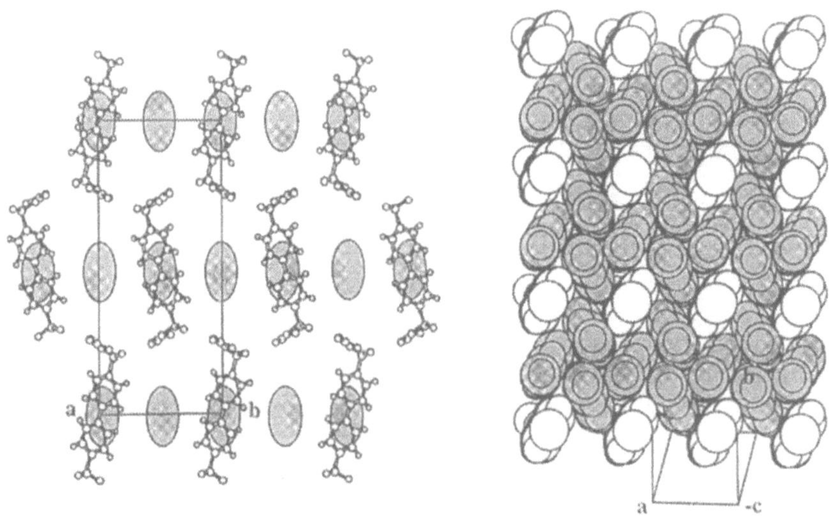

Figure 7. Molecular packing diagrams illustrating the brick framework of $(\mathbf{G})_2(\mathbf{V})$•1.5(1,4-divinylbenzene). (left) View along 1-D pores. Shaded ovals represent 1,4-vinylbenzene molecules. (right) Space-filling diagram depicting the arrangement of guests in the pores. The **GS** ribbons are horizontal. The **G** ions and sulfonate oxygens of the top layer have been omitted so that the guests (shaded) can be viewed.

prove to be continuous as the guest network suggests, we anticipate they will have porosity reflecting the pillar dimensions.

4. Summary

The synthesis of these nanoporous networks demonstrates that solid state architectures can be predictable if modules that maintain their structural integrity with changes in ancillary functional groups are employed. The use of the robust 2-D **GS** networks reduces the crystal engineering problem to the last remaining dimension. In the case of the porous frameworks described above, this enables rational adjustment of pillar structure and nanopore dimensions. The robustness of the **GS** networks can be attributed to their ability to adapt to the steric demands of ancillary functional groups and different guest molecules without loss of the 2-D H-bond network. The **GS** networks adapt by (i) flexing of hydrogen bonds joining ribbons in the **GS** sheet (ii) switching between the quasihexagonal and shifted ribbon networks, (iii) tilting of the pillars so that the network can shrink around the guest molecules, which is made possible by the flexibility of the (**G**)N-H···O(**S**) H-bonds, (iv) adjusting the conformations of the pillars to optimize packing in the galleries, and (v) forming either bilayer or brick architectural isomers. Additionally, the use of dense 2-D H-bonded networks such as the **GS** sheets prevents network interpenetration that commonly frustrates the formation of voids in molecular crystals. These networks provide a platform for fundamental studies of solid state structure, and we anticipate that analogous disulfonate pillars which have specific functionality will have potential as host lattices for optoelectronics materials, molecular separations, and topochemically directed polymerizations.

Acknowledgements. The authors gratefully acknowledge D. Britton, W. Ojala and V. Young, Jr. for crystallographic services and the National Science Foundation for financial support.

References

1. Lehn, J. M., Mascal, M., DeCian, A., and Fisher, J. J. (1992) Molecular Ribbons from Molecular Recognition Directed Self-Assembly of Self-Complementary Molecular Components, *J. Chem. Soc. Perkin Trans. 2*, 461 - 463.
2. Fan, E., Yang, L., Geib, S. J., Stoner, T. C., Hopkins, M. D., and Hamilton, A. D. (1995) Hydrogen-bonding control of molecular aggregation - Self-complementary subunits lead to rod-shaped structures in the solid state *J. Chem. Soc. Chem. Commun.* 1251-1252.
3. Zerkowski, J. A., MacDonald, J. C., Seto, C. T., Wierda, D. A., and Whitesides, G. M. (1994) New Varieties of Crystalline Architecture Produced by Small Changes in Molecular Structure in Tape Complexes of Melamines and Barbiturates, *J. Am. Chem. Soc.* 116, 4305 - 4315.
4. Lehn, J.-M., Mascal, M., DeCian, A., and Fisher, J. J. (1990) Molecular Recognition Directed Self-Assembly of Ordered Supramolecular Strands by Cocrystallization of Complementary Molecular Components, *J. Chem. Soc. Chem. Commun.* 479-481.

5. Aakeroy, C. B. (1997) Crystal Engineering: Strategies and Architectures, *Acta. Cryst.* **B53**, 569 - 586.
6. Simard, M., Su, D., and Wuest, J. D. (1991) Use of Hydrogen Bonds to Control Molecular Aggregation. Self-Assembly of Three-Dimensional Networks with Large Chambers, *J. Am. Chem. Soc.* **113**, 4696.
7. Ermer, O. (1988) Fivefold-Diamond Structure of Adamantane-1,3,5,7-tetracarboxylic Acid, *J. Am. Chem. Soc.* **111**, 3747 - 3754.
8. Zaworotko, M. J. (1994) Crystal Engineering of Diamondoid Networks, *Chem. Soc. Rev.* 283-288.
9. Russell, V. A., Etter, M. C., and Ward, M. D. (1994) Layered Materials by Molecular Design: Structural Enforcement by Hydrogen Bonding in Guanidinium Alkane- and Arenesulfonates, *J. Am. Chem. Soc.* **116**, 1941 - 1952.
10. Russell, V. A., Etter, M. C., and Ward, M. D. (1994) Guanidinium Para-Substituted Benzenesulfonates: Competitive Hydrogen Bonding in Layered Structures and the Design of Nonlinear Optical Materials, *Chem. Mater.* **6**, 1206 - 1217.
11. Etter, M. C. (1991) Hydrogen Bonds as Design Elements in Organic Chemistry, *J. Phys. Chem.* **95**, 4601 - 4610.
12. Etter, M. C., MacDonald, J. C., and Bernstein, J. (1990) Graph Set Analysis of Hydrogen-Bond Patterns in Organic Crystals, *Acta Cryst.* **B46**, 256 - 262.
13. Bernstein, J., Davis, R. E., Shimoni, L., and Chang, N.-L. (1995) Patterns in Hydrogen Bonding: Functionality and Graph Set Analysis in Crystals, *Angew. Chem. Int. Ed. Engl.* **34**, 1555-1573.
14. Russell, V. A., Evans, C. C., Li, W. and Ward, M. D. (1997) Nanoporous Molecular Sandwiches: Pillared Two-Dimensional Hydrogen-Bonded Networks with Adjustable Porosity, *Science* **276**, 575-579.
15. Swift, J. A., Russell, V. A., and Ward, M. D. (1997) Organoporous Hosts with Adjustable Molecular Environments, *Adv. Mater.* in press.
16. Swift, J. A., Reynolds, A. M. and Ward, M. D. manuscript in preparation.
17. Swift, J. A., Pivovar, A. M., Reynolds, A. M. and Ward, M. D. (1997) Template Induced Architectural Isomerism in Porous Molecular Lattices, manuscript submitted.
18. Packing fractions were calculated by Connolly surfaces using Cerius^2 molecular modeling software (version 1.6)
19. The guest-free packing fractions of bilayer structures $\mathbf{(G)_2(V)}$•naphthalene, $\mathbf{(G)_2(V)}$•*p*-bromotoluene and $\mathbf{(G)_2(V)}$•*p*-xylene are 0.52, 0.52 and 0.53, respectively.
20. The linear packing density of the nitrobenzene guest molecules along the π-stacks (2 molecules per 7.68Å repeat) is similar to that observed in the low-temperature crystal structure of pure nitrobenzene (1 molecule per 3.86Å repeat). See Trotter, J. (1959) The Crystal Structure of Nitrobenzene at -30°C, *Acta. Cryst.* **12**, 884-888.

RECOGNITION OF CONVEX FULLERENE MOLECULES BY CONCAVE COMPLEXING AGENTS. STRUCTURE AND REACTIVITY.

H.B. BUERGI
Laboratory of Crystallography
University of Bern
Freiestr. 3,
CH-3012 BERN, Switzerland

1. Introduction

Molecular recognition is mediated by various types of noncovalent interactions, including relatively strong metal-ligand bonds, electrostatic interactions, directional hydrogen bonds, donor-acceptor interactions, anisotropic interactions between polarized and polarizable atoms and also the relatively weak and isotropic dispersion interactions. Mostly recognition is controlled by some combination of these and it is often difficult to assess their relative importance[1]. The role played by the weaker, isotropic interactions is best studied in the absence of the stronger, anisotropic ones. Fullerenes, aromatic hydrocarbons and their hybrids would thus seem ideal for studying weak interactions because they contain mainly hydrogen and sp^2-hybridized carbon atoms.

In order to achieve molecular recognition in the absence of strong directional interactions, two factors require special attention[2]: (1) the complementary fragments of recognition partners have to be of a certain size to ensure a sufficient number of intermolecular contacts, i.e. strong overall binding. (2) In the absence of directional interactions, geometric shape and shape complementarity are especially important.

Shape may be considered at different scales: A representation of the van der Waals surface of a molecule emphasizes the bumps and hollows produced by individual atoms. If atomic detail is neglected the overall molecular shape comes to the fore. Given the highly developed art of organic synthesis, it is possible to build objects with relatively simple overall shapes, such as discs, bowls and spheres. The molecular bumps and hollows of these are typically five or more times larger than atomic ones. At the atomic scale the problem of recognition resembles that of finding complementary individuals in a bag of potatoes. At the molecular scale recognition may be compared with accommodating table-ware in the limited space of a cupboard. The former problem is quite difficult, the latter is relatively easy. Again, fullerenes, aromatic hydrocarbons and their hybrids are convenient objects for study, because they come in a range of sizes and in a variety of relatively simple shapes.

G. Tsoucaris (ed.), Current Challenges on Large Supramolecular Assemblies, 145–155.

2. Crystal packing

The study of molecular packing patterns in crystals is one approach towards a better understanding of molecular recognition. Conversely, an understanding of recognition is helpful for designing crystal structures from molecular structures and from the conditions of crystallization, as well as for engineering the physical and chemical properties of crystals[1,3].

A study of molecular packing patterns in crystalline polynuclear aromatics has indicated that two types of interactions are important (Fig. 1)[4]: (1) weak electrostatic bonds between the small positive charges in the C-H perimeter of one molecule and the small permanent and induced negative charges in the carbon core of its neighbours[5]. These bonds are found between parallel molecules, which are shifted by various degrees relative to each other (~1.5 - 3.5 Å), and, more importantly, between molecules which are inclined to each other (~50 - 90 deg). (2) dispersion interactions, especially those between the carbon cores of stacks of parallel (but displaced) molecules. Although the details of the observed crystal structures are the result of a delicate balance between these interactions, the basic interaction types seem undisputed.

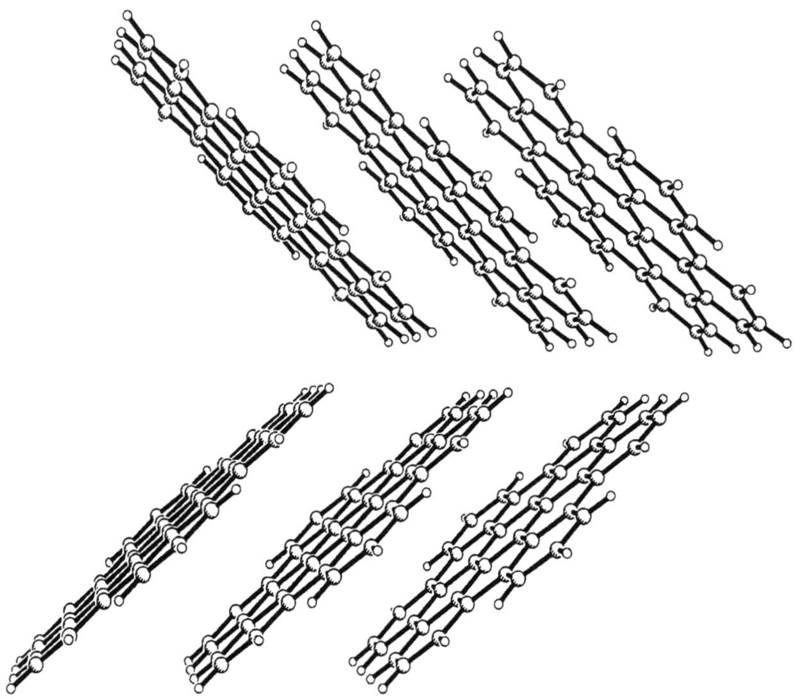

Figure 1. Types of intermolecular interactions in hexabenzocoronene[26]: (1) H...C interactions within and between stack of molecules; (2) C...C stacking interactions. The packing of hexabenzocoronene is an example of the so called γ-structure[4], characterized by a large sideways shift of molecules within stacks and large inclination between neighbouring stacks.

2.1 PLANAR VS. CURVED AROMATIC MOLECULES

Many polynuclear aromatic molecules are essentially planar. In the presence of sterically congested fragments they become warped. Examples include molecules containing the helical benzo[c]phenanthrene fragment. Introducing five-membered rings between aromatic fragments can also result in steric congestion. An example is decacyclene, $C_{36}H_{18}$, whose shape can be described as a monkey saddle (Fig. 2)[6]. Condensing seven benzene rings into a circle leads to the saddle shaped (7)-circulene[7]. Five-membered rings surrounded by five six-membered rings produce bowl-shaped molecules, such as $C_{20}H_{10}$ (corannulene)[8], $C_{22}H_{10}$ (cyclopentacorannulene)[9] and $C_{36}H_{12}$[10] (Fig. 2). Twelve five-membered rings connected through six-membered rings are necessary to obtain the spheroidal fullerenes.

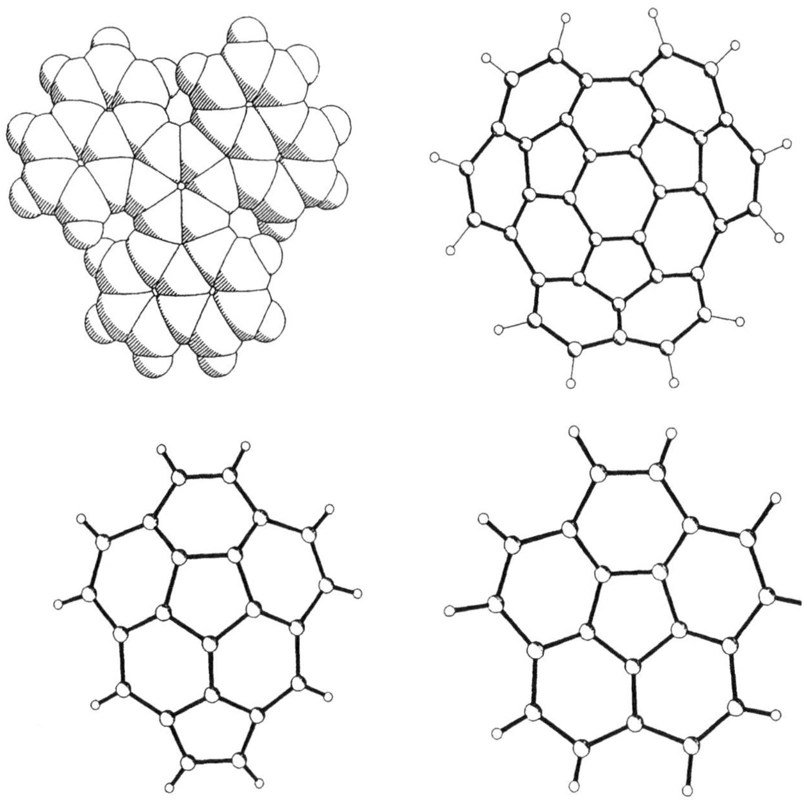

Figure 2. Clockwise from top left: steric congestion in $C_{36}H_{18}$ (decycyclene), the bowl shaped molecules $C_{36}H_{12}$, $C_{20}H_{10}$ (corannulene) and $C_{22}H_{10}$ (cyclopentacorannulene).

In all these molecules the atomic distribution in space may be approximated by curved surfaces[11]. Surfaces with positive curvature are called convex, with negative curvature they are called concave. For a sphere or spherical cap the curvature of the outer, convex surface is simply 1/R, i.e. inversely proportional to the sphere radius R. The curvature of the inner, concave surface is -1/R. Shape complementary of molecules depends on their van der Waals surfaces with radii $R_1 + R_1^*$ and $R_2 - R_2^*$, where R_i is the radius of the surface described by the atomic nuclei and R_i^* is the relevant van der Waals radius. Optimal complementarity requires $1/(R_1 + R_1^*) = 1/(R_2 - R_2^*)$. Except for planar objects ($R_1 = R_2 = \infty$), this condition is rarely fulfilled exactly. In particular, the shape complementarity of identical warped molecules is never perfect and needs to be balanced against other types of interactions. In any case, matching convex and concave molecular surfaces is an element that can be exploited in molecular recognition.

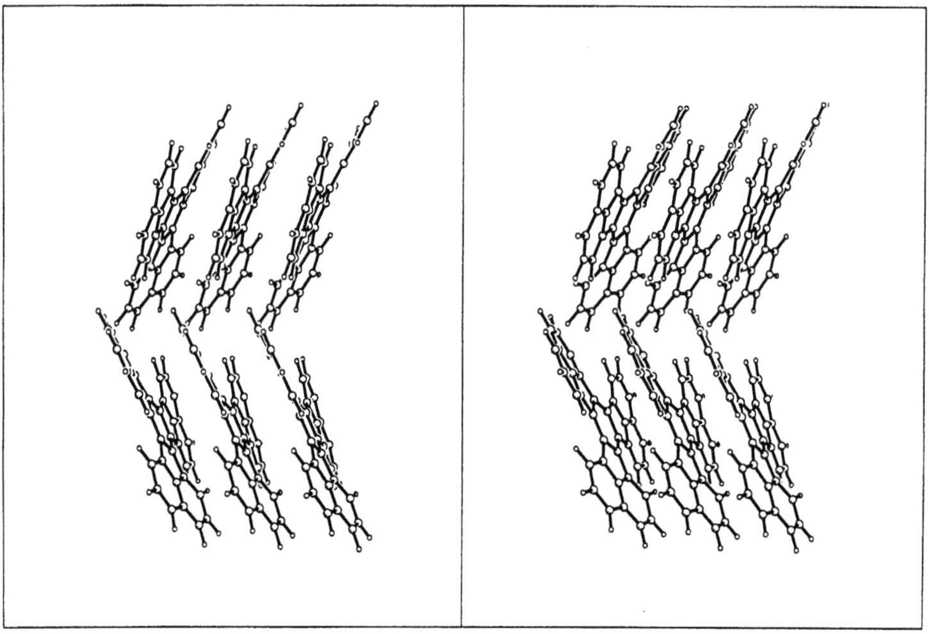

Figure 3. Decacyclene packs in the so called β-structure, characterized by a small sideways shift of molecules within stacks and a small inclination between neighboring stack (stereo representation, compare with Fig. 1).

It has been observed that extended planar aromatic molecules form the so called γ-structure (Fig. 1), whereas most weakly nonplanar aromatic molecules pack in the so called β-structure (Fig. 3). Both packings show molecular stacks with interplanar spacings of ~3.5 Å. In the β-structure the translational repeat distance of the molecules is shorter (< 4Å) and the displacement parallel to the molecular planes is correspondingly smaller[4]. This can be interpreted as a compromise between shape complementarity of the nonplanar molecules and other interactions, H...C in

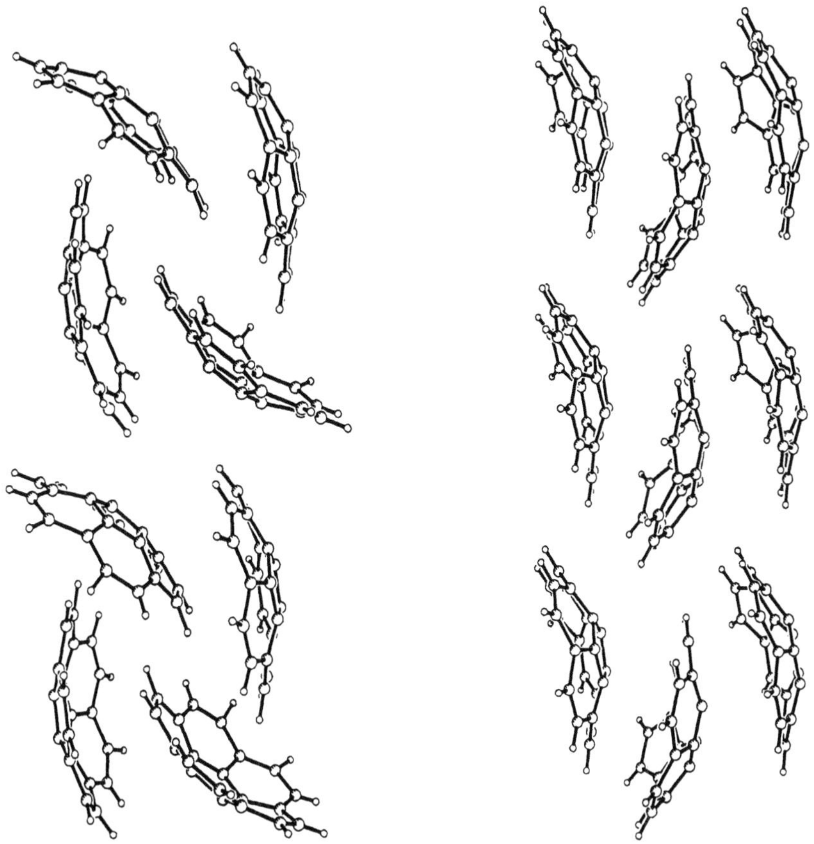

Figure 4. Left: centrosymmetric arrangement of four pairs of independent molecules in the crystal structure of corannulene. Right: a layer of cyclopentacorannulene molecules with polar stacks running horizontally.

Figure 5. Section of the polar crystal structure of $C_{36}H_{12}$.

particular. A different compromise is seen in corannulene[8]. The two crystallographically independent molecules match parts of their complementary surfaces, but because of their bowl shape they are now rotated, rather than slided, with respect to each other to allow for some H...C interactions (Fig. 4). A similar motif is found in cyclopentacorannulene, although the overall packing is different[9] (Fig. 4). Note the polar stacks of molecules and the polar layer of stacks. Neighboring layers are related to that shown in Fig. 4 by centers of inversion, i.e. the crystal structure as a whole is apolar. In the case of $C_{36}H_{12}$[10], another molecular bowl, the molecules are nicely stacked by translation in spite of imperfect shape complementary: the curvature of the concave, inner surface is about three times as large as that of the convex, outer one. However, the rim of 12 C-H groups interacts with the cap formed by the 12 central carbon atoms of the next molecule. The shortest C-C distances are about 3.4 Å, the translational repeat is 5.2 Å, in the same range as found for stacked planar aromatics. The stacks of cups are polar and hexagonally close packed with all stacks showing the same direction of polarity (Fig. 5). This arrangement can be understood in terms of the contact interface between stacks. The rims of the cups are inclined to each other in a way that is reminiscent of the inclination between stacks of planar hydrocarbons (Fig. 1). This allows them to engage in the same type of H...C interactions. A combination of curved molecular surface and H...C interactions thus leads to a macroscopically polar crystal structure.

2.2 FULLERENE COCRYSTALS

The carbon atoms of C_{60} lie on the surface of a sphere with a radius of ~3.5 Å. Packing C_{60} molecules is geometrically straight forward, but far from optimal energetically. As expected for identical spheres, C_{60}-molecules arrange in hexagonal or cubic close packing. Shape complementarity is poor because the molecules are convex. Their van der Waals diameter of ~10 Å implies, that here the octahedral and tetrahedral cavities always found in close packings, have diameters of as much as ~ 4 and ~ 2 Å, respectively. The octahedral cavity is large enough to accommodate an NH_4^+ ion, the tetrahedral ones can host alkali cations, as found in the superconducting C_{60} phases. The poor shape complementarity correlates with an unusually small sublimation energy of 0.67 kcal mol^{-1} per carbon atom[12]. For C_{70}, which is less curved on average, the sublimation energy is 0.93 kcal mol^{-1} per carbon atom[13]. In view of this, it is not too surprising that C_{60} forms cocrystals with a large number of solvents and other molecules. The Cambridge Crystallographic Data Base[14] list 26 such compounds. As an example we mention the solvate $C_{60} \cdot 4\ C_6H_6$, the only pure hydrocarbon cocrystal for which a complete structure determination is available[15]. In this crystal structure the basic cubic close packing pattern of C_{60} is maintained, but modified by deleting one out of every three rows of C_{60} molecules (along the [110]-direction). The channels which result give access to all of the cavities and most of the empty space is filled by benzene molecules (Fig. 6). As an aside we note that the arrangement of the solvent molecules in the channels is very regular and reminiscent of the packing of benzene in its high pressure modification with its characteristic T-shaped benzene...benzene

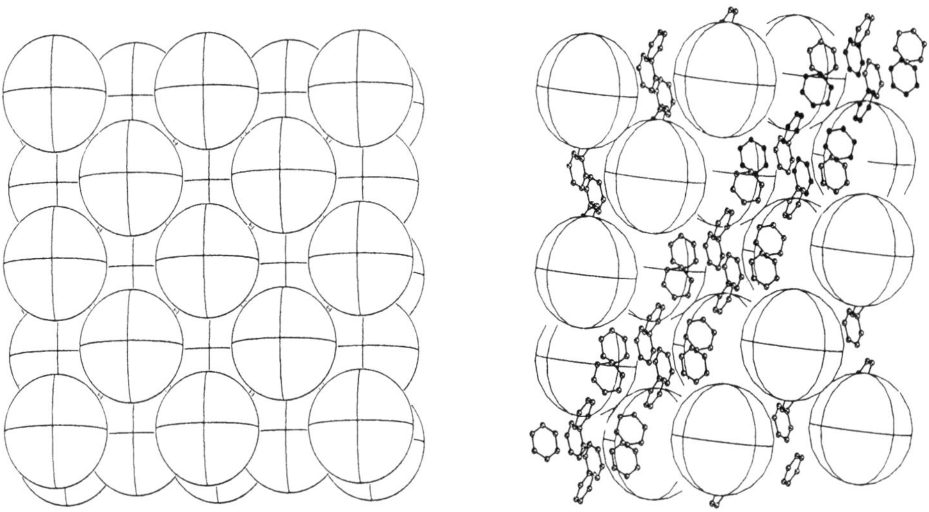

Figure 6. Molecular packing of C_{60} (cubic modification, left) and $C_{60} \cdot 4C_6H_6$ (right). The latter is obtained from the former by deleting one out of three rows along the positive diagonal and filling the resulting channels with benzene molecules. C_{60} molecules are shown as spheres. The motif of eight benzene molecules marked with black spheres is also found in the ambient and high pressure modifications of benzene

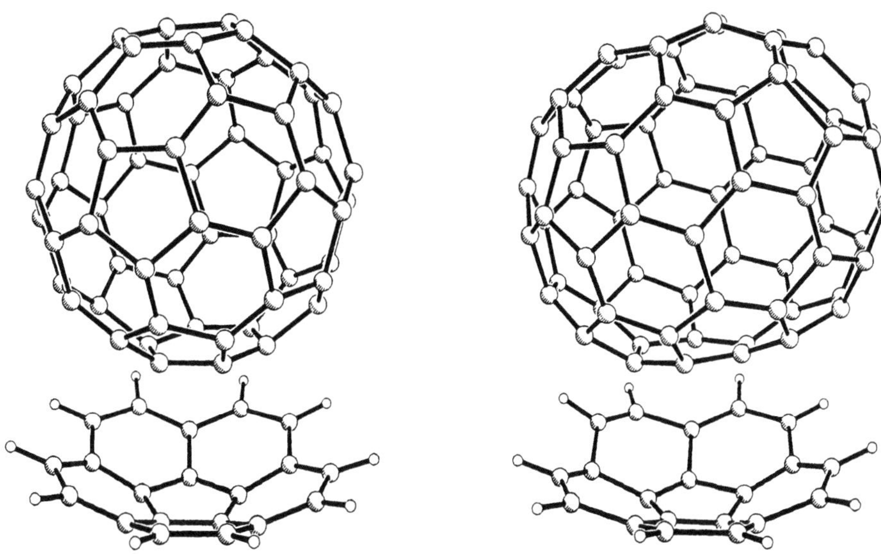

Figure 7. Association of corannulene with C_{60} and C_{70} in the solid state.

interactions. The packing density is about 25% lower, however, and still about 5% below that in the related ambient pressure modification.

If the benzene rings in the previous example are preorganized to some degree with the help of covalent bonds, molecular complexes may be expected. Calixarenes and related compounds form such complexes with $C_{60,}$ both in the solid and in solution[16]. Although not a pure hydrocarbon, we mention the example of a trimethyl-diiodocalix[5]arene[17]. Equilibrium constants for 1:1 association with C_{60} in benzene and toluene solution are found to be ~2000 mol^{-1} at room temperature. In the solid C_{60} is embraced by two calixarenes, each of them displaying a bowl-shaped conformation.

An even higher degree of preorganisation is provided by the rigid corannulene molecule, which forms 1:1 cocrystals with both C_{60} and C_{70}. It contacts the fullerenes through its concave surface as expected (Fig. 7). In the case of C_{70}, corannulene has a choice to associate with the more curved polar or the less curved equatorial regions of the ellipsoidal C_{70} molecule[18]. It chooses a location which is inbetween, but closer to the equator than the poles. This seems to allow the relatively flat C-H rim of corannulene to engage in better H...C interactions.

2.3 FULLERENE ADDITION COMPOUNDS

Several carbene and Diels-Alder addition products of C_{60} have been synthesized and characterized crystallographically. Many of these contain both convex and concave parts and are thus potentially self complementary. In bisanisylmethano derivatives of C_{60} the two anisyl groups form a mirror symmetric concave cleft which coordinates to the convex surface of a neighbouring methano fullerene[19] (Fig. 8). Since 1,1-diphenyl fragments prefer a chiral, twofold symmetric propeller conformation[20], the fit is induced and made possible by a rotation of the two phenyl groups by ~30 deg.

A more rigid and somewhat more open concave fragment is found in anthracene adducts of C_{60}. In the crystal structure of the monoadduct[21] the molecules are arranged in linear stacks, the intermolecular contacts being between the anthracene cleft and the spherical surface of the next molecule. The binding energy has been estimated from empirical interaction potentials[22] to be ~10 kcal mol^{-1} (Fig. 9). In the bisadduct[23] with two anthracene fragments at opposite ends of C_{60}, the molecules form square, two-dimensional sheets. Each molecule recognizes two neighboring C_{60}-fragments through its anthracene groups and is being recognized by two anthracene fragments, one from each of two additional neighbors (Fig. 8). Analogous patterns have been observed in cyclohexadiene mono- and tris-adducts of C_{60}[21, 24]. Here the cyclohexene-fragments act as concave contact sites. The tris-adduct forms hexagonal sheets. Each molecule makes contacts to three neighbouring C_{60}-fragments through its cyclohexene fragments. Its C_{60} part is being contacted by the cyclohexene fragments of three additional neighbours.

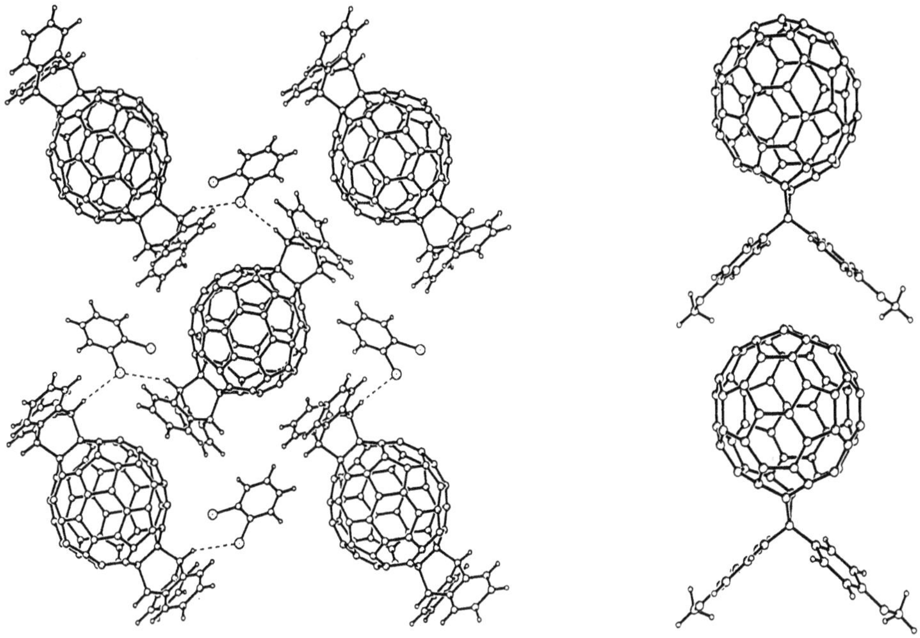

Figure 8. Self recognition of 1,1-diphenyl methano fullerenes (left) and of an anthracene bisadduct of C_{60} (right).

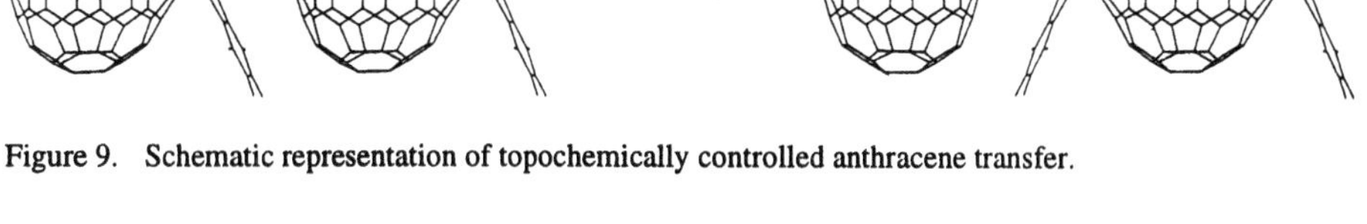

Figure 9. Schematic representation of topochemically controlled anthracene transfer.

3. A thermal topochemical reaction

As mentioned above, the anthracene monoadduct forms linear stacks, ...C_{60} - $C_{14}H_{10}$... C_{60} - $C_{14}H_{10}$... . Heating it to 180°C for a few minutes produces C_{60} and $H_{10}C_{14}$ - C_{60} - $C_{14}H_{10}$ in a one to one ratio and with ~96% conversion[19] (Fig. 9). The reaction can proceed through a sequence of thermally allowed cycloreversion / cycloaddition steps which are assisted by the parallel alignment of the dienophilic 6,6-bonds of the two C_{60}-fragments participating in the reaction. The driving force for this topochemical reaction is probably due to an increase in entropy, brought about by almost free rotation of C_{60} in the product crystal structure. The high degree of conversion indicates cooperativity, the nature of which is not understood at present.

4. Conclusion

Analysis of crystal structures containing fullerenes, nonplanar aromatic hydrocarbons and fullerene derivatives with aromatic substituents, suggests that weak interactions can mediate molecular recognition provided that their number is optimized by shape complementarity[25]. Some of the structures show interesting properties, e.g. macroscopic polarity or topochemical reactivity. These observations, although mostly serendipitous, may serve as guidelines for engineering crystal structures containing extended aromatic hydrocarbon fragments.

References

1. Desiraju, G.R. (1997) Designer crystals: intermolecular interactions, network structures and supramolecular synthons, *Chem. Commun.* **16**, 1475-1482.

2. Lehn, J.M. (1995) *Supramolecular Chemistry*, Verlag Chemie, Weinheim.

3. König, O., Bürgi, H.B., Armbruster, Th., Hulliger, J., Weber, Th., (1997) A Study in Crystal Engineering: Structure, Crystal Growth and Physical Properties of a Polar Perhydrotriphenylene Inclusion Compound, *J. Am. Chem. Soc.*, in press.

4. Gavezzotti, A., Desiraju, G.R. (1988) A Systematic Analysis of Packing Energies and Other Packing Parameters for Fused-Ring Aromatic Hydrocarbons, *Acta Cryst.* **B44,** 427-434.
Desiraju, G.R., Gavezzotti, A. (1989) Crystal Structures of Polynuclear Aromatic Hydrocarbons, Classification, Rationalization and Prediction from Molecular Structure, *Acta Cryst.* **B45,** 473-482.

5. Williams, D.E., Xiao, Y. (1993) Benzene, Naphthalene and Anthracene Dimers and their Relation to the Observed Crystal Structures, *Acta Cryst.* **A49**, 1-10.

6. Ho, D.M., Pascal, R.A. Jr. (1993) Decacyclene: A Molecular Propeller with Helical Crystals, *Chem. Mater.* **5**, 1358-1361.

7. Sygula, A., Folsom, H.E., Sygula, R., Abdourazak, A.H., Marcinow, Z., Fronczek, F.R., Rabideau, P.W. (1994) Bowl Stacking in Curved Polynuclear Aromatic Hydrocarbons: Crystal and Molecular Structure of Cyclopentacorannulene, *J. Chem. Soc., Chem. Commun.* 2571-2572.

8. Hanson, J.C., Nordman, C.E. (1976) The Crystal and Molecular Structure of Corannulene, $C_{20}H_{10}$, *Acta Cryst.* **B32**, 1147-1153.

9. Yamamoto, K., Harada, T., Okamoto, Y., Chikamatsu, H., Nakazaki, M., Kai, Y., Nakao, T., Tanaka, M., Harada, S., Kasai, N. (1988) Synthesis and Molecular Structure of [7]Circulene, *J. Am. Chem. Soc.* **110**, 3578-3584.

10. Forkey, D.M., Attar, S., Noll, B.C., Koerner, R., Olmstead, M.M., Balch, A.L. (1997) Crystallographic Characterization of the Molecular Structure and Solid State Packing of the Fullerene-Shaped Hydrocarbon, $C_{36}H_{12}$, *J. Am. Chem. Soc.* **119,** 5766-5767.

11. Andersson, S., Jacob, M. (1977) *The Mathematics of Structures*, R. Oldenburg Verlag, München.

12. Pan, C., Sampson, M.P., Chai, Y., Hauge, R.H., Margrave, J.L. (1991) Heats of Sublimation from a Polycrystalline Mixture of C_{60} and C_{70}., *J. Phys. Chem*, **95**, 2944-2946.

13. Beckhaus, H.-D., Verevkin, S., Rüchardt, Ch., Diederich, F., Thilgen, C., ter Meer, H.-U., Mohn, H., Müller, W. (1994) C_{70} ist stabiler als C_{60}: experimentelle Bestimmung der Bildungswärme von C_{70}, *Angew. Chem.* **106**, 1033-1035.

14. Allen, F.H., Kennard, O. (1993) 2D Search and Research Using the Cambridge Structural Database, *Chem. Des. Automat. News*, **8**, 31-37.

15. Bürgi, H.B., Restori, R., Schwarzenbach, D., Balch, A.L., Lee, J.W., Noll, B.C., Olmstead, M.M. (1994) Nanocrystalline Domains of a Monoclinic Modification of Benzene Stabilized in a Crystalline Matrix of C_{60}, *Chem. Mater.* **6**, 1325-1329.

Balch, A.L., Lee, J.W., Noll, B.C., Olmstead, M.M. (1993) Disorder in a Crystalline Form of Buckminsterfullerene: $C_{60}·4C_6H_6$, *Chem. Commun.* 56-58.

Meidine, M.F., Hitchcock, P.B., Kroto, H.W., Taylor, R., Walton, D.R.M. (1992) Single Crystal X-Ray Structure of Benzene-solvated C_{60}, *Chem. Commun.* 1534-1537.

16. Steed, J.W., Junk, P.C., Atwood, J.L. (1994) Ball and Socket Nanostructures: New Supramolecular Chemistry Based on Cyclotriveratrylene, *J. Am. Chem. Soc.* **116**, 10346-10347.

Atwood, J.L., Barnes, M.J., Gardiner, M.G., Raston, C.L. (1996) Cyclotriveratrylene polarisation assisted aggregation of C_{60}, *Chem. Commun.*, 1449-1450.

Barbour, L.J., Orr, G.W., Atwood, J.L. (1997) Supramolecular intercalation of C_{60} into a calixarene bilayer - a well-ordered solid-state structure dominated by van der Waals contacts, *Chem. Commun.*, 1439-1440.

17. Haino, T., Yanase, M., Fukazawa, Y. (1997) Ein supramolekularer Komplex von C_{60} mit Calix[5]aren - Struktur im Kristall und in Lösung, *Angew. Chem.* **109**, 288-290.

18. Bürgi, H.B., Olmstead, M.M., Pilkington, M., Abeln, D., Pichierri, F., unpublished.

19. Osterodt, J., Nieger, M., Vögtle, F. (1994) First X-Ray Determination of Cyclopropane Structure in Methanofullerenes, *Chem. Commun.* 1607-1608.

Tezuka, Y., Kawasaki, N., Yajima, H., Ishii, T., Oyama, T., Takeuchi, K., Nakao, A., Katayama, C. (1996) Bis(4-methoxyphenyl)methano[60]fullerene with a Methano Bridge at the 6-6-Ring Junction, *Acta Cryst.* **C52**, 1008-1010.

Djojo, F., Herzog, A., Lamparth, I., Hampel, F., Hirsch, A. (1996) Regiochemistry of Twofold Additions to [6,6]Bonds in C_{60}: Influence of the Addend-Independent Cage Distortion in 1,2-Monoadducts, *Chem. Eur. J.* **2**, 1537-1547.

20. Rappoport, Z., Biali, S.E., Kaftory, M. (1990) Application of the Structural Correlation Method to Ring-Flip Processes in Benzophenones, *J. Am. Chem. Soc.* **112**, 7742-7748.

Klebe, G. (1994) Mapping common molecular fragments in crystal structures to explore conformation and configuration space under the conditions of a molecular environment, *J. Mol. Struct.* (Theochem) **308**, 53-89.

21. Kräutler, B., Müller, T., Maynollo, J., Gruber, K., Kratky, C., Ochsenbein, P., Schwarzenbach, D., Bürgi, H.B. (1996) Eine topochemisch kontrollierte, regiospezifische Fulleren-Bisfunktionalisierung, *Angew. Chem.* **108**, 1294-1296.

22. Filippini, B., Gavezzotti, A. (1993) Empirical Intermolecular Potentials for Organic Crystals: the '6-exp' Approximation Revisited, *Acta Cryst.* **B49**, 868-880.

23. Kräutler, B, Müller, T., Ochsenbein, P., Bürgi, H.B., unpublished.

24. Bürgi, H.B., Pilkington, M., Abeln, D., unpublished.

25. Bürgi, H.B. (1996) On Molecular Recognition in Fullerene Chemistry, in: *Crystallography of Supramolecular Compounds*, G. Tsoucaris et al. (eds.), Kluwer Academic Publishers, Dordrecht, 31-44.

26. Goddard, R., Haenel, M.W., Herndon, W.C., Krüger, C., Zander, M. (1995) Crystallization of Large Planar Polycyclic Aromatic Hydrocarbons: The Molecular and Crystal Structures of Hexabenzo[*bc,ef,hi,kl,no,qr*]coronene and Benzo[1,2,3-*bc*:4,5,6-*b'c'*]dicoronene, *J. Am. Chem. Soc.* **117**, 30-41.

The Structure Of Solvates In The Solid State - Systems Model For Heteromolecular Association ?

Janusz Lipkowski [1], Konstantin Udachin [2], Yurij Simonov [3] and Dymitry Soldatov [2]

[1] Institute of Physical Chemistry, Polish Academy of Sciences, Kasprzaka 44, 01 224 Warszawa, Poland

[2] Institute of Inorganic Chemistry, Siberian Branch of Russian Academy of Sciences, Lavrentyeva 3, 630090 Novosibirsk, Russia

[3] Institute of Applied Physics, Moldavian Academy of Sciences, Grosula 5, Kishinev, Moldova

It happens quite often during chemical preparation work that products of chemical reactions are obtained in the form of "solvates" (or "adducts", etc.) in which solvent and/or other foreign molecules co-crystallize with the "main" component. The large amount of structural information which has been accumulated over the last decades provides interesting data on the role played by inclusion of molecular species in the crystalline state.

In the present paper solvation of several chemically different species are presented and discussed from the crystallochemical standpoint. Solvation 'by' and 'of' *coordination complexes* and *hydrate* structures in which the guest component is a molecule or quarternary alkylammonium salt, crown polyether or its molecular complex are demonstrated. The lattices are quite different from a chemical point of view (polar and hydrophilic water molecules forming the host lattice in hydrates, and hydrophobic metalloorganic complexes in Werner clathrates). Nevertheless, some common phenomena may be observed on their inclusion compound formation.

G. Tsoucaris (ed.), Current Challenges on Large Supramolecular Assemblies, 157–171.

Among the large varieties of different clathrate structures there are two main types which may be defined as, respectively, "skeletal" and "packing" architectures. In skeletal clathrates it is both customary and convenient to represent their structures as assemblies of polyhedra constructed from, say, tetrahedral OH_4 units (water host structures, each H_2O molecule donating two and accepting two hydrogen bonds) [12]. There is a tendency in nature to build regular pentagonal dodecahedra from the tetrahedral units. The resulting spatial structure, containing some extra cages (in addition to the dodecahedra representing small cavities) is unstable without absorbed guest species filling the voids in the structure.

Hydrates

Clathrate (or, semiclathrate) type hydrates formed with alkylammonium halides are relatively well known. However, only recently some topologically new compounds of the series have been studied by x-ray diffractometry.The example shown in Figure 1 represents a rather open

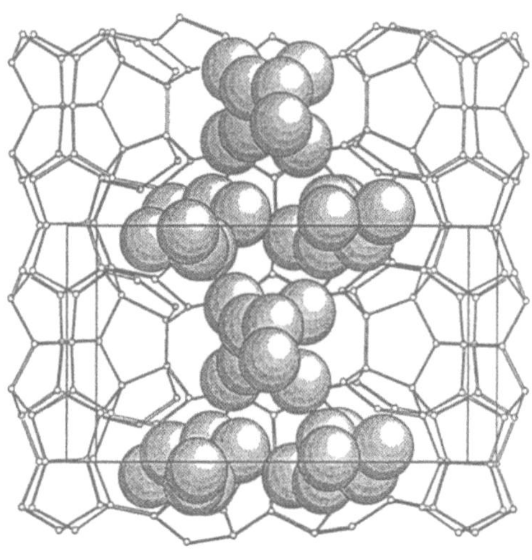

Figure 1. The structure of tetraethylammonium fluoride hydrate (1 : 10.5 mol/mol)

structure in which guest cations (tetraethylammonium) are arranged along open channels in the water framework. This structure may be considered a sort of *zeolite-like* structure built of water molecules [3].

The tetra-n-propylammonium fluoride hydrate [4] shows a layered structure (Figure 2) in which the three-dimensional intermolecular bonding between water molecules is preserved only partially, i.e. within the layer formed by H_2O molecules. Hydrate structures in which only a two-dimensional pattern of the host water is observed are known as well (see e.g. the structure of the tetraethylammonium fluoride pentahydrate [5]).

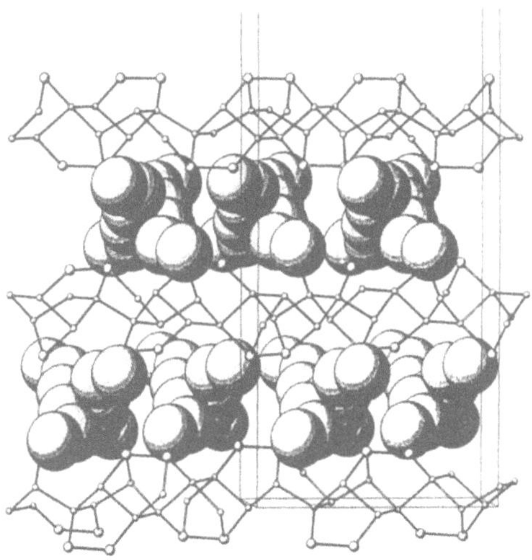

Figure 2. The layered structure of tetra(n-propyl)ammonium fluoride hydrate.

In a recently studied series of hydrate structures of the crown-type molecular receptors it has been found that some of the supramolecular patterns mentioned above can also be found in these compounds. Although there is no contribution of ionic guest-lattice interactions, as is the case of alkylammonium salt hydrates, in the crown/water structures there is more guest-host interaction than clathration forces alone. As may clearly be seen by inspecting Figures 3 through 5, there is some hydrogen bonding between water and the crown species.

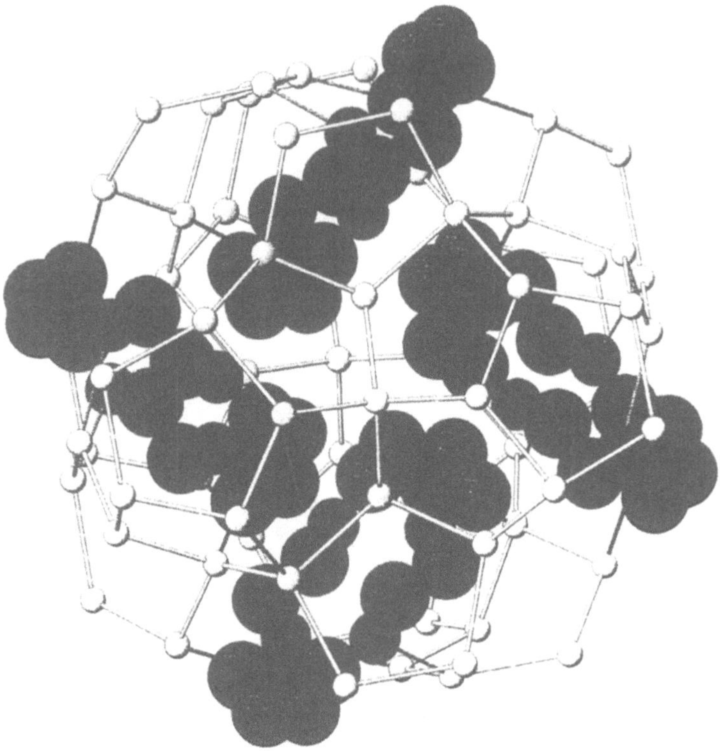

Figure 3. The structure of 1,10-diaza-18-crown-6 nonahydrate [6].

When the internal crown ether receptor site is blocked by complexation of another species, (e.g. by an aminoacid or peptide moiety), then water-crown hydrogen bonding is no longer feasible. Still high hydration of the complexes is possible.

Supramolecular complexes formed by 18-crown-6 and 1,10-diaaza-18-crown-6 and small peptides show remarkable tendency to form hydrates in the solid state. These may contain as much as several water molecules per molecule of the complex and the overall supramolecular architecture may consist of a network of water molecules around the crown-peptide complex supramolecule.

The structure of the glycylglycine/water/18-crown-6 ternary compound [7], which is illustrated in Figure 4, shows two distinct types of the water-complex interaction. One is 'typical' hydrogen bonding between water molecules and the hydrophilic end groups of the peptide (i.e. the carboxylic end group, the peptide existing in the zwitterionic form in the structure). The other is *hydrophobic hydration*, i.e. clathration of the hydrophobic part of the complex by the supramolecular assembly of water molecules.

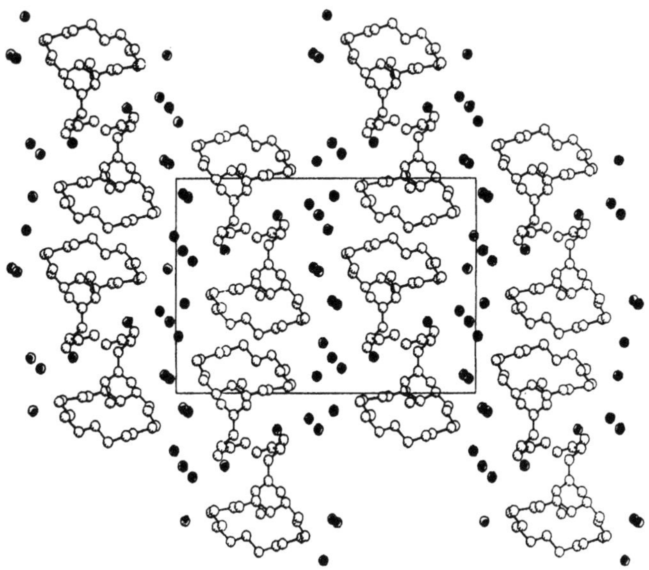

Figure 4. The structure of glycylglycine/water/18-crown-6 ternary compound.

Analogous situation arises in the hydrate structures of crown-ammonium salt complexes; a suitable example of the polyhydrate [18-crown-6]•[$C_3H_7NH_3F$]•[$10H_2O$] [8] being illustrated below (Fig. 6). As in semihydrate hydrates of ammonium salts alone, water molecules and fluoride anions connect with each other *via* hydrogen bonds. In the example given below they form tetra- penta- and heptagons which in turn make up layers.

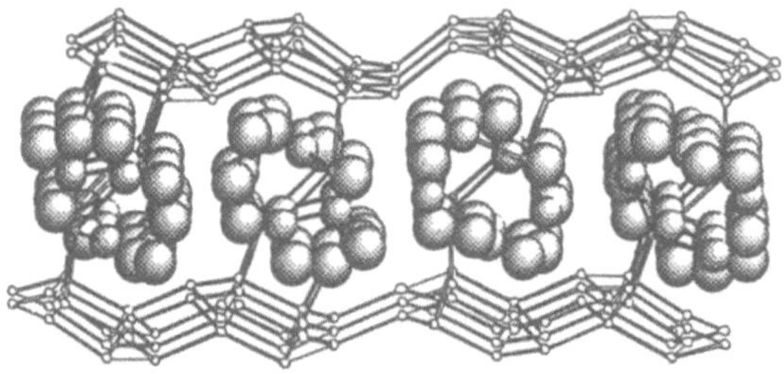

Figure 5. The structure of 15-crown-5 octahydrate.

Entirely different picture is obtained in the ternary complex formed by 18-crown-6, methylammonium fluoride and water (1 : 1 : 5). In this hydrate [5] the molecules of the crown ether and the complexed cations $[CH_3NH_3]^+$ make columns along the *c* axis (Fig. 7). The channels consisting of water molecules (Fig. 7b) are also arranged along the *c* axis. The channels consist of water molecules not bound to the molecules of the crown ethers neither to the fluoride anions which are included inside the channels and are disordered around the hexad axis (not shown in the Figure).

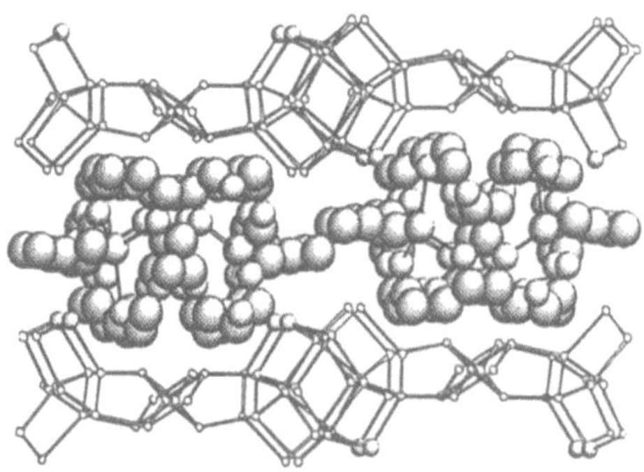

Figure 6. Hydrate 18-crown-6/$C_3H_7NH_3F$/10•H_2O.

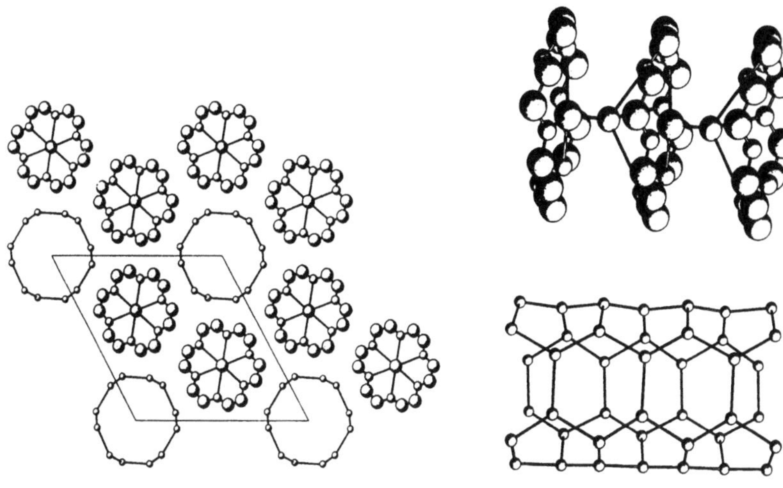

Figure 7. View of the channel structure in 18-crown-6/$CH_3NH_3F/5{\bullet}H_2O$: (a) along c axis, and (b) approximately along the *a* axis. Hydrogen atoms and fluoride anions are omitted.

Hydration of crown ethers performed from acidic media, containing boron, silicon or germanium fluoride species, leads to formation of interesting structural patterns in which different modes of hydration coexist [9]. Water may play the role of bridgeing unit as in the example given in Figure 9 but also may build up a supramolecular structure in which some hydrophobic hydration play a role. In the solid complex formed by diprotonated diaza-18-crown-6 with silicon pentafluoride anions (Fig. 8) water molecule is coordinated to silicon and donates two hydrogen bonds: one to crown oxygen and the other to a fluorine atom of the neighbouring anion. The crystal structure of trihydrate of monoaza-18-crown-6 pentafluoride contains three water molecules which play different roles: one is coordinated to the pentafluoride anion, the second forms a bridge between thus formed hexacoordinated $[SiF_5OH]^-$ moiety and the crown, and the third one is coordinated by the protonated macrocycle ring (Fig. 9). In the crystal structure of tetrafluorosilicate of 18-crown-6 (Fig. 10) water molecules also are both coordinated by the anion and the crown macrocycle; the latter one serving also as a bridge to the water coordinated in the anion.

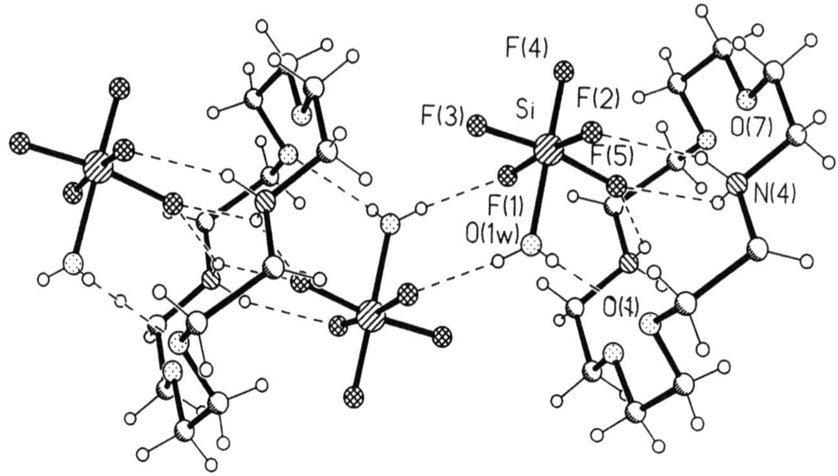

Figure 8. Illustration of fragment of the chain structure in the complex of diprotonated diaza-18-crown-6 with silicon pentafluoride anions and water (1 : 2 : 1).

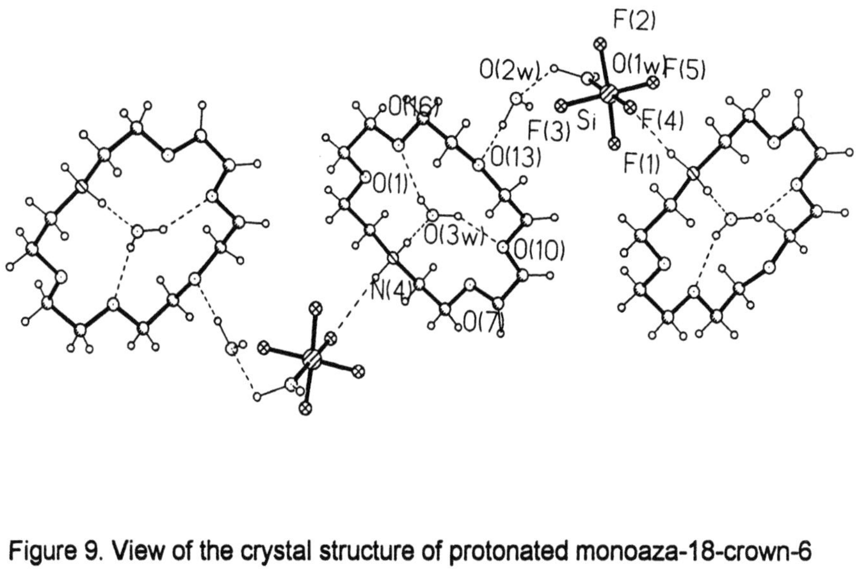

Figure 9. View of the crystal structure of protonated monoaza-18-crown-6 complexed with pentafluoride and water (1 : 1 :3)

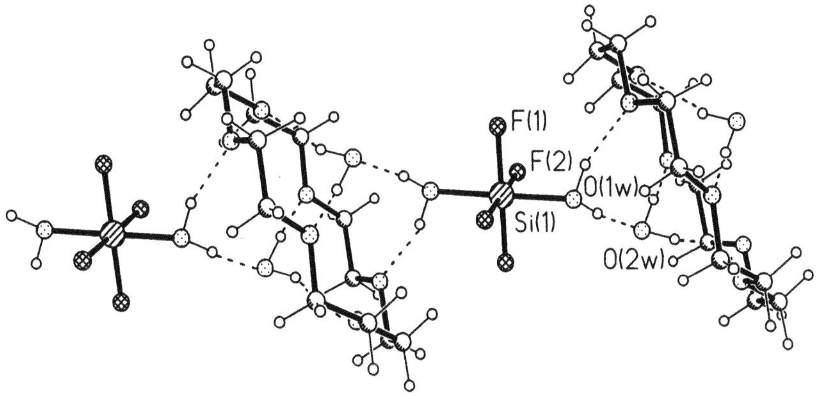

Figure 10. The chain architecture of the dihydrated form of tetrafluorosilicate of 18-crown-6

Hydration of chemically different species may, perhaps, follow similar geometric architecture. A convenient example may be taken from a recent study on hydration of zirconium acetylacetonate (Figure 11). As much as 10 water molecules surround the hydrophobic outer sphere of the zirconium complex thus forming hydrophobic hydration assembly [10]. Some of them form, additionally, rather weak hydrogen bonds to oxygen atoms of the acac ligands.

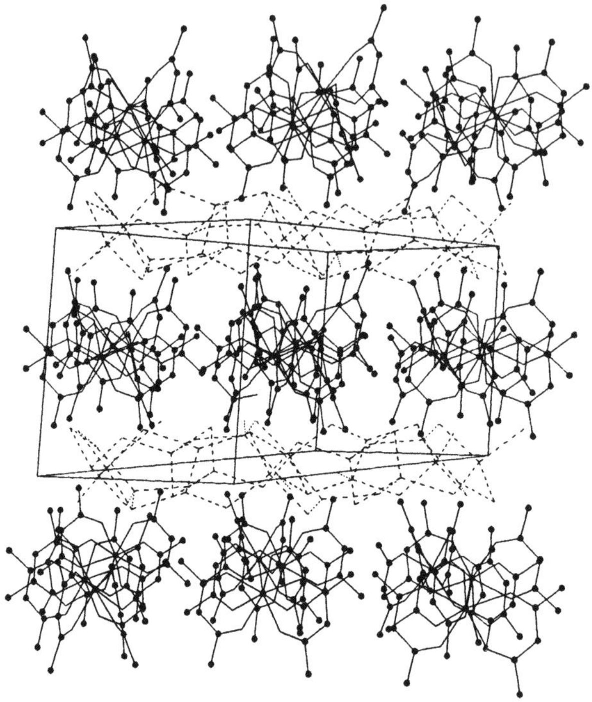

Figure 11. Molecular packing in the structure of $Zr(acac)_4$ decahydrate.

Werner clathrates

Suitable examples of the packing clathrate structure type are the series of the, so called, Werner clathrates [11]. These are products of the co-crystallization of some coordination complexes with either solvent molecules or some other guest compounds. The structures thus formed and their physicochemical behaviour are strongly dependent upon the size and shape of the guest molecule(s) present in the voids of the host crystalline lattice, as well as the formally assigned structure type based on the packing of the host molecules. The series is represented by the general formula $M^{II}X_2A_4$, where M stands for a divalent metal cation, X - monovalent anionic and A - monodentate neutral ligand. The complexes display outstanding ability to form broad range of inclusion - type compounds. Their crystal structures depend upon the guest, its content in the solid, and upon thermodynamic conditions of preparation. In several cases more than just one crystalline phase is formed in the system under investigation, depending on external physicochemical parameters. E.g. the $Cu(NCS)_2$(4-methylpyridine)$_4$ complex, when crystallized from 4-methylpyridine below 5° C, forms blue crystals containing 2 moles of included solvent per mole of the complex but the crystals quickly transform into green solid of the composition 2/3 : 1 when heated above that temperature or if exposed to traces of water [12].

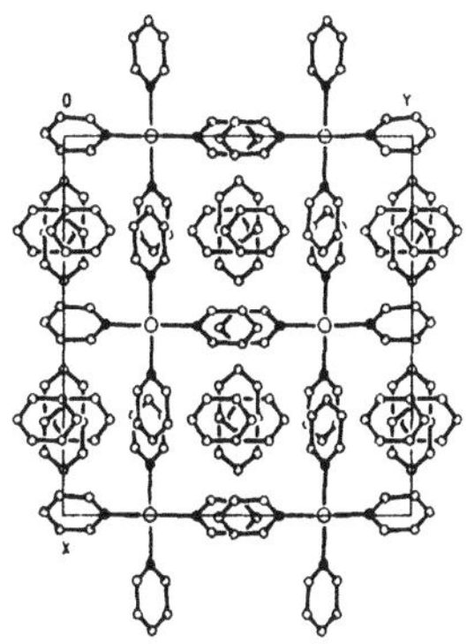

Figure 12. Example of an inclusion compound formed by the NiX_2py_4 complex [13]

The cage structure of the m-bromonitrobenzene clathrate might behave like a channel-type structure if the guest molecules would have appropriate size and shape. The same holds for organic zeolite in which some guest species are held firmly as in a cage structure while the other guest species can easily diffuse in and out. The porous structure of the organic zeolite host displays a pronounced contraction-dilation ability according to the amount of guest material absorbed in its intracrystalline pores. The minimum volume of an *organic zeolite* unit cell is attained when no guest is absorbed. Such *empty* clathrate structures, as well as some partially filled lattices, although unstable thermodynamically, show significant kinetic stability which can be studied experimentally. The organic zeolite structure of this host complex can vary its molar volume from 472 cm^3 (empty structure) to 547 cm^3 (m-xylene clathrate). This difference is greater than the lattice volume expansion during the $\alpha \rightarrow \beta^\circ$ phase transformation (431 to 472 cm^3 ; β° denoting the *empty* phase while α stands for the non-clathrated crystalline modification of the host material). Host lattice dilation may cause significant effects on guest diffusion kinetics in the solid and on the guest-lattice interaction energy and the partial molar enthalpy of clathration.

Since the dependence is primarily dependent on the amount of the given component and <u>not</u> on the lattice volume, one may assume that the absorbed guest molecule "knows" whether the adjacent cage is filled or empty since the energy of its interaction with the lattice is "sensitive" to the degree of filling of the cavities. Molecular modeling calculations[14] have shown that there are no significant guest-guest interactions in the experimentally studied examples. Thus, the mechanism of the dependence seems to function *via* guest-lattice rather than guest-guest interactions. Put in other words, the host structure adapts itself in order to accommodate a guest molecule in such a way that the next absorbed molecule of the same kind will find the structure more suitable for absorption, and so on. It may be considered a sort of *induced fit* behaviour in solid inclusion compounds. This would seem reasonable assuming that an absorbed molecule can "feel" not only if its adjacent neighbouring cavities are filled or empty, and also "recognizes" if the cavities contain similar or different guest species. This being so, one would expect a dependence of the partial molar enthalpy of clathration of a guest on the qualitative composition of the enclathrated mixture, as given above.

One may expect larger enthalpy effects when, instead of isomeric guest molecules, a mixture of species which are dissimilar in chemical and geometrical parameters is used for clathration. Such is indeed the case, as was observed in the examples of simultaneous clathration of 4-methylpyridine and a solvent (methanol or acetone). The enthalpy of clathration of 4-methylpyridine (the partial molar value) varies from 2.8

kcal mol^{-1} (in ternary Mepy/acetone clathrate) to 4.8 kcal mol^{-1} (in MePy clathrate) [15].

Recent x-ray and thermodynamic studies on some of the systems discussed above [16] have shown that the non-stoichiometric character of sorption may also be observed at high guest concentration, or, put in other words, sorption does not stop at a given 'maximum' sorption capacity of a given host structure. Such 'excess' sorption may lead to formation of a sort of precursor structure for host phase transformation which produces structures of higher porosity, e.g. layered one from a channel structure.

Discussion

As demonstrated above, it seems important to consider clathration as a sort of collective interaction at the multimolecular level. Thus, clathration may be assumed as the distinguishing factor contributing to complete pattern of solvent-solute interactions.

Presently, it is quite common practise to use the terms *clathrate*, and inclusion or addition *molecular complexes* for two- or multicomponent molecular crystals in which there is no apparent chemical interaction between the components. In other words, *clathration* is used instead of *solvation* if only there is no specific, chemical interaction between the *solvent* and *solute*.

The ordered arrangements of these species comprised of "main" and "solvent", or "host" and "guest" components, contain important information on intermolecular interactions. Crystalline "inclusion" or "clathrate" compounds provide unique opportunities to study model structures which are likely to occur in liquid solutions [17].

Clathrate or inclusion-type compounds may also serve as a convenient matrix to study molecular species which do not form monocomponent single crystals. The idea of contact stabilization (Dyadin) comes from this observation and has recently proven [18] to be very useful in investigating coordination complexes in their less stable conformations.

References

[1] Jeffrey G.A. (1984), Hydrate inclusion compounds, in J.L. Atwood, J.E.D. Davies and D.D. MacNicol (eds.) *Inclusion Compounds*, Academic Press, London, vol. 1, ch. 5, pp. 135-190, and references therein.

[2] Davidson D.W. (1973), Clathrate hydrates, in F. Franks (ed.) *Water: A Comprehensive Treatise*, Plenum Press, New York, vol. 2, ch. 3, pp. 115-203.

[3] K.A. Udachin, J. Lipkowski - "Channel polyhydrate $(C_2H_5)_4NF \bullet 11H_2O$" - *Mendeleev Commun.* (1996), 3, 92-93

[4] Lipkowski J., Luboradzki R., Udachin K. and Dyadin Yu. (1992), A layered clathrate hydrate structure of tetrapropylammonium fluoride, *J. Inclusion Phenomena and Molecular Recognition in Chem.* **13**, 295.

[5] K.A. Udachin, J. Lipkowski - unpublished data.

[6] K.A. Udachin, J. Lipkowski - "Polyhydrate 1,10-Diaza-18-crown-6•$8.25H_2O$" - *Mendeleev Commun.* (1995), 3, 237-238

[7] Lipkowski J., Kulikov O.V. and Zielenkiewicz W. (1992) Complexation of peptides with crown ethers. Structure and thermal behaviour of ternary compounds: 18-crown-6/glycylglycine/water and 18-crown-6/ L-α-alanyl-L-α-alanine/water, *Supramolecular Chem.* **1**, 73-79.

[8] K.A. Udachin, J. Lipkowski - "Layered polyhydrate 18-crown-6•$C_3H_7NH_3F \bullet 10H_2O$" - *Mendeleev Commun.* (1996), 195-196

[9] Yu.A. Simonov, M.S. Fonari, J. Lipkowski, V.O. Gelmboldt, E.V. Ganin - "Inclusion Compounds: the Products of the Interaction of Silicon, Germanium and Boron Fluorides with Crown Ethers" - *J. Incl. Phenom. Mol. Recogn. Chem.* , **24**, (1996), 149-161

[10] J. Lipkowski, K. Udachin, J. Narbutt and P. Staszewski - "On supramolecular architecture water molecules form in hydrophobic hydrates", XVII IUCr Congress, Seattle, Washington, USA, C-275 (1966)

[11] Lipkowski J., Werner Clathrates, in J.L. Atwood, J.E.D. Davies, D.D. MacNicol and F. Voegtle (eds.) *Comprehensive Supramolecular Chemistry*, Pergamon, vol. 6, ch. 20, pp. 691-714.

[12] K. Kislykh, Yu.A. Dyadin, D.V. Soldatov, J. Lipkowski - unpublished data

[13] D.V. Soldatov, Yu.A. Dyadin, J. Lipkowski, K. Suwińska - "Structure and supramolecular nature of the compounds $[MPy_4(NCO)_2] \bullet 2Py$ (M = Mn, Fe, Co, Ni, Cu, Zn, Cd)" - *Mendeleev Commun.* (1997), 3, 100-102

[14] Starzewski P. (1982), Calorimetric study of clathration phenomena (*Pol.*), PhD thesis, Warszawa.

[15] Lipkowski J., Zielenkiewicz A., Hatt J. and Zielenkiewicz W. (1989) , Calorimetric and thermogravimetric study of competitive clathration in the β-$Ni(NCS)_2(4\text{-}MePy)_4$ host, *J. Inclusion Phenomena and Molecular Recognition in Chem.* **7**, 519-527.

[16] A.Yu. Manakov, J. Lipkowski, K. Suwińska, M. Kitamura - "New Crystal Structures of β-$[Ni(NCS)_2(4\text{-methylpyridine})_4]$ Clathrates with Furan, Tetrahydrofuran, Methylene Chloride, Benzene + Ethanol and Methylcellosolve as Guest Molecules" - *J. Incl. Phenom. Mol. Recogn. Chem.* **26**(1-3), (1996), 1-20

[17] J. Lipkowski - "Clathration and Solvation of Molecules" - in G. Tsoucaris, J.L. Atwood and J. Lipkowski (Eds.) *Crystallography of Supramolecular Compounds,* NATO ASI Series, Series C: Mathematical and Physical Sciences - vol. 480, Kluwer Academic Publishers, Dordrecht/Boston/London, 1996, pp. 265-283

[18] Yu.A. Dyadin, D.V. Soldatov, V.A. Logvinenko, J. Lipkowski - "Contact Stabilization of Host Molecules During Clathrate Formation: The Pyridine-Zinc Nitrate and the Pyridine-Cadmium Nitrate Systems" - *J.Coord.Chem.,* **37**(1-4), (1996), 63-75

THE ROLE OF CHARGE ASSISTED C-H$^{\delta+}$---O$^{\delta-}$ AND C-H$^{\delta+}$---F$^{\delta-}$ HYDROGEN BONDS IN ORGANOMETALLIC CRYSTALS

D. BRAGA AND F. GREPIONI
Dipartimento di Chimica G. Ciamician, University of Bologna
Via Selmi 2, 40126 Bologna, Italy

Abstract

Weak hydrogen bonding interactions between organometallic cations and organic and inorganic anions have been discussed. It has been shown that charge assisted C-H---O hydrogen bonds can be used in conjunction with strong O-H---O bonds to prepare organic organometallic crystals in which the organic molecule (usually an organic acid) self assembles into three-dimensional networks around organometallic cations. Hydrogen bonding interactions of the C-H---F type have also been investigated in organometallic salts containing PF_6^- and BF_4^- anions. These weak interactions are reinforced by the opposite charge present on the ionic components, and have been shown to be responsible for the crystal stability and the phase transitional behaviour of some organometallic systems.

1. Introduction

We are currently exploring crystal engineering strategies to bring together organic and organometallic molecules or ions *via* a combination of weak and strong hydrogen bonding interactions.[1] The interest stems from the idea that crystals containing organic and organometallic molecules or ions may possess physical, chemical and structural features that are intermediate between those of typical molecular solids and those of transition metal complexes and bulk metal.[2]

The involvement of transition metal atoms in crystal build-up has a manifold effect. The topology of ligand-metal bonding interactions, for instance, provides a powerful means to organize in space extramolecular links (e.g. hydrogen bonding and polar groups). The metal atoms can be chosen for their direct participation in extramolecular interactions and/or for their capacity of tuning the acid/base behavior of ligands. These aspects have been discussed in other reports.[3]

G. Tsoucaris (ed.), Current Challenges on Large Supramolecular Assemblies, 173–191.

An important consequence of the variable oxidation state of transitional metal atoms and of the possibility of using ionic ligands is that many organometallic building blocks are ions. The ionic charge has, of course, a strong influence on interactions of electrostatic nature, such as the hydrogen bond.

This contribution is devoted to a discussion of organometallic architectures based on hydrogen bonds. In particular we will discuss a design strategy to obtain organic-organometallic supramolecular aggregates in the solid state by means of a selective use of strong hydrogen bonds between the organic component and of weak hydrogen bonds between organic and organometallic components, with emphasis on the role of charge assisted C-$H^{\delta+}$---$O^{\delta-}$ hydrogen bonds. In the second part of the paper the effect of C-$H^{\delta+}$---$F^{\delta-}$ interactions on the stability and phase transitional behavior of some organometallic crystals will be addressed.

Research in the field of organic solid state chemistry and crystal engineering has been carried out for the past three decades with remarkable results.[4] Crystal engineering conforms strictly to the paradigm of supramolecular chemistry, *i.e.* the intelligent utilization of non-covalent interactions to make aggregates that function differently from the individual components.[5] This logical connection with supramolecular chemistry has given further impetus to the field. Modern crystal engineering can be viewed as a bridge between supramolecular and materials chemistry.[4]

2. Construction of Organometallic Solids *via* C-$H^{\delta+}$---$O^{\delta-}$ Interactions and Strong Hydrogen Bonds.

The interaction of choice in most crystal engineering strategies is the hydrogen bond, because it joins strength to a certain degree of directionality.[5] Crystal design strategies which make use of strong hydrogen bonds between "conventional" donors, such as the -COOH, -OH, -CONRH etc. are being widely exploited to make nanoporous systems that promise to be extremely interesting for enclathration, selective adsorption and catalysis.[6] Such strong binders, however, are often not available in organometallic crystals for the simple reason that these groups, being strong Lewis bases, tend to interact directly with the metal atoms giving rise to dative bonds and complexation. The organometallic crystal engineer needs, therefore, to make use of weaker interactions such as C-H---O bonds.[7] Researchers are beginning to realize that it is possible to *compensate weakness with number* in devising crystal synthesis strategies. One way is that of utilizing a large number of C-H donors of tunable polarity and/or a large number of O-acceptors of tunable basicity. Furthermore, the use of transition metal atoms permits easy access to charged species. The ionic charge may provide, in turns, a simple way to strengthen weak electrostatic interactions such as C-H---O bonds.[8]

These considerations constitute the basis of our strategy to the preparation of organic-organometallic crystalline aggregates by means of cooperative strong O-H---O and weak C-H---O hydrogen bonds. The synthetic procedure is founded on the reaction between organic or inorganic molecules possessing acidic protons and organometallic molecules that can be easily oxidized to yield organometallic hydroxides, such as $[(\eta^6\text{-arene})_2Cr][OH]$ (arene = $\eta^6\text{-}C_6H_6$ $\eta^6\text{-}C_6H_5Me$) and $[(\eta^5\text{-}C_5H_5)_2Co][OH]$ prepared *in situ*. The result of this reaction are supramolecular salts formed *via* hydrogen bonding interactions.[9]

For example, when $[(\eta^6\text{-arene})_2Cr][OH]$ (arene = $\eta^6\text{-}C_6H_6$ $\eta^6\text{-}C_6H_5Me$) is reacted with 1,3-cyclohexanedione (CHD) the cations $[(\eta^6\text{-arene})_2Cr]^+$ (arene = $\eta^6\text{-}C_6H_6$ and $\eta^6\text{-}C_6H_5Me$) can be encapsulated in aggregates of the type $[(\eta^6\text{-}C_6H_6)_2Cr]^+[(CHD)_2]^-\cdot(CHD)_2$ and $[(\eta^6\text{-}C_6H_5Me)_2Cr]^+[(CHD)_2]^-$.[9]

An inspection of the factors responsible for the stability of the crystalline edifice and for the interaction between organometallic cations and organic superanions showed us that, since strong donor/acceptor hydrogen bonding groups are present only on the organic acidic species, these tend to self aggregate, leaving "unused" acceptors sites to interact with the only donors available on the organometallic partners, namely the large number of $C_{(sp^2)}$-H systems present on the metal bound arenes. Since the C-H---O interaction is between the organic anion and the organometallic cation, it is *assisted by the charge difference,* resulting in $C\text{-}H^{\delta+}\text{---}O^{\delta-}$ separations shorter than those usually observed in organometallic crystals.[11]

As mentioned above, a useful organometallic component is also provided by $[(\eta^5\text{-}C_5H_5)_2Co]$. Cobaltocene is readily oxidized by oxygen, in water or THF solution, to the yellow cobalticinium cation. If the oxidation, and the consequent formation of the hydroxide $[(\eta^5\text{-}C_5H_5)_2Co]^+[OH]^-$ occurs in the presence of an organic acid, this is deprotonated leading to the formation of organic/organometallic salts.[12] The crystalline products of chemical formulae $[(\eta^5\text{-}C_5H_5)_2Co]^+[(D,L\text{-}HTA)(D,L\text{-}H_2TA)]^-$ and $[(\eta^5\text{-}C_5H_5)_2Co]^+[L\text{-}HTA]^-$ have been obtained by utilizing D,L-tartaric acid and L-tartaric acid, respectively. The two crystalline materials are respectively constituted of hexagonal and tetragonal superanions molded around the cobalticinium cations, as shown in Figure 1.

In both crystals the cobalticinium cation is encapsulated within organic honeycomb-type frameworks. In the case of the D,L-tartrate the organic superanion has stoichiometry $\{[(D,L\text{-}HTA)(D,L\text{-}H_2TA)]^-\}_n$ generated by the aggregation of dimeric monoanions $[(D,L\text{-}HTA)(D,L\text{-}H_2TA)]^-$ formed as a consequence of the loss of one proton every two tartaric acid molecules. In the chiral crystal obtained by using L-tartaric acid the anionic structure is not based on hexagonal channels but on square ones (see Figure 1b). The channels are formed by $L\text{-}HTA^-$ ions, interwoven in spirals which wrap around the cobalticinium cations.

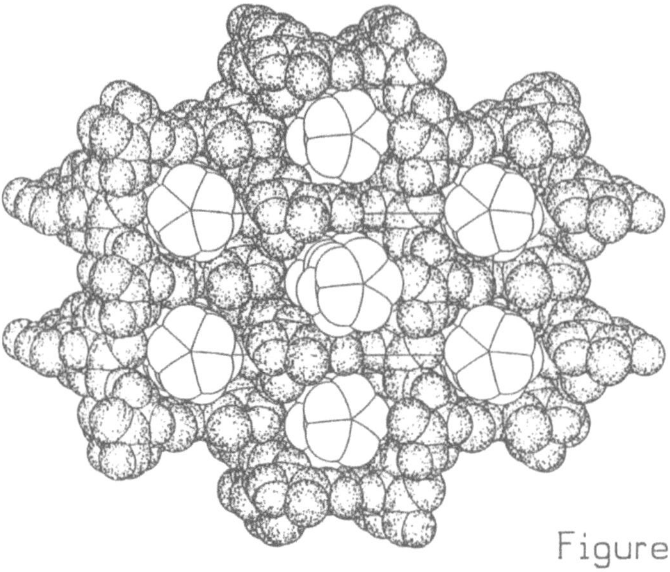

Figure 1a

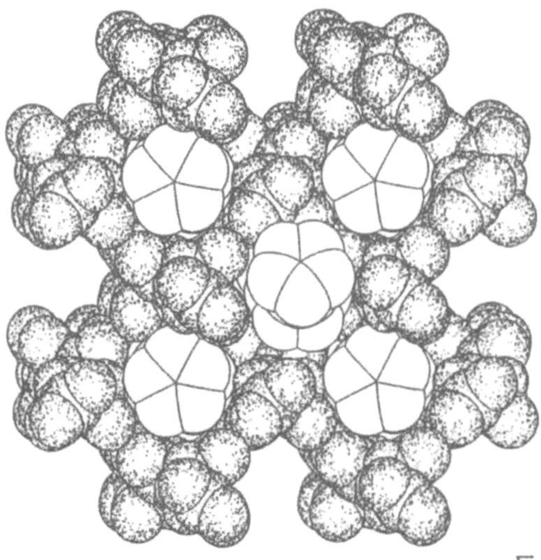

Figure 1b

Figure 1. Supramolecular organization in crystalline $[(\eta^5\text{-}C_5H_5)_2Co]^+[(D,L\text{-}HTA)(D,L\text{-}H_2TA)]^-$ (a) and $[(\eta^5\text{-}C_5H_5)_2Co]^+[L\text{-}HTA]^-$ (b).

The hydrogen bond between the tartaric acid moieties in both organic frameworks is of the *negatively-charged* type[13] [O-H---$O^{(-)}$] formally between one neutral -COOH group and one -COO^- unit. The O-H---$O^{(-)}$ bonds have length 2.434 Å and 2.440/2.413 Å for the two independent L-tartaric acid units present in the chiral crystal. The honeycomb frameworks are then generated *via* O-H---O bonds involving the carboxyl and hydroxyl oxygen acceptors.

As pointed out before, the interaction between the anionic framework and the encapsulated $[(\eta^5\text{-}C_5H_5)_2Co]^+$ cations is based on C-H---O hydrogen bonds between the cyclopentadienyl ligands of the cations and the unused lone pairs of the organic framework. (C)H---O distances are observed (after normalization of the C-H lengths to neutron values) in the range 2.10-2.35(1) Å, *viz.* at values that fall towards the lower limit for interactions of this type. This is a clear indication that the electrostatic interaction C-$H^{\delta+}$---$O^{\delta-}$ is reinforced by the difference in charge between hosts and guests.

The solid state structure of the ferricinium trichloroacetate salt $[(\eta^5\text{-}C_5H_5)_2Fe]^+[(CCl_3COO)(CCl_3COOH)]^-$ was reported by Boeyens.[14] Figure 2 shows the close structural analogy with crystalline $[(\eta^5\text{-}C_5H_5)_2Co]^+$[(D,L-HTA)(D,L-$H_2$TA)]$^-$. The supramolecular $\{[(CCl_3COO)(CCl_3COOH)]^-\}_n$ framework is formally constituted of one CCl_3COO^- ion and of one neutral CCl_3COOH molecule organized in a honeycomb-type structure encapsulating the ferricinium cations, *viz.* the stoichiometry is the same as for crystalline $[(\eta^5\text{-}C_5H_5)_2Co]^+$[(D,L-HTA)(D,L-$H_2$TA)]$^-$. There are some remarkable features that deserve a close examination. The interaction between the CCl_3COO^- ion and the neutral CCl_3COOH molecule is of the same type as observed for the other superanions. The bond linking the two moieties of the $[(CCl_3COO)(CCl_3COOH)]^-$ unit is, therefore, a negatively charge-assisted O-H---O^- hydrogen bond [2.463 Å] of the type discussed above. These hydrogen bonds define two opposite edges of the distorted hexagonal frame, while the other four edges are generated *via* interlocking and stacking of the CCl_3 moieties. The shortest Cl---Cl distance is 3.518 Å. The ferricinium cation adopts an orientation that allows optimization of the C-H---O hydrogen bonds between the cyclopentadienyl ligands and the O-atoms of the acid which are only present on the sides of the channel.

The problem of whether the Cl---Cl separations (which are shorter than the sum of van der Waals radii) represent attractive interactions or are the result of anisotropy of the chlorine atomic shape is still being debated.[15] Many organic molecular crystals present Cl---Cl distances in the range 3.2-3.4 Å, and a very short distance of 3.27 Å is also present in solid chlorine.[15]
Cl---Cl distances have been observed as short as 3.070(1) Å in crystalline ClF.[16] The analogies and differences between the crystal structures of the D,L-tartrate and

that of the trichloroacetate represent an important manifestation of a supramolecular analogy between hydrogen bonds and Cl---Cl interactions.[17]

If the stoichiometric ratio between $[(\eta^5\text{-}C_5H_5)_2Co][OH]$ and D,L-tartaric acid in the acid-base reaction is changed from 1:2 to 1:1, the hydrated crystalline salt $[(\eta^5\text{-}C_5H_5)_2Co]^+[D,L\text{-}HTA]^-\cdot H_2O$ is obtained. The tartrate ion forms with one solvate water molecule a layered structure (see Figure 3). As in the cases discussed previously the organic acid anion interacts with itself via -COOH/-COO$^-$ bonds [2.479(8) Å, compare with the values discussed above]. The water molecules form O-H---O hydrogen bonds involving two hydroxyl groups of two neighboring tartrate anions. Two hydroxyl groups belonging to two tartrate ions and two water molecules form tetramer-like $(\text{-OH})_4$ hydrogen bonded units of the type commonly observed for alcohols.[18]

The O---O separation in the meso-tartaric acid salt Na[(M-HTA)(M-H_2TA)][19] [2.448 Å] is strictly comparable to the values discussed above. A hydrogen atom is situated on the inversion center relating the two moieties [O-H 1.224(2) Å], although the aggregation of tartaric acid and tartrate units does not constitute an infinite three-dimensional network.

3. The Role of Charge Assisted C-H---F Interactions and the Phase Transitional Behaviour of Some Organometallic Crystals

The organometallic cations utilized for organic-organometallic crystal synthesis had all been characterized earlier as hexafluorophosphate salts in which the most common interaction is between the peripheral hydrogen atoms and the fluorine atoms bound to phosphorus. Based on our previous experience with C-H---O bonds the extension to C-H---F was quite natural. C-H---F(-P) interactions are expected to be weak because of the low acidity of the C-H system, but may be reinforced, as in the cases discussed above, by the electrostatic contribution arising from the opposite polarity on the anions and cations. In other words interactions of the C-$H^{\delta+}$---$F^{\delta-}$ type between organometallic cations and the fluorine atoms of the PF_6^- anion are possible.[20] As in the case of C-H---O the number of such interactions is also potentially very large (compared with compounds carrying conventional strong donors such as -COOH, -OH etc.) and may well result in an important *collective* contribution to cohesion.

Furthermore, PF_6^- as well as BF_4^- are among the most commonly employed anions for crystallization of inorganic and organometallic cations. In the course of this study we have found that the simple metallocenium cations $[(C_5H_5)_2Co]^+$, $[(C_5H_5)_2Fe]^+$ form polymorphic modifications that undergo phase transitions on changing the temperature. The phase transitional behaviour is quite intriguing and will be briefly described below.[21]

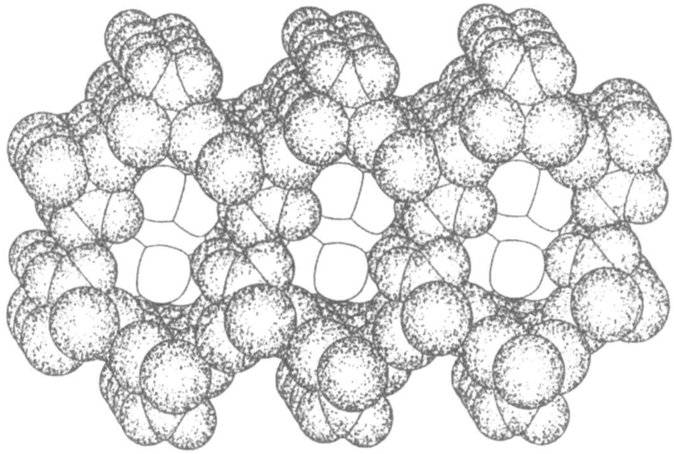

Figure 2

Figure 2. Space-filling representation of crystalline $[(\eta^5\text{-}C_5H_5)_2Fe]^+[(CCl_3COO)\ (CCl_3COOH)]$ (coordinates from ref. 14).

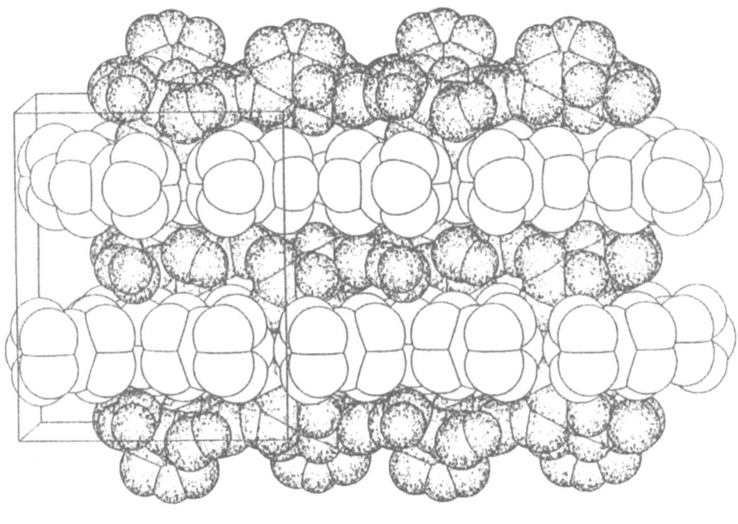

Figure 3

Figure 3. The layered structure of organometallic cations and organic/water moieties in the hydrated crystalline salt $[(\eta^5\text{-}C_5H_5)_2Co]^+[D,L\text{-}HTA]^- \cdot H_2O$.

A further reason of interest in the participation of F-atoms in hydrogen bonds arises from the recent report by Dunitz and Taylor on the participation of organic fluorine in intermolecular hydrogen bonds reported.[22] Database analysis and theoretical calculations have been used to demonstrate that fluorine, when covalently bound to carbon, rarely forms hydrogen bonds with conventional hydrogen bond donors, including O-H, N-H and C-H groups.

Before discussing the intermolecular interactions, however, it is useful to recall the phase transitional behavior of crystalline $[(C_5H_5)_2Co][PF_6]$ and in $[(C_5H_5)_2Fe][PF_6]$. Both systems undergo two phase transitions, indicated as Form-I⇔Form-II and Form-I⇔Form-III, Form-I being the room temperature structure, within narrow temperature intervals (see Table 1). The phase relationship in the case of the cobalt complex is shown schematically in Figure 4. Importantly, these three crystal phases have been fully characterized by single crystal X-ray diffraction at variable temperature on the same crystal specimen.

$[(C_5H_5)_2Co][PF_6]$	Form-I⇔Form-II 252K	Form-I⇔Form-III 314K
$[(C_5H_5)_2Fe][PF_6]$	Form-I⇔Form-II 213 K[23]	Form-I⇔Form-III 347K

TABLE 1. Phase transitions in crystalline $[(C_5H_5)_2M][PF_6]$ (M = Co, Fe)

The low temperature phase transition relates two ordered crystal phases, whereas the high temperature one leads from the ordered room temperature crystals to semi-disordered high temperature ones, in which the anions retain almost the same position in the crystal edifice of the cobalt salt, while the Cp-ligands of the iron and cobalt organometallic cations, and the anions of the iron salt become disordered. In the case of cobalt the Form-I⇔Form-III transition has been demonstrated to be fully reversible on the same crystal sample as the low-temperature Form-I⇔Form-II transformation.

The intriguing reversibility of the order-disorder transition has been explained as the result of the relationship between static and dynamic disorder. Form-III is cubic only on average over space, but remains monoclinic on average over time, i.e. the disorder observed in the high temperature cubic phase is not degenerate, so that the two crystallographically independent sites retain their difference although, on average over space, the crystal symmetry becomes pseudo-cubic. In fact, the cubic structure of $[(C_5H_5)_2Co][PF_6]$ can be correlated to a lower symmetry monoclinic one which is almost isomorphous. Space-filling projections of Form III in the monoclinic and cubic settings are compared in Figure 5.

The crystal structures of $[(C_5H_5)_2Co][PF_6]$ and $[(C_5H_5)_2Fe][PF_6]$ can be compared with that of $[(C_6H_6)_2Cr][PF_6]$, which can ideally be obtained by substituting benzene for the cyclopentadienyl ligand. As a matter of fact the crystal architecture is retained essentially unchanged, *viz.* the three crystals are *quasi-*

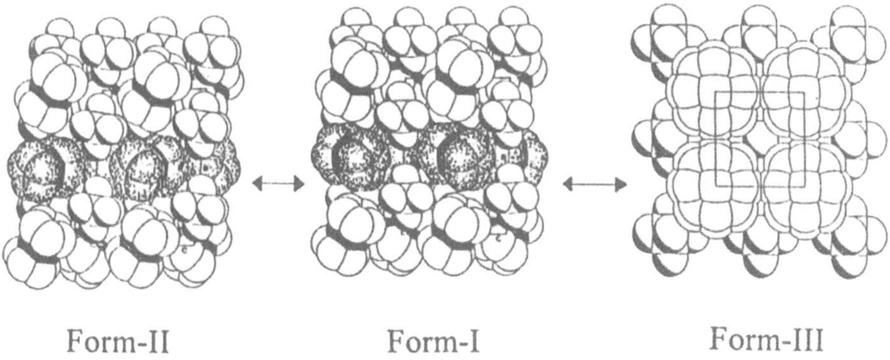

Figure 4

Figure 4. Space-filling representation of the phase transitions in crystalline $[(C_5H_5)_2Co][PF_6]$.

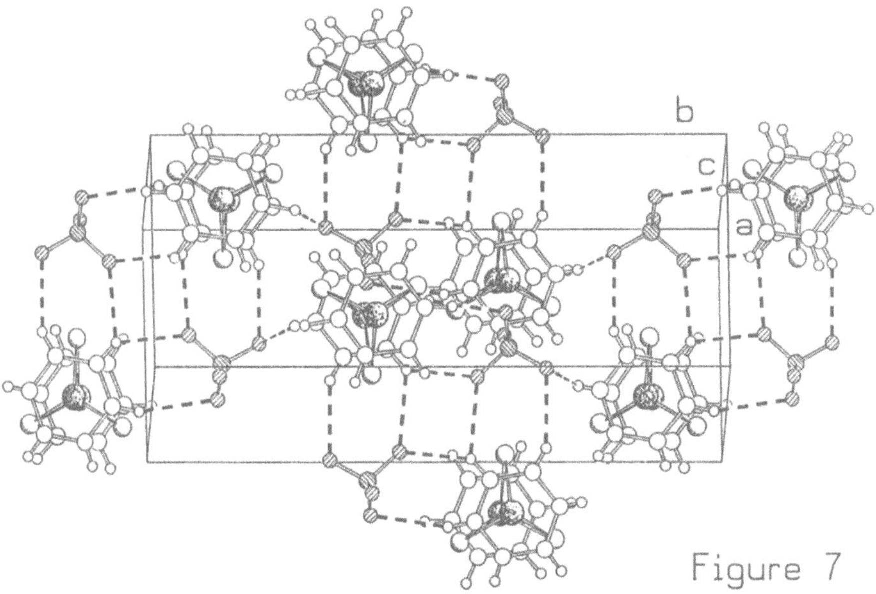

Figure 7. Intermolecular C-H---F(B) interactions between cations and anions in crystalline$[(\mu\text{-}Cl)_3\{(\eta^6\text{-}C_6H_5Me)Ru\}_2][BF_4]$.

Figure 5a

Figure 5b

Figure 5. Space-filling projections of the structure of $[(C_5H_5)_2Co][PF_6]$ (Form III) in the monoclinic (a) and cubic (b) settings.

isomorphous, in spite of the structural differences between the organometallic cations. The two independent cations in crystalline $[(C_6H_6)_2Cr][PF_6]$ (see Figure 6) have the same relative arrangement observed in the low temperature phases of $[(C_5H_5)_2Co/Fe][PF_6]$ (Form-II) but a different stacking sequence.

Let us now compare the three crystal structures at the various temperatures in terms of C-H---F(P) distances. In Form-I and Form-II of $[(C_5H_5)_2Fe][PF_6]$ there are two and five C-H---F distances between 2.3 and 2.4 Å, respectively, while all C-H---F(P) distances are longer than 2.42 Å in crystals of $[(C_5H_5)_2Co][PF_6]$. In crystalline $[(C_6H_6)_2Cr][PF_6]$, on the other hand, distances are as short as 2.3 Å. Whether these differences reflect the effect on the C-H acidity of the metal atoms (Co, Fe, Cr) is difficult to state with confidence. However, it is interesting to note that shorter and more numerous C-H---F distances are present in $[(C_6H_6)_2Cr][PF_6]$ which does not undergo a phase change, and in $[(C_5H_5)_2Fe][PF_6]$ whose intermediate phase is much more stable than in the case of the cobalt complex.

The crystal structures of the complexes $[(\mu\text{-Cl})_3(\eta^6\text{-arene})Ru\}_2]$ $[BF_4]$ (arene = C_6H_6, and C_6H_5Me)[24] provide other examples of the participation of C-H systems in C-H---F interactions between anions and cations. In $[(\mu\text{-Cl})_3\{(\eta^6\text{-}C_6H_5Me)Ru\}_2][BF_4]$ the fluorine atoms of the anion form C-H---F interactions in the range 2.32-2.45 Å (see Figure 7). C-H---F distances in crystalline $[(\mu\text{-Cl})_3\{(\eta^6\text{-}C_6H_6)Ru\}_2][BF_4]$ are in the range 2.36-2.45 Å

In order to put this analysis in a broader perspective, the CSD[25] has been searched for intermolecular H-bonds of the C-H---$F^{(\delta-)}$ type involving the PF_6^- and the BF_4^- anions. The hydrogen bonds involving nitrogen and oxygen as donors have been discussed elsewhere.[20] Since the number of compounds containing C-H groups is very large, two separate searches were carried out. In the first only C-H fragments for which the C atom is directly bound to the transition metal atom were considered. In the second one no restrictions were applied to the C atom. The number of compounds obtained is 149 (247 fragments) and 831 (6094 fragments) for the PF_6^- anion and 149 (249 fragments) and 614 (4210 fragments) for the BF_4^- anion, in the first and second searches, respectively.

Scattergrams of (C)H---F(P) angles *versus* (C)H---F(P) distances and of C-H---F(B) angles *versus* (C)H---F(B) distances are shown in Figure 8. Although the sample population is different in the two cases, the two types of interactions follow the normal trend observed for hydrogen bonds, *i.e.* as the distance between hydrogen and acceptor atoms decreases, the C-H---F vectors becomes straighter and the bond approaches linearity. Beside this general behavior, there is a large number of bonds that fall below the van der Waals cutoff distances and are clearly indicative of specific and directional interactions involving donors and acceptors. Comparison of the two types of search also shows that there is no appreciable

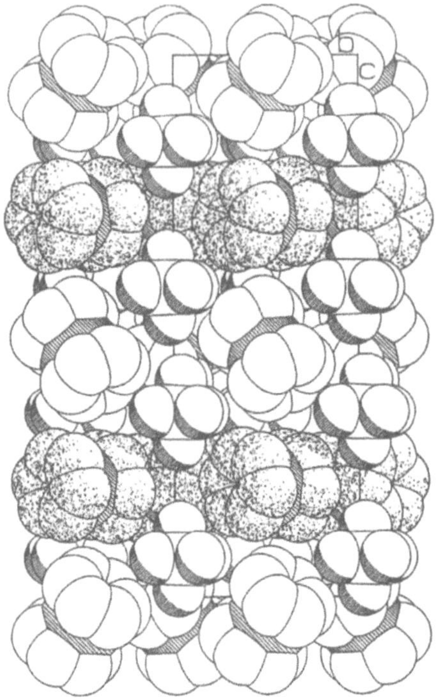

Figure 6

Figure 6. Space-filling representation of the structure of $[(C_6H_6)_2Cr][PF_6]$.

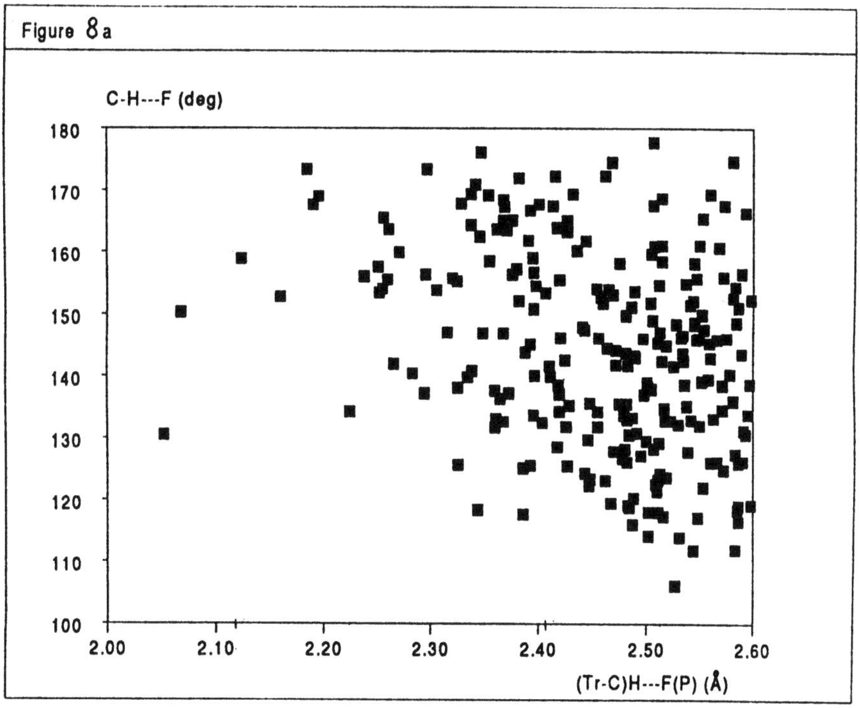

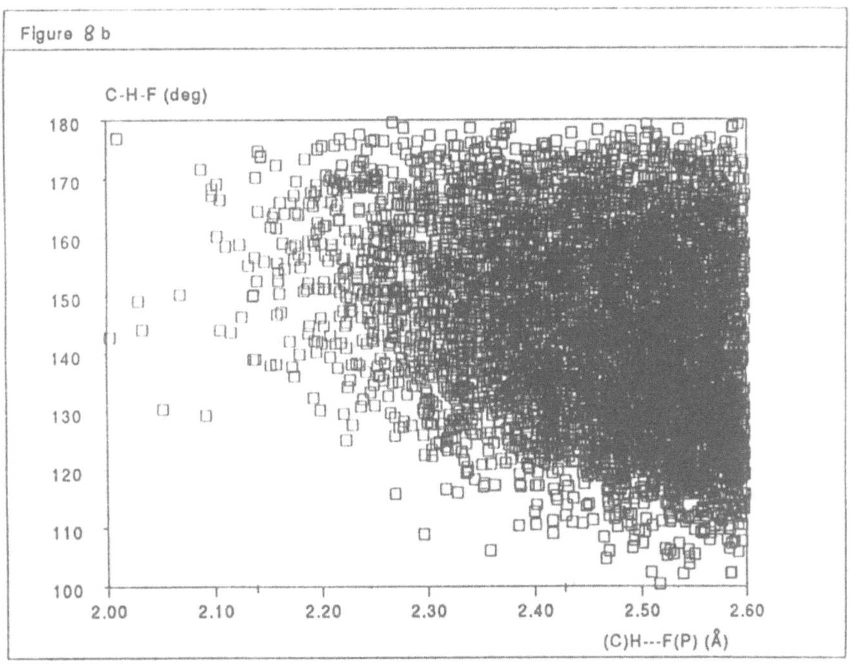

Figure 8. Scattergrams of C-H---F(P) angles *versus* (Tr-C)H---F(P) (a) and (C)H---F(P) (b) distances and of C-H---F(B) angles *versus* (Tr-C)H---F(B) (c) and (C)H---F(B) (d) distances.

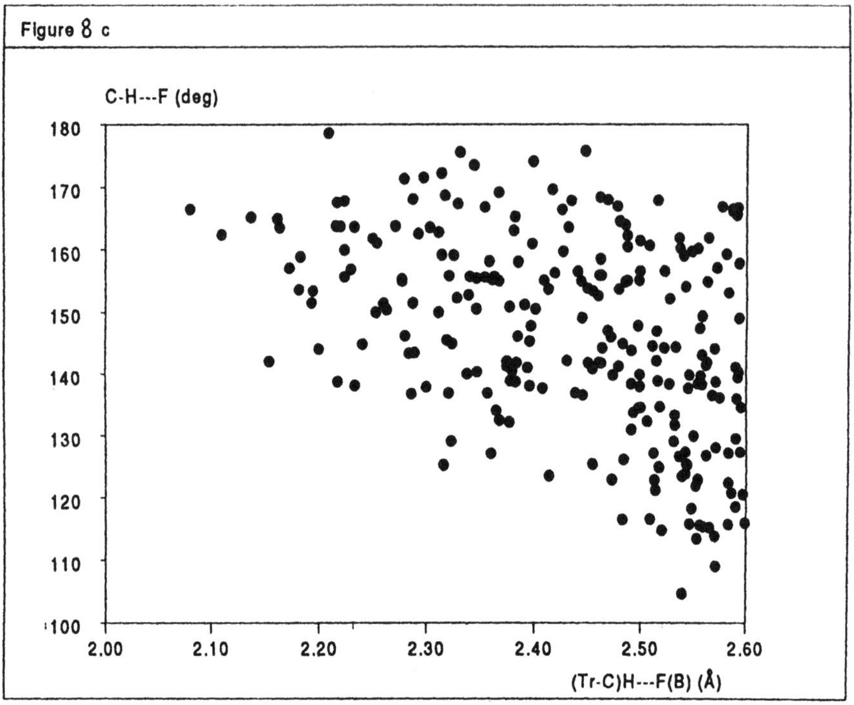
Figure 8 c
C-H---F (deg)
180
170
160
150
140
130
120
110
100
2.00
2.10
2.20
2.30
2.40
2.50
2.60
(Tr-C)H---F(B) (Å)

Fig. 8 (c) and (d)

effect of the direct bonding to transition metal atoms on the geometry of the interaction, as the two populations have the same distributions. Interestingly, while the short contacts constituting the first 10% of the total number of contacts (see histograms in Figure 9) fall in the range 2.00 - 2.32 Å in the case of PF_6^-, they fall in a narrower range in the case of BF_4^-, the range being 2.00-2.24 and 2.00-2.28 Å for the reduced and extended searches, respectively. We see two possible explanations for this difference: (i) the presence of a larger number of short contacts in the case of fluorine bound to boron than in the case of flurine bound to phosphorous may reflect the higher polarity of the F-atom in the formed anion than in the latter (the anion charge is shared by four versus six fluorine atoms), and (ii) the smaller BF_4^- anion can approach the C-H donors more efficaciously than the bulkier PF_6^-.

4. Conclusions

We have discussed in this contribution the role of weak intermolecular interactions in determining the stability of organometallic crystals and their behaviour with temperature. The main outcome of this work is essentially twofold:

i) weak interactions can be reinforced by the different polarity between anions and cations (charge assistance). The effect of charge assistance is particularly important in the case of fluorine. While organic fluorine do not frequently form intermolecular hydrogen bonds with hydrogen bonding donors, this is not so when fluorine belongs to inorganic anions such as PF_6^- and BF_4^-. These anions, in fact, appear to form a profusion of interactions with the surrounding donors, in particular the C-H systems which are abundant in organometallic crystals. Therefore, the intrinsic weakness of these interactions is compensated by the presence of a large number of donors (all hydrogen atoms of carbocycles such as π-bonded benzene or cyclopentadienyl ligands can form bonds) and by the difference in polarity between the inorganic anions which accept and the organometallic cations which donate hydrogen bonds.

ii) weak interactions can be used when there is a large number of accepting sites. Lone pairs on electronegative atoms (or π-electron clouds on alkynes, arenes, etc.[26]) can be satisfied by weak donors when strong donors are not available. This is quite advantageous in organometallic crystal engineering for the same reasons stated above: there is always a great number of weak donors available for intermolecular links. If the choice of building blocks in the crystal synthesis strategy is such that strong donors and acceptors are present only on the organic partner, while the organometallic molecule or ion can only provide a shape template and weak bonds, than the organic molecules or ions (as the tartrate ions discussed in the first part) are "forced" to self aggregate thus forming three-dimensional organic frameworks in which the organometallic moiety is

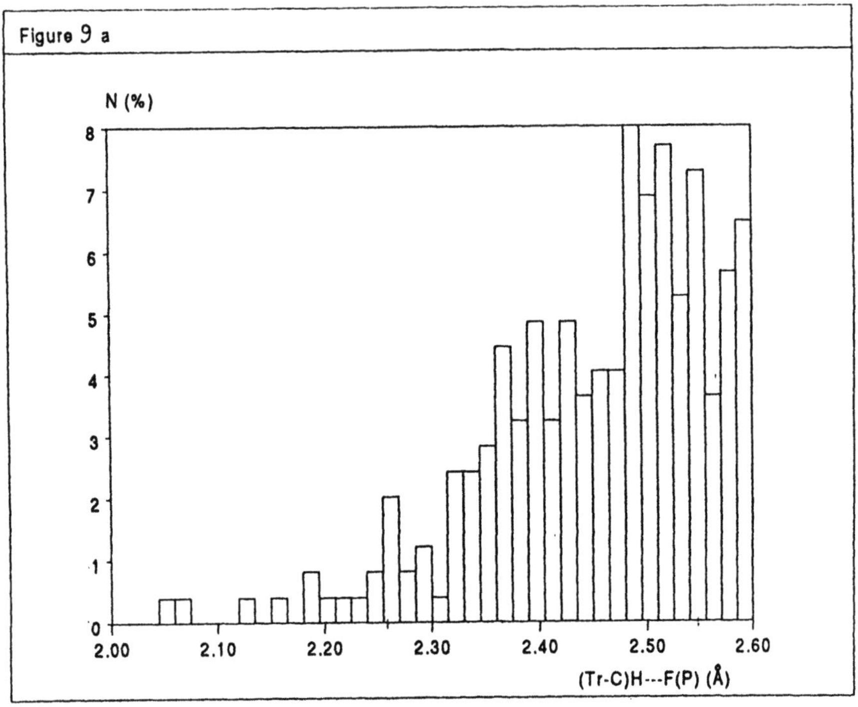

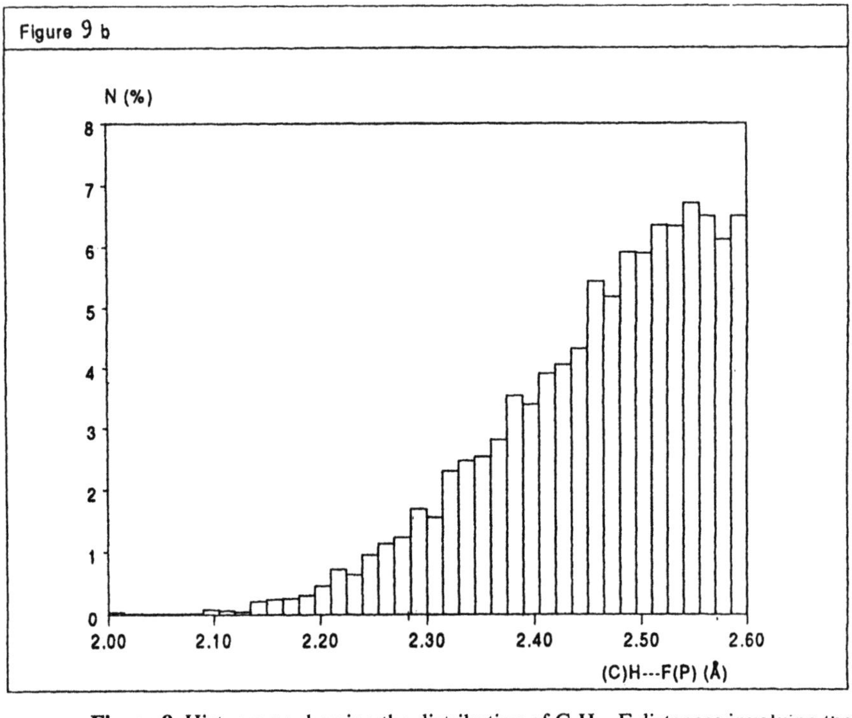

Figure 9. Histograms showing the distribution of C-H---F distances involving the PF_6^- (a and b) and the BF_4^- (c and d) anions. Scattergrams a and c refer to C-H fragments for which the C atom is directly bound

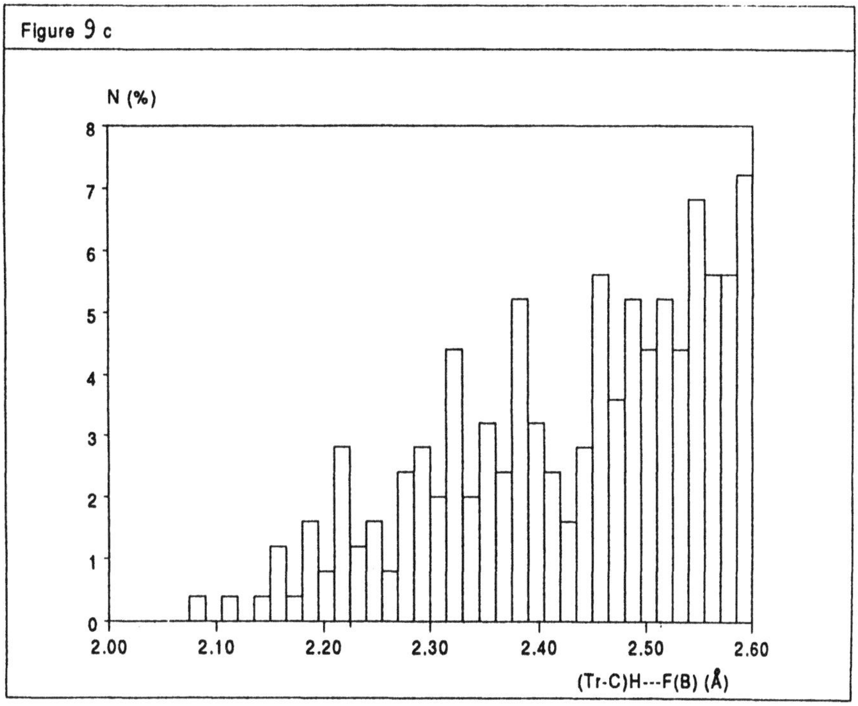
Figure 9 c
N (%)
8
7
6
5
4
3
2
1
0
2.00
2.10
2.20
2.30
2.40
2.50
2.60
(Tr-C)H---F(B) (Å)

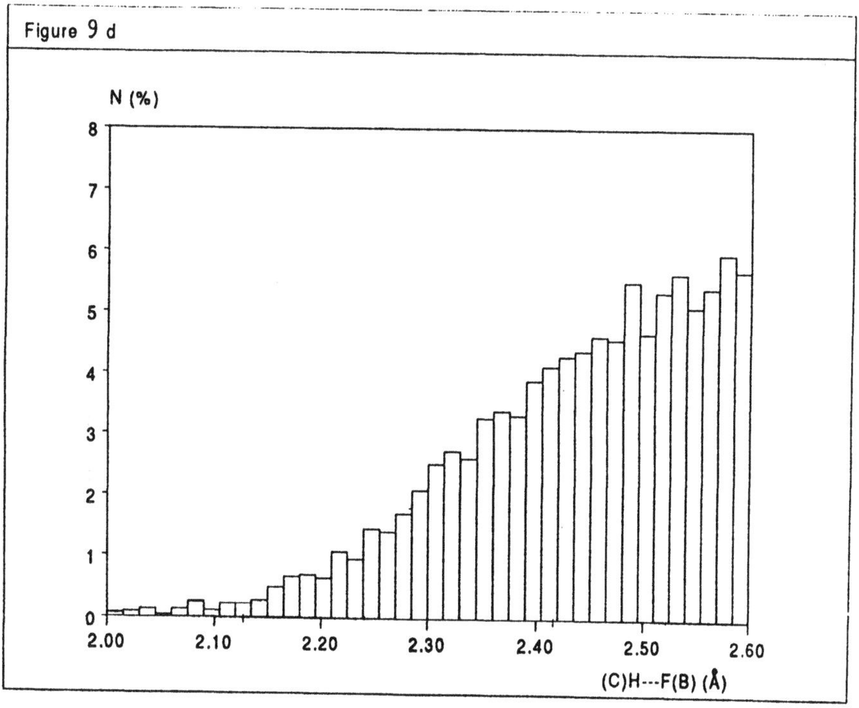
Figure 9 d
N (%)
8
7
6
5
4
3
2
1
0
2.00
2.10
2.20
2.30
2.40
2.50
2.60
(C)H---F(B) (Å)

Fig. 9 (c) and (d)

encapsulated. This provides a simple route towards preparation of chiral crystals as enatiomerically pure organic acids can be used.

Acknowledgment

The work discussed here has seen the contribution of numerous collaborators who share the authorship of the papers quoted throughout this article. We acknowledge financial support by the University of Bologna (project "*Intelligent Molecules and Molecular Aggregates*", 1995), by MURST, by CIBA-Geigy for an ACE fellowship, and by DAAR-CRUI (Vigoni Programme). The ERASMUS student exchange program "Crystallography" is also acknowledged.

References

1. (a) Braga, D., Grepioni, F., Byrne, J. J., and Wolf, A. (1995) *J. Chem. Soc. Chem. Commun.* 1023. (b) Braga, D., Costa, A. L., Grepioni, F., Scaccianoce, and L., Tagliavini, E. (1996) *Organometallics* **15**, 1084. (c) Braga, D., Angeloni, A., Grepioni, and F., Tagliavini, E. (1997) *J. Chem. Soc. Chem. Commun.* 1447.
2. Braga, D., Grepioni, F., and Desiraju, G. R. *Chem. Rev.* in press
3. (a) Braga, D., and Grepioni, F. (1997) *Comments Inorg. Chem. 19*, 185. (b) Braga, D., and Grepioni, F. *Coord. Chem. Rev.* in press
4. Desiraju, G. R. (1989) Crystal Engineering: The Design of Organic Solids, Elsevier, Amsterdam.
5. (a) Jeffrey, G. A., and Saenger, W. (1991) Hydrogen Bonding in Biological Structures, Springer-Verlag, Berlin. (b) Aakeröy, C. B., and Seddon, K. R. (1993) *Chem. Soc. Rev.* 397. (c) Brammer, L., Zhao, D., Ladipo, F. T., and Braddock-Wilking, J. (1995) *Acta Cryst.* **B51**, 632.
6. (a) Hulliger, J. (1994) *Angew. Chem. Int. Ed. Engl. 33*, 143. (b) Bowes, C. L., and Ozin, G. A. (1996) *Adv. Mater.* **8**, 13. (c) Zaworotko, M. J. (1997) *Nature* **386**, 220. (d) Yaghi, O. M., Guangming, L., and Li, H. (1995) *Nature* **378**, 703. (e) Ball, P. (1996) *Nature* **381**, 648.
7. (a) Desiraju, G. R. (1986) *Acc. Chem. Res. 19*, 222. (b) Desiraju, G. R. (1996) *Acc. Chem. Res.* **29**, 441. (c) Steiner, T. (1997) *Chem. Commun.* 727. (d) Steiner, T. (1996) *Cryst. Rev.* **6**, 1. (e) Braga, D., and Grepioni, F. (1997) *Acc. Chem. Res.* **30**, 81.
8. Braga, D., and Grepioni, F. in W. Gans and J.C.A. Boeyens (eds.), Intermolecular Interactions, Plenum, New York, in press.
9. Braga, D., Costa, A. L., Grepioni, F., Scaccianoce, L., and Tagliavini, E. (1997) *Organometallics* **16**, 2070.
10. (a) Allerhand, A., and von Rague' Schleyer, P. (1963) *J. Am. Chem. Soc* **85**, 1715. (b) Steiner, T. (1996) *Cryst. Rev.* **6**, 1.
11. Braga, D., Grepioni, F., Biradha, K., Pedireddi, V. R., and Desiraju, G. R. (1995) *J. Am. Chem. Soc.* **117**, 3156.
12. Braga, D., Angeloni, A., Grepioni, F., and Tagliavini, E. (1997) *Organometallics.* in press.
13. (a) Meot-Ner (Mautner), M. (1984) *J. Am. Chem. Soc.* **106**, 1257. (b) Meot-Ner (Mautner), M., and Sieck, L.W. (1986) *J. Am. Chem. Soc.* **108**, 7525.
14. Allen, C. C., Boeyens, J. C. A., and Neuse, E. W. (1987) *S. Afr. J. Chem.* **40**, 17.
15. (a) Dunitz, J., and Taylor, R. (1997) *Chem. Eur. J.* **3**, 89.(a) Mirsky, K., and Cohen, M. D. (1978) *Chem. Phys.* **28**, 193. (b) Sarma, J. A. R. P., and Desiraju, G. R. (1986) *Acc. Chem. Res.* **19**, 222. (c) Price, S. L., Stone, A. J., Lucas, J., Rowland, R. S., and Thornley, A. E. (1994) *J. Am. Chem. Soc.* **116**, 4910.

16. Boese, R., Boese, A. D., Blaeser, D., Antipin, Y. M., Ellern, A., and Seppert, K. (1997) *Angew. Chem. Int. Ed. Engl.* **36**, 1489.
17. Navon, O., Bernstein, J., and Khodorkovsky, V. *Angew. Chem. Int. Ed. Engl.* 1997, **36**, 601.
18. Brock, C. P., and Duncan, L. L. (1994) *Chem. Mater.* **6**, 1307. Braga, D., Grepioni, F., Sabatino, P., and Desiraju. G. R. (1994) *Organometallics* **13**, 3532. Walther, D., Klettke, T., Schmidt, A., Görls, H., and Imhof, W. (1996) *Organometallics* *15*, 2314. Braga, D., Grepioni, Walther, D., Klettke, T., Schmidt, A., Görls, H., and Imhof, W. *Organometallics* in press.
19. Blankensteyn, A. J. A. R., and Kroon, J. (1986) *Acta Cryst.* **C42**, 291.
20. Braga, D., Cojazzi, G., Grepioni, F., Scully, N., and Draper, S.M. *Organometallics*, in press.
21. Braga, D., Scaccianoce, L., Grepioni, F., and Draper, S.M. (1996) *Organometallics* **15**, 4675.
22. Dunitz, J., and Taylor, R. (1997) *Chem. Eur. J.* **3**, 89.
23. (a) Daniel, M. F., Leadbetter, A. J., Meads, R. E., and Parker, W. G. (1978) *J. Chem. Soc. Faraday Trans.* **74**, 456, (b) Sato, K., Katada, M., Sano, H., and Konno, M. (1984) *Bull. Chem. Soc. Jpn.* **57**, 2361
24. Grepioni, F., Braga, D., Dyson, P., Johnson, B. F. G., Sanderson, F.M., Calhorda, M. J., and Veiros, L. (1995) *Organometallics* **14**, 121.
25. Allen, F. H., Davies, J. E., Galloy, J. J., Johnson, O., Kennard, O., Macrae, C. F., and Watson, D. G. (1991) *J. Chem. Inf. Comp. Sci.* **31**, 204.
26. Braga, D., Grepioni, F., and Tedesco, E. *Organometallics* submitted.

WEAK HYDROGEN BONDS IN SUPRAMOLECULAR SYNTHESIS

ASHWINI NANGIA and GAUTAM R.DESIRAJU

School of Chemistry, University of Hyderabad,

Hyderabad 500 046, India

ABSTRACT The importance of C-H···O and C-H···π hydrogen bonds in the assembly of supramolecular architectures is highlighted in three recent examples. The predictable construction of target supramolecular synthon mediated through multi-point recognition between complementary donor and acceptor groups shows the robustness of the synthon. Retrosynthetic analysis of a trigonal non-centrosymmetric network structure leads to the engineering of an octupolar non-linear optical crystal. A comparison of two nearly identical crystal structures reveals subtle differences which indicate that C-H···π interactions exhibit the characteristics of weak hydrogen bonds.

1. Introduction

The weak hydrogen bond, typified by the C-H···O interaction is of relevance in the assembly of supramolecular crystalline architectures [1,2] At the outset, it must be emphasised that the term 'hydrogen bond' is applied to this interaction because it has many of the attributes and characteristics of the strong hydrogen bond, such as dependence on the acidity and basicity of donor and acceptor groups and cooperative properties that enhance interaction strength [3,4]. The fall-off of interaction

G. Tsoucaris (ed.), Current Challenges on Large Supramolecular Assemblies, 193–208.

effectiveness is much more gradual for weak hydrogen bonds than it is for van der Waals interactions and in general, these represent distinct situations that need not be confused with each other. Even as the shrinking grey area between the weak hydrogen bond and the van der Waals interaction is particularly contentious, it appears that there is a structural continuum between these limiting possibilities. Here we discuss some examples where the weak hydrogen bond is used in crystal engineering.

2. Multipoint Recognition

Crystal engineering with conventional, or strong, O-H···O and N-H···O hydrogen bonds may appear sufficiently reliable, but it is in reality incomplete if weak intermolecular interactions are not considered [5]. The advantage of using the weaker interactions is that the variety of compounds that can be used to construct supramolecular motifs and patterns is significantly increased. Among such interactions, C-H···O hydrogen bonds have attracted considerable attention. Database studies have resulted in a better understanding of the nature of these interactions and this in turn has led to their utilisation in the construction of supramolecular synthons [6] through a consideration of complementarity of functional groups. To summarise, C-H···O hydrogen bonds can be used quite efficiently in the design of supramolecular synthons and crystal structures. However, owing to the inherent weakness of these interactions, multi-point rather than single-point recognition is the preferred strategy.

We have aimed at the design of the three-point C-H···O recognition synthon **1** and at an analysis of its robustness. Synthon **1** is a mimic of the well-known synthon **2** that is constructed purely with strong hydrogen bonds. Towards these ends, the crystal structures of complexes **5a-g** have been examined [7]. Among these structures, synthon **1** was found to occur in complexes **5a-e**.

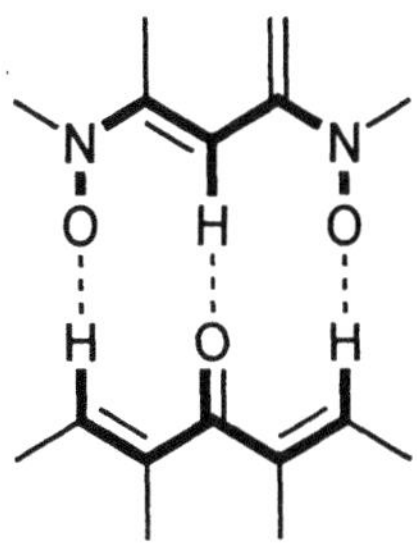

1

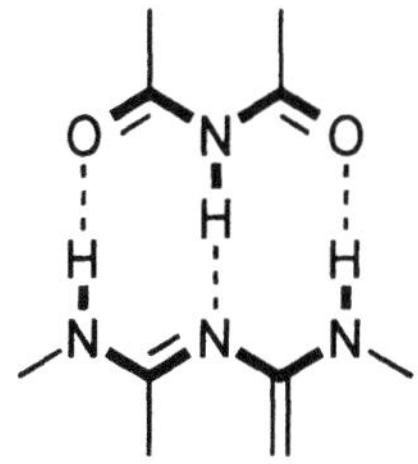

2

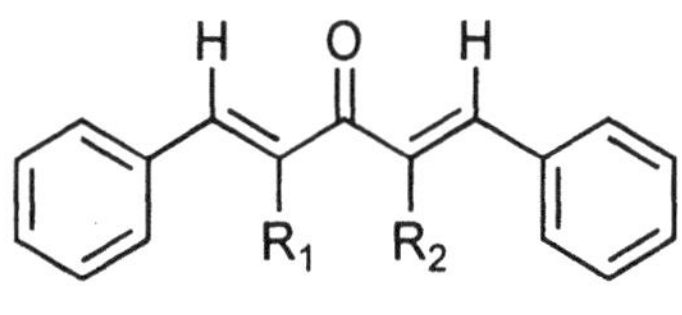

3

a: $R_1 = R_2 = H$
b: $R_1 = R_2 = (CH_2)_2$
c: $R_1 = R_2 = (CH_2)_3$
d: $R_1 = R_2 = CH{=}CH$

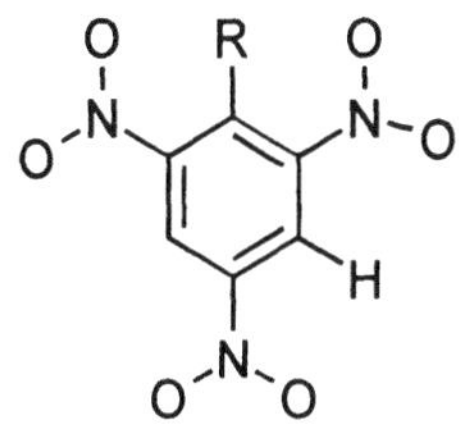

4

a: R = H
b: R = Cl
c: R = OH

5

a: 3a:4a (1:2)	**d: 3b:4b (1:2)**
b: 3b:4a (1:2)	**e: 3d:4b (1:2)**
c: 3c:4a (1:2)	**f: 3b:4c (1:1)**
	g: 3a:4c (1:1)

The choice of trinitrobenzene as one of the components to form synthon **1** is due to the high acidity of its aromatic C-H and the ability of the nitro groups to form C-H···O hydrogen bonds. The choice of dibenzylidene ketones as the second component is made by matching complementary groups. Compound **4a** forms a 2:1 crystalline complex **5a** with dibenzylideneacetone **3a**.

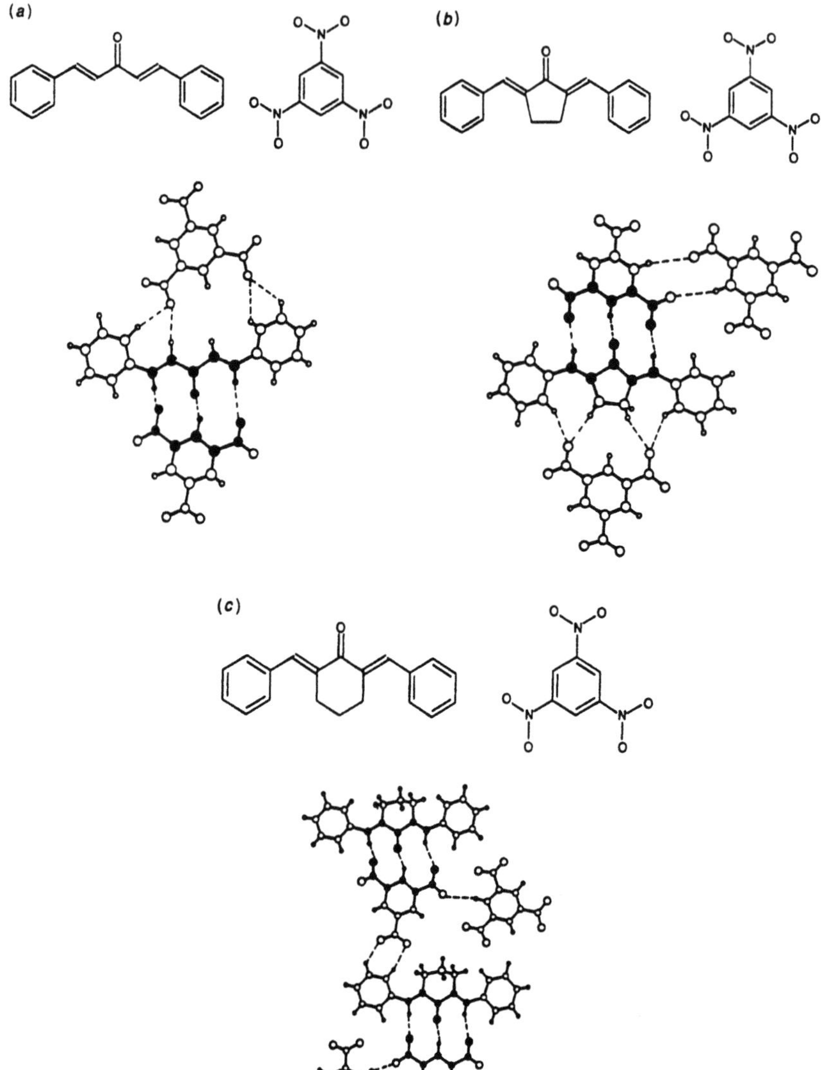

Figure 1. Structures of complexes **5a**, **5b** and **5c** showing synthon **1** (highlighted) and other C-H···O hydrogen bond arrangements. Notice that while synthon**1** is present in all three structures, the other C-H···O pattern varies.

The expected synthon **1** is observed in **5a** and the overall crystal structure is fortified by additional C-H···O hydrogen bonds on the other side of the carbonyl group (Fig. 1a). In order to avoid this alternate C-H···O hydrogen bonded pattern and also to assess the robustness of synthon **1**, we considered complexes of **4a** with **3b** and **3c** instead of **3a**. Compound **4a** also forms 2:1 crystalline complexes **5b** and **5c** with **3b** and **3c**, respectively. While synthon **1** is found in both these complexes, the alternate C-H···O hydrogen bonded pattern which involves the hydrocarbon side of the of the molecule is found only in complex **5b** (Fig. 1b,c). The presence of the alternate motif on the hydrocarbon side in complexes of **3a** and **3b** is attributed to the higher acidity of vinylic sp^2 and allylic sp^3 C-H groups, respectively, which enables them to make the three-centre motif consisting of vinylic/allylic C-H, aromatic C-H and nitro oxygen atoms. The observation that the occupancies of the two orientations of **3b** in complex **5b** are exactly 0.5 and 0.5 indicates that synthon **1** and the alternate C-H···O hydrogen bonded pattern are of comparable significance.

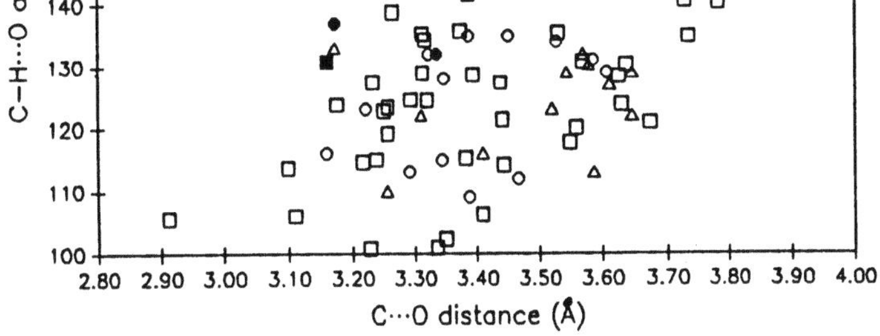

Figure 2. Scatterplot of C-H···O interactions in complexes **5a** (O), **5b** (Δ) and **5c** (□). Filled symbols represent C-H···O hydrogen bonds that contribute to synthon **1**. Note that all these filled symbols are in the region corresponding to strong hydrogen bonding.

To analyse the C-H···O hydrogen bonds in complexes **5a-c**, plots of the C···O distances *versus* the C-H···O angles were obtained (Fig. 2). Circles, triangles and squares represent the hydrogen bonds in complexes **5a**, **5b** and **5c** respectively. The C-H···O hydrogen bonds that are part of synthon **1** are shown as filled symbols. From this plot, one notes that the C-H···O hydrogen bonds within synthon **1** are shorter and more linear, in other words they constitute the the essence of these crystal structures. This observation strengthens the idea that the significance of a C-H···O hydrogen bond increases if it is part of a multpoint synthon.

In order to assess the robustness of synthon **1** in the presence of other functional groups like Cl and OH, complexes **5d-g** were prepared. Compound **4b** forms 2:1 crystalline complexes **5d** and **5e** with ketones **3b** and **3d**, respectively. The presence of synthon **1** in complexes **5d** and **5e** indicates that the Cl group does not interfere in the formation of **1**. On the other hand, the C-H···O hydrogen bonded synthon **1** is absent in the crystal structures of complexes **5f** and **5g** that contain the strong hydrogn bonding OH group in the molecular component **4c**. Thus, there is a limit to synthon robustness and an appreciation of these limits is advantageous in the design of newer crystal structures within the same family.

3. Towards Functionalised Solids

Crystal engineering, that is solid state supramolecular synthesis, is increasingly turning towards the design of functionalised materials [8,9]. Non-Linear Optical (NLO) properties with such applications as Second Harmonic Generation (SHG) or Electrooptic (EO) modulation have attracted much interest [10]. Recently, a new class of SHG-active substances, namely, octupolar molecules have been proposed and shown to display significant NLO behaviour at the molecular level. The engineering problem in this case at the crystalline or supramolecular level amounts to steering the structure of an appropriately substituted trigonal molecule **6** towards the trigonal, non-centrosymmetric network structure 7 characterised by specific interactions between unlike groups in the molecular skeleton rather than towards the hexagonal, centrosymmetric network **8** characterised by close approaches between like groups. Of course, the majority of trigonal molecules **6** adopt neither structure 7 nor **8** but some trivial close-packed structure [11] and this renders the engineering problem only more challenging.

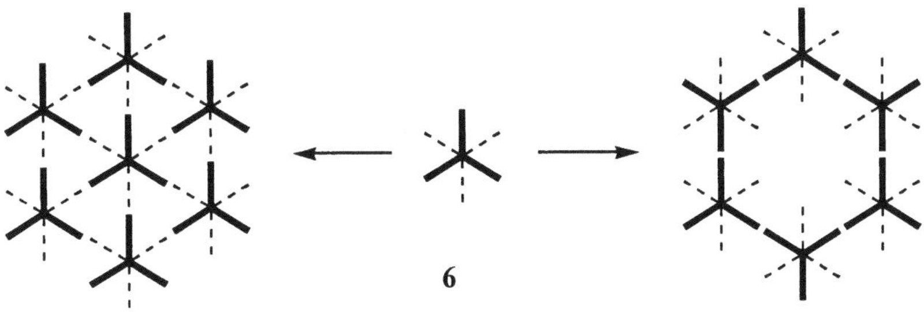

7 **6** **8**

In the crystal structure of the 1:1 complex of trimethyl isocyanurate, **9** and 1,3,5-trinitrobenzene **4a** the respective components are found in distinct layers, each possessing the trigonal symmetry of 7. Specifically, the molecules of **9** are linked via C-H···O mediated supramolecular synthon **12** that may be readily identified in network 7.

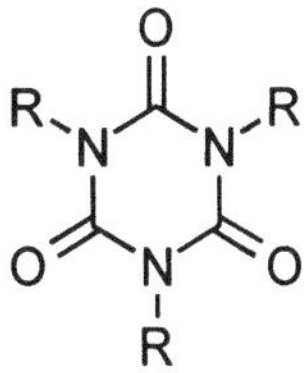

9: R = CH_3
10: R = CH_2CH_3
11: R = $CH_2C_6H_5$

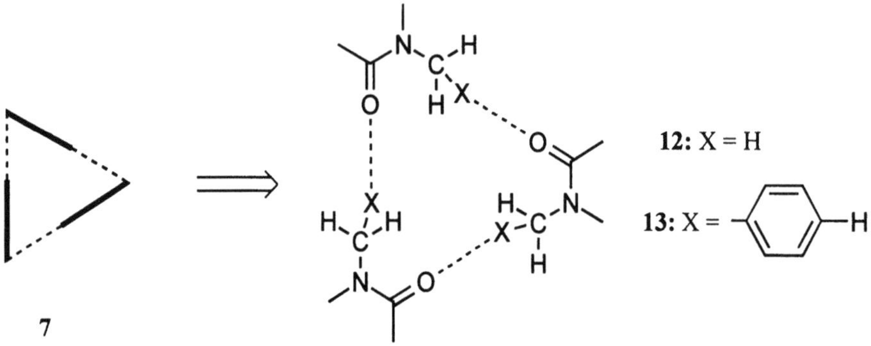

7

In general, single component crystals are preferred to molecular complexes for NLO applications because of issues connected with material purification, crystal growth and optical characterisation in both solution and the solid state. Therefore we further examined the symmetrical isocyanurates **9-11**, all of which have alternating C-H···O donors and acceptors in the molecular structure. Such an alternation is an essential prerequisite for the formation of network **7**. In the crystal structures of trimethyl- and triethyl isocyanurates **9** and **10**, the trigonal network **7** is not found. The very feeble acidity of the sp^3 C-H groups probably leads to non-adoption of the desired structure, based as it is on C-H···O hydrogen bonds [12].

Keeping such carbon acidity considerations in mind, tribenzyl isocyanurate, **11** was next examined [13]. The crystal structure of **11** (Fig. 3) shows that the desired non-centrosymmetric structure has been obtained. The structure is characterised by synthon **13** which is the *tris*-phenylogous extension of synthon **12**. The molecule of **11** is far from planar. With respect to the central heterocyclic ring, two benzyl groups are oriented in one direction whilst the third points in the other direction leading to an overall 'chair' shape. The layer structure in **11** is corrugated as shown in Fig. 4, with two distinct intralayer C-H···O hydrogen bonds. Molecules translated along [001] are connected by interaction **j** in Fig. 3 (d 2.97 Å, θ 145°) to form chains. Adjacent chains are n-glide related and connected by interaction **k** (d 2.51 Å, θ 142°). The chains are additionally interconnected by herringbone interactions (dotted lines in Fig. 3).

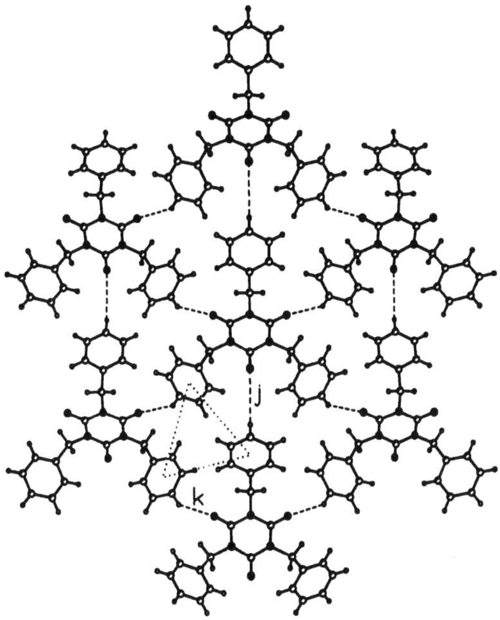

Figure 3. A view of the crystal structure of tribenzyl isocyanurate **11** down [010] to show the trigonal network structure. C-H···O hydogen bonds (**j**,**k**) and herringbone interactions (dotted lines) are indicated. Notice the occurrence of supramolecular synthon **13** and the overall non-centrosymmetry of the network.

Figure 4. Space-filling view of the network structure shown in Fig. 3. Notice the corrugation in the layer.

The non-centrosymmetric nature of crystalline **11** was confirmed by a visible SHG powder signal at 1.064 μm of the order of 0.1 x urea. The nonlinear characterisation of **11** at the molecular level was performed by Harmonic Light Scattering (HLS) measurements in solution. The nonlinear $\beta = \sqrt{\langle\beta^2\rangle}$ HLS coefficient measured at 1.064 μm in $CHCl_3$ is $(10 \pm 2) \times 10^{-30}$ esu, which is comparable to that of the classical dipolar pNA molecule under the same measurement conditions. The octupolar structure of **11** was confirmed by a HLS depolarisation measurement and the depolarisation ratio $D = 0.65$ is close to the theoretical value of 2/3 for purely octupolar molecular structures.

In the above retrosynthetic strategy, a trigonal, non-centrosymmetric target network is identified and analysed through supramolecular synthons to the molecular components. Further, optimisation of the donor acidity in the symmetrical isocyanurates through the phenyl C-H group leads to the formation of the desired octupolar nonlinear crystal mediated through C-H···O hydrogen bonds.

4. The C-H···π Hydrogen Bond

A number of recent studies have commented on the C-H···π interaction [14]. The term 'hydrogen bond' has been used explicitly in this context. In a recent study, we have noted that 2,3,7,8-tetraphenyl-1,9,10-anthyridine, **14** forms 1:1 solvates with toluene and chlorobenzene [15]. The crystal structures of these solvates, **15** and **16**, were determined at low temperature and to good precision. While the two structures are very similar, with nearly the same unit cell dimensions and space group symmetry, there are small differences with respect to the positioning of the solvent molecules. Though small, these differences are significant and show that the C-H···π interactions found in these structures have the characteristics of weak hydrogen bonds.

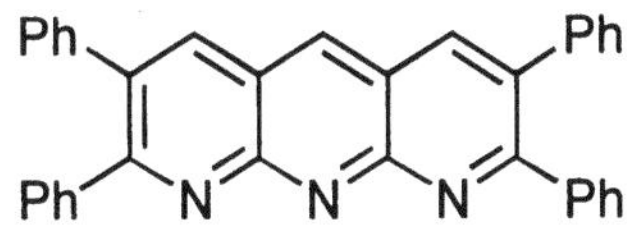

14

15: 14.toluene
16: 14.chlorobenzene

Figures 5 and 6 show the crystal structures of solvates **15** and **16**. The positioning of the heterocyclic molecule in the two unit cells is identical and so are the gross packing features. The ordered toluene and chlorobenzene molecules in **15** and **16** are situated in nearly identical pockets. However, the Me- and Cl- substituents point in almost opposite directions in the two cases. This is unusual. When chloro-methyl exchange is obtained, the two substituents behave in an isosteric fashion with only shape and size factors controlling the packing. The switching around of the Me- and Cl- groups in **15** and **16** is indicative of the fact that electronic factors are involved in stabilisation of the solvent molecule in one or both cases.

A very short C-H···π interaction of 2.54(2) Å (all H-atom positions normalised in this and subsequently discussed contacts) in structure **15** (**a** in Fig. 5) is noteworthy. This interaction is clearly of the H···ring centroid type ($\theta = 151^{\circ}$), the distances from the H-atom to the six ring atoms of the toluene molecule lying in the range 2.73-3.05 Å. Accompanying contact **a**, is a much longer contact of 3.01 Å ($\theta = 126^{\circ}$, **b** in Fig. 5) so that the toluene molecule may be be said to be positioned at unequal distances from adjacent glide-related molecules of anthyridine **14**. In contrast, the two corresponding contacts, **g** (2.61(2) Å, $\theta = 143^{\circ}$) and **h** (2.62(2) Å, $\theta = 147^{\circ}$) in structure **3** (Fig. 6) are significantly longer than contact **a**. However, unlike the toluene molecule in **15**, the chlorobenzene molecule in **16** is disposed at nearly equal distances from the adjacent molecules of **14**.

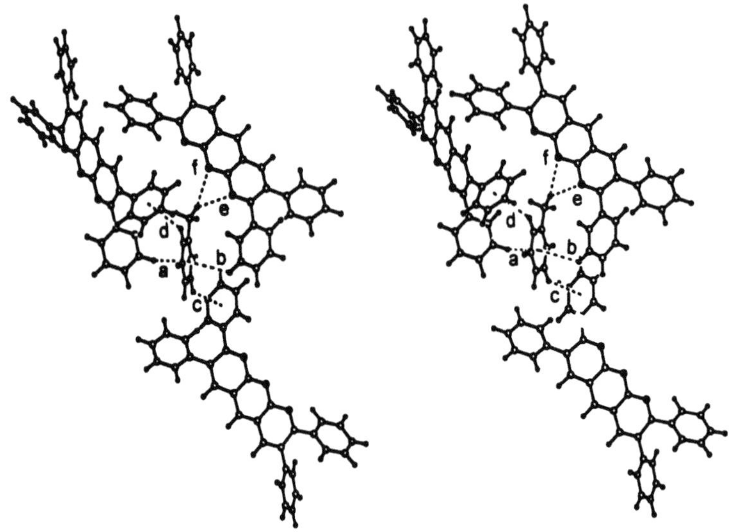

Figure 5. Stereoview of the crystal structure of solvate**15** approximately down [010] showing the binding of the toluene molecule. The anthyridine molecules are inversion- and glide related. Interactions**a-f** are indicated. Notice the cooperative scheme of C-H···π hydrogen bonds and the positioning of the methyl H-atom between the two heterocyclic N-atoms. Interaction **a** is unusually short.

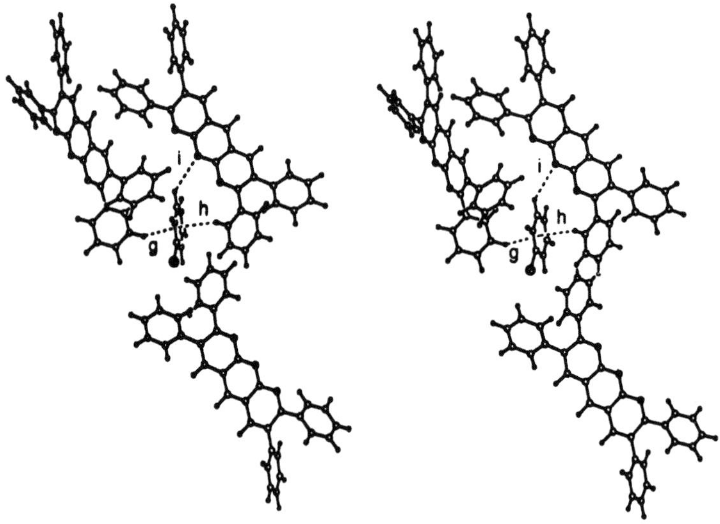

Figure 6. Stereoview of the crystal structure of solvate**16** showing the binding of the chlorobenzene molecule. Interactions **g-i** are indicated. Contrast this with Fig. 5. Notice that the Cl-atom is not involved in any specific intermolecular contact.

Comparing solvates **15** and **16**, the shortness of contact **a** with respect to **g** and **h** also correlates with the electron-rich character of the aromatic ring in toluene *vis-a-vis* chlorobenzene. Additionally, given that there would be little steric problem were the toluene molecule to be placed symmetrically between the two approaching C-H groups, one may state with confidence that the C-H···π interaction **a** is stabilised by the electron rich aromatic ring in toluene. Our observations here adduce evidence that the C-H···π interaction has structural properties of weak hydrogen bonds, rather similar to O-H···π and N-H···π interactions.

Hydrogen bonds formed by carbon acids and π-bases are soft and this in turn can lead to cooperative patterns of interactions. The weak hydrogen bond **a** is able to activate the C-H groups in the toluene molecule at least to the extent that the long contacts **c**-**f** are possible. Interactions **c** (2.90 Å, $\theta = 140^\circ$) and **d** (3.01 Å, $\theta = 140^\circ$) formed by the aromatic C-H groups are of the C-H···π type while interactions **e** (2.76 Å, $\theta = 146^\circ$) and **f** (2.83 Å, $\theta = 161^\circ$) formed by the methyl C-H groups are of the C-H···N type. In contrast, there is just one extra interaction of the C-H···N type in the unactivated structure **16** (**i**; 2.87 Å, $\theta = 139^\circ$). All these interactions may make only small contributions to the crystal binding energy but they help in defining the structures of the solvates. While it has always been recognised that C-H···π interactions in highly activated systems such as terminal acetylenes are of the hydrogen bond type, this study shows that even moderately activated systems such as tolyl rings participate in interactions that may be considered to be weak hydrogen bonds formed by soft acids and soft bases.

5. Conclusions

The examples discussed in this paper show that weak hydrogen bonds have many of the properties of strong hydrogen bonds. In the context of supramolecular synthesis, they may be employed in cases where interference from stronger interactions is minimal. In such an event, the weak hydrogen bond may be manipulated appropriately for crystal structure design. The C-H···O hydrogen bond is the archetype of the weak hydrogen bond. Interactions such as the C-H···π hydrogen bond are less well-documented. With our increasing knowledge of these lesser-known interactions, they may be used in future in crystal engineering.

6. Acknowledgements

We acknowledge our collaborations with Professor R.Boese (Essen), Dr.H.L.Carrell (Philadelphia) and Professor J.Zyss (Paris). The experimental work described here was carried out by K.Biradha, S.Brasselet, C.J.Carrell, N.N.L.Madhavi and V.R.Thalladi. Financial assistance from the Department of Science and Technology, Government of India is acknowledged.

7. References

1. Desiraju, G.R. (1997) Designer crystals: intermolecular interactions, network structures and supramolecular synthons, *Chem. Commun.* 1475-1482.
2. Philp, D. and Stoddart, J.F. (1996) Self-assembly in natural and unnatural systems, *Angew. Chem. Int. Ed. Engl.* **35** 1155-1196.
3. Desiraju, G.R. (1996) The C-H···O hydrogen bond: structural implications and supramolecular design, *Acc. Chem. Res.* **29** 441-449.
4. Steiner, T. (1997) Unrolling the hydrogen bond properties of C-H··O interactions, *Chem. Commun.* 727-734.
5. Desiraju. G.R. (ed.) (1996) *Perspectives in Supramolecular Chemistry: The Crystal as a Supramolecular Entity*, Vol. 2, Wiley, Chichester.
6. Desiraju, G.R. (1995) Supramolecular synthons in crystal engineering - A new organic synthesis, *Angew. Chem. Int. Ed. Engl.* **34** 2311-2327.
7. Biradha, K, Nangia, A., Desiraju, G.R., Carrell, C.J. and Carrell H.L. (1997) C-H··O hydrogen bonded multi-point recognition in molecular assemblies of dibenzylidine ketones and 1,3,5-trinitrobenzenes, *J. Mater. Chem.* **7** 1111-1122 (Special issue on Molecular Assemblies and Nanochemistry).
8. Gavezzotti A. (1996) Organic crystals: engineering and design, *Curr. Opin. Solid State Mater. Sci.* **1** 501-505.
9. Desiraju, G.R. (1997) Crystal engineering: solid state supramolecular synthesis, *Curr. Opin. Solid State Mater. Sci.* **2** 451-454
10. Ledoux, I. and Zyss, J. (1997) in *Novel Optical Materials and Applications,* ed. Khao, I.C., Simoni, F. and Umeton, C., Wiley, Chichester, 1-48.
11. Britton, D. (1997) 1,3,5-Trichlorotricyanobenzene complexes with mesitylene (1/1) and benzene (1/3), *J. Chem. Crystallogr.* **27** 405-412.
12. Thalladi, V.R., Katz, A.K., Carrell, H.L., Nangia, A. and Desiraju, G.R. (1997) Trimethyl isocyanurate and triethyl isocyanurate, *Acta. Crystallogr.* **C53** 0000, in the press.
13. Thalladi, V.R., Brasselet, S., Bläser, D., Boese, R., Zyss, J., Nangia, A. and Desiraju, G.R. (1997) Engineering of an octupolar nonlinear crystal: tribenzyl isocyanurate, *Chem. Commun.* 1841-1842.

14. Nishio, M., Umezawa, Y., Hirota, M., and Takeuchi, Y. (1995) The CH/π interaction: significance in molecular recognition, *Tetrahedron* **51** 8665-8701.

15. Madhavi, N.N.L., Katz, A.K., Carrell, H.L., Nangia, A. and Desiraju, G.R. (1997) Evidence for the characterisation of the C-H···π interaction as a weak hydrogen bond: toluene and chlorobenzene solvates of 2,3,7,8-tetraphenyl-1,9,10-anthyridine, *Chem. Commun.* 1953-1954.

MOLECULAR TECTONICS : AN APPROACH TO ORGANIC NETWORKS

MIR WAIS HOSSEINI
Université Louis Pasteur
Laboratoire de Chimie de Coordination Organique
Institut Le Bel,
F-67000 Strasbourg, France

1. Introduction

Molecular solids are defined by the chemical nature of their molecular components and by their interactions with respect to each other in the crystalline phase. Whereas molecular chemistry deals with the design and the synthesis of the individual units, their assembly is governed by concepts of supramolecular chemistry [1].

The construction of large size molecules (10^{-6}-10^{-3} m scale) or hyper molecules with predicted and programmed structure may hardly be envisaged through stepwise classical synthesis using covalent linkages. However, the preparation of such higher-order materials may be attained through iterative process based on self-assembly [1] of individual tectons (from Greek TEKTON, builder) [2]. This particular area of research, called molecular tectonics [3], is currently under active investigation [4,5].

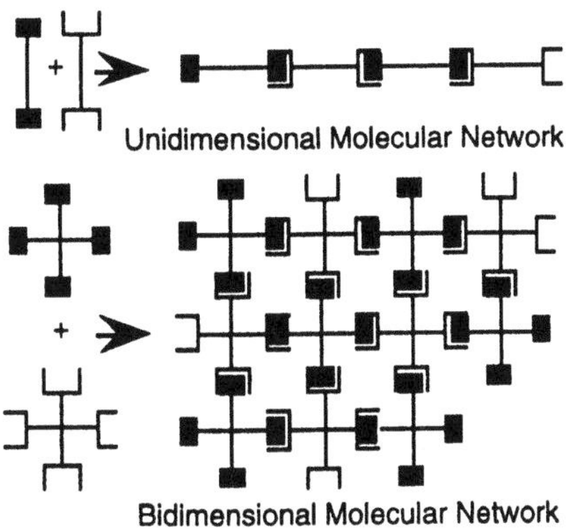

Fig. 1 Schematic representation of mono- and di-dimensional networks based on assembling of a self-complementary tectons

One of the most important aspect of molecular tectonics appears to be the fact that the assembling process is based on molecular recognition processes operating at the level of the complementary tectons. In other terms, the operational concepts in molecular tectonics are those defined for molecular recognition [6], as well as topological features allowing iteration.

The manufacture of molecuar networks may be achieved by an iterative assembling process of either a self-complementary tecton (one component system composed of a unique motif), or of several complementary modules (multi component system). In the later case the most viable approach would engage two complementary tectons (Figure 1).

In order to achieve an iterative assembling process, the building blocks or

G. Tsoucaris (ed.), *Current Challenges on Large Supramolecular Assemblies*, 209–219.

tectons must fulfil both structural and energy criteria. In particular, the complementary tectons must recognise each other (molecular recognition) and furthermore should allow the repetition of the recognition pattern (iteration).

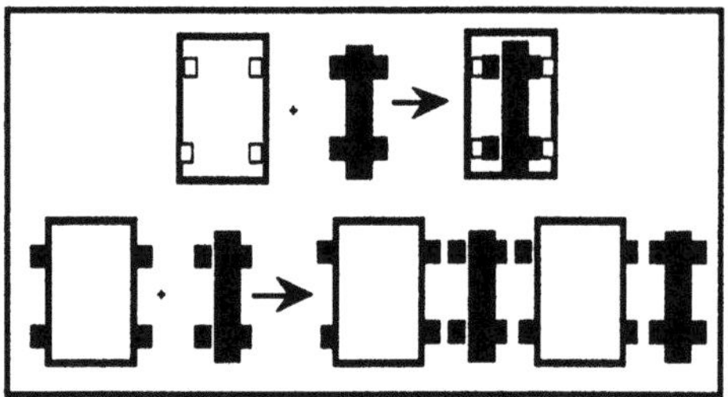

Fig. 2 Schematic representation of the formation of an endo-molecular complex (top) and of a one-dimensional network resulting from an iterative assembly of one exo- and one endo-tectons

These two requirements lead to molecular modules possessing connecting points or interaction sites located in a divergent fashion (Figure 2). Whereas an endo molecular receptor (endo-receptor), by conception, is designed to form only discrete molecular complexes with a selected exo-substrate, an exo-receptor may form molecular networks with appropriate exo-substrates.

2. Design of molecular networks based on weak interactions

The synthesis of solids based on iterative assembling of individual complementary tectons still remains a challenge. A strict control of the self-assembly of molecular modules in the solid state should lead to structurally strictly controlled molecular networks (one-, two- and three-dimensional solids).

Although in principle any type of weak interactions may be used to self-assemble complementary tectons, so far, we have focused our attention on hydrogen bonding and on van der Waals interactions. Only results obtained in the latter case will be presented here.

2.1. Linear molecular arrays using van der Waals interactions

The chemistry of inclusion complexes [7], *i.e.* the inclusion of a substrate within the cavity of a receptor molecule is an established area.

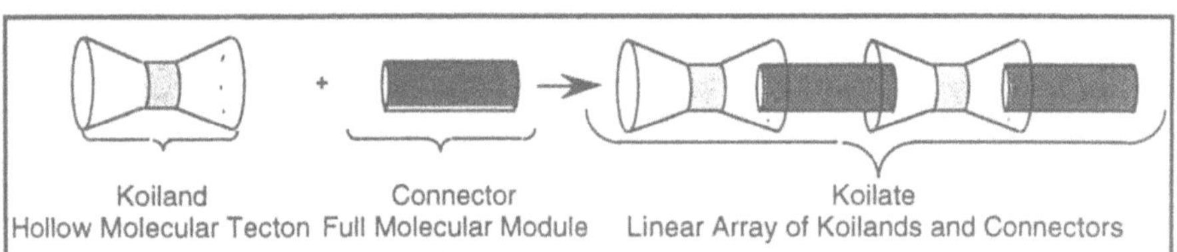

Fig. 3 Schematic representation of a linear array (koilate) resulting from interpenetration of a convex molecule (connector) into the cavity of a concave receptor (koiland)

One may extend the concept of inclusion in solution to the construction of new networks in the solid state. For example, a rigid and compact direceptor possessing two divergent cavities (linear *koiland* , from Greek koilos : hollow), and a linear connector, possessing two extremities capable each to be included within the cavities of the direceptor, may be assembled in the solid state, by an iterative process through non-covalent van der Waals interactions leading thus to a linear molecular array which may be called linear *koilate* [8,9] (Figure 3).

For the construction of linear koilates, the design of a koiland possessing two divergent cavities with an angle of 180° between them is crucial. For this purpose, calix[4]arene seems to be an interesting unit since, it possess both hydrophobic and hydrophilic sites[10]. The *p-tert*-butylcalix[4]arene **1** (Figure 4), in its favoured cone conformation below 20°C in chloroform [11-13] and in the solid state [14,15] offers a genuine hydrophobic pocket surrounded by four *p-tert*-butyl aryl moieties. This basket type cavity was proved to accommodate a variety of neutral substrates such as chloroform [13], toluene [14], anisole [15], xylene [16] and benzene [16]. Complexation of Cs^+ through metal-π-aromatic interactions has also been reported for **1** [17] in its cone conformation.

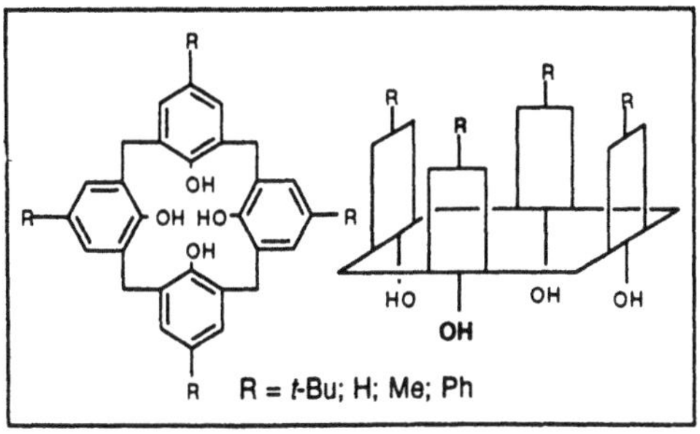

Fig. 4 calix[4]arenes (left) and a schematic representation of their cone conformation (right)

The fusion of two *p-tert*-butylcalix[4]arenes **1** in its cone conformation by two silicon atoms leads to a hollow brick (koiland) **2** [8,9] (Figure5a,6). Other similar compounds using titanium [18], niobium [19] and aluminium [20] have been reported.

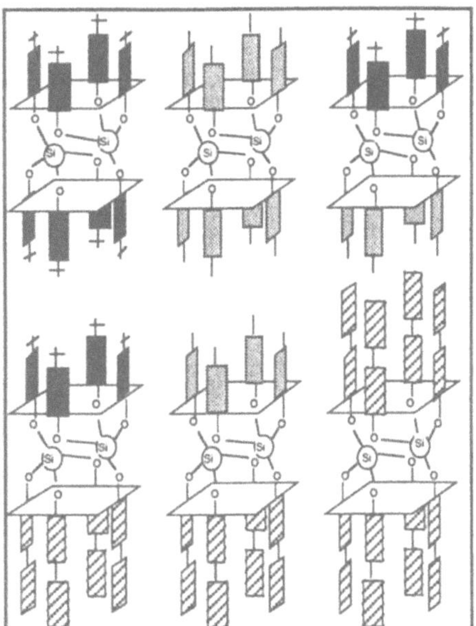

Fig. 5a Schematic representation of koilands resulting from the fusion of two identical or different calix[4]arenes

As stated above, calixarenes seem particularly well suited for the design of the hollow modules since one may control both the entrance and the depth of their preorganised cavity by the nature of the substituant R at the *para* position, *i.e.* H, CH_3, $C(CH_3)_3$, Ph (Figure 4). The synthesis of different koilands (Figure 5) based on the double fusion of two *p*-methylcalix[4]arenes, and two *p*-phenylcalix[4]arenes with two silicon atoms was achieved (22). In order to control both the dimensionality and the directionality of koilates, one may use non-symmetrical koilands. Such hollow modules may be obtained by two different strategies. The differentiation of the two cavities may be achieved by using two different cavities (Figure 5). Mixed compounds based on the double fusion of two different calix units have been prepared [23].

Another alternative to the formation of non-symmetrical koilates may be the use of two different metals. Thus, the heterobinuclear dimer **4** containing both a Si and a Ti atoms was prepared [9]. The non-symmetric nature of **4** is based on the induced difference between the two calix units by their coordination to two different metals with different elctronegativity. Indeed, one of the calix is triply coordinated to a Si atom whereas the other unit is triply coordinated to a Ti atom (Figure 5). In order to prepare **4** by a directed strategy, compound **3** was prepared. For **3**, due to possible free rotation around the Si-O bond, different rotamers may be expected. However, NMR studies revealed that, due to inter calix H-bonding between the hydroxy groups, the face to face (syn) rotamer was the preferred one (Figure 5). Treatment of the isolated **3** with $TiCl_4$ produced, in high yield, the heterobinuclear compound **4**.

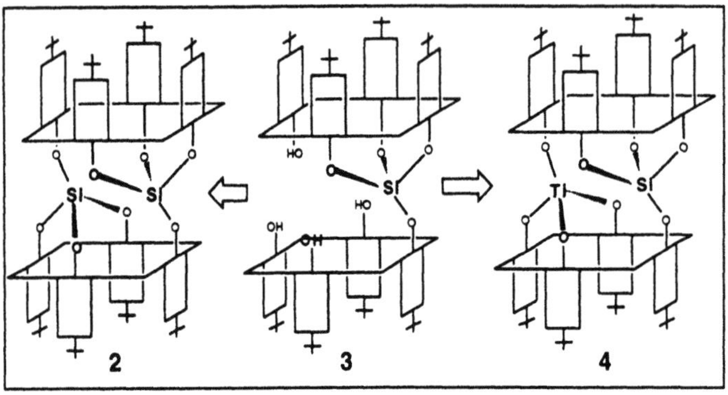

Fig. 5 Schematic representation of koilands based on: the fusion of two p-tert-butylcalix[4]arene 1 by two Si atoms 2, the fusion of two tert-butylcalix[4]arene by one Si atom 3, or by both Si and Ti atoms 4.

In the solid state, the structure of **2** was investigated by X-ray analysis. Compound **2** as shown in Figure 6 was indeed a centro-symmetric silicon aryloxy dimer consisting of 2 Si and 2 calixarenes. Both calixarene units were in cone conformation, thus, presenting two divergent cavities. The coordination geometry around both Si atoms was tetrahedral. Each of these two tetrahedrons was composed of 3 oxygen atoms belonging to one of the two calixarenes and the fourth one to the other calix unit.

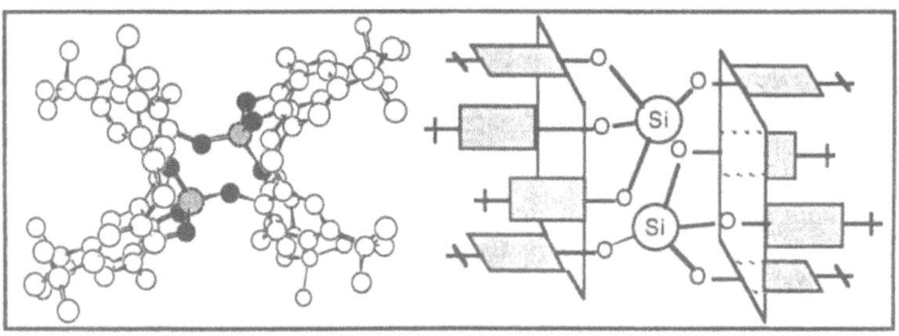

Fig. 6 Schematic representation of 2 composed of two calixarenes fused by two silicon atoms yielding a double basket type receptor (right), and its crystal structure (left).

As shown in the figure 6, in the solid state, the hollow ditopic receptor **2** indeed possesses two divergent cavities. In order to ascertain the possibility of double inclusion, binuclear complexes composed of one compound **2** and two molecules of anisol or *p*-xylene were prepared and their structures investigated by X-ray analysis (Figure 7). The crystal study showed that in both cases, each cavity accommodates, in a inclusive fashion, an anisol or a *p*-xylene molecule. Both substrates penetrate the cavity of **2** through their CH_3 group. It is worth noting that anisol, due to its reduced dimension and its bent shape, is not a well suited connector for bridging the hollow modules **2**, and thus leading to the linear koilate.

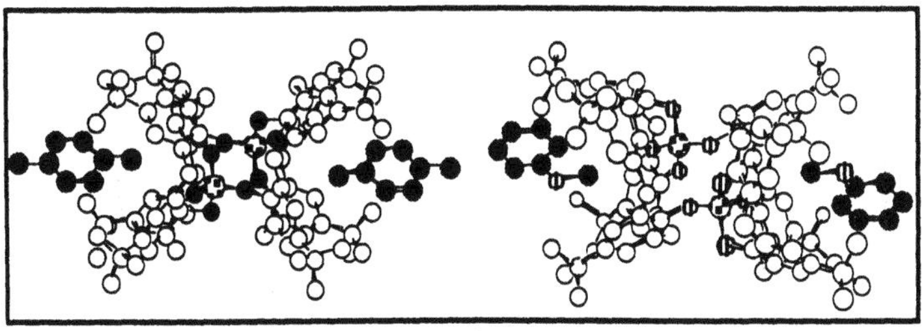

Fig. 7 *X-ray structure of the binuclear inclusion complexes formed between 2 and p-xylene (left) and anisol (right)*

However, for *p*-xylene, although due to the discrete nature of the complex, *i.e.* the absence of interconnection between the discrete bi-nuclear complexes, the linear koilate was not formed, nevertheless, the aligned positioning, with respect to each other, of binuclear complexes in the solid appeared to be rather interesting and encouraging.

Our aim was the preparation of linear arrays resulting from an iterative inclusion of both extremities of connectors within the cavities of koilands (Figure 8).

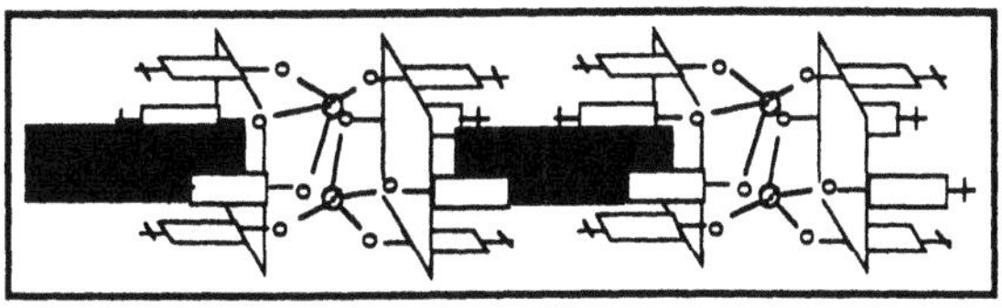

Fig. 8 *Schematic representation of the formation of koilates resulting from an iterative assembling process between hollow molecular modules and connectors capable of penetrating the cavity of the ditopic receptors*

Hexadiyne, a rod type molecule, possessing both the required linear geometry and a sufficient distance between its two terminal CH_3 groups ($d_{C\text{-}C}$ = 6.65 Å) leads to the formation of the desired linear koilate (Figure 9) [21]. The latter crystallises from a $CHCl_3$/MeOH solution of koiland **2** and hexadiyne in large excess. The X-ray analysis revealed that in the crystal, in addition to **2** and hexadiyne present in 1/1 ratio, two $CHCl_3$ molecules were also present.

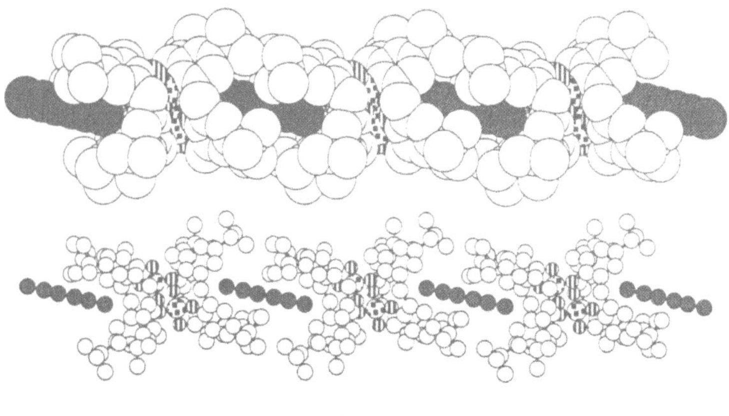

Figure 9: *X-ray structure of the koilate formed between 2 and hexadiyne. space filling (top) and balls & sticks (bottom) representations.*

The lattice is composed of linear arrays of koilates formed between koilands **2** and hexadiyne as connectors. Each connector bridges two consecutive koilands by penetrating their cavities through its terminal CH_3 groups. The methyl groups of hexadiyne are deeply inserted into the cavity of the koiland. Indeed, the shortest distance between the CH_3 group of the connector and the aromatic carbon atom of the koiland is 3.64 Å. It is worth noting that the hexadiyne molecules are extremely well encapsulated by two consecutive koilands **2** forming a sandwich, the shortest distance between the CH_3 groups of the *tert*-butyl moieties of **2** being 3.51 Å.

2.2. Functionalisation of koilands

Calix[4]arene, in addition to the features stated above, presents further advantages by being readily modified at the *para* position by a large variety of substituents bearing functionalities. In particular, one may introduce at the *para* position functional groups such as allyl moieties [24] (Figure 10). The design of the new koiland **6** (Figure 11) possessing functional groups at both faces of the molecule was based on the double fusion of the parent compound **5** (Figure 10) with two silicon atoms [25]. The doubly fused compound **16** was obtained after column chromatography in 53 % yield upon treatment of compound **4** in dry THF by NaH followed by addition of $SiCl_4$.

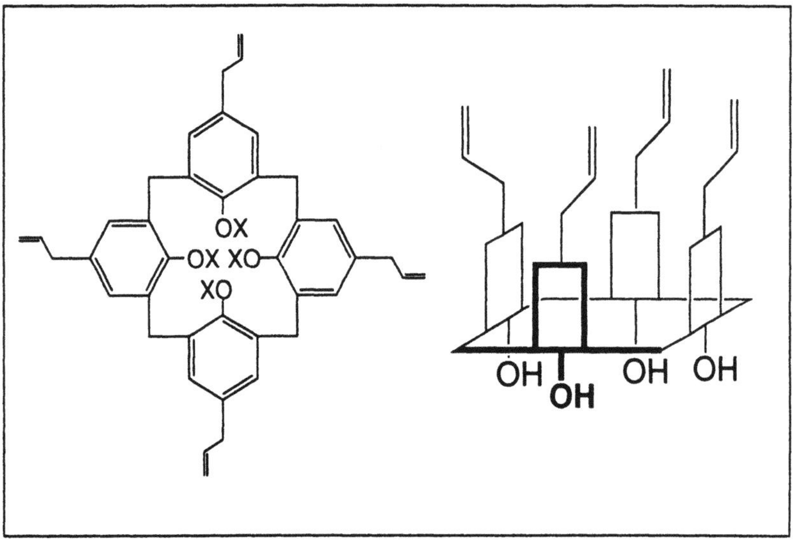

***Figure 10**: p-allylcalix[4]arene 5 (left) and a schematic representation of its cone conformation (right).*

Since it has been previously observed that *p*-xylene, a rigid, compact and ditopic molecule, forms in the solid mononuclear and binuclear inclusion complexes with *p-tert*-butylcalix[4]arene and with the koiland **2** respectively, the formation of 1-D network using compound **6** as the koiland and *p*-xylene as the connector was investigated (Figure 11).

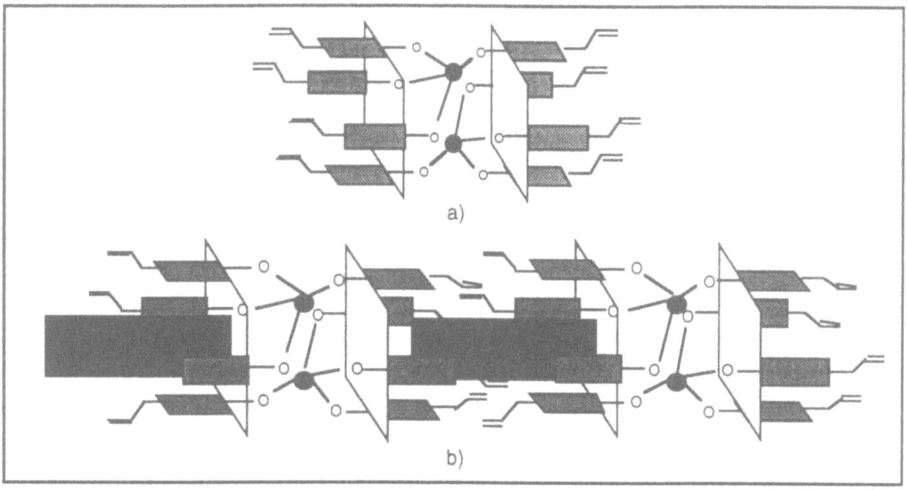

***Figure 11**: Schematic representations of the koiland 6 and of its α-network in the presence of a connector molecule.*

Suitable crystals were obtained from a mixture of compound **6** and *p*-xylene in excess and *i*PrOH at room temperature. The X-ray analysis showed the following features (Figure 12). The crystals were composed exclusively of **6** and *p*-xylene disposed in a alternate fashion. Both the connector and the koiland were centrosymmetric, the latter offering two divergent cavities. The coordination geometry around the silicon atoms was tetrahedral. As predicted, a linear koilate was formed through the interconnection of two cavities belonging to consecutive koiland **6** with the *p*-xylene molecule. In the crystal one could identify an assembling core which could be defined as the inclusion of a CH_3 group of the connector within a cavity of **6**. The shortest C-C distance of 3.63 Å between the CH_3 group of the connector and one of the carbon atoms belonging to one of the phenolic group at the bottom of the cavity indicated a high degree of inclusion. The formation of the α-network resulted from a single translation of the assembling core.

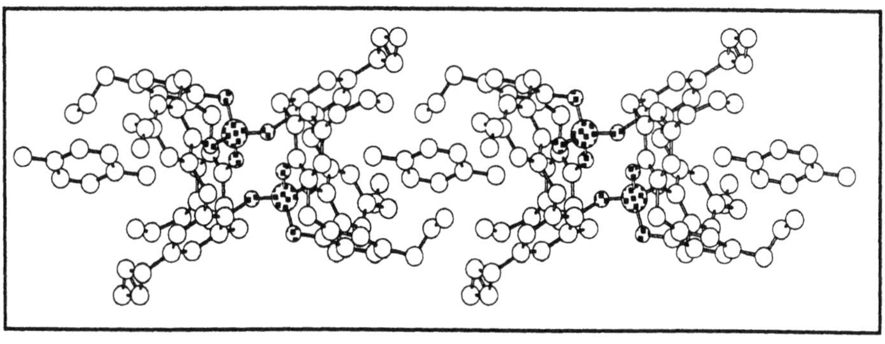

***Figure 12**: A portion of the crystal structure of the α-network formed between the koiland **6** and p-xylene.*

Although the formation of the α-network was predicted, the X-ray analysis revealed another unexpected feature. Due to the large number of rotations available for the allyl moieties, one can not in fact predict their conformations in the crystal lattice. Interestingly, when looking at lateral packing of consecutive linear koilates, it was observed that among the eight allyl groups present in each koiland **6**, two of them were oriented centrosymmetrically in a peculiar manner. Indeed, two alkene moieties belonging to two consecutive koilands were located perfectly parallel to each other with a alkene-to-alkene distance of 4.52 Å between them. Although the observed distance seems slightly to long, the proximity of the two double bonds suggests the possibility to interconnect covalently the koilates by performing a [2+2] cycloaddition reaction in the solid sate.

It is interesting to notice that although, using compound **2** and hexadiyne, the first fully characterised linear koilate was obtained in the solid state (see above), one could nevertheless argue that, due to the presence of $CHCl_3$ molecules in the lattice, the formation of the koilate was governed by solvent molecules. The formation of the koilate with compound **6** and *p*-xylene precludes this objection since in that case the linear koilate was formed in the solid without the presence of any solvent molecules in the lattice.

2.3. Formation of koilates by self-inclusion

As demonstrated above, koilands may lead in the solid state either to discrete molecular complexes in the presence of molecules acting as stoppers (Figure 13a) or to linear molecular networks in the presence of appropriate connectors (Figure 13b). Another alternative route to the formation of linear koilate may be based on the use of self-complementary koilands possessing simultaneously two divergent cavities and connecting moieties (Figure 13c). Compound **6** fulfil these requirements. Indeed, the latter possesses two preorganised cavities resulting from the fusion of two calix units in cone conformation, as well as allyl groups at both faces of the molecule which may act as connectors.

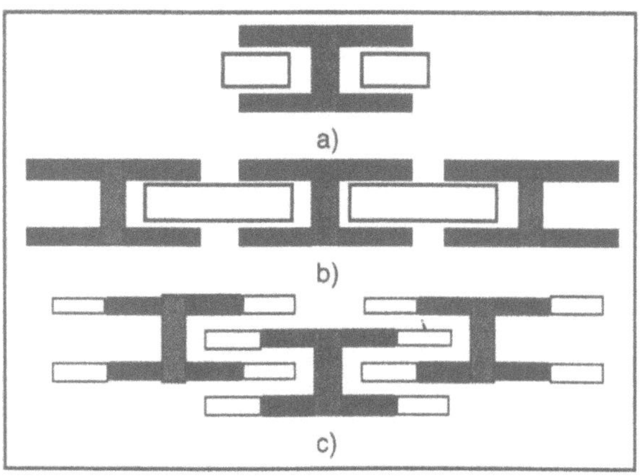

***Figure 13** : Schematic representation of discrete binuclear inclusion complexes formed between koilands and stopper molecules (a) or of infinite linear molecular networks formed by koilands in the presence (b) and absence of (c) connector molecules.*

A study based on X-ray diffraction [26] on single crystals obtained under various conditions for the compound **6** revealed indeed the formation of an α network in the solid state. Upon slow liquid-liquid diffusion at room temperature of EtOH, into a $CHCl_3$ solution of compound **6**, suitable large, colourless, air stable at rt., mono-crystals of the same morphology (rhombic) were obtained. The X-ray analysis revealed that the crystal was exclusively composed of compound **6**. The latter, offering two divergent cavities, was centrosymmetric. As in the case of other koilands presented above, the coordination geometry around the silicon atoms was tetrahedral. In the crystalline phase, the self-complementary koiland **6** formed a linear koilate (Figure 14) through the interconnection of two cavities belonging to consecutive koilands **6** by van der Waals interactions. The assembling core leading by a single translation to the α-network, could be identified as the inclusion of one of the four allyl groups located at one of the two faces of the molecule **6** by one of the two cavities of the consecutive self-complementary compound **6**. The shortest C-C distance between the terminal CH_2 of the allyl group acting as connector and carbon atoms of one of the aromatic moieties of the cavity belonging to the next unit varied from 3.63 to 3.92 Å, indicating a rather high degree of inclusion. Interestingly, when crystals of compound **6** were grown from other solvent systems such as toluene/EtOH, benzene/EtOH, $CHCl_3$/*i*PrOH and toluene/*i*PrOH, in all cases, crystals of the same morphology and stability were obtained and their investigation by X-ray (lattice parameters) confirmed that they were the same.

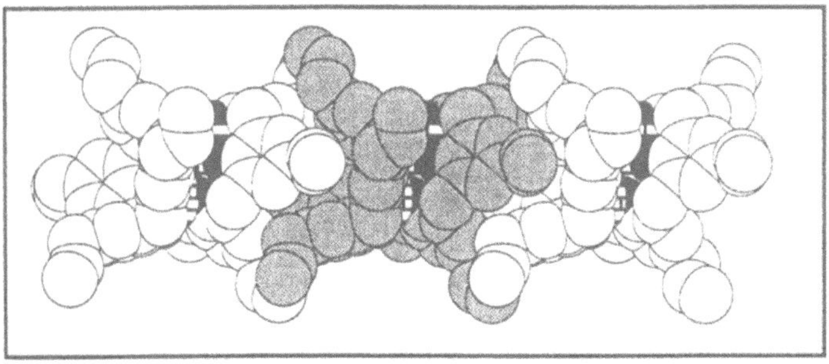

Figure 14: *A portion of the crystal structure of the α-network formed by self-inclusion of the self-complementary koiland* ***6***.

However, when crystals of **6** were grown by the same technique from a CH_2Cl_2/EtOH mixture, two types of crystals, one with a rod-type morphology and the other with rhombic shape, were observed. Whereas the crystals with rhombic morphology were stable at rt. and proven by X-ray analysis to be of the same type as those mentioned above, the rod-shape crystals were unstable outside the solution. Their analysis by X-ray diffraction revealed the presence of **6** and 3 molecules of CH_2Cl_2. In the lattice two types of discrete inclusion complexes were present and therefore no network based on inclusion processes could be identified (Figure 15). The two types of units were a binuclear (Figure 15a) and a tetranuclear (Figure 15b) CH_2Cl_2 inclusion complexes. For the binuclear complex, both cavities of compound **6** were occupied by a CH_2Cl_2 molecule. For the tetranuclear specie, among the four CH_2Cl_2 present, as in the case of the binuclear complex, two of them occupied the two preorganised cavities, whereas the remaining solvent molecules were located within the two "pockets" resulting from the positioning of the allyl groups. The difference between the two discrete complex observed in the lattice may be assigned to two different conformations adopted by the allyl groups.

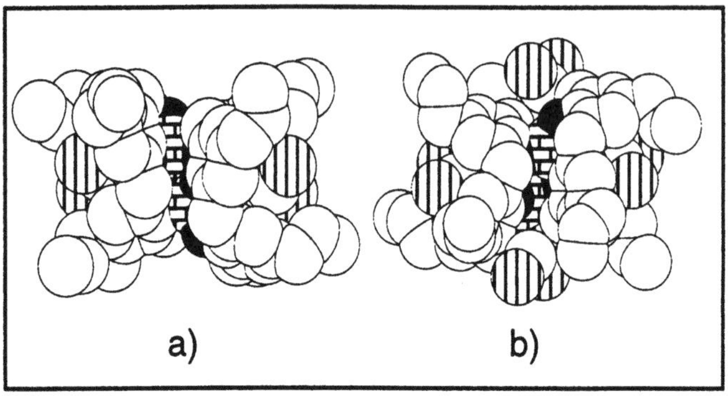

Figure 15: *X-ray structures of the discrete bi- (a) and tetra-nuclear (b) inclusion complexes formed by the linear koiland **6** and CH_2Cl_2. For the sake of clarity, the hydrogen atoms are not presented.*

It is worth noting that in the case of the self-complementary compound **6**, depending on composition of the solution, all three situations (Figure 13), *i.e.* discrete inclusion complexes and 1-D networks formed either by interconnection of consecutive units by connectors or by self-interconnection, were obtained.

3. Conclusion

In conclusion, it has been demonstrated that hollow molecular modules may be assembled in the solid state into 1-D molecular arrays either by self inclusion or by interconnection using connector molecules. Thus, inclusion processes based on van der Waals interactions may be used as design principle for the generation of molecular networks in the solid state. Although at the present state of affairs, this approach remains essentially structural, by using molecular building blocks possessing specific magnetic, photonic or electric properties, one may extend this strategy to the preparation of functional solids. Work along these lines is under progress.

References

1. Lehn, J.-M. **(1995)** *Supramolecular Chemistry, Concepts and Perspectives*, VCH, Weinheim; Desiraju, G. R. **(1995)***Angew. Chem. Int. Ed. Engl.*, *34*, 2311; see also Tsoucaris, G. *et al.*, Ed.,**(1996)** *Crystallography of Supramolecular Compounds*, Kluwer; Lindsey, J. S. **(1991)** *New. J. Chem.*, *15*, 153; Lawrence, D. S., Jiang, T., Levett, M. **(1995)** *Chem. Rev.*, *95*, 2229; Stoddart, J. F., Philip, D. **(1996)** *Angew. Chem. Int. Ed. Engl.*, *35*, 1155.
2. Simard, M., Su, D., Wuest, J.D. **(1991)** *J. Amer. Chem. Soc.*, *113*, 4696.
3. Mann, S. **(1993)** *Nature*, *365*, 499.
4. Lehn, J.-M. **(1990)** *Angew. Chem. Int. Ed. Engl.*,*29*, 1304.
5. Whitesides, G. M., Mathias, J. P., Seto, C. T. **(1991)** *Science*, *254*, 1312.
6. Lehn, J.-M. **(1973)** *Struc. Bonding.*, *16*, 1; *ibid*, **(1978)** *Pure Appl. Chem.*, *50*, 871.
7. Lehn, J.-M.**(1980)** *Leçon Inaugurale*, Collège de France, Paris.
8. Delaigue, X., Hosseini, M. W., De Cian, A., Fischer, J., Leize, E., Kieffer, S., Van Dorsselaer, A. **(1993)** *Tetrahedron Lett.*,*34*, 3285.
9. Delaigue, X., Hosseini, M. W., Leize, E., Kieffer, S., Van Dorsselaer, A. **(1993)** *Tetrahedron Lett.*, *34*, 7561; Delaigue, X., Hosseini, M. W., Graff, R., Kintzinger, J.-P., Raya, J.**(1994)** *Tetrahedron Lett.*, *35*, 1711.

10. Gutsche,C.D. (**1989**) *"Calixarenes"*, Monographs in Supramolecular Chemistry, Ed. J.F. Stoddart, R.S.C., London; J. Vicens and V. Böhmer, Eds. (**1991**) *Calixarenes. A Versatile Class of Macrocyclic Compounds*, Kluwer Academic Publishers, Dordrecht; Böhmer, V. (**1995**) *Angew. Chem. Int. Ed. Engl.*, *34*, 713.
11. Kämmerer, H., Happel, G., Caesar, F. (**1975**) *Makromol. Chem.*, *176*, 3317.
12. Gutsche, C. D., Bauer, L. J. (**1985**) *J. Amer. Chem. Soc.*, *107*, 6052.
13. Gutsche, C. D., Dhawan, B., No, K. H., Muthukrishnan, R. (**1981**) *J. Amer. Chem. Soc.*, *103* , 3782.
14. Andreetti, G. D., Ungaro, R., Pochini, A. (**1979**) *J. Chem. Soc., Chem. Commun.*, 1005.
15. Ungaro, R., Pochini, A., Andreetti, G. D., Domiano, P. (**1985**) *J. Chem. Soc., Perkin Trans. II*, 197.
16. Coruzzi, M. ; Andreetti, G. D.; Bocchi, V.; Pochini, A., Ungaro, R. (**1982**) *J. Chem. Soc. , Perkin Trans. II*, 1133.
17. Harrowfield, J. M., Ogden, M. I., Richmond, W. R., White, A. H. (**1991**) *J. Chem. Soc., Chem. Commun.*, 1159.
18. Olmstead, M. M., Sigel, G., Hope H.; Xu, X., Power, P. P. (**1985**) *J. Amer. Chem. Soc.*, *107*, 8087.
19. Corazza, F., Floriani, C., Chiesti-Villa, A., Guastini, C. (**1990**) *J. Chem. Soc., Chem. Commun.*, 1083.
20. Atwood, J. L., Bott, S. G., Jones, C., Raston, C. L. (**1992**) *J. Chem. Soc., Chem.Commun.*. 1349.
21. Hajek, F., Graf, E., Hosseini, M. W., Delaigue, X., De Cian, A., Fischer, J.(**1996**) *Tetrahedron Lett.*, *37*, 1401.
22. Hajek, F., Graf, E., Hosseini, M. W. (**1996**) *Tetrahedron Lett.*, *37*, 1409.
23 Hajek, F., Graf, E., Hosseini, M. W. , unpublished.
24. C. D. Gutsche, J. A. Levine (**1982**), *J. Amer. Chem. Soc.*, *104*, 2653.
25. F. Hajek, E. Graf, M. W. Hosseini, A. De Cian, J. Fischer (**1997**), *Angew. Chem. Int. Ed. Engl.*, *36*, 1760.
26. F. Hajek, E. Graf, M. W. Hosseini, A. De Cian, J. Fischer (**1998**), *Mat. Res. Bull.*.,in press.

MD SIMULATIONS ON IONS AND IONOPHORES AT A LIQUID-LIQUID INTERFACE: FROM ADSORPTION TO RECOGNITION

F. BERNY, N. MUZET, R. SCHURHAMMER, L. TROXLER AND G. WIPFF *

UMR 7551, Institut de Chimie,
4, rue B. Pascal,
67 000 Strasbourg (France)

Abstract. Molecular dynamics (MD) simulations provide important insights into the interfacial behaviour of ions and ionophores (**L**) at liquid-liquid interfaces. We report computer experiments on the water / chloroform interface represented explicitely, starting with hypothetical situations where the solutes (ions, **L** free, **L** complexed) are either right at the interface, or immersed in one phase. The demixion of "homogeneous" binary water / chloroform mixtures containing species involved in the assited ion extraction process has also been simulated. The results demonstrate the interfacial activity of free and complexed ionophores, which contrasts with the hydrophilicity of uncomplexed metal cations, or of small anions like Cl^- or NO_3^-. The picrate or ClO_4^- anions, often used in extraction experiments, display a high interfacial activity, as do quaternary ammonium cations or the $As\phi_4^+ B\phi_4^-$ salt. Beyond the interfacial behaviour of monomeric solutes, collective properties are observed when the concentration is increased. Some ionophores like 18-crown-6 display a high concentration at the interface, in equilibrium with molecules diluted in the organic phase. Others like TBP or TPTZ remain completely adsorbed at the interface. The results have implications concerning the mechanism of assisted ion transfer, interfacial electrochemistry, and supramolecular organization induced by the asymmetrical nature of the liquid-liquid interface.

Keywords: liquid-liquid extraction; interface; ionophore; calixarene; counterion effect; solvation; molecular dynamics; immiscible liquids; phase separation, organization.

G. Tsoucaris (ed.), Current Challenges on Large Supramolecular Assemblies, 221–248.

I- INTRODUCTION

Two immiscible liquids are separated by an interfacial region, whose unique properties are stressed to markedly determine the local concentration, conformation, orientation, and reactivity of ionic or neutral solute species [1-4]. In this paper, we focus on the interfacial behaviour of species involved in the assisted ion transfer from an aqueous to an organic phase, a problem which has been the cornerstone of supramolecular chemistry [5, 6]. In extraction experiments, the source phase containing the hydrophilic ions is mixed with an organic phase containing water insoluble extractant molecules ("ionophore" or "ligand", hereafter noted **L**) which complex ions. The nature of the mixed phase is ill-defined, as it may range from a perfectly mixed homogeneous system to droplets or microemulsions involving the two liquid systems. Then, phase separation takes place, due to gravitational forces or centrifugation, and differences in surface tension between the two liquids. In transport experiments, the ions are transported by carrier molecules **L** through an organic liquid phase or an organic membrane, from the aqueous source phase to the aqueous receiving one [7-9]. At some stage, the cation is captured by the ionophore and extracted into the organic phase. The precise mechanism of ion capture and extraction is still unclear but it is stressed that the interfacial region between water and the organic solvent plays a crucial role [10-12]. Computational chemistry may contribute to understand what happens at the liquid-liquid interface in the assisted ion extraction [13].

There have indeed always been tight connections between theoretical and supramolecular chemistry. In the early seventies, quantum mechanical studies defined the "supermolecule approach" to study the elementary non-covalent interactions which hold interacting molecules together. Since, computational chemistry has achieved spectacular progress, which allows to routinely calculate the interactions between small model systems "in the gas phase"[14], as well as to simulate solutes in condensed phases [15]. Today, molecular dynamics simulations ("MD") provide dynamic pictures of systems of several thousands of atoms for timescales ranging from 100 ps to several ns [16]. Monte Carlo simulations can also provide statistical views of supramolecular systems in solution [17]. Examples of simulations at liquid-liquid or liquid-air interfaces are cited in ref. 18-21 (for ionic solutes) and in ref. 22-24 (for neutral solutes). For ionophores we simulated at the water- chloroform interface, see ref. 25-32.

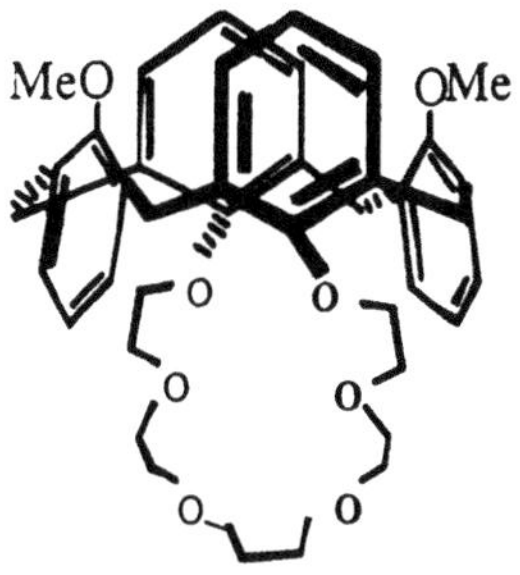

CalixC6

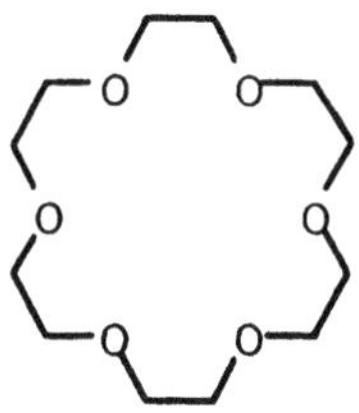

18C6

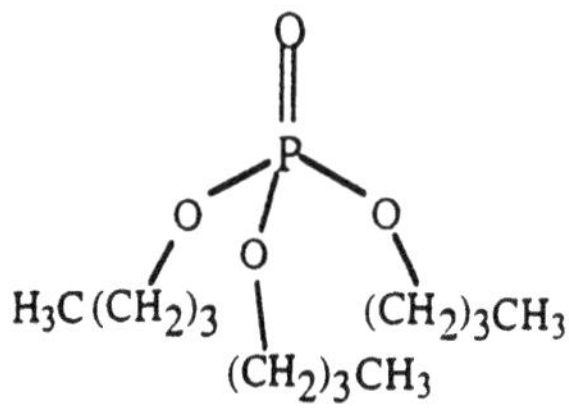

TBP

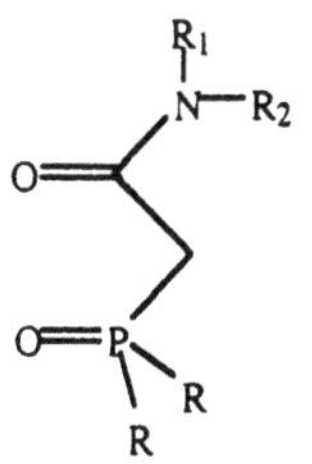

CMPO

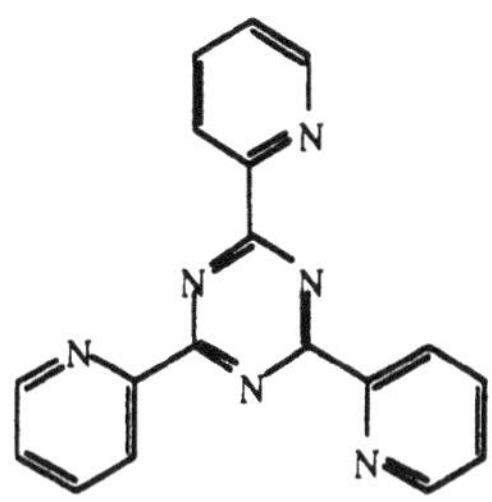

TPTZ

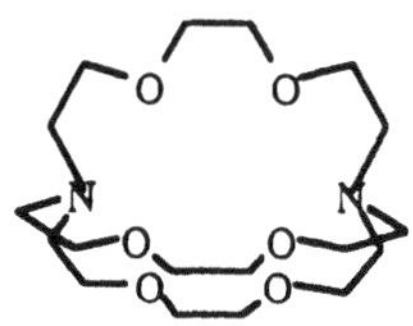

222

Chart 1: Examples of simulated extractant molecules.

This paper reports results of MD simulations performed in our laboratory on structural and energy features of ions and ionophores at the water chloroform interface. We first consider uncomplexed ions or salts, of different hydrophilic characters: (i) the uncomplexed X^- anions (halides and ClO_4^-) in the absence of counterion, and in the presence of K^+ or NMe_4^+ counterions; (ii) the more concentrated solutions of the $(K^+Cl^-)_{20}$ salt and the $(NTMA^+Cl^-)_4$ salt of a lipophilic quaternary ammonium cation; (iii) the $As\phi_4^+B\phi_4^-$ salt, of high interest in the context of electrochemical [33] or thermodynamical [34] measurements. Then, we consider monomers of extractant molecules of macrocyclic type (calixarenes,18-crown-6, the *222* cryptand, naturally occuring ionophores like valinomycin) or acyclic ones like TBP, CMPO or TPTZ (Chart 1), which are shown to display a high interfacial activity. The effect of ligand concentration is illustrated, in relation with the possible formation of "monolayers" or aggregates, involved in the assisted ion transfer from the aqueous to the organic phase. Finally, we report a "computer demixion experiment" where the time evolution of a binary water - chloroform mixture containing a complexed ionophore is followed as a function of time.

As organic liquid, we chose chloroform because it is used experimentally in extraction processes. On the computational side, there are potentials available for chloroform, fitted on the pure liquid properties. It is more polar than solvents like CCl_4 or benzene and simpler to simulate than flexible solvent molecules like alkanes, or their derivatives.

II- METHODS

1- Energy representation of the system. The system is represented via the computerized "ball and stick model", where the potential energy U is calculated by a classical force field model. We use the modified AMBER4.1 software [35] with the following representation of U:

$$U = \Sigma_{bonds} K_r (r-r_{eq})^2 + \Sigma_{angles} K_\theta (\theta-\theta_{eq})^2 + \Sigma_{dihedrals}\Sigma_n V_n(1+\cos n\phi) + \Sigma_{i<j} (q_iq_j/R_{ij} - 2\varepsilon_{ij}(R_{ij}^*/R_{ij})^6 + \varepsilon_{ij}(R_{ij}^*/R_{ij})^{12})$$

The bonds and bond angles are treated as harmonic springs and a torsional term is associated to the dihedral angles. The interactions between atoms separated by at least three bonds are described by pairwise additive 1-6-12 contributions, without explicit account for polarization or many body effects. The van der Waals parameters involving unlike atoms are calculated by the Berthelot-Lorentz combination rules. Most parameters were taken from the AMBER force field [36] and from our previous studies in pure homogeneous solvents [28, 37]. The alkali cation parameters [38], and those of halide anions were fitted to reproduce their relative hydration energies. The charges on ClO_4^-, Pic^-, $B\phi_4^-$ and $As\phi_4^+$ were fitted from ESP potentials. The solvents were represented explicitely, using the TIP3P model for water [39] and the OPLS model for chloroform [40] where CH is represented in the united atom approximation. In order to reduce the computational costs, the non-bonded interactions were calculated within a residue based cutoff of 12Å in most cases.

The MD simulations started with random velocities, and the temperature was controlled by coupling to a thermal bath with a relaxation time of 0.2 ps.

All C-H, O-H, H···H, C-Cl and Cl···Cl "bonds" were constrained with SHAKE, using a time step of 1 fs.

Additional computational tests have been performed, using the "Particle Mesh Ewald" summation method, as implemented in AMBER4.1, or the reaction field correction [41].

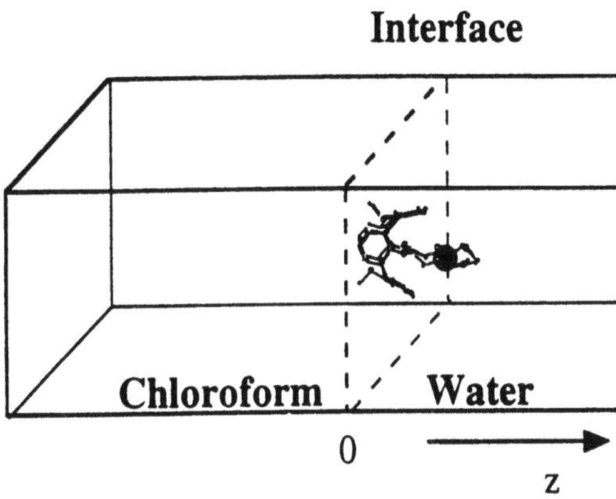

Chart 2: Representation of the simulation box.

2- The solvent system. The solvent system consists of two adjacent cubic boxes of pure water / chloroform liquids (Chart 2 and Table 1). After immersion of the solute near the interface, each system was first energy minimized and equilibrated at 300K under a constant pressure of 1 atm, to ensure that both liquid phase reached densities close to their experimental values (of 1.0 and 1.49, respectively) "in the bulk region". Most simulations use periodic boundary conditions in the three dimensions ("3D"), where the system corresponds to alternating slabs of water and of chloroform. In other simulations, periodicity has been applied along the x and y directions only ("2D"), while the Z-edges of the box correspond to a liquid - air interface, where solvent evaporation is prevented by using a restraining potential (see details in ref 42).

Solute	Box Size ($Å^3$)	Solvent $N_{chl} + N_{wat}$	2D / 3D	(Mol/l)	Time (ps)
F^-	26*26*(26+26)	130 + 568	2D	0.05	150
$NMe_4^+ ClO_4^-$	31*31*(28+29)	202 + 903	2D	0.03	250
$K^+ ClO_4^-$	30*28*(27+27)	175 + 755	2D	0.04	150
$(K^+Cl^-)_{20}$	34*32*(36+36)	279 + 1237	2D	0.43	1000
$(NTMA^+ Cl^-)_4$	38*35*(39+32)	365 + 1340	2D	0.07	250 - 500
$As\phi_4^+ B\phi_4^-$	38*38*(34+28)	364 + 1333	3D	0.02	400
*222*K^+Pic^-	23*33*(45+24)	252 + 808	3D	0.03	850
$(18C6)_{10}$	33*33*(44+27)	251 + 808	3D	0.21	1000
$(18C6.K^+Pic^-)_6$	33*33*(42+33)	325 + 1088	3D	0.12	1000
$(18C6)_6/(K^+Pic^-)_6$	34*34*(42+27)	262 + 923	3D	0.12	1000
222	25*24*(34+27)	140 + 538	3D	0.04	600
$(TPTZ)_{24}$	44*44*(36+26)	457 + 1477	2D	0.33	500
CalixC6.Na^+Pic^-	42*38*(31+30)	363 +1564	3D	0.03	450+720

Table 1: Simulation conditions for the different solutions.
Box size $V_x*V_y*(V_{z\text{-chlor}}+V_{z\text{-wat}})$. Number of chloroform and water molecules. Periodicity (2D / 3D) of the system. Concentration of the solute (mol/l) and simulated time (ps).

3- Analysis of results. The results were analyzed with our DRAW and MDS software from the trajectories which were saved every 0.25 ps. The precise position of the interface was recalculated at each step, and defined as in ref. [26] by the intersection between the density curves of the two liquids.

III - RESULTS

Because of the dynamics nature of the system, it is not easy to depict their evolution during the simulations. Some of them remain close to their initial configuration. For others, major changes may take place, concerning the position of the solute species. In the following, we select a few snapshots concerning the initial and "final" states, with a particular focus on the location of the solutes. As described in ref. 26, the interface between the aqueous and the organic liquids is more or less flat, of about 10 Å wide.

1- Simulation of X^- anions at the interface (no counterion).

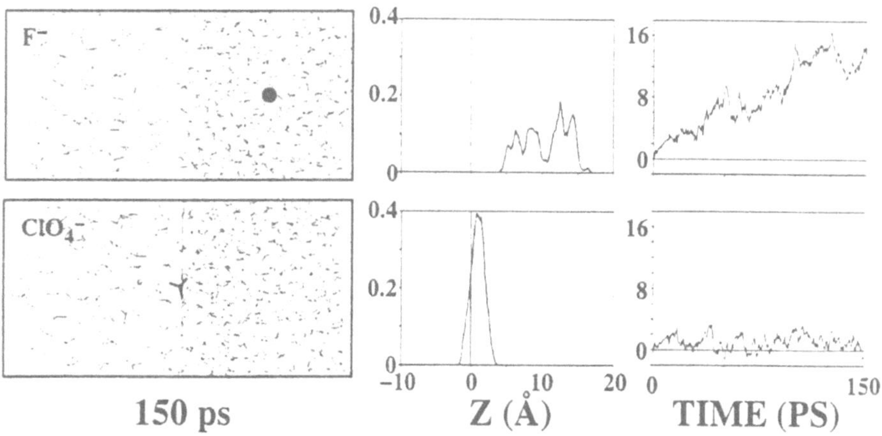

Figure1: The X^- anions at the interface. Snapshots of the solutes after 150 ps of simulation (*left*), density profiles obtained from the last 100 ps (*middle*), and distances X^-...interface as a function of simulation time (*right*).

One century ago, Hofmeister established a classification of anions, in relation with their effect on the solubilization or denaturation of proteins [43]:

$$SO_4^{2-} < F^- < Cl^- < ClO_4^- < SCN^-$$

As it was stressed that this sequence also relates to differences in interfacial behaviour [44], we decided to compare the time evolution of hypothetical situations, where X^- was initially set right at the interface, and simulated consistently for 150 ps. The results of Figure 1 show two extreme cases: the most hydrophilic F^- anion migrates rapidly from the interface to water, in contrast to ClO_4^- which remains adsorbed at the interface during the whole simulation. Similarly, the Picrate ("Pic-") or BPh_4^- anions (see next) remain at the interface, like phospholipidic or fatty acid anions. It is thus noteworthy that tetrahedral ions like ClO_4^- or BPh_4^- are surface active, although they do not possess, like classical surfactants, a "polar head + hydrophobic tail".

2- Simulation of $X^- M^+$ salts at the interface. Role of the cation and of the concentration on the interfacial behaviour.

2.1 The $K^+ ClO_4^-$ and $NMe_4^+ ClO_4^-$ salts. We performed a systematic simulation study of $X^- M^+$ salts, first with K^+, then with the less hydrophilic NMe_4^+ cation as counterion [45], as the accompanying cations may markedly modify the interfacial behaviour of anions, and vice versa. The results depend on the interplay between cation - anion interactions and "solvation forces". In all cases, the ion pair migrates from the interface to water, but different situations were found, ranging from intimate ion pairs to fully separated ion pairs, where the ions display different relationships with respect to the interface. An example is given below with the $K^+ClO_4^-$ and $NMe_4^+ClO_4^-$ salts.

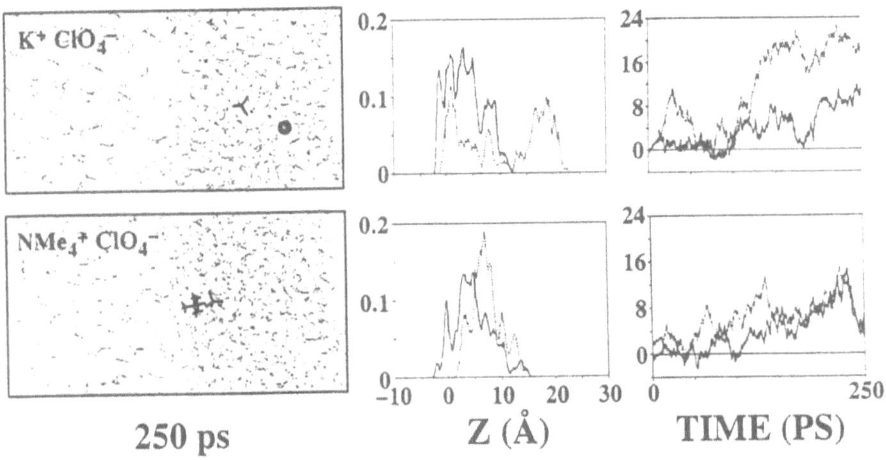

Figure 2:The $K^+ ClO_4^-$ and $NMe_4^+ ClO_4^-$ salts at the interface. Snapshots after 250 ps of simulation (*left*), density profiles obtained from the last 100 ps (*middle*), and distances ion···interface as a function of time (*right*). Solid lines correspond to ClO_4^- and dotted lines to the cations.

In both cases, the ClO_4^- anion leaves the interface. At the beginning of the simulation, K^+ and ClO_4^- oscillate at about 8 Å from each other, ClO_4^- being close to the interface and K^+ deeper in water. Then, the ions separate further, but ClO_4^- remains closer to the interface than K^+. In the case of the $NMe_4^+ClO_4^-$ salt, the two ions always remain in loose contact at a N···Cl distance of about 8 Å, and oscillate at similar short distances from the interface (about 4 Å, after 250 ps). In the case of the NMe_4^+ F^- pair, the anion is more remote from the interface than NMe_4^+. Other examples are given in ref [45]. Thus, all salts simulated behave somewhat differently, as far as the nature of the ion pair, and the relationship with the interface are concerned.

2.2 The $(K^+Cl^-)_{20}$ salt. It is generally considered that small cations are more repulsed than anions of similar size by the interface [46, 47]. This led us to simulate the $(K^+Cl^-)_{20}$ salt for 1 ns, starting with "two grids of alternating K^+ and Cl^- ions", right at the interface (Figure 3). Dissolution takes place readily, driven by migration of the ions to water. The snapshots of Figure 3 indicate that both cations and anions are spread in the water slab, none achieving contacts with the interface. This is confirmed statistically by the density profiles (Figure 3), which peak at about 15 Å for both ions. The density of ions close the interface is almost zero. However, at 3Å, a tiny shoulder is observed for Cl^-, while the density of K^+ ions is strictly zero.

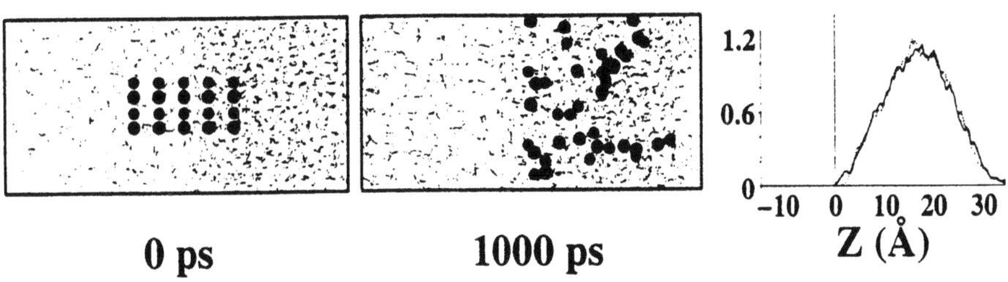

Figure 3: The $(K^+ Cl^-)_{20}$ salt at 0 ps, and after 1 ns of MD. *Right*: density profiles of the K^+ and Cl^- species (anions are represented with solid lines and cations with dotted lines), as a function of the Z coordinate (defined in Chart 2), calculated between 0.15 and 1 ns of simulation.

2.3 The $(NTMA^+Cl^-)_4$ salt at the interface. Salts of quaternary amonium cations have been used in extraction processes, in relation with the high interfacial activity of these cations [48]. On the computational side, there have been a number of simulations of monolayers, micelles, or single quaternary ammonium ions [49-55] involving the water interface.

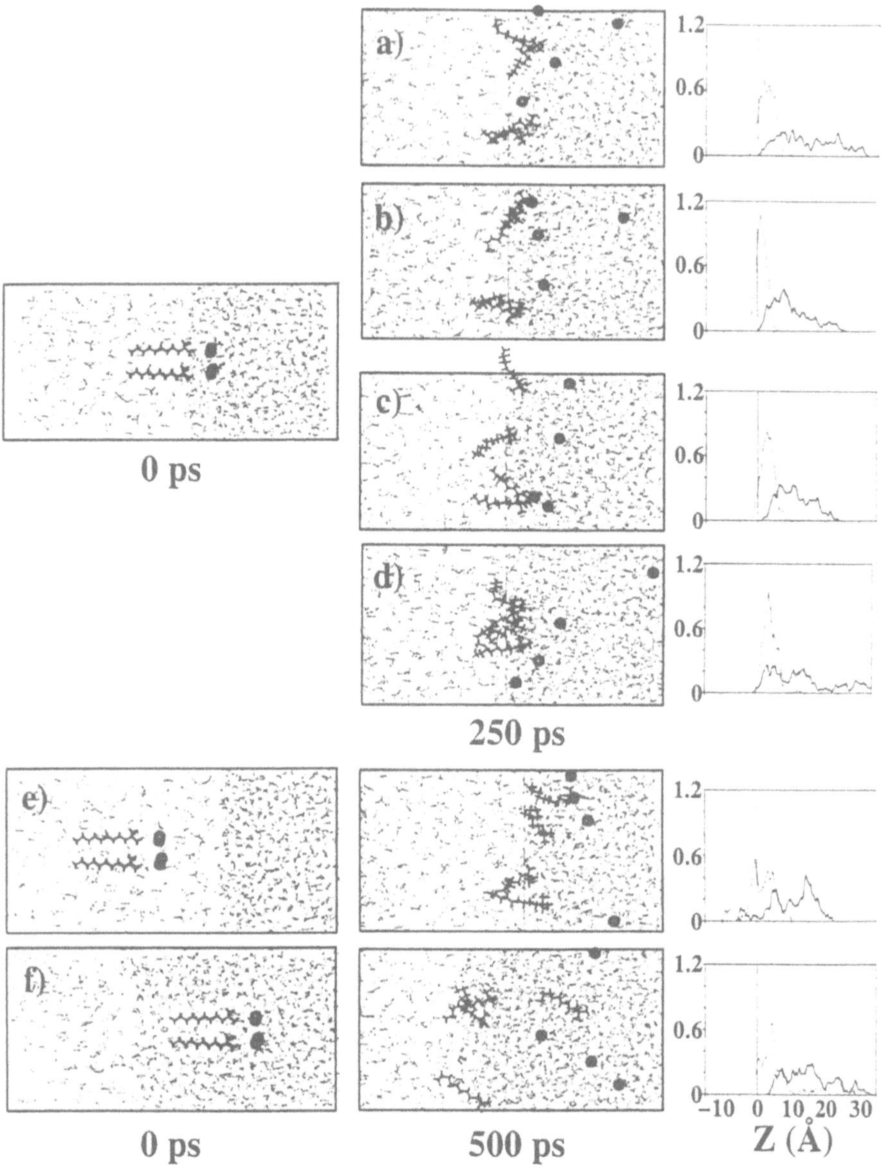

Figure 4: The $(NTMA^+ Cl^-)_4$ salt at the interface. From *left* to *right*: snapshots at 0 ps, at the end of the simulation and ions density profiles (Cl^- anions are represented with solid lines and N^+ atoms with dotted lines) obtained from the last 200 ps. Calculations were performed with a 12.5 Å cutoff (*a*; *e*; *f*), with a 15 Å cutoff (*b*), with the Ewald summation +12.5 Å cutoff (*c*), with the reaction field + 15 Å cutoff (*d*). The solute sits initialy at the interface (*a* to *d*), in chloroform (e) or in water (*f*).

We simulated $NTMA^+$ (Nonyl-trimethylammonium $(H_3C)_3NC_9H_{19}^+$), with atomic charges chosen to mimic a "polar head" of +1 charge, flanked by a neutral hydrophobic tail (C_8H_{17}). Three initial positions of the salt were considered (Figure 4), where the ions are placed respectively at the interface, in the chloroform and in the water phase. It can be seen that the outcome of the three simulations is qualitatively similar: the four Cl^- anions are completely immersed in water, while the $NTMA^+$ cations are at, or close to the interface, which is therefore positively charged. No cation is fully immersed in the organic phase. Instead, we sometimes find one or two cations in water, close to the Cl^- counterions. As expected, the polar heads of $NTMA^+$ interact more with water than the alkyl tails do. Although there are no intimate ion pairs between Cl^- and $NTMA^+$, the ions display some attractions, as found at the end of one simulation, where the Cl^- ions are attracted by the positively charged interface.

The organization of ionic species at the interface results, in principle, from "short range", a well as from "long range" electrostatic interactions. This question was tested on the $(NTMA^+Cl^-)_4$ system via four simulations which started at the interface, using respectively (i) a cutoff of 12.5 Å, (ii) a cutoff of 15 Å, (iii) a 12.5 Å cutoff + Ewald summation, or (iv) a 15Å cutoff + Reaction Field correction. With the four treatments of electrostatics, similar arrangements are found after 250 ps: the $NTMA^+$ cations are right at the interface, next to a "diffuse layer" of Cl^- anions. These layers correspond respectively to a sharp density peak for the N^+ atoms and to a broad peak for the Cl^- anions (Figure 4). It is thus stressed that the position of ionic species at the interface, at the concentration studied here, is not critically determined by the "long range" electrostatic forces.

2.4 The $As\phi_4^+$ $B\phi_4^-$ salt at the interface. It is generaly considered that bulk cations and anions of same charge display similar transfer properties from an aqueous to an organic medium. According to the TATB "extrathermodynamic hypothesis" [34], both $As\phi_4^+$ (tetraphenylarsonium) and $B\phi_4^-$ (tetraphenylborate) ions should have the same free energy of transfer from water to chloroform. This implies that they should display similar interfacial properties. Our simulations of the $As\phi_4^+$ $B\phi_4^-$ salt at the water / chloroform interface show that this is not the case (Figure 5).

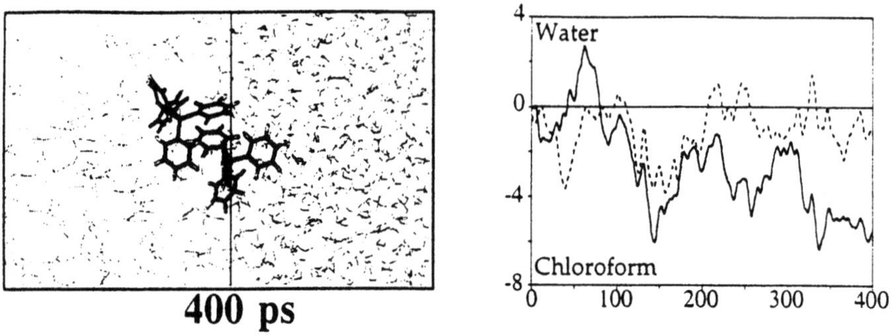

Figure 5: The $As\phi_4^+$ $B\phi_4^-$ salt simulated at the interface. Snapshot after 400 ps. Distances between the As^+ (*full line*) and B^- (*dotted line*) atoms from the interface as a function of time (ps).

The $B\phi_4^-$ anion remains anchored at the interface, somewhat on the chloroform side, during the whole simulation. This contrasts with $As\phi_4^+$, which moves from the interface to chloroform, remaining in loose contact with $B\phi_4^-$. Thus, the computations reveal that $As\phi_4^+$ is more hydrophobic than $B\phi_4^-$. When simulated in *pure* water, the cation interacts less with water than the anion (-85 and -121±8 kcal/mol, respectively). They also display different interaction energies with the *pure* chloroform liquid (-65 and -87 ±6 kcal/mol, respectively). The discrimination between the two ions by solvent is thus larger in water than in chloroform. These differences in solvation properties are important for the interpretation of electrochemical [56] and of thermodynamical data. They also suggest that in more concentrated systems of small X^- and M^+ ions of similar size, the $As\phi_4^+$ X^- bilayer differs from a $B\phi_4^-$ M^+ bilayer.

3- Simulation of single ionophores, free or complexed, at the interface.

Our first simulations involving interfaces concerned ionophores **L** free or complexed by cations [26]. As these molecules are not soluble in water, but are soluble in organic liquids or membranes, we naively thought that, when placed initially at the border between the two liquids, they should diffuse to chloroform. This turned out not to be the case, whatever the conformation of **L**, or the nature of its substituents. For instance, two conformers of the 222 cryptand, of zero dipole moment, remained at the interface, on the choroform side [26]. For other systems, we investigated methodological aspects concerning the temperature of the liquids, the choice of (N,V,T) / (N,P,T) thermodynamic ensembles to perform the simulations, the representation of chloroform molecules [28]. In all cases, the extractant molecule **L** adsorbed on the chloroform side of the interface, with the polar moities solvated by a few water molecules ("water fingers"). Conversely we checked that, in the same simulation conditions, the hydrophic *n*-butane molecule migrates from the interface into chloroform, while hydrophilic ions like Cl^- are expelled from the interface into water.

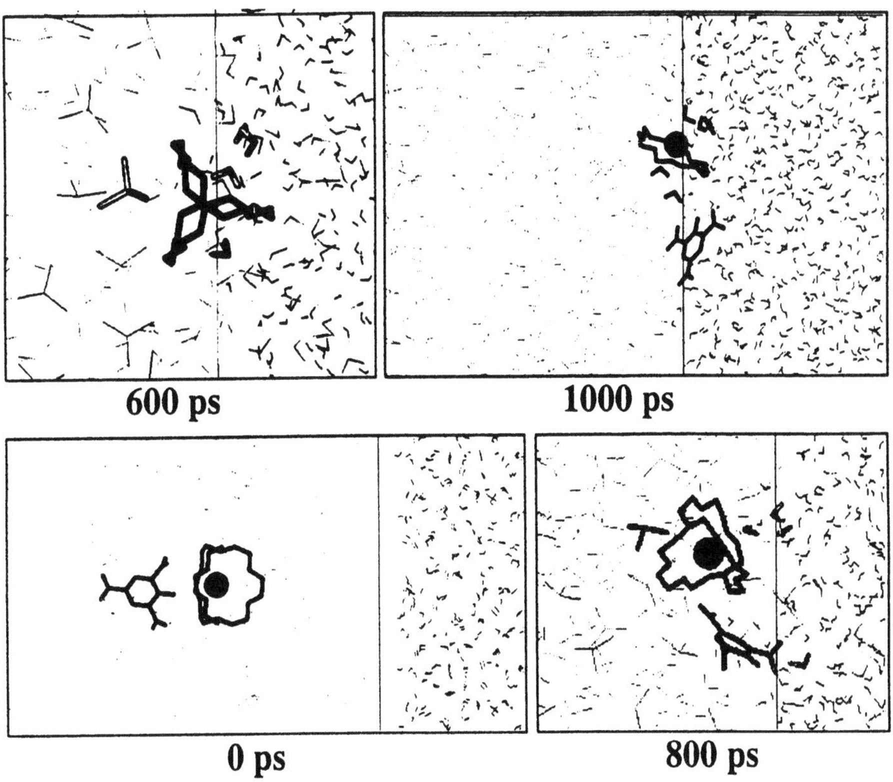

Figure 6: *Top:* the *222* cryptand after 600 ps and the *18C6*K^+Pic^- complex after 1 ns. *Bottom*: the $222K^+Pic^-$ cryptate at 0 ps (in chloroform) and after 800 ps

Complexed ionophores were calculated to also display a strong affinity for the interface. For instance the *222*K^+Pic^- cryptate, initially immersed in chloroform, at about 12 Å from the interface, finally diffused back to the interface where it remained trapped [30] (see Figure 6). A similar result was obtained for the *calixC6*·Cs^+Pic^- complex [28]. Adsorption of other $\mathbf{L}M^+$ complexes (**L** = calix[4]arene$^-$ anion [25]; calix[4]arene-tetraamide [26]; *18C6* [32], valinomycine [57], calix[4]arene-crown6 [37] or calix[4]arene-*bis*crown6 [27], podants [30]) with alkali cations was observed, in the absence as well as in the presence of counterions. Some anions like Pic^- were found to strongly adsorb at the interface, while more hydrophilic ones like Cl^- or NO_3^- moved to the bulk water phase.

More quantitative insights into the energetic features of interface crossing were obtained by the "PMF" (Potential of Mean Force) calculations corresponding to the departure of ions and of ionophores from the interface to the water and chloroform phases, respectively [42]. We considered the uncomplexed Cs^+Pic^- and Cs^+Cl^- salts, the uncomplexed *calixC6* ionophore, and the *calixC6*Cs^+ complex, in the presence of Cl^- / Pic^- counterions. Details are reported in ref. 42. These free energy profiles fully support trends observed in the MD simulations: for *calixC6* free or complexed, the energy curves display a flat minimum on the chloroform side of the interface. Migration to chloroform is easier than to water, and the *calixC6* Cs^+ complex is more surface active than *calixC6* uncomplexed. This contrasts with the free Cs^+ ion whose departure from the interface to water corresponds to a slightly downhill process, while migration to chloroform is quite energy demanding, although facilitated by Pic^-, compared to the Cl^- counterion.

Most of the simulations on complexed ionophores were performed with alkali cations as guests, but more charged ions, like UO_2^{2+}, Sr^{2+} [29], or Eu^{3+} [58] were also considered. As the cation charge increases, the complex may become more hydrophilic and adopt a different relationship with respect to the interface. For instance, the $Eu(NO_3)_3\mathbf{L}_3$ complexes (**L** = diamide ligands), migrate from the interface to water, and finally dissociate [58]. Thus, the interfacial behaviour relates to the hydrophobic / hydrophilic balance of the solute, which is difficult to estimate, as it depends on its total charge, its size, volume, shape and precise conformation and orientation of its polar / apolar groups [30].

4- Assemblies of ionophores simulated at the water / chloroform interface: from monolayers to diluted aggregates. Implications for the assisted ion process.

The high affinity of free or complexed ionophores for the interface, and the fact that they are more soluble in organic solvents than in water, led us to simulate more concentrated solutions, to test whether spontaneous migration to the organic phase would take place. The concentration was increased more or less arbitrarily, without performing systematic studies. Interestingly, different situations were obtained, depending on the nature of the solute.

4.1 Assemblies of uncomplexed ionophores at the interface.

TBP and TPTZ (see Chart 1) are used to extract lanthanides and actinides from nuclear waste solutions. Decamers of TBP molecules were found to adsorb and form a monolayer at the interface (see details in ref. 31). Three MD simulations were performed starting with different arrangements of $(TBP)_{10}$ (either in a "normal layer", where the $O_{(P)}$ oxygens point towards water; or an "inversed layer" where $O_{(P)}$'s point toward chloroform; or a "spherical aggregate"). They all led to a similar situation where the TBP's sit at the interface, pointing their $O_{(P)}$ atoms towards water, "ready to capture" an approaching cation. Thus, although TBP is more soluble in chloroform than in water, it does not migrate from the interface to chloroform. Similarly a $(TPTZ)_{24}$ aggregate, initially placed at the interface, remained there, with the TPTZ molecules stacked more or less over each other, instead of diffusing to chloroform (Figure 7).

A different situation is found with *18C6* as solute (see ref. 32 for details). The simulation starting with the $(18C6)_{10}$ decamer right at the interface, leads, after 1 ns, to a complex situation. Most of the crowns remained at the interface, hydrogen bonded to water molecules. Others moved to chloroform, changing their conformation from D_{3d} to C_i. A few water molecules followed these crowns to the organic phase ("water dragging effect" [59]). The concentration of *18C6* is higher at the interface than in the bulk organic phase, without however forming a monolayer. Interestingly, we notice that the *18C6* molecules which remain at the interface retain their D_{3d} form, which has a zero dipole moment. Thus, as pointed out above for anions and for the 222 cryptand, the *surface activity does not necessarily relate to the intrinsic asymmetry of the solute.*

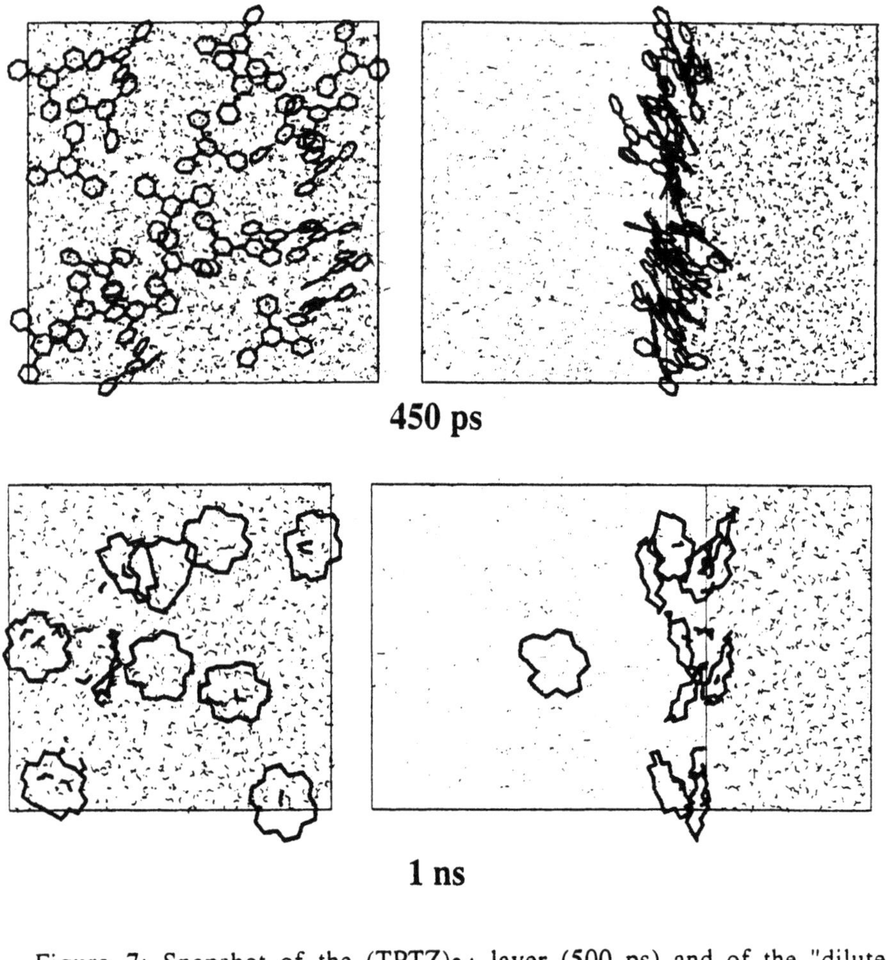

Figure 7: Snapshot of the $(TPTZ)_{24}$ layer (500 ps) and of the "diluted" $(18C6)_{10}$ aggregate (1 ns) at the interface. Orthogonal views.

Another related important feature concerns the *preorganization of the ionophore at the interface for ion capture.* Indeed, the D_{3d} form of *18C6*, like the "K" form of cryptand *222* (D_3 symmetry), which is the most surface active conformer, is also preorganized to capture K^+, the cation which is selectively complexed and extracted.

4.2 Assemblies of complexed ionophores at the interface: towards the extraction process. We simulated the *18C6*K^+Pic^- complexes at the interface, starting with two different arrangements [32]. The first one (A) started with the hypothetical situation where the *18C6*K^+ inclusion complexes, stacked with Pic^- counterions, are right at the interface. The second one (B) mimicked the "extraction experiment", and started with the uncomplexed *18C6* molecules in the chloroform phase, while the K^+Pic^- ions were in water, at about 12 Å from the interface. Simulations A and B did not converge after 1 ns to similar situations concerning the relationship between the solutes and the interface. However, a number of clear trends were observed.

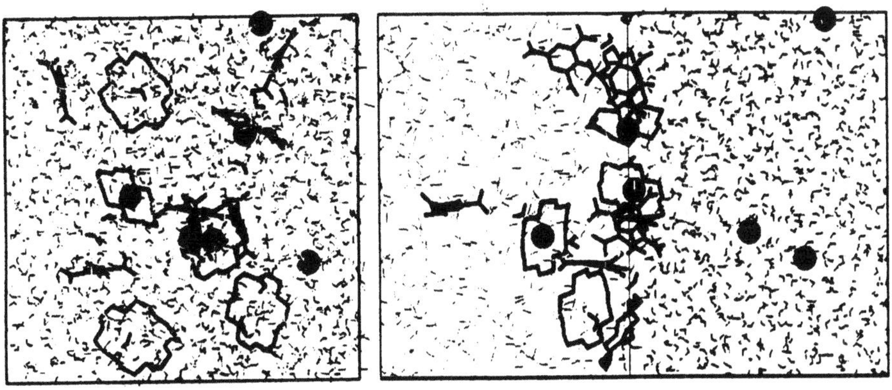

$(18C6.K^+ Pic^-)_6$

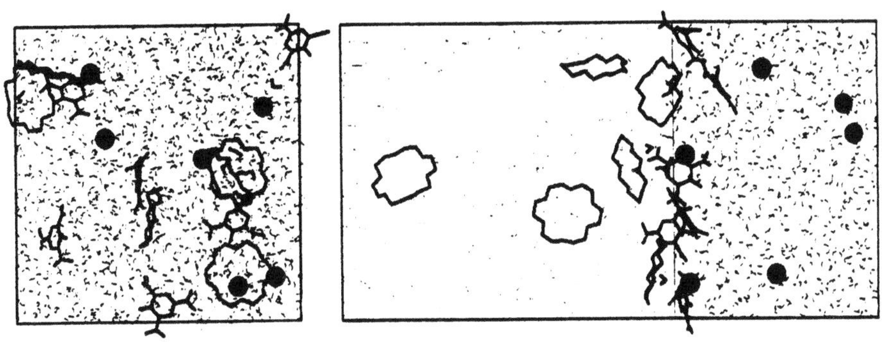

$(18C6)_6$–chlor $(K^+ Pic^-)_6$–wat

Figure 8: *Top*: the $(18C6K^+Pic^-)_6$ aggregate, after 1 ns (the simulation started with the solute right at the interface). *Bottom*: "simulated extraction experiment" of $(K^+Pic^-)_6$ by $(\mathit{18C6})_6$. At 0 ps, the K^+Pic^- ions were in water, and the *18C6* molecules in chloroform. See text and ref. 32.

First, *increasing the concentration of the solute modifies its interfacial behaviour.* A monomeric *18C6*K^+Pic^- solute remains of inclusive type, and anchored at the interface (Figure 6). This contrasts with simulation A on (*18C6*$K^+Pic^-)_6$ which leads to a complex equilibrium between some *18C6*K^+ complexes at the interface, others which migrated to chloroform, and others which decomplexed. The K^+ decomplexed cations are captured by water, while the decomplexed crowns remain at the interface. The remarkable result, in relation with the assisted ion transfer, concerns the *extraction of a 18C6K+ complex to chloroform* (Figure 8), seemingly driven by a Pic^- anion. The calculations reveal indeed the *crucial role of the counterions*. We first notice that no Pic^- anion sits in water, while some K^+ cations do. Most Pic^- anions, adsorbed at the interface, create a negative potential which likely pushes some of them to the organic phase. Figure 8 shows how the *18C6*K^+ extracted complex, surrounded by one H_2O molecule at each face, follows a Pic^- anion, hydrogen bonded with one H_2O.

After 1 ns of the "extraction simulation" B, an equilibrium is found involving *18C6* molecules which have moved to the interface, others remaining in chloroform, and the Pic^- anions which all sit at the interface. Due to the induced negative potential, some K^+ cations moved from the bulk water to the interface, from which they would, otherwise, be "repulsed". This situation thus highly facilitates the ion capture, as seen in Figure 8. Some K^+ cations, next to interfacial Pic^- anions, seem ready, via a least motion pathway, to be captured by adjacent *18C6* ionophores.

Thus, although the spontaneous complexation process, followed by extraction to the organic phase, is not observed in the computer output at the timescales simulated, these simulations reveal an essential feature of the interfacial behaviour of ions and ionophores. They display a clear local organization, induced by the asymmetrical nature of the interface. It may extent from the formation of a well structured monolayer, to complex equilibria involving different local concentrations of the partners: anions, ionophores free and complexed, and cations. It is remarkable that *the resulting organization markedly facilitates the ion extraction by the ionophore, from the aqueous to the organic phase.*

5- Simulation of the demixion of binary solutions containing salts or ionophores

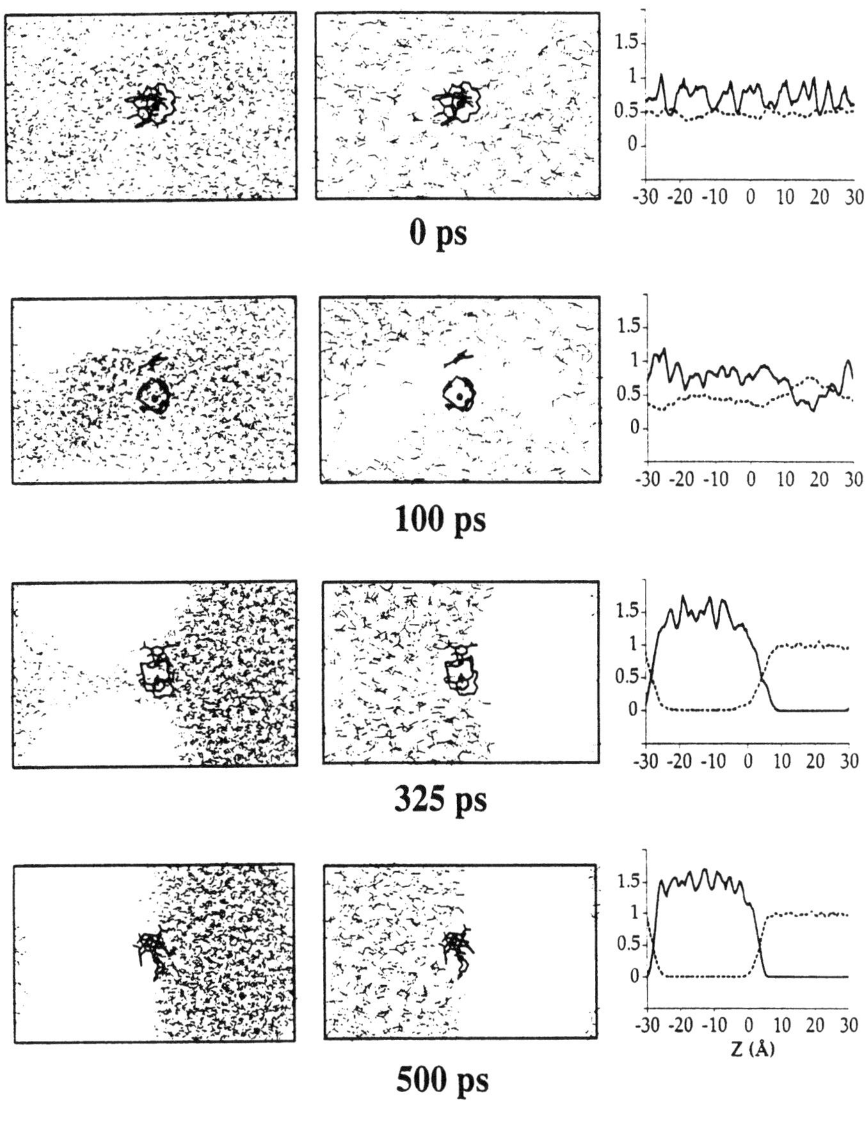

Figure 9: Demixion of a binary water / chloroform mixture containing the *CalixC6*Na^+ Pic^- complex as solute. Snapshots at 0 ps, 100, 325 and 500 ps. For a purpose of clarity, we display separately the water (*left*) and chloroform (*middle*) solvents. The solvent density curves (chloroform: *full line*; water: *dotted line*) as a function of Z are shown on the right side.

All simulations reported above started with a prebuilt interface between two adjacent liquid phases. However, in relation with the liquid-liquid extraction experiment, where the two liquid phases are mixed, we decided to simulate the demixion of "perfectly mixed" binary water - chloroform mixtures, containing different types of solutes, involved in the assisted ion extraction process. The results are described in ref. 60, with free salts (Cs^+Pic^-, $Cs^+NO_3^-$) or complexes of *calixC6* (with Cs^+Cl^- or with Cs^+Pic^-) as solutes. An illustration is given in Figure 9, where the two liquid components are shown separately side by side, for clarity. The solute is the *calixC6*Na^+ complex, with Pic^- as counterion.

At the beginning of the simulation (0 ps), the two liquids were homogeneously mixed, reaching densities about half of their values in the bulk pure liquids. Rapidly, at the early stages of the simulation, water aggregated around the polar moities (the crown ether fragment and the Na^+ cation, the Pic^- anion), while chloroform molecules concentrated around the aromatic groups of *calixC6*. After 100 ps, two pockets of water molecules formed, surrounded by chloroform, while the *calixC6*Na^+Pic^- solute sat at the border region between the two liquids. At 325 ps, the two liquids were almost completely separated, connected by a few water molecules only. After additional 175 ps, they formed two "planar interfaces", which are in fact equivalent due to the 3D periodicity used in the simulations. The remarkable result is that *the calixC6*Na^+ *complex finally adsorbed at the interface, somewhat more on the chloroform side than on the water side.* This situation is nearly identical to the one obtained from a simulation starting with the two adjacent liquid phases. This is quite satisfactory from the computational point of view as it demonstrates that the simulated adsorption is not an artefact caused by the choice of the initial state. More basically, these results, taken together with the adsorption observed similarly for *calixC6* uncomplexed [37], suggest that *ion capture and recognition take place at the interface.* This is confirmed by free energy perturbation simulations on the Na^+ / Cs^+ binding *at the interface* [61].

CONCLUSION

Computer simulations do not constitute, *per se*, a tool for structure elucidation. However, with a given representation of the energetics ("Hamiltonian") of a precisely defined system, they allow to investigate complex situations, which cannot be clearly pictured at the microscopic level from "real experiments". "Computer MD experiments" allow to test the lifetime and stability of selected hypothetical situations, or the effect of selected "hamiltonians". As far as liquid interfaces are concerned, the simulated results are in qualitative agreement with conclusions drawn from "macroscopic experiment". For instance, according to surface tension measurements [2], small ions are "repelled by the surface", whereas ionophores, free or complexed, are adsorbed [62-68]. This has been discussed previously [25-32,37]. Interfacial electrochemistry also allows to quantify energetic features of interface crossing by free and complexed salts [33, 69-71]. As far as ion transport is concerned, the interpretation of kinetic data on related systems is consistent with an interfacial complexation process [72]. Spectroscopic data also provide structural insights into the precise arrangement of molecules at interfaces [73].

The computational results deserve some words of caution concerning the energy representation of the system, and particularly of electrostatics when dealing with highly charged solutes. The precise interactions between unlike solvents have to be accounted for precisely, which may require explicit account of polarization effects [74]. The behaviour at longer timescales than the ones simulated has to be explored. Physical questions concerning the size effect of the simulated system, concentration, salting out and synergistic effects, remain to be investigated. It becomes clear however that the problem of ion extraction, revisited from the point of view of the interfacial mechanism, raises the question of *supramolecular organization and recognition at the interface*. The free and complexed extractant molecules are highly *concentrated* or *ordered* at the interface, as do lipophilic ions. Organization at the interface, different from organization in pure solvents (and *a fortiori* in the solid state) may relate to special functions like ion recognition, signal transduction, or photophysical processes. Thus, computer simulations like those performed on membranes or membrane mimetics [21, 24, 50, 53, 75-83], micelles [49, 84-89], solvent mixtures [90-97], should play an increasing role to provide dynamical microscopic models, whose status may range from simple mental pictures, to representation of "the reality".

Acknowledgements. The authors are grateful to CNRS IDRIS for allocation of computer time, to PRACTIS for support, and to E. Engler for software developments. LT thanks EEC (F14W-CT0022 contract) for a grant. FB and NM thank the French Ministry of Research for a grant.

References

(1) Aveyard, R.; Vincent, B. "Liquid-liquid Interfaces: in Isolation and in Interaction" *Progress Surf. Science* **1977**, *8*, 59-102.

(2) Adamson, A. W. *Physical Chemistry of Surfaces. 5th ed.*; Wiley: **1990**.

(3) Rosen, M. J. *Surfactants and Interfacial Phenomena*; Wiley: New York, **1989**; 431 .

(4) Kazarinov, V. E. *The Interface Structure and Electrochemical Processes at the Boundary between two Immiscible Liquids.*; .; Springer: Berlin, **1987**; .

(5) Lehn, J. M. "Design of Organic Complexing Agents. Strategies towards Properties." *Struct. Bonding* **1973**, *161*, 1-69.

(6) Lehn, J.-M. *Supramolecular Chemistry. Concepts and Perspectives*; VCH: Weinheim New York, **1995**; .

(7) Izatt, R. M.; Clark, G. A.; Bradshaw, J. S.; Lamb, J. D.; Christensen, J. J. "Macrocyle Facilitated Transport of Ions in Liquid Membrane Systems" *Separation and Purification Methods* **1986**, *15*, 21-72.

(8) Behr, J. P.; Kirch, M.; Lehn, J. M. "Carrier-Mediated Transport through Bulk Liquid Membranes: Dependence of Transport Rates and Selectivity on Carrier Properties in a Diffusion-Limited Process." *J. Am. Chem. Soc.* **1985**, *107*, 241-246.

(9) Visser, H. C.; Reinhoudt, D. N.; de Jong, F. "Carrier-mediated Transport through Liquid Membranes" *Chem. Soc. Rev.* **1994**, 75-81.

(10) Danesi, P. R. "Separation of Metal Species by Supported Liquid Membranes" *Separation Science Tecnology* **1985**, *19*, 857-894.

(11) Danesi, P. R.; Chirizia, R.; Coleman, C. F. "*The Kinetics of Metal Solvent Extraction*" in *Critical Reviews in Analytical Chemistry*; Campbell, B.; CRC Press, Boca Raton, Florida, **1980**; pp 1.

(12) Moyer, B. A. "*Basic Principles of Extraction and Liquid-Liquid Systems Employing Crown Ethers and Related Metal-Ion Receptors*" in *Comprehensive Supramolecular Chemistry. Vol. 1 Molecular Recognition: Receptors for Cationic Guests*; Atwood, J. L.; Davies, J. E. D.; McNicol, D. D.; Vögtle, F. Lehn, J.-M.; Pergamon, **1996**; pp 325-365.

(13) Watarai, H. "What's happening at the Liquid-Liquid Interface in Solvent Extraction Chemistry?" *Trends in Analytical Chemistry* **1993**, *12*, 313-318.

(14) Feller, D. "Ab initio Study of M^+: 18-crown-6 Microsolvation" *J. Phys. Chem.* **1997**, *101*, 2723-2731.

(15) Wipff, G. *Computational Approaches in Supramolecular Chemistry*; Kluwer Acad. Publ.: Dordrecht, **1994**.

(16) van Gunsteren, W. F.; Berendsen, H. J. C. "Computer Simulation of Molecular dynamics : Methodology, Application and Perspectives in Chemistry." *Angew. Chem. Int. Ed. Engl.* **1990**, 992-1023.

(17) Jorgensen, W. L.; Duffy, E. M. "Structure and Binding for Complexes of Rebeks Acridine Diacid with Pyrazine, Quinoxaline, and Pyridine from Monte Carlo Simulations with an All-atom Force Field." *J. Am. Chem. Soc.* **1994**, *116*, 6337-6343.

(18) Benjamin, I. "Chemical Reactions and Solvation at Liquid Interfaces: a Microscopic Perspective" *Chem. Rev.* **1996**, *96*, 1449-1475.

(19) Schweighofer, K. J.; Benjamin, I. "Transfer of Small Ions Across the Water/1,2-Dichloroethane Interface" *J. Phys. Chem.* **1995**, *99*, 9974-9985.

(20) Torrie, G. M.; Valleau, J. P. "Double Layer Structure at the Interface between two Immiscible Electrolyte Solutions." *J. Electroanal. Chem.* **1986**, *106*, 69-79.

(21) Wilson, M. A.; Pohorille, A. "Mechanism of Unassisted Ion Transport across Membrane Bilayers." *J. Am. Chem. Soc.* **1996**, *118*, 6580-6587.

(22) Sokhan, V. P.; Tildesley, D. J. "MD Simulation of the Non-Linear Optical Susceptibility at the Phenol/Water/Air Interface" *Faraday Discuss.* **1996**, *104*, 193-208.

(23) Chipot, C.; Wilson, M. A.; Pohorille, A. "Interactions of Anesthetics with the Water-Hexane Interface. A Molecular Dynamics Study." *J. Phys. Chem. B* **1997**, *101*, 782-791.

(24) Guba, W.; Kessler, H. "A Novel Computational Mimetic of Biological Membranes in MD simulations" *J. Phys. Chem.* **1994**, *98*, 23-27.

(25) Varnek, A.; Sirlin, C.; Wipff, G. "*Interaction of the t-butyl-calix[4]arene⁻ Anion with Alkali Cations in Vacuo, in Water, in Acetonitrile and in Chloroform: Host-guest Complexes or Exo Counterions ? MD and FEP Computer Investigations on the Cs^+ Binding Affinity*" in *Crystallography of Supramolecular Compounds*; Tsoucaris, G. Ed. Kluwer Acad. Publ., Dordrecht, **1995**; pp 67-100.

(26) Wipff, G.; Engler, E.; Guilbaud, P.; Lauterbach, M.; Troxler, L.; Varnek, A. "Free Calixarene and Cryptand Ionophores at the Water/Chloroform Interface." *New J. Chem.* **1996**, *20*, 403-417.

(27) Varnek, A.; Wipff, G. "Theoretical Calculations of Extraction Selectivity: Alkali Cation Complexes of Calix[4]-bis-crown6 in Pure Water, Chloroform, and at a Water/chloroform Interface" *J. Comput. Chem.* **1996**, *17*, 1520-1531.

(28) Lauterbach, M.; Wipff, G. "*Liquid-liquid Extraction of Alkali Cations by Calix[4]crown Ionophores: Conformation and Solvent Dependent Na^+/Cs^+ Binding Selectivity: A MD FEP Study in Pure Chloroform and MD Simulations at the Water /chloroform Interface*" in *Physical Supramolecular Chemistry*; Echegoyen, L. Kaifer, A.; Kluwer Acad. Publ, Dordrecht, **1996**; pp 65-102.

(29) Guilbaud, P.; Wipff, G. "MD Simulations on $UO_2^{2+}$and Sr^{2+} Complexes with CMPO Derivatives in Aqueous Solution and at a Water/Chloroform Interface." *New J. Chem.* **1996**, *20*, 631-642.

(30) Varnek, A.; Troxler, L.; Wipff, G. "Adsorption of Ionophores and of their Cation Complexes at the Water/Chloroform Interface: a MD Study of the *222* Cryptand and of Phosphoryl-containing Podants"" *Chem. Eur. J.* **1997**, *3*, 552-560.

(31) Beudaert, P.; Lamare, V.; Dozol, J.-F.; Troxler, L.; Wipff, G. "Theoretical Studies on TBP: MD Simulations in Vacuo, in Water, and at a Water / Chloroform Interface" *Solvent Exraction and Ion Exchange* **1997**, *in press.*

(32) Troxler, L.; Wipff, G. "Molecular Dynamics Studies on the Interfacial Behaviour of 18-crown-6 and of 18-crown-6 K^+ Pic^- Aggregates at the Water-Chloroform Interface" *Anal. Sciences* **1997**, *in press.*

(33) Girault, H. H.; Schiffrin, D. J. "*Electrochemistry of Liquid-Liquid Interfaces*" in *Electroanalytical Chemistry*; Bard, A. J.; Dekker, New York, **1989**; pp 1-141.

(34) Marcus, Y. "The Thermodynamics of Solvation of Ions" *J. Chem. Soc. Farad. Trans. 1* **1987**, *83*, 2985-2992.

(35) Pearlman, D. A.; Case, D. A.; Cadwell, J. C.; Seibel, G. L.; Singh, U. C.; Weiner, P.; Kollman, P. A. *AMBER4*; University of California: San Francisco, **1991**.

(36) Cornell, W. D.; Cieplak, P.; Bayly, C. I.; Gould, I. R.; Merz, K. M.; Ferguson, D. M.; Spellmeyer, D. C.; Fox, T.; Caldwell, J. W.; Kollman, P. A. "A Second Generation Force Field for the Simulation of Proteins, Nucleic Acids, and Organic Molecules" *J. Am. Chem. Soc.* **1995**, *117*, 5179-5197.

(37) Wipff, G.; Lauterbach, M. "Complexation of Alkali Cations by Calix[4]crown Ionophores: Conformation and Solvent Dependent Na^+ / Cs^+ Binding Selectivity: A MD FEP Study." *Supramol. Chem.* **1995**, *6*, 187-207.

(38) Åqvist, J. "Ion-Water Interaction Potentials Derived from Free Energy Perturbation Simulations." *J. Phys. Chem.* **1990**, *94*, 8021-8024.

(39) Jorgensen, W. L.; Chandrasekhar, J.; Madura, J. D. "Comparison of Simple Potential Functions for Simulating Liquid Water." *J. Chem. Phys.* **1983**, *79*, 926-936.

(40) Jorgensen, W. L.; Briggs, J. M.; Contreras, M. L. "Relative Partition Coefficients for Organic Solutes from Fluid Simulations." *J. Phys. Chem.* **1990**, *94*, 1683-1686.

(41) Smith, P. E.; van Gunsteren, W. F. "*Methods for Evaluation of Long Range Electrostatic Forces in Computations of Molecular Systems*" in *Computer Simulations of Biomolecular Systems*; van Gunsteren, W. F. ; Weiner, P. K. Wilkinson, A. J.; ESCOM, Leiden, **1993**; pp 182-212.

(42) Lauterbach, M.; Engler, E.; Muzet, N.; Troxler, L.; Wipff, G. "Migration of Ionophores and Salts through a Water-Chloroform Liquid-Liquid Interface: Computer MD-PMF Investigations." *J. Phys. Chem.* **1997**, *101*, 000

(43) Hofmeister, F. "On the Understanding of the Effect of Salts. Second report. On Regularities in the Precipitating Effect of Salts and their Relationship to their Physiological Behavior." *Naunyn-Schmiedebergs Archiv für Experimentelle Pathologie und Pharmakologie* **1888**, *24*, 247-260.

(44) Collins, K. D.; Washabaugh, M. W. "The Hofmeister Effect and the Behaviour of Water at Interfaces" *Quater. Rev. Biophys.* **1985**, *18*, 323-422.

(45) Berny, F.; Wipff, G. *Unpublished.*

(46) Benjamin, I. "Theory and Simulation of Solvation and Chemical Reactions at Liquid Interfaces" *Acc. Chem. Res.* **1995**, *28*, 233-239.

(47) Randles, J. E. "Ionic Hydration and the Surface Potential of Aqueous Electrolytes" *Discuss. Faraday Soc.* **1957**, *24*, 194-199.

(48) Popov, A. N. "*Counterions and Adsorption of Ion-Exchange Extractants at the Water/Oil Interface*" in *The Interface Structure and Electrochemical Processes at the Boundary Between two Immiscible Liquids*; Kazarinov, V. E.; Springer Verlag, **1987**; pp 179-205.

(49) Böcker, J.; Brickmann, J.; Bopp, P. "MD Simulation Study of an *n*-Decyltrimethylammonium Chloride Micelle in Water." *J. Phys. Chem.* **1994**, *98*, 712-717.

(50) Böcker, J.; Schlenkrich, M.; Bopp, P.; Brickmann, J. "MD Simulation of a n-Hexadecyltrimethylammonium Chloride Monolayer" *J. Phys. Chem.* **1992**, *96*, 9915-9922.

(51) Karaborni, S.; Esselink, K.; Hilbers, P. A. J.; Smit, B.; Karthäuser, J.; van Os, N. M.; Zana, R. "Simulating the Self-Assembly of Gemini (Dimeric) Surfactants." *Science* **1994**, *266*, 254-256.

(52) Smit, B.; Hilbers, P. A. J.; Esselink, K. "Computer Simulations of Simple Oil/Water/surfactants Systems." *Tenside Surf. Det. 30* **1993**, *4*, 287-294.

(53) van der Ploeg, P.; Berendsen, H. J. C. "Molecular Dynamics Simulation of a Bilayer Membrane" *J. Chem. Phys.* **1982**, *76*, 3271.

(54) Ahlström, P.; Berendsen, J. C. "A MD Study of Lecithin Monolayers" *J. Chem. Phys.* **1993**, *97*, 13691-13702.

(55) van Buuren, A. R.; Marrink, S.-J.; Berendsen, J. C. "Characterization of Aqueous Interfaces with Different Hydrophobicities by MD" *Colloids and Surfaces* **1995**, *102*, 143-157.

(56) Shao, Y.; Stewart, A. A.; Girault, H. H. "Determination the Half Wawe Potential of the Species limiting the Potential Window. Measurements of Gibbs Transfer Energies at the Water/1,2-Dichloroethane Interface" *J. Chem. Soc. Faraday Trans.* **1991**, *87*, 2593-2597.

(57) Lauterbach, M.; Wipff, G. *Unpublished*

(58) Giovannino, L.; Wipff, G. *Unpublished.*

(59) Fan, W.; Tsai, R.-S.; El Tayar, N.; Carrupt, P.-A.; Testa, B. "Solute-water Interactions in the Organic Phase of a Biphasic System. 2. Effects of Organic Phase and Temperature on the "Water Dragging" Effect" *J. Phys. Chem.* **1994**, *98*, 329-333.

(60) Muzet, N.; Wipff, G. "MD Simulations on the Demixion of Binary Water-Chloroform Mixtures Containing Ionophoric Solutes" *Submitted.*

(61) Muzet, N.; Wipff, G. *Unpublished.*

(62) Ishikawa, Y.; Kunitake, T.; Matsuda, T.; Otsuka, T.; Shinkai, S. "Formation of Calixarene Monolayers which Selectively Respond to Metal Ions" *J. Chem. Soc. Chem. Commun.* **1989**, 736-737.

(63) Danesi, P. R.; Chiarizia, R.; Pizzichini, M.; Saltelli, A. "Extraction Kinetics of 18-crown-6 and DB18-crown-6" *J. Inorg. Nucl. Chem.* **1978**, *40*, 1119-1123.

(64) Sagert, N. H.; Lee, W.; Quinn, M. J. "The Adsorption of tri-*n*-Butylphosphate at the *n*-Dodecane-Water Interface" *Can. J. Chem.* **1979**, *57*, 1218-1223.

(65) Carlà, M.; C. M. C. Gambi; Baglioni, P. "Adsorption Properties of Cryptand *222* at the Charged Mercury-Solution Interface" *J. Phys. Chem.* **1996**, *100*, 11067-11071.

(66) Dei, L.; Casnati, A.; Nostro, P. L.; Baglioni, P. "Selective Complexation by *p-tert*-Butylcalix[6]arene in Monolayers at the Water-Air Interface." *Langmuir* **1995**, *11*, 1268-1272.

(67) Dei, L.; Casnati, A.; Nostro, P. L.; Pochini, A.; Ungaro, R.; Baglioni, P. "Complexation Properties of *p-tert*-Butylcalix[6]arene Hexamide in Monolayers at the Water-Air Interface." *Langmuir* **1996**, *12*, 1589-1593.

(68) Nostro, P. L.; Casnati, A.; Bossoletti, L.; Dei, L.; Baglioni, P. "Complexation Properties of Calixarenes in Langmuir Films at the Water-Air Interface." *Colloids and Surfaces A* **1996**, *116*, 203-209.

(69) Senda, M.; Kakiuchi, T.; Osakai, T. "Electrochemistry at the Interface between two Immiscible Electrolyte Solutions" *Electrochimica Acta* **1991**, *36*, 253-262.

(70) Koryta, J. "Ion Transfer across Water/Organic Boundaries and Analytical Applications" *Ion Selective Electrode Rev.* **1983**, *5*, 131-164.

(71) Vanysek, P. "Charge Transfer Processes on Liquid/Liquid Interfaces: The First Century" *Electrochim. Acta* **1995**, *40*, 2841-2847.

(72) Danesi, P. R. "*Solvent Extraction Kinetics*" in *Principles and Practices of Solvent Extraction*; Rydberg, J.; Musikas, C. Chopin, G. R.; M. Dekker, Inc., New York,, **1992**; pp 157-207.

(73) Eisenthal, K. B. "Liquid Interfaces Probed by Second-Harmonic and Sum-Frequency Spectroscopy." *Chem. Rev.* **1996**, *96*, 1343-1360.

(74) Chang, T.-M.; Peterson, K. A.; Dang, L. X. "MD Simulations of Liquid, Interface, and Ionic Solvation of Polarizable CCl_4" *J. Chem. Phys.* **1995**, *103*, 7502-7513.

(75) Benjamin, I. "Mechanism and Dynamics of Ion Transfer across a Liquid-Liquid Membrane" *Science* **1993**, *261*, 1558-1560.

(76) Marrink, S.-J.; Berendsen, H. J.-C. "Simulation of Water Transport through a Lipid Membrane." *J. Phys. Chem.* **1994**, *98*, 4155-4168.

(77) Pohorille, A.; Wilson, M. A. "MD Studies of Simple Membrane-Water Interfaces: Structure and Functions in the Beginning of Cellular Life" *Origins of Life and Evolution of the Biosphere* **1995**, *25*, 21-46.

(78) Pohorille, A.; Wilson, M. A. "Excess Chemical Potential of Small Solutes across Water-Membrane and Water-Hexane Interfaces." *J. Chem. Phys.* **1996**, *104*, 3760-3774.

(79) Pohorille, A.; Cieplak, P.; Wilson, M. A. "Interactions of Anesthetics with the Membrane-Water Interface" *Chem. Phys.* **1996**, *204*, 337-345.

(80) Schlenkrich, M.; Nicklas, K.; Brickmann, J. "A MD Study of the Interface between a Membrane and Water" *Ber. Bunsenges. Phys. Chem.* **1994**, *94*, 133-145.

(81) Stouch, T. R. "Lipid-Membrane Structure and Dynamics Studied by All-atom MD Simulations of Hydrated Phospholipid Bilayers" *Molecular Simulation* **1993**, *10*, 317-345.

(82) Sheng, Q.; Schulten, K. "Molecular Dynamics Simulation of Immobilized Artifical Membranes" *J. Phys. Chem.* **1995**, *99*, 11018-11027.

(83) Venable, R. M.; Zhang, Y.; Hardy, B. J. "Molecular Dynamics Simulations of a Lipid Bilayer and of Hexadecane : An Investigation of Membrane Fluidity." *Science* **1993**, *262*, 223-225.

(84) Jönsson, B.; Edholm, O.; Teleman, O. "MD Simulations of a Sodium Octanoate Micelle in Aqueous Solution" *J. Chem. Phys.* **1986**, *85*, 2259-2271.

(85) Karabomi, S.; O'Connell, J. P. "MD Simulations of Model Micelles. 4. Effects of Chain Length and Head Group Characteristics" *J. Phys. Chem.* **1990**, *94*, 2624-2631.

(86) Shelley, J.; Watanabe, K.; Klein, M. "Simulation of a Sodium Dodecylsulfate Micelle in Aqueous Solution" *Int. J. Quant. Chem.: Quantum Biol. Symp.* **1990**, *17*, 103-117.

(87) Smit, B.; Hilbers, P. A. J.; Esselink, K.; Rupert, L. A. M.; van Os, N. M. "Structure of a Water/Oil Interface in the Presence of Micelles: a Computer Simulation Study" *J. Phys. Chem.* **1991**, *95*, 6361-6368.

(88) Wallin, T.; Linse, P. "Monte Carlo Simulations of Polyelectrolytes at Charged Micelles. 2. Effects of Linear Charge Density." *J. Phys. Chem.* **1996**, *100*, 17873-17880.

(89) Woods, M. C. "Internal Structure of a Model Micelle via Computer Simulation. 2. Spherically Confined Aggregates with Mobile Head Groups" *J. Phys. Chem.* **1986**, 1875-1885.

(90) de Leeuw, S. W.; Smit, B.; Williams, C. P. "Molecular Dynamics Studies of Polar/Nonpolar Fluid Mixtures. I. Mixtures of Lennard-Jones and Stockmayer Fluids." *J. Chem. Phys.* **1990**, *93*, 2704-2714.

(91) Ferrario, M.; Haughney, M.; McDonald, I. R.; Klein, M. L. "Molecular-Dynamics Simulation of Aqueous Mixtures: Methanol, Acetone, and Ammonia." *J. Chem. Phys.* **1990**, *93*, 5156-5166.

(92) Kovacs, H.; Kowalewski, J.; Laaksonen, A. "Molecular Dynamics Simulation of Liquid Mixtures of Acetonitrile and Chloroform." *J. Phys. Chem.* **1990**, *94*, 7378-7385.

(93) Laaksonen, A.; Kusalik, P. G.; Svishchev, I. M. "Three-Dimensional Structure in Water-Methanol Mixtures." *J. Phys. Chem. A* **1997**, *101*, 5910-5918.

(94) Nakanishi, K. "Computer Simulations on Aqueous Solutions of some Non-Electrolytes" *Chem. Soc. Rev.* **1993**, 177-182.

(95) Wallqvist, A. "Pressure Dependence of Methane Solvation in Aqueous Mixture and the Relation to the Structure of Liquid Water." *J. Chem. Phys.* **1992**, *96*, 1655-1656.

(96) Skaf, M. S.; Ladanyi, B. M. "Molecular Dynamics Simulation of Solvation Dynamics in Methanol-Water Mixtures." *J. Phys. Chem.* **1996**, *100*, 18258-18268.

(97) DeBolt, S. E.; Kollman, P. A. "Investigation of Structure, Dynamics, and Solvation in 1-Octanol and its Water-Saturated Solution: Molecular Dynamics and Free-Energy Perturbation Studies" *J. Am. Chem. Soc.* **1995**, *117*, 5316-5340.

SELF-ORGANIZATION IN MOLECULAR CRYSTALS, LIQUIDS AND SOLUTIONS: COMPUTER STUDIES

A.Gavezzotti
Dipartimento di Chimica Strutturale e Stereochimica Inorganica, University of Milano, Milano, Italy

G.Filippini
Centro CNR per lo Studio delle Relazioni tra Struttura e Reattività Chimica, University of Milano, Milano, Italy

1. Crystals

The study of intermolecular interactions in molecular systems has seen great advancements in the last decades, thanks to the detailed structural information made available by X-ray diffraction studies in crystals. Although recognition and systematization of the geometrical patterns in crystal structures has provided a number of clues on the basic mechanisms of molecular recognition that lead to the formation of condensed phases, theories are still in a rudimentary stage.

I.C. Paul and D.Y. Curtin, two pioneers of the field, stated fifteen years ago [1]: "We seem to be some time away from being able to control or even to predict with real assurance the packing a compound will adopt when it crystallizes, or the relationship of the crystal symmetry to molecular structure". The semantic accuracy of this statement, in its careful distinction between control and the much less ambitious task of prediction, should be appreciated. To be more specific, one could say that a basic accomplishment for the chemically oriented crystallographer would be to provide an answer to some or all of the following questions, given only molecular constitution and connectivity:

1) will this compound, if properly purified, crystallize at all, in a temperature range reasonably close to normal organic chemistry

G. Tsoucaris (ed.), Current Challenges on Large Supramolecular Assemblies, 249–263.

operations? What will be the appearance of the crystal crop? What could be the best method - growth from solution, from melt, from vapor?
2) are polymorphs possible, meaning that more than one crystal structure will frequently appear under ordinary conditions?
3) will this molecule, during crystallization, form clathrates or inclusion complexes? If yes, what are candidate guest molecules?
4) can one predict the density and sublimation enthalpy of the crystal(s)? Can these quantities be related to dissolution properties?
5) are there any special intermolecular recognition points, and can one expect from these the appearance of some partial packing motifs - rings, chains, layers, herringbones?
6) can the translational repetitions and the intermolecular symmetry (ideally, unique cell dimensions and space group) be predicted?
7) eventually, can the molecular conformation, as well as the location and orientation of the molecule within the crystal cell, be predicted with good accuracy, yielding a result comparable to that of a single crystal structure determination?
If answering these questions would be "prediction", the ultimate question, a good definition of crystal chemistry, refers to "control": can one steer the answers to all of the above questions by changing in a rational way the chemistry of the molecule?

Static computer simulations (that is, using empirical formulations for the potential, but neglecting kinetic energy) can provide estimates of heats of sublimation, of enthalpy differences between polymorphs, and of lattice vibrational frequencies and entropies. The calculation of some thermodynamic properties of crystals is easily feasible, using structural data from X-ray work, (see e.g. ref. [2]). The crystal structure can be guessed, with moderate success, by recursive crystal construction algorithms (see e.g. ref. [3]), under the action of suitably calibrated potentials (see e.g. ref. [4]).

2. Molecular dynamics

On a more fundamental level, however, understanding and predicting crystal structures requires also control of their liquid - melt or solution - precursors. One is therefore brought to a consideration of the complete phase behaviour of an organic molecule, and, for any liquid state, static calculations are clearly inadequate. In principle, molecular dynamics (MD) simulations allow the study of all stages of recognition and condensation, in terms of structure, thermodynamics and kinetics. There are however, as could have been anticipated, quite a few difficulties:

1) the force field problem. Parameterization of a force field for dynamic simulations is even more problematic than for static lattice energy calculations. Since all force fields are incomplete, transferability is a problem not only over different compounds, but over different phases of the same compound, as the dynamics of an isolated molecule is different from that of the liquid or crystal phases. Moreover, especially for solvent-solute interactions, mutual polarisation is presumably one of the main driving forces in the dynamical evolution, but no simple and sound method for empirically accounting for polarisation is apparently available;

2) the sampling problem. Derivation of thermodynamic averages and of kinetic pathways requires a robust sampling of phase space. This depends in a subtle way on many factors; for example, the same force field on the same system may produce different trajectories on different computers, because of infinitesimal rounding errors, cumulated over tens of thousands of iterative steps. Complete sampling would do away with this difficulty, but would require an infinite sampling time. Besides, even on the fastest available computers, simulation times on chemically significant systems (10,000 atoms) are limited to the order of nanoseconds; this time span is long for the lattice dynamics in the solid state, and is adequate for diffusion in a liquid phase, but no one knows for sure if it is long enough for nucleation (presumably, yes) and for growth of elementary nuclei (presumably not); it is borderline for phase transitions among solids, and for freezing and ordering in molecular clusters; it is orders of magnitude too short for melting. Given the easy manipulation of temperature in an MD run, supercooling a liquid to change artificially the nucleation rate, or superheating a crystal to speed up melting, by tens or even hundreds of degrees, is just a matter of computer input. Simulated annealing, which allows jumping over barriers, and then cooling down to new states, may help;

3) the design-interpretation problem. Seldom recognized, the importance of this conceptual step increases exponentially with computing power. MD is so ingenious that its results have a captivating tendency to be mistaken for truth, while they just reflect a simulation. Just as ordinary chemistry requires the design of the appropriate chemical experiment to prove a certain hypothesis, so computational chemistry requires the design

of an appropriate computational experiment, and a sound and skeptical interpretation of its results, otherwise one is just piling up an inordinate amount of uninterpretable numerical data.

3. Nucleation

The development of a formal theory of molecular events at phase boundaries is quite a demanding task. The following may be preliminary steps in that direction.

Consider a chemical system consisting of a number of different molecular species, A_i, each molecule A_{ij} with an unperturbed potential field U_i* and at infinite distance from any other molecule A_{kl}. As the volume of the system is reduced, molecules begin to interact, and polarisation distorts the molecular fields from U_i* to U_i. Under the action of the fields, molecules are driven together to aggregate, against random diffusion motions; a cluster can be defined as an aggregate of molecules whose centers of mass librate around an equilibrium position, and which are in a collective potential energy valley (as distinct from a saddle point), that is, no thermal libration may disrupt the ensemble without expense of energy. Such a cluster should have a pseudo-crystalline structure, since it can be expected that larger and larger aggregates should be more and more ordered, in agreement with the close-packing principle. Fluxionality within the aggregate would produce instead pseudo-liquid aggregates, or droplets; this possibility will not be further considered here.

Aggregation of these pseudo-crystalline clusters can be written as:

$$[N_i A_i : N_j A_j] + A_i \rightarrow [(N_{i+1})A_i : N_j A_j] \qquad (1)$$

where the N's are numbers of molecules of each species, and square brackets denote a cluster in the above sense. We define molecular recognition as one elementary step in this equation: the attractive potential adds a molecule of species i to a cluster composed of different amounts of molecules of each molecular species. The advantage of this definition is that it focusses on the intrinsic energetic aspects of the association process; the disadvantage is that it is not readily applicable to systems in which all species and clusters are surrounded by a solvent. In that case, thermodynamics and kinetics are complicated by the fact that solvation and desolvation energies, entropies and velocities must be taken into account. Even more problematic is the application of this definition to liquids: in a homogeneous bulk liquid there may be no clusters at all (a gas-like liquid), but, with varying temperature and chemical conditions, several degrees of clustering may occur. For example, it is certain that the structure of a liquid far from the freezing point is quite different from that of the same liquid in a pre-freezing state. A perfect, ordered crystal is a stable cluster in itself, where (paradoxycally, according to some different definitions) molecular recognition no longer occurs.

If interactions with the medium are neglected in a first approximation, nucleation of an organic compound from its solution involves only one molecular species, A, and equation 1 can be rewritten as

$$[nA] + A \rightarrow [(n+1)A] \qquad (2)$$

with n going from 1 to a few units or a few tens. The thermodynamics and kinetics of recognition are in principle different for each step, since the structure and potential (U* versus U) of the cluster depend on n.

Crystal growth may just be described by equation 2 with n very large, and actually approaching Avogadro's number. This means assuming that even in the ultimate stages, when the crystal has already taken macroscopic dimensions, growth may proceed by addition of single molecules, rather than by sintering of large clusters of different sizes. Possibly, the single-molecule species A could be replaced by aggregates of a few molecules, bound by strong forces like hydrogen bonds, and already present in solution. This is most likely the case, for example, for alcohols, in whose crystals multimolecular asymmetric units are very frequent [5].

Clusters may be structurally flexible: that is, in between elementary aggregation steps, equation 2, rearrangement steps may occur:

$$[nA]_j \leftrightarrow [nA]_k \qquad (3)$$

where the index denotes a cluster "conformation", or a particular arrangement of molecules within the cluster. In the nucleation stage, the number of these arrangements must be small, otherwise fluxional clusters proper of the liquid state would obtain; in the growth stage, n is very large, and equation 3 might represent the migration of one last incoming molecule through the surfaces, steps and kinks of the nucleus, on its way to its proper location.

The enthalpy of addition (recognition enthalpy) can be approximated by the difference in energy between the clusters and isolated molecules:

$$\Delta H^{+} = E[(n+1)A] - E[nA] - E(A) \quad (4)$$

while the rearrangement enthalpy can be estimated as the difference in energy between two conformations of the cluster:

$$\Delta H_{jk} = E[nA]_k - E[nA]_j \quad (5)$$

Both enthalpy differences could be estimated by empirical potential fields, if the structures of the clusters were known. But even in this very favorable and unlikely case, confidence in the results would be limited by the lack of polarisation in the force field, which sticks to U_i* and cannot evolve into U_i.

As concerns entropy differences, the following considerations apply. ΔS^{+} will include a negative configurational contribution, due to joining two fragments into one. The new cluster has however some more low-frequency vibrational degrees of freedom, for intermolecular librations. Assuming that internal degrees of freedom do not change very much, dispersion over these new vibrational levels could raise the thermal entropy of the new cluster, yielding a positive contribution to ΔS^{+}. It is tempting, although hardly justifiable, to assume cancellation here. ΔS_{jk} is expected to be negligible, just as it is for crystal polymorphs [7].

A severe limitation to the discussion of cluster stability in terms of free energies is the fact that these systems are not infinitely uniform, and therefore the usual thermodynamic criteria (minimum absolute free energy at constant T and P) do not apply in a straightforward way. Clusters have a finite size and well defined boundaries, and therefore contributions from surface effects should be included for a proper treatment. Sorting out these contributions in force field calculations is not easy, so that no simple methods are available for their computational estimation.

Nucleation and growth, as depicted by the above equations, are not activated processes, since there is no energy barrier to be surmounted when a molecule is attracted, downhill in energy, by a cluster. This obviously does not correspond to reality, since both processes are markedly temperature-, pressure- and volume-dependent: observed barriers must therefore arise from heat and mass transfer and from rearrangement and/or migration. The release of a molecule from a cluster should be activated, at least, by the sticking energy, ΔH^{+}, but a correlation to solubility is hampered by the fact that dissolution kinetics seems to be mainly diffusion-controlled.

The theoretical study and the computer simulation of the above phenomena may proceed through molecular dynamics. A computational box for this task can be prepared by randomly inserting a few solvate molecules into a solvent bath, and by monitoring the aggregation in time.

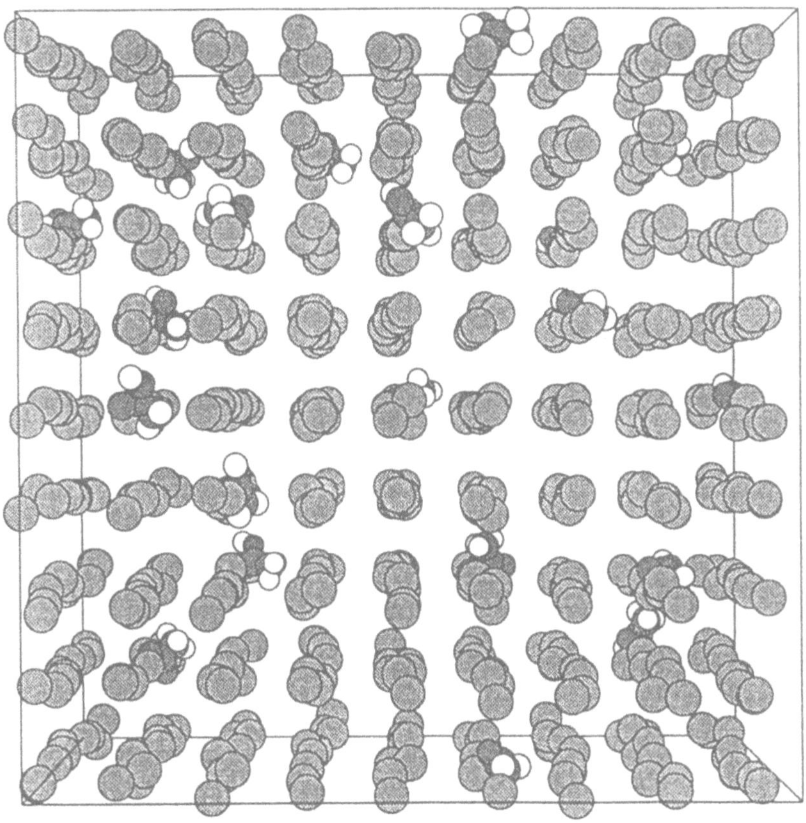

Figure 1. The computational box with 709 carbon tetrachloride (large circles) and 20 acetic acid molecules [7].

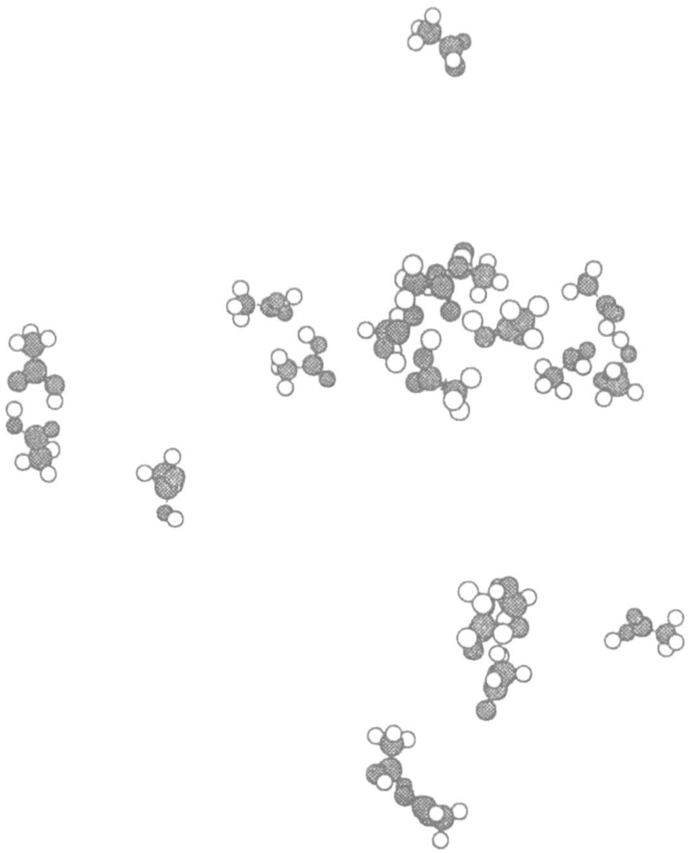

Figure 2. The acetic acid molecules after a 500 ps MD run within the box shown in Figure 1 [7].

Figures 1 and 2 show the results of such a computational experiment for acetic acid in carbon tetrachloride [7]; recognition and clustering are clearly observed, but the way from there to a true simulation of pre-crystallization phenomena seems to be still a long one. One never knows how long the dynamic run should last before the results can be considered positively realistic. MD runs on isolated clusters of small organic molecules have shown convergence to the actual crystal structure [8]; extensive computational tests have been performed on ensembles of spherical objects or rare gas atoms (see for example ref. [9]), but these have little resemblance to actual molecular systems. The interaction of solvate urea molecules with crystal faces [10] and the nucleation and growth of *n*-alkane crystals [11] have been simulated. One is always struggling with timescale, with constraints posed by system size, system boundaries and approximate potentials, as well as with possible numerical artifacts: for example, rounding errors generate undesired drifts in the computational box, which could obscure, if not carefully screened out, subtle details of diffusion [10]. Analysis of the results requires the definition of adequate geometrical and energetic parameters and averages to identify and monitor the onset of order and disorder: for example, how does one describe clustering and nucleation with just a few time-dependent parameters?

The final crystal structure must however provide some clues on the nuclei from which it was generated and grown. A promising computational experiment could be the extraction, from the actual crystal structure, of different aggregates of, say, 10-30 molecules, which could then be probed for stability by solvating them and running MD simulations. Those clusters which would not disrupt could then be computationally grown by MD addition of further solvate molecules.

Some progress along these lines has been made with the computational study [12] of the crystal polymorphism of tetrolic (methylacetylene carboxylic) acid. Crystals of its two phases show a cyclic dimer motif and a catemer motif [13]. Two tetrolic acid molecules were placed in a cyclic dimer configuration within a box of 226 carbon tetrachloride solvent molecules, and the dynamic behaviour of the dimer was observed by an MD simulation of about 2000 ps. Events involving cleavage of one of the two hydrogen bonds, and thus leading to the formation of catemer precursors, were found to take about 10% of the simulation time. Similar, but more sparse, catemer events were found [14] for the dimer of 2-pyridone, whose known crystal phase has the dimer motif. Figure 3 shows typical MD results.

A computational experiment to mimic crystal dissolution could be run in a computational box consisting of a slab cut out of a crystal structure and wetted by a layer of solvent, monitoring the detachment of crystalline molecules and their diffusion into the bulk solution by MD. However, given a typical dissolution velocity of 1 mmole/(liter minute) for an organic compound in water, in a computational box of typical size (say,

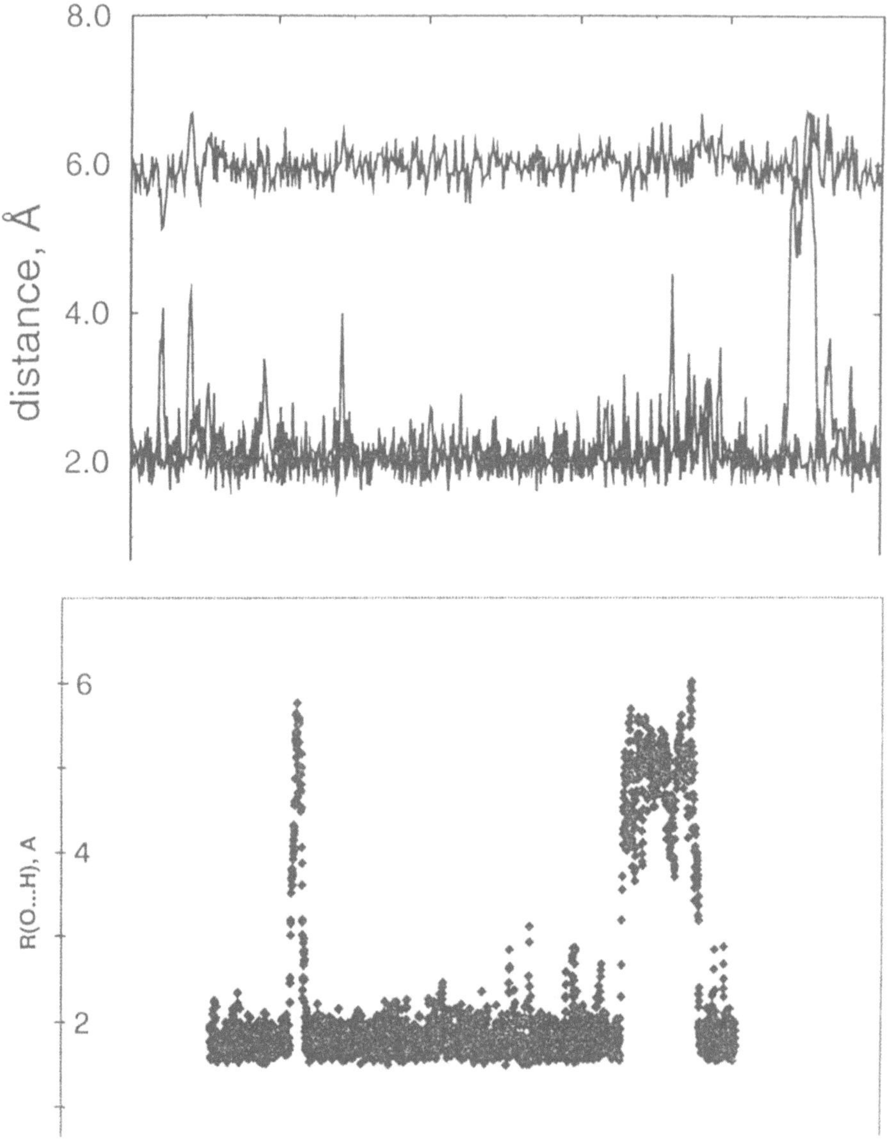

Figure 3. Above: 2-pyridone cyclic dimer, distance between centers of mass and O...H distances during a 100 ps MD run. Below: tetrolic acid cyclic dimer, O...H distances during a 250 ps MD run. Peaks to 5- 6 Å in the O...H distance denote breaking of one hydrogen bond.

100 Å edge) one should observe one bulk solvate molecule per minute, a timescale which is 8-9 orders of magnitude longer than presently affordable in MD simulations. This gap could presumably be bridged by an appropriate choice of solvate and solvent, as well as by computationally overheating the system.

4. Freezing and melting

"Everybody would like to know more about two of the most common phenomena in physical science - the melting of solids and the inverse process, their crystallization.... how does a whole crystal being warmed know that the point is approaching when each of its atoms or molecules is to forsake imprisonment on the crystal lattice for the freedom to move about at will?" [15]. A very good question indeed....

The state of affairs can be summarized by reviewing two papers [16, 17] on the subject. The first states that "The microscopic mechanism by which a crystalline solid forms in the freezing of a pure liquid is far from understood"; the second, 12 years later, concurs: "Although crystal nucleation has been the subject of extensive experimental and theoretical study, our understanding of crystal nucleation at the microscopic level is still limited". In the formation of a solid, the development of critical nuclei is the rate-determining step, since growth is much faster. The free energy separation between solid and liquid depends on the difference in bulk chemical potentials, but the barrier to nucleation is likely to be mostly determined by the surface free energy of the crystallites. The time a crystallite spends at the top of that barrier is of the order of nanoseconds, and this hampers experimental studies; computer simulation is then the best tool. Again, however, one runs into a timescale problem: the structure of the nuclei is not known, and they must be somehow produced in the simulation. Actual nucleation rates being of the order of one nucleus/(ml second), the observation of a nucleation event would require an astronomical time in a typical computational box. Extensive computational supercooling is therefore a must; presumably however, instead of artificially decreasing the kinetic energy, one could manipulate the potential function by artificially increasing its attractive parts. Apparently, the only definite conclusion one can draw so far about homogeneous nucleation is that it does occur, i.e., crystallization originates in localized regions rather than globally.

Nucleation for crystallization could be described by the equilibrium:

$$(m+n)A \leftrightarrow [mA] + nA \qquad (6)$$

with m+n of the order of Avogadro's number, and m of the order of a few units to a few tens. A microscopic cluster, [mA], somehow "dissolved" within the bulk liquid, nA, could evolve and grow by the following steps:

$$[mA]_j \rightarrow [mA]_k \quad \text{and} \quad [mA] + A \rightarrow [(m+1)A] \qquad (7)$$

Melting should somehow involve the collapse of the crystal structure, or the macroscopic cluster [(m+n)A]. No doubt, molecular librations have to do with that, but basic theories, like the Lindemann mechanism, have very few chances of being applicable to much other than simple spheres. In actual molecular crystals, premelting phenomena, or rearrangements and diffusion within the crystal structure with partial preservation of long-range order, are very likely, and could be symbolized as

$$[(m+n)A]_j \rightarrow [(m+n)A]_k \qquad (8)$$

Incidentally, equation (8) is also a schematization of polymorphic transitions, which however very seldom (if ever) proceed by a direct crystal-to crystal pathway. Actual melting is almost certainly preceded by some local loss of order, of which surface melting is the most likely form. This would split the crystal in two regions, one still crystal-like and one liquid-like, the former merging eventually into the second:

$$[(m+n)A] \rightarrow [mA] + nA \qquad (a)$$

$$[mA] \rightarrow mA \qquad (b)$$

(9)

In these equations, both m and n have macroscopic dimensions. Needless to say, the detailed mechanisms underlying these processes are still quite obscure.

The solid-liquid equilibrium can in principle be studied by MD, but the timescale problem is devastating if one looks at the equilibrium in equation (6) from left to right; computational overheating can make the right-to-left pathway more practical. Quite a few simulations have been carried out on the Lennard-Jones fluid, a test system presumably with little resemblance to a real molecular system, although the essentials of the process could be the same. Besides, dP/dT is so large that instabilities may easily arise in the computations, so that some simulations are carried out at constant volume (NVT), rather than at constant pressure (NPT); this condition is rather unrealistic for a solid-liquid transition.

The equilibrium thermodynamics and density of pure, separate liquids and crystals can be simulated rather easily (Table 1), at constant or variable temperature [18]. For liquid benzoquinone, in spite of simulation temperatures being well below the melting point, no trace of nucleation was observed over 200 ps. Computational supercooling encourages crystallization on one side, for obvious reasons, but also reduces rotational and translational diffusion, producing a glassy state in which molecules cannot cross the diffusional barriers. A sound computational experiment could be the MD equilibration of a supercooled liquid, followed by its computational seeding with small crystalline clusters, extracted from the actual crystal structure and strategically located within the bulk of the liquid. A subsequent MD treatment would hopefully see the preformed or symmetry-enforced clusters templating the further nucleation and growth steps.

Table 1. Results of MD simulations on liquids (from ref. [18]).

compound	ΔH_{vap}, kJ/mole		density, g/cc
	calc	exptl	
formic acid	32.8	41.4	1.138
maleic anhydride	43.8	50.6	1.289
acetamide	51.8	58.6	1.034
furan	28.3	27.7	0.938
succinimide	67.0	68.6	1.291
benzoic acid	69.9	63.8/68.2	1.157
coumarin	68.6	63.6	1.198

References

1. Paul,I.C. and Curtin,D.Y. (1981) Chemical Consequences of the Polar Axis in Organic Solid-State Chemistry, *Chem.Revs.* **81**, 525-541.
2. Gavezzotti, A. and Filippini, G. (1997) Energetic Aspects of Crystal Packing, in Gavezzotti, A. (ed.), *Theoretical Aspects and Computer Modeling of the Molecular Solid State*, Wiley, Chichester, Ch. 3.
3. Gavezzotti, A. (1997a) Crystal Symmetry and Molecular Recognition, in Gavezzotti, A. (ed.), *Theoretical Aspects and Computer Modeling of the Molecular Solid State*, Wiley, Chichester, Ch. 1.
4. Gavezzotti, A. and Filippini,G. (1994) Geometry of the Intermolecular X-H...Y (X,Y=N,O) Hydrogen Bond and the Calibration of Empirical Hydrogen-Bond Potentials, *J.Phys.Chem.* **98**,4831-4837.
5. Brock, C.P. and Duncan, L.L. (1994) Anomalous Space-Groups Frequencies for Monoalcohols C_n H_m OH, *Chem.Mater.* **6**, 1307-1312.
6. Gavezzotti, A. and Filippini, G. (1995) Polymorphic Forms of Organic Crystals at Room Conditions: Thermodynamic and Structural Implications, *J.Am.Chem.Soc.* **117**,12299-12305.
7. Gavezzotti, A. (1997c). Unpublished results.
8. Bartell, L.S. (1995) Nucleation Rates in Freezing and Solid-State Transitions. Molecular Clusters as Model Systems, *J.Phys.Chem.* **99**, 1080-1089.
9. Agrawal, R. and Kofke, D.A. (1995) Thermodynamic and structural properties of model systems at solid-fluid coexistence. II. Melting and sublimation of the Lennard-Jones system, *Molec.Phys.* **85**, 43-59.
10. Boek, E.S, Briels, W.J. and Feil, D. (1994) Interfaces between a Saturated Aqueous Urea Solution and Crystalline Urea: A Molecular Dynamics Study, *J.Phys.Chem.* **98**, 1674-1681
11. Esselink, K., Hilbers,P.A.J. and van Beest, B.W.H. (1994) Molecular dynamics study of nucleation and melting of *n*-alkanes, *J.Chem.Phys.* **101**,9033-9041.
12. Gavezzotti,A., Filippini,G., Kroon,J., van Eijck,B.P. and Klewinghaus, P. (1997) The Crystal Polymorphism of Tetrolic Acid ($CH_3C{\equiv}CCOOH$): A Molecular Dynamics Study of Precursors in Solution, and a Crystal Structure Generation, *Chem.Eur.J.* **3**, 893-899.

13. Benghiat,V. and Leiserowitz,L. (1972) Molecular Packing Modes. Part VI. Crystal and Molecular Structures of Two Modification of Tetrolic Acid, *J. Chem. Soc. Perkin II*, 1763-1768.
14. Gavezzotti, A. (1997b) Computer simulations of organic solids and their liquid-state precursors, *Faraday Discuss.* 106, xxxx-xxxx.
15. Maddox, J. (1995) Colloidal crystal model real world, *Nature*, **378**, 231.
16. Harrowell,P. and Oxtoby,D.W. (1984) A molecular theory of crystal nucleation from the melt, *J.Chem.Phys.* **80**, 1639-1646.
17. ten Wolde, P.R., Ruiz-Montero, M.J. and Frenkel, D. (1996). Numerical calculation on the rate of crystal nucleation in a Lennard-Jones system at moderate undercooling, *J.Chem.Phys.* **104**, 9932-9947.
18. Gavezzotti, A., Filippini, G. and van Eijck ,B.P.(1997). To be published.

GAS HYDRATES AS SELF-ASSEMBLED TEMPLATED LATTICES

Towards Time-resolved Studies of Clathrate Hydrate Formation

JOHN A. RIPMEESTER, IGOR L. MOUDRAKOVSKI AND
CHRISTOPHER I. RATCLIFFE
Steacie Institute for Molecular Sciences
National Research Council of Canada
Ottawa, Ontario, Canada K1A 0R6

1. Introduction

The theme of this NATO ARW is the challenge posed by the design, synthesis and characterization of large supramolecular assemblies. On the structural characterization side, the advent of the CCD-detector equipped diffractometer has revolutionized the way that crystallographic analysis is carried out. One example is the solution of the bromine hydrate structure some 150 years after its existence was first reported[1], and another is the double rosette structure published recently[2]. Significant advances are also being made in solution techniques such as NMR spectroscopy and gas-phase methods such as mass spectrometry. In this setting of improved instrumentation and characterization methodology it is again of interest to examine the possibilities of studying the processes that lead to self-assembly and that connect the product to possible pre-cursor states. Developments in the NMR spectroscopy of noble gases has made it particularly attractive to focus on the longstanding and ever important problem of gas hydrate formation[3].

1.1 BACKGROUND

The simple (two-component) clathrate hydrates are probably the best understood examples of templated, self-assembled lattices. About 120 guests ranging in size from argon to isobutane are known to form either the cubic structure I or structure II clathrates[4 5]. If double hydrates are also taken into account the size range is expanded to include molecules the size of adamantane and 1,2 dimethylcyclohexane. These form the hexagonal str. H hydrate. Each hydrate family has unit cell parameters that are very similar, and has compositions that are variable within reasonably well-known limits (Table 1). Recent interest in hydrates has been sparked by the vast quantities of natural gas stored in off-shore hydrate deposits that exist in many locations on the continental slopes[6].

Over the years there has been much interest in the processes by which hydrates form: a solid lattice is assembled from liquid water and a gaseous guest that is often quite insoluble in water[7]. Alternatively, hydrates will form directly from solid ice and gaseous guest. It is also of great technical interest to learn how to prevent hydrates from forming, as solid hydrate plugs are known to form in pipelines used for the transmission of gas and

G. Tsoucaris (ed.), Current Challenges on Large Supramolecular Assemblies, 265–272.

crude oil, often with disastrous results[8]. Today, much effort has been expended on so-called kinetic inhibitors that slow the growth of hydrate crystals after nucleation. Thus, the study of processes such as nucleation, crystal growth and inhibition of crystal growth, from a molecular point of view, are of critical importance.

Table 1. Summary of the clathrate hydrate structural and composition data.

Hydrate Structure	Lattice Symmetry	Unit Cell Parameters (Å)	Ideal* Stoichiometry
Str. I	Pm3n	12.0	$2X \cdot 6Y \cdot 46H_2O$
Str.II	Fd3m	17.0	$16X \cdot 8Y \cdot 136H_2O$
Str.H	P6/mmm	a = 12.2 c = 10.0	$6X \cdot 5X' \cdot Y \cdot 34H_2O$

*X and X' are small cage sites, Y is the large cage site

1.2 MACROSCOPIC KINETIC STUDIES

In the past, most of the information on hydrate kinetics has come from gas uptake measurements, where gas at a pressure greater than the decomposition pressure of the hydrate at the temperature of measurement is admitted to a container of vigorously stirred water. The gas uptake is then followed as a function of time as the hydrate nucleates and the solid hydrate phase grows. In such studies it is often difficult to separate the actual kinetic parameters from the factors that are limited by transport, eg gas transport to the interfacial regions where hydrate growth occurs, and the removal of the heat of

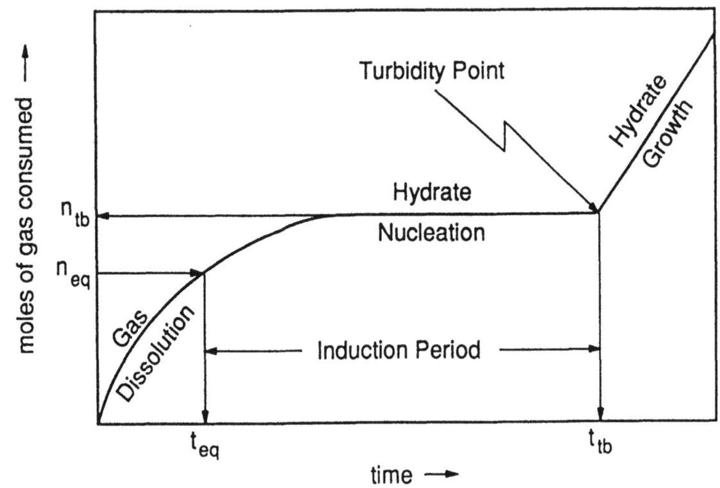

Figure 1. Macroscopic kinetic study of hydrate formation. A typical plot of gas consumed as a function of temperature at constant temperature and pressure (after ref. 7).

crystallization. Some of these problems were addressed by Barrer et al[9], who studied the direct reaction of solid ice with gas in a system where the solid ice/hydrate was ground continuously to provide fresh surface for reaction. Various stages of the reaction (initial gas dissolution, induction period associated with nucleation and a period of rapid hydrate growth) can be recognized in the gas uptake *versus* time curve (fig.1). These have been interpreted in terms of a number of models, although there is little or no confirmation of their validity since they require information on a molecular scale.

2. Experimental Considerations

What are likely to be the best approaches for obtaining information about transient states? Diffraction will be of somewhat limited use[10] as it can give definitive information only after crystal phases have nucleated. Techniques sensitive to local order are likely to be able to give significant amounts of information. This includes the various spectroscopic methods as well as methods that provide information on radial distribution functions in solution or for putative precursor states.

Table 2. Hydrate cages, cage symmetries, cage sizes and ^{129}Xe chemical shift data

Structure	cage type	symmetry	radius/Å^{a}	δ^{Xe}_{iso} ^{b}/ppm	Δ^{Xe} ^{c}/ppm	η^{Xe} ^{d}
I	5^{12}(D)	$\underline{m}3$	2.50	242	0	
I	$5^{12}6^{2}$(T)	42m	2.93	152	21	
II	5^{12}(D)	$\underline{3}$m	2.50	231	16	
II	$5^{12}6^{4}$(H)	43m	3.28	80	0	
H	5^{12}(D)	mmm	2.45	231	13.7	0.70
H	$4^{3}5^{6}6^{3}$(D')	62m	2.40	212.4	31.4	
H	$5^{12}6^{8}$(E)	6/mm	3.8	---------	------	

^{a} estimates for the cages roughly approximated as spheres from X-ray diffraction data; ^{b} isotropic chemical shift; ^{c} $\Delta=\delta_{iso} - \delta_{zz}$ chemical shift anisotropy; ^{d} asymmetry parameter $\eta = (\delta_{yy} - \delta_{xx}) / (\delta_{iso} - \delta_{zz})$; the δ_{ii} are the chemical shift tensor elements; the ^{129}Xe shifts are referenced to xenon gas at zero pressure. The error estimates on both δ_{iso} and Δ^{Xe} are ±1 ppm and ±0.04 on η^{Xe}.

Other considerations include the ease with which time resolved studies can be carried out, the time scales that are accessible, and the certainty with which data can be interpreted, eg, spectral assignments and quantitation in vibrational spectroscopy[11].

2.1 XENON NMR SPECTROSCOPY

Over the years we have used ^{129}Xe NMR spectroscopy as a technique to study structural aspects of the clathrate hydrates[12]. In fact, this technique played a large part in identifying a new structure known as structure H hydrate[13]. Briefly, each guest site, commonly known as a hydrate "cage", gives the xenon guest in that cage its own spectral signature: the isotropic chemical shift reflects the cage size and the anisotropic chemical shift reflects the cage symmetry (fig. 2, table 2). There is every expectation that in the future the xenon shifts will be interpretable directly in terms of the xenon - hydrate cage

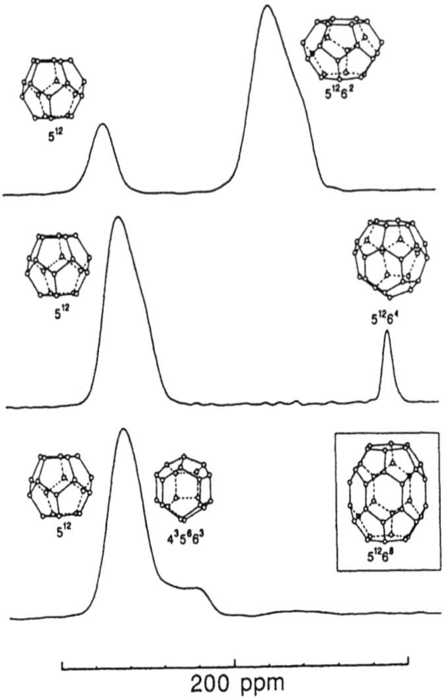

Figure 2. Xe NMR spectra observed for structure I, II and H hydrate; the cages are shown as insets. The x^y descriptor refers to the fact that the cage has y faces made up out of polygons with x sides.

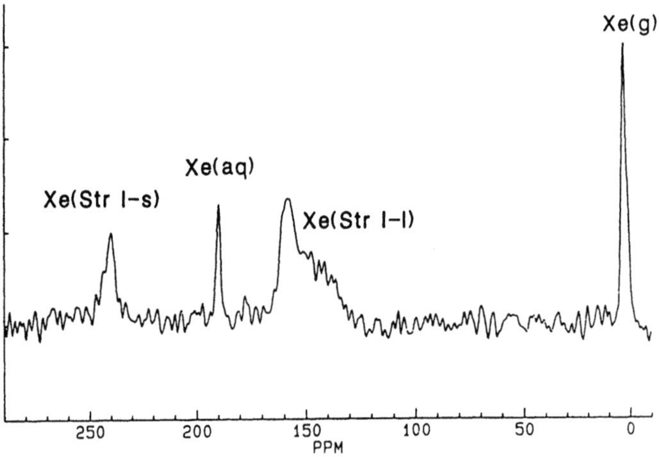

Figure 3. ^{129}Xe NMR spectrum of xenon structure I hydrate in equilibrium with water and xenon gas. The signals can be attributed to xenon in the two cages of str. I hydrate, xenon dissolved in water, and xenon gas.

potential function. Because of the large ^{129}Xe chemical shift dispersion, Xe NMR spectroscopy has the potential for keeping track of all the xenon signals even in the presence of a number of phases (fig. 3). However, it has been difficult to use ^{129}Xe NMR spectroscopy as a quantitative technique to study processes because of the long times usually required for data collection. This is due to the low inherent sensitivity of the NMR technique and the long relaxation times of the xenon atom inside the hydrate cages. This can be alleviated somewhat by the use of cross-polarization techniques that improve the signal to noise ratios, but at the cost of quantitative intensity information.

2.1.1 *Low temperature approaches*

Several approaches have been tried to utilize Xe NMR as a site-specific and structure-specific probe for local structure formation in hydrate precursor materials despite the long data acquisition times. Some of these involve the slowing down of the processes to be studied in order to accommodate the long data acquisition times. One way this can be done is to produce intimate mixes of the guest and host materials, then to anneal the mixture under an appropriate temperature program, and to collect data during the "cold" periods of the cycle. The intimate mix of guest and host is prepared at very low temperature (well below the glass transition) by co-deposition of metered amounts of the vapours at low temperature. In our case, this is accomplished by using the tail of a displex refrigerator that is kept at 10 - 20K. Gram-sized quantities can be produced in about 8 hrs., harvested under liquid nitrogen, and stored for subsequent study. It is advantageous to use complementary methods to study the development of order in such materials: NMR for local order and X-ray powder diffraction for long range crystalline order. Preliminary reports of such experiments have appeared[5].

Another low temperature approach that has considerable appeal concerns the hyperquenching of aqueous solutions of hydrate formers. Hyperquenching is the extremely rapid freezing (10^6K/sec) of small, nebulized drops of the solution. Again, the product of such a process can be harvested at low temperature and studied on timescales that correspond to correspond to data acquisition times for standard methods such as diffraction and NMR spectroscopy. Some work in this area is in progress and will be reported in the near future.[14]

2.1.2 *NMR Spectroscopy and Hyperpolarized Noble Gases*

An alternative to slowing down kinetic processes to the time scale of the experiment is to increase the sensitivity of the experiment so that the processes can be studied directly on their characteristic time scale. A development that has made this possible is the use of optical polarization to enhance the sensitivity of the NMR experiment. For the ^{129}Xe isotope, signal enhancements of up to 10^4 have been reported[15], and this increases the prospects for real time observation of kinetic processes tremendously.

Hyperpolarized xenon can be produced by optically polarizing the electron spins in vapourized Rb by laser pumping in a magnetic field and transferring the polarization to the nuclear spins on xenon atoms through a transient binary complex that exists in the gas phase (fig.4). Polarization levels of up to 40% have been achieved as compared to ~20 ppm achievable through polarization in a high-field NMR magnet. The hyperpolarized state persists long enough so that the xenon can be transferred to appropriate samples for observation by NMR spectroscopy.

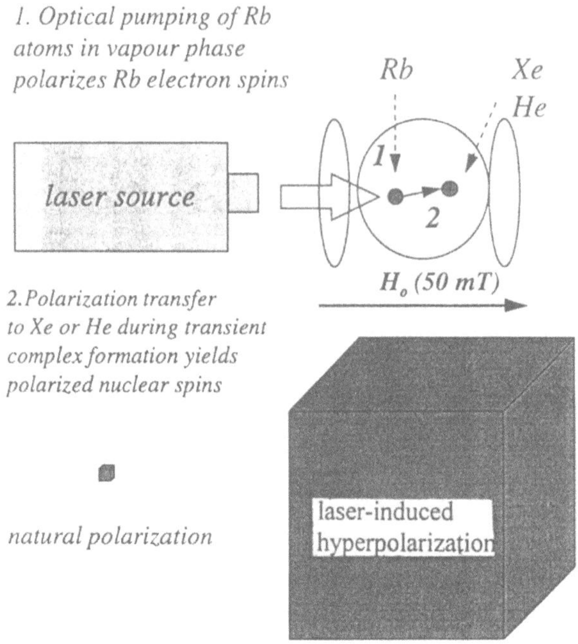

Figure 4. Schematic representation of the hyperpolarization of noble gases and the increase in sensitivity that can be expected under optimum conditions. The optical polarization process is carried out in a weak magnetic field H_0 with circularly polarized radiation (rubidium D_1 line) .

We have followed the formation of clathrate hydrates on the surface of powdered ice with hyperpolarized xenon (figs. 5, 6). Time resolution down to ~ 40 msec can be achieved with excellent signal to noise ratios, thus allowing the observation of site-specific kinetic processes. The reaction appears to go to completion well within the lifetime of hyperpolarized xenon gas (the slow decay of all signals is due to the finite lifetime of the hyperpolarized state) . The gas line at 0 ppm shows a slow decrease with time after initial exposure, although there is little to see of the two xenon sites in the solid hydrate. This seems to correspond to the initial period of gas dissolution and the nucleation time as shown in figure 1. Formally one can say that an induction period then also exists in the molecular scale studies, thus suggesting that it may be possible to réfine the macroscopic hydrate formation models as we learn to analyze this molecular scale data. It is also hoped that information will become available on the composition of the hydrate as a function of time, and that the induction time can be studied systematically as a function of temperature and pressure. This will help to identify the driving force for nucleation. There also are indications that the formation of hydrate on the surface of ice will have potential in testing the reactivity of ice surfaces. Measurable parameters involved in the process include the induction time as well as the rate and the yield of hydrate formation. For example, the action by which antifreeze proteins inhibit surface growth is thought to occur via a molecular recognition mechanism on specific crystal faces.

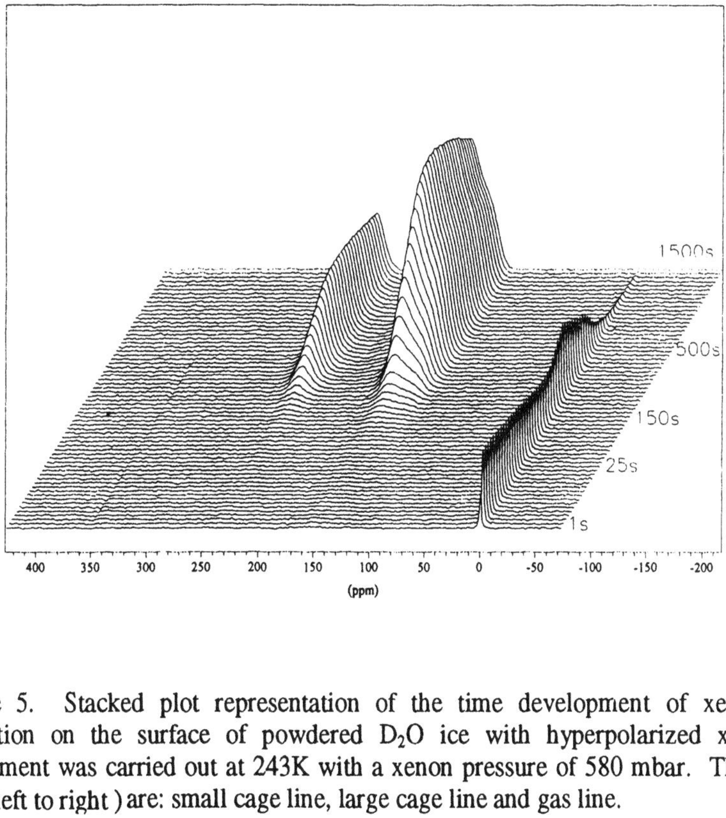

Figure 5. Stacked plot representation of the time development of xenon hydrate formation on the surface of powdered D_2O ice with hyperpolarized xenon. The experiment was carried out at 243K with a xenon pressure of 580 mbar. The features (from left to right) are: small cage line, large cage line and gas line.

3. Conclusions

In this short presentation, which is very much a progress report, we have shown that there are excellent prospects for making major advances in understanding processes such as self-assembly in the two phase solid – gas systems that form clathrate hydrates. There is every reason to expect that time-resolved studies with molecular-level resolution will be possible from a number of experimental perspectives. As with all complex problems, the best models are likely to be derived from strategies that employ complementary experimental approaches.

References

[1] Udachin, K. A., Enright, G. D., Ratcliffe, C. I., Ripmeester, J. A. (1997), The Crystal Structure, Composition and Morphology of Bromine Hydrate, *J. Am. Chem. Soc.* **119**, 11481-11486

[2] Timmerman, P., Vreekamp, R. H., Hulst, R., Verboom, W., Reinhoudt, D. N., Rissanen, K., Udachin, K. A., Ripmeester, J., (1997) Non-covalent assembly of functional groups on Calix[4]arene Molecular Boxes, *Chem.. Eur. J.* **3**, 1823 - 1832

[3] Proceedings of the 1st Int'l Conference on Natural Gas Hydrates, (1994) *Ann. N. Y. Acad. Sci.* **715**

[4] Davidson, D. W., Clathrate Hydrates (1974) in *Water. A Comprehensive Series*, Ed. Felix Franks., Plenum, Vol. 2

[5] Ripmeester, J. A., Ratcliffe, C. I., Klug, D. D., Tse, J. S., (1994) Molecular perspectives on Structure and Dynamics in Clathrate Hydrates, *Ann. N. Y. Acad. Sci.* **715**, 161 - 176

[6] Kvenvolden, K. A., (1994) Natural Gas Hydrate Occurrence and Issues, *Ann. N. Y. Acad. Sci,* **715**, 232 - 246

[7] Bishnoi, P. R., Natarajan, V., Kalogerakis, N. (1994) A unified description of the kinetics of hydrate nucleation, growth and decomposition, *Ann. N. Y. Acad. Sci.,* **715**, 311-322

[8] Various papers, (1994) *Ann. N. Y. Acad. Sci,* **715**

[9] Barrer, R. M., Ruzicka, D. J., (1962) Non-Stoichiometric Clathrate compounds of water, Part 4., *Trans. Farad. Soc.* **58**, 2262 - 2271

[10] Koh, C. A., Savidge, J. L., Tang, C. C., (1996) Time resolved in-situ experiments on the crystallization of natural gas hydrates, *J. Phys. Chem.* **100**, 6412-6414

[11] Sum, A.K., Burruss, R. C., Sloan, Jr., E. D., (1997) Measurement of clathrate hydrates via Raman spectroscopy, *J. Phys. Chem.* **101**, 7371-7377

[12] Ripmeester , J. A., Ratcliffe, C. I., (1990) ^{129}Xe NMR studies of clathrate hydrates: New guests for str. II and str. H., *J. Phys. Chem.* **94**, 8773

[13] Ripmeester, J. A., Tse, J. S., Ratcliffe, C. I., Powell, B. M., (1986) A new clathrate hydrate structure, *Nature* **311**, 142

[14] Tulk, C. A., Klug, D. D., McLaurin, G., Ripmeester, J. A., unpublished

[15] Pietrass, T., Gaede, H. C., Bifone, A., Pines, A, Ripmeester, J. A., (1995) Monitoring clathrate hydrate formation on ice surfaces with optically polarized xenon, *J. Am. Chem. Soc.* **117**, 7520

STUDY BY ELECTROSPRAY MASS SPECTROMETRY OF LARGE DENDRITIC SUPRAMOLECULAR COMPLEXES

Leize Emmanuelle[1], Moucheron Cécile[2], Dupont-Gervais Annick[1], Kirsch Andrée[2] and Van Dorsselaer Alain[1].

[1]LSMBO, Université Louis Pasteur, CNRS URA 31, Faculté de Chimie, 1 rue Blaise Pascal, 67008 Strasbourg, France.
[2]Service de Chimie Organique Physique, Université Libre de Bruxelles, 50 avenue F.D. Roosevelt, 1050 Bruxelles, Belgique.

The use of electrospray (ES) for the ionization of a very wide variety of biochemical and chemical compounds with low or high molecular masses is now well established in mass spectrometry (MS). Now, ESMS is also usually used in the field of supramolecular chemistry to characterize and to follow the formation of complexes and their intermediates in solution. [1 , 2]

In the case of some polynuclear transition metal complexes of dendritic nature, an unequivocal characterization is still missing, mainly due to stereogenic problems. Indeed, because of the numerous diastereoisomers in polynuclear compounds composed of polyazaaromatic Ru(II) complexes, the characterization by NMR is of limited use. For these artificial nanostructures, ESMS is the only analytical tool allowing their unambiguous characterization.

The aim of this work is the characterization by ESMS of a series of novel dendritic polymetallic complexes of Ruthenium (II). The largest complex with seven Rutheniums (II) has a molecular weight of 5602.2 Da. [3]

Three problems were encountered for the characterization by ESMS of this type of complexes :

1. The low amounts of sample available needed the use of a nanospray source. Then it was necessary to optimize the nanospray capillary for the use of organic solvents (acetonitrile, dichloromethane).
2. The low stabilitiy of these compounds in the gas phase made necessary the use of very mild ES interface conditions (low extraction cone voltage Vc and capillary exit voltage).
3. The partial decomposition of the counter-ions PF6- in the solution complicated the ES mass spectra since no clear series of multiply-charged ions could be detected.

For the four complexes studied, all the peaks observed in the ESmass spectra corresponded to the expected pure Ru(II) metallic complex. Ions were obtained by the loss of the counter-ions. For the monometallic, dimetallic and trimetallic complexes, the counter-ions were PF6- and the spectra were simple. The heptametallic complex was associated with different types of counter-ions resulting from the

G. Tsoucaris (ed.), Current Challenges on Large Supramolecular Assemblies, 273–274.

quite complex and high resolution analysis has been necessary for the ions identification. The stability of the complexes in the gas phase was shown as decreasing with the increase of their nuclearity.

ESMS brought clearcut evidences for the nuclearity of this series of Ru(II) dendritic complexes. Now, ESMS could be regarded as extremely valuable to prove the existence of very large polynuclear transition metal complexes, and paves the way to future design of artificial nanostructures with unambiguous characterization.

1. Fenn et al., Science, 1989, 246, 64 ; Fenn et al., Mass Spectrom. Rev., 1990, 9, 37.
2. Bitsch et al., J. Am. Chem. Soc., 1991, 113, 4023 ; Leize et al., J. Chem. Soc., Chem. Commun., 1993, 990 ; Marquis-Rigault et al., Inorg. Chem., 1996, 354, 2307 ; Leize et al., J. Mass Spectrom., 1996, 31, 537.
3. Ortmans et al., Inorg. Chem., 1995, 34, 3695 ; Moucheron et al., J. Am. Chem. Soc., 1997, 118, 12834.

ADVANCED CRYSTALLOGRAPHIC METHODS: EXPERIMENTAL ELECTRON DENSITIES AND THE STUDY OF EXCITED STATE STRUCTURE

PHILIP COPPENS,
Chemistry Department, State University of New York at Buffalo, Buffalo NY 14260-3000

Abstract

As techniques for structure determination are further perfected, other crystallographic applications, of relevance for our understanding of the electronic and geometric structure of solids and molecules, merit attention. The first is the mapping of charge densities in crystals, based on accurate X-ray diffraction intensities, usually collected at low temperatures and with significant redundancy in the data sets. New developments, including area detectors and novel software, are opening the possibility of routine determination of the charge distribution in crystals. The second application discussed here, which is in a much less advanced state, exploits the time structure of the third-generation synchrotron sources through combined laser/X-ray experiments, in which a laser-triggered process is probed by the X-ray photon beam. Such photo-crystallographic experiments will allow the study of transient species in crystals.

1. Introduction

A quick glance at the literature shows that much of the structural insight into Supramolecular Chemistry would not have been possible without the tremendous advances in structure determination techniques of the past decades. Many new compounds being synthesized have hundreds of atoms in the crystallographic unit cell, and are commonly solved routinely by standard programs. The development of methods to rapidly solve even larger structures is treated in other chapters of this volume.

As techniques for structure determination are further perfected, other crystallographic applications, of relevance for our understanding of the electronic and geometric structure of solids and molecules, merit attention. The first is the mapping of charge densities in crystals, based on accurate X-ray diffraction intensities, usually collected at low temperatures and with significant redundancy in the data sets. The recent advent of area detectors and the development of much easier to use software, such as the program package XD,[1] has raised the prospect of routine determination of the charge distribution in crystals.

The second application we will discuss is at a much less advanced state, and made possible by the very bright, pulsed X-ray beams that are now becoming available. The time structure of the third generation synchrotron sources such as the European Synchrotron Source, the Advanced Photon Source in the US, and SPRING-8 in Japan lends itself to combined laser-X-ray experiments, in which a laser-triggered process is probed by the X-ray photon beam. By exposing a small sample crystal to an exciting laser pulse, just before it is being probed by the X-ray photons, photoinduced processes in crystals can be studied. Given a reversible process, the excitation may be repeated until a sufficiently strong signal is obtained.

G. Tsoucaris (ed.), Current Challenges on Large Supramolecular Assemblies, 275–285.

2. The Measurement of Charge Densities by X-ray Diffraction

2.1. BACKGROUND

The ability to measure the experimental charge distribution in crystals from the intensities of the scattered X-rays was realized almost immediately after the discovery of X-ray diffraction. Debye[2] stated as early as 1915, only three years after the seminal experiment by von Laue and collaborators,

'It seems to me that experimental study of the scattered radiation, in particular from light atoms, should get more attention, since along this way it should be possible to determine the arrangements of the electrons in the atoms'

Later statements by prominent crystallographers, however, were much more pessimistic.[3] The goal became more realistic as a result of technical developments of the 1960's and 70's, including the design of new diffractometers, automation of data collection, much better low temperature techniques, advances in computing power and software. In the last few years data collection techniques have been further revolutionized by the capability of area detectors to rapidly provide large amounts of quite accurate intensity data, and by the development of program systems specifically designed for Charge Density Analysis.[1]

Comparison of the experimentally obtained electron densities with theoretical results provides a test for both techniques, which have very different limitations. Many such comparisons have now been done with often very good agreement. Discrepancies have often been traced to either experimental flaws, shortcomings in the basis set or correlation effects in the theoretical density, or simply to the fact that most calculations are performed on isolated molecules, while the experiment gives information on the solid.

The ability to obtain information on a molecule in a crystalline environment is a major advantage of the experimental technique. From the experimental density other physical properties, such as electrostatic moments, the electrostatic potential and lattice energies can be derived, and insight in molecular interactions can be obtained. Since experimental charge densities are only now becoming routinely accessible, such applications are few, but increasing.

2.2. METHODS[4]

The simplest way to illustrate the effect of bonding on the charge density is by subtraction of the density corresponding to ground state spherically averaged atoms, referred to as the promolecule density, from the total density as mapped by X-rays. The resulting *deformation density* shows accumulations in the bond and lone pair regions of a molecule and in regions of crystal field-stabilized orbitals of low spin transition metal complexes. However, for bonds between electron-rich atoms such as O and F, this may not be the case. The reason is that such atoms have on average more than one electron per atomic orbital, so that subtraction of a spherically averaged atom leads to a deficiency in the bonding region. Various alternative reference states have therefore been suggested. A properly oriented 'prepared' atom, for example, will have only one electron in the orbital that participates in the formation of the covalent bond. When such atoms are used as the reference state, overlap density appears in all covalent bonds. However, the choice of a 'prepared' atom introduces a measure of ambiguity as several alternatives have been used.

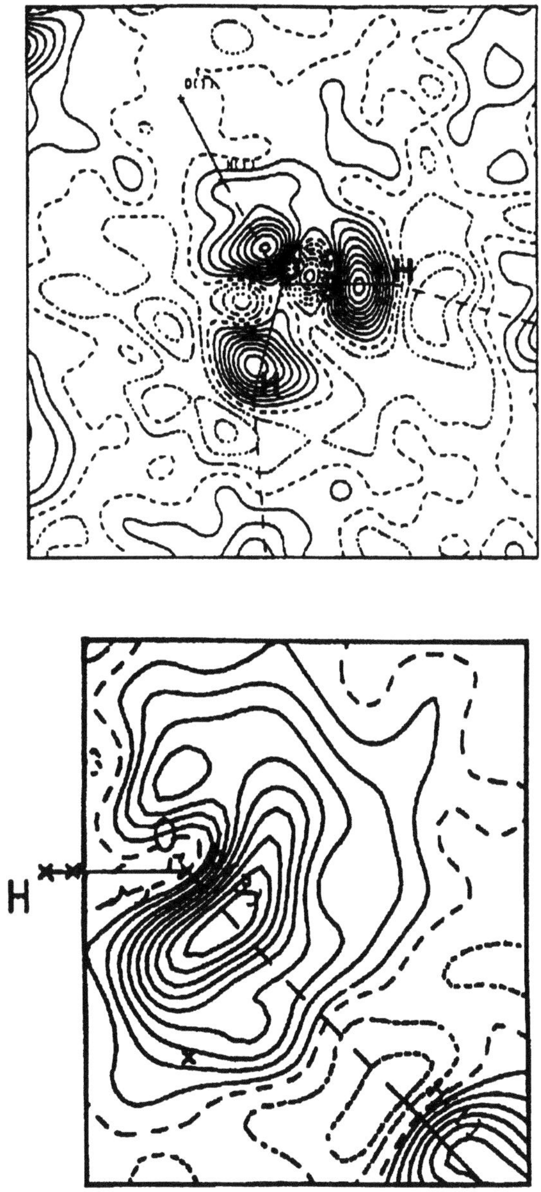

Figure 1. The water molecule in oxalic acid. Top: the deformation density in the plane of the molecule. Bottom: in a section perpendicular to the plane of the molecule. Contours at $0.05e\text{Å}^{-3}$. Zero contour broken. Negative contours dash-dotted. The hydrogen bond is indicated by the broken line.[5]

The deformation density is most easily obtained for centrosymmetric crystals, for which the phase problem is negligible, though special techniques applicable to acentric space groups have been developed.[6] An early picture of the deformation density in the water molecule in oxalic acid dihydrate is shown in Fig. 1. It shows the overlap density in the O···H bonds, and the lone pair density at the back of the oxygen atom. A section oriented perpendicular to the plane of the water molecule reveals that the lone pair density is polarized in the direction of the hydrogen donor atom of the short hydrogen bond (O···O = 2.48 Å). This is one of the first demonstrations of the effect of intermolecular interactions on the charge density. It is supported by a theoretical calculation on the water dimer.[7]

A different technique, based on least-squares refinement of the experimental structure factor amplitudes is now common. The structure factor expression of the *multipole formalism* is based on an aspherical-atom algorithm that also accounts for the effect of bonding on the atomic density. Thus, the atom-centered functions also represent the two-center overlap density between atoms. Because of this, Stewart introduced the term *pseudo-atom* to describe the sum of the density fragments assigned to a particular atom.[8]

The algorithm expresses the atomic density as an expansion of spherical harmonic functions centered on the nucleus. In the Hansen and Coppens formulation,[9] the pseudo-atom density is defined as

$$\rho_{atom}(\mathbf{r}) = P_{core}\rho_{core}(r) + P_{valence}\kappa^3\rho_{valence}(\kappa r) + \sum_{l=0}^{l_{max}} \kappa'^3 R_l(\kappa' r) \sum_{m=0}^{l} P_{lm\pm} d_{lm\pm}(\theta,\phi) \quad (1)$$

The populations P^v, $P_{lm\pm}$ and radial expansion/contraction parameters κ and κ' are the new *charge density variables* in the least-squares refinement which uses aspherical form factors corresponding to the atoms defined by (1). The charge density variables are refined together with the conventional positional and thermal parameters. The angular density functions $d_{lm\pm}$ are identical in shape to the atomic orbitals used in quantum chemistry, but describe the density distribution rather than the wave function.

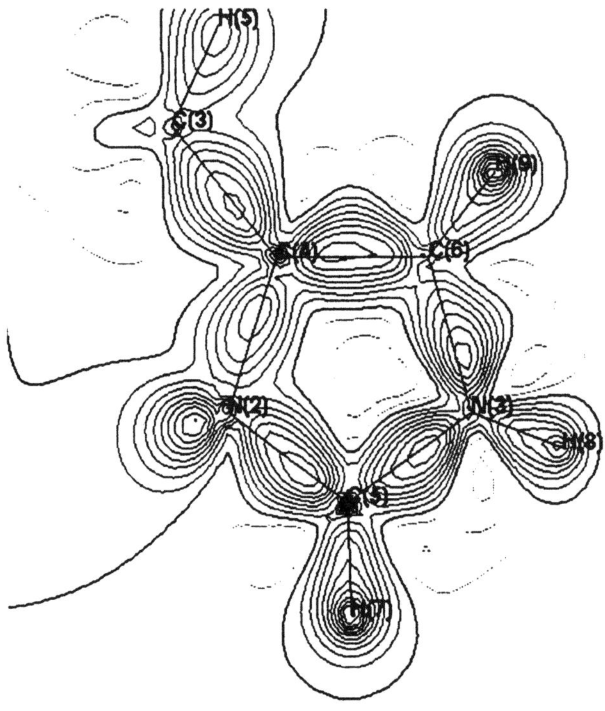

Figure 2. a) Model deformation map in the plane of the imidazole ring of DL-histidine. Contours at $0.1 e\text{Å}^{-3}$. Zero contour heavy, negative contours light (ref. 17).

Their radial dependence, and that of the valence density function $\rho_{valence}$, are multiplied by the κ and κ' expansion/contraction parameters respectively.

Combined, the refined parameters and the density functions provide an analytical description of the density in the crystal. In principle, this is a description of the static density, as the effect of the thermal smearing of the density has been explicitly accounted for by the temperature factors in the scattering formalism. Thus, within the limitations of that formalism, the thermal motion has been deconvoluted from the experimental density. An example of a model-based deformation density, for the imidazole plane of the histidine molecule is shown in Fig. 2

The populations and shape of the atom-centered spherical harmonic density functions in (1) are directly related to the electrostatic atomic moments. The total valence charge follows from the l=0 monopolar function, the atomic dipole moments from the three dipolar functions with l=1, the quadrupole moments from the functions with l=2, etc. It is often sufficient to truncate the expansion at the l=3, octupolar level for light atoms, and at the l=4, hexadecapolar, level for transition metal atoms in molecular complexes. In extended solids with atoms located at highly symmetric sites, functions with higher quantum numbers may be needed.

The electrostatic moments of a molecule in a crystal can be derived by summing the moments of the constituent atoms. Dipole and quadrupole moments obtained by X-ray diffraction up to 1992 have been summarized by Spackman.[10] Not unexpectedly, the moments for a molecule in a solid are generally larger than those calculated theoretically for isolated molecules, in agreement with recent theoretical results for molecules in solution.[11] For small molecules with conjugated double bonds in polar solvents, dipole moments enhancements can be pronounced.[12]

In crystals, the enhancement will be both a function of the polarisability of the molecule, and of the packing arrangement. The water molecule has a dipole moment of 1.84 D in the gas phase, while reported X-ray values cluster around 2.25 D. For comparison, a theoretical calculation gives 2.23 – 2.73 D for ice VIII and a number of prototype structures.[13] In crystals of urea, the molecular dipole moment is calculated to increase from a value of 5.14 D for the isolated molecule, to a value of 7.04 D in the crystal.[14] An early X-ray value[15] of 5.4 - 5.7 D needs to be reevaluated. For somewhat larger molecules, such as amino acids, the matrix effects are much larger when the molecule contains conjugated bonds or aromatic rings. For example, while for L-alanine the X-ray value of 12.9 D is close to the theoretical isolated molecule result of 12.4 D, [16] for the histidine molecule in crystals of DL-histidine, the X-ray and isolated molecule values are 17.4 and 15.9 D, respectively.[17] Such environment-induced changes in the charge distribution are of crucial importance, and must be accounted for to obtain reliable force fields for the calculation of protein folding.[18] Because the X-ray experiment gives information on the charge distribution in the crystal, it can be exploited for the quantitative evaluation of the intermolecular interactions that are the basis for supramolecular chemistry.

A powerful way of studying bonding is by topological analysis of the total density, based on the *Atoms In Molecules* (AIM) theory of Bader.[19] The topological analysis quantifies the interactions between atoms in terms of parameters such as the density of the critical points, the Laplacian at the critical point and the ellipticity of the bond path at the critical point. Since the analysis is density-based, it is a logical sequence to the X-ray experiment. Noise in the experimental density before modeling interferes with topological analysis, but this is not true for the analytical model density, which is now being used extensively for this purpose.

2.3. TOPOLOGICAL ANALYSIS OF WEAK INTERACTIONS

It is well recognized that electrostatic interactions, including hydrogen bonding, play a key role in determining the relative orientation of molecules in solids. They contribute a major part of the crystal-lattice energy, in particular for polar solids.

Crystal design can be put on a more quantitative basis by identifying the energy and force contributions of each of the intermolecular pairwise atomic interactions. In the topological AIM theory, such interactions are identified by the (3,-1) critical points of the charge density along the bond paths linking the atoms. Wile the density along the bond path goes through a minimum, contraction of the density in the plane perpendicular to the bond path gives a maximum in this plane, leading to a saddle point, described in the topological theory as a (3,-1) critical point. Some (3,-1) intermolecular critical points in the crystal of DL-histidine are listed in Table 1. In all but the weakest interactions, the electron density at the bond critical point exceeds that of the densities based on spherical atoms. The density at the critical point can be related to the kinetic energy density with a functional due to Abramov.[20] The example illustrates that energies related to intermolecular interactions can be obtained from the charge density. Though intramolecular bonding has been studied extensively, the topological analysis of intermolecular interactions is still in its infancy.

Table 1. Topological analysis of some intermolecular interactions in DL-histidine[21]

Interaction	R†	ρ(b)††	ρ(b)-I	ρ(b)-II	G†††
N-H...O	1.88	0.24	0.23	0.19	0.87
C-H...O	2.48	0.07	0.06	0.06	0.39
C-H...O	2.88	0.03	0.03	0.03	0.21

† distance in Å; †† column 3, density at the critical point in the experimental density; column 4, as 3 but in the promolecule density; column 5 as 3, but for density of charged spherical atoms with charge transfer as determined by experiment. ††† kinetic energy density(kJ/mol) at the critical point as derived from the experimental density according to Abramov.

In an alternative analysis of intermolecular interactions based on the charge density, the experimental atomic multipole expansions can be directly used to evaluate the electrostatic interactions between atoms or groups of atoms in adjacent molecules. An appropriate expression for the electrostatic interaction between two non-overlapping charge distributions is due to Buckingham.[22]

The application of experimental charge densities to the development of molecular force fields is still tentative, and will require careful analysis of the multitude of experimental charge densities that are now becoming available. Analysis of the weak interactions between molecules and the dependence of the charge distribution on the molecular environment is one a physical-chemistry frontiers of Supramolecular Chemistry. A better understanding of the strength of molecular interactions may be expected to lead to a more quantitative approach to crystal design.

3. Photoinduced changes in crystals and the study of transient species

3.1. BACKGROUND

In the late fifties and sixties Schmidt and coworkers carried out a series of pioneering studies on photochemical reactions in crystals. In work spanning two decades, they showed that the nature of the products of a solid reaction is topochemically-controlled by the geometry of the crystal structure.[23] One prime example is the photoinduced cycloaddition of cinnamic acids, which was found to lead to centrosymmetric or mirror-symmetric dimers, depending on the packing of the molecules in the crystal being irradiated.

The dramatic technical advances that have occurred since that time have put at our disposal highly intense pulsed synchrotron sources, pulsed lasers, fiber optic light guides, sub-nitrogen temperature cryostats, helium gas-flow systems and area detectors. In research initiated several years ago, we are exploiting the new technology in low-temperature diffraction studies of long-lived metastable states of transition metal nitrosyl complexes.[24] X-ray Crystallography has been traditionally used for the study of ground-state structure, but the elimination of this limitation is now within reach. The aim is to develop methods for the determination of the geometry, and eventually the charge density of transient light-induced species with lifetimes of μs and less.[25] To accomplish this aim, a pulsed source with suitable time-structure and high brightness is required. Third generation synchrotron storage rings, which are now increasingly available, are well suited for this purpose.

3.2. EXPERIMENTAL STRATEGY

The Advanced Photon Source (APS) at Argonne National Laboratory provides a pulsed beam with a frequency of about 300 kHz if one the ring is filled with one bunch of electrons, or with a group of closely-spaced bunches. Each electron bunch produces a light flash of about 100 ps length.[26] A *superbunch* of, for example 20 bunches, will thus provide a semi-continuous flash of about 1.9 ns.

The process to be studied may be triggered by a laser pulse of the proper wavelength, typically in the 300-600 nm range. By suitable synchronization of the laser and synchrotron pulses, an instantaneous non-equilibrium concentration can be established, and probed before significant decay occurs. Such pump-probe experiments have now been performed to study processes like the photodissociation and recombination of myoglobin and carbon monoxide.[27,28] In these experiments, a polychromatic Laue technique is used with a single synchrotron bunch of 60 ps length. By varying the delay between the exciting pump pulse and the probe pulse, the reversible dynamic process can be followed on a very fine time scale, and a "motion picture" of the reaction can be obtained. For experiments that aim at detailed atomic resolution or better, monochromatic methods have significant advantages over the polychromatic Laue technique. If the process is reversible, a stroboscopic experiment can be carried out, in which counts are accumulated over a number of synchrotron photon pulses, each preceeded by an exciting laser pulse (Fig. 3). The 'light-on' period is to be followed by a reference measurement carried out during a light-off period.

Such a differential experiment eliminates many systematic errors due to long-range fluctuations in intensity or other instabilities. It therefore gives superior sensitivity to the light-induced changes. As the repeat rate of pulsed lasers is significantly shorter than that of the synchrotron, either a fast pulse selector, or a detector with very fast read-out is needed to eliminate pulses not preceeded by a laser flash.

For such experiments to be successful, the number of photons in the exciting laser pulse must be comparable to the number of molecules in the crystal. This is well within reach of current technology. A 5 kHz Nd-YAG laser with a 50mJ pulse energy gives, for example, about 10^{14} photons per pulse at 355nm. For a typical complex, 10^{14} molecules will aggregate into a crystal of a linear dimension of about 40 μ. This is typical for samples used in many conventional synchrotron diffraction experiments. Working with small sample crystals has the further advantage that illumination is more uniform, and heat dissipation to the surrounding is enhanced. A reasonably uniform illumination throughout the crystal requires that only a fraction of the incident beam be absorbed. Thus, only part of the incident photons will be utilized. To counter this effect, the number of active molecules can be reduced further by embedding the

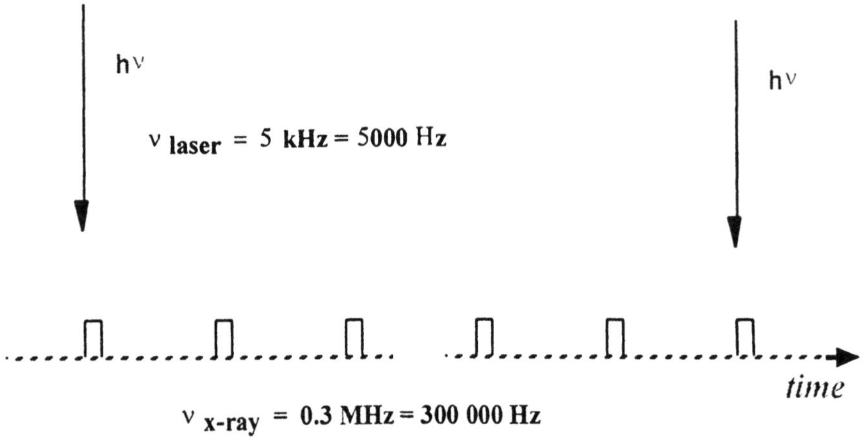

Figure 3. Schematic drawing of a stroboscopic experiment at the synchrotron source. The small rectangles represent the X-ray flashes at the pulsed source. The frequency indicated refers to the Advanced Photon Source at Argonne National Laboratory.

molecules as hosts in a spectroscopically inert matrix. The advantages are that i) the number of molecules to be excited to obtain a given conversion percentage is reduced, ii) concentration quenching due to exciton-exciton annihilation is diminished, and iii) choice of a transparent host will increase the transparency of the solid to the exciting light, and thus assure more uniform illumination.

3.3. SELECTION OF A HOST SYSTEM.

It has long been recognized that in photochemistry 'organized and constrained' media may have significant advantages over isotropic liquids.[29] Molecules can be held in fixed orientations that favor specific processes, and may inhibit deactivation of excited states through bimolecular interactions.

For the photocrystallographic experiments, a first condition is that the host has no absorption in the wavelength range to be used to excite the guest molecules. This eliminates most hosts with several conjugated bonds, including molecules containing larger condensed ring systems, while transition-metal containing host are also likely to be less suitable. Secondly, heavy atoms, which are strong X-ray absorbers, must be avoided. Third, the guest molecules must be ordered, which general implies a cavity of low symmetry, of a size matched to that of the intended guest. If the molecules are located in channels, as in the perhydroterphenylene inclusion complexes,[30] another type of disorder may occur. If a fixed height relationship between adjacent columns is lacking, i.e. the columns are not in registry, the diffraction pattern of the guest will consist of diffuse lines. In that case the amount of structural information that can be extracted is severely limited.

Microporous inorganic solids, such as zeolites and aluminophosphates, are known to incorporate organic guests including *p*-nitroaniline,[31] but will only be suitable if disorder of the guest molecules can be avoided. Cyclodextrin inclusion complexes offer attractive possibilities, as a wide range of molecules of interest have been included in the hydrophobic cavities of α- β- and γ-cyclodextrins. Although disorder is often observed in cyclodextrin inclusion complexes, there are many exceptions.[32]

The photochemistry of cyclodextrin complexes in solution has been studied extensively. It offers examples of both enhancement and quenching of the fluorescence of guest molecules, as well as changes in fluorescence lifetimes, and shifts in the emission spectra. In some cases the fluorescence is quenched by a chemical reaction which itself is an interesting subject of study.

Many other host systems can be selected, some of much greater complexity than those mentioned here. Additional studies are needed to explore further possibilities.

3.4. WHICH TYPES OF COMPLEXES ARE CANDIDATES FOR TIME-RESOLVED EXPERIMENTS ?

a) Transition metal complexes

Many transition metal complexes have excited triplet states with long lifetimes. Prime examples are complexes of rhodium studied more than twenty years ago by Crosby and coworkers. The light-induced triplet state of $[RhBr_2(d^5\text{-pyridine})_4]Br$, for example, has a lifetime of 1.5 μs in the solid state. With such long lifetimes, even conventional sources offer possibilities, as we have recently demonstrated in a series of test experiments carried out in preparation for the synchrotron studies.[33] In these experiments, a steady state concentration of excited molecules was maintained during a period of the order of 20 ms. In the synchrotron experiments, on the other hand, no equilibrium is to be established, and much shorter lifetime-species can be studied.

Examples of transition metal complexes with photoinduced excited triplet states having μs lifetimes are the binuclear metal d^8-d^8 complexes of rhodium and platinum. They show unusually reactivity, and are, in solution, intermediates in metal-catalyzed photoreactions of small molecules and organic substrates.[34] Since for d^8-d^8 complexes the excitation corresponds to a dσ* to pσ transition, very large contractions of the metal-metal distance occur if the bridging ligands are sufficiently flexible. An example is $Rh^2(dimen)_4^{2+}$ (dimen = 1,8-diisocyanomenthane) for which metal-metal bond shortening from 4.8 Å to about 3.2 Å upon excitation to a 21 μs lifetime state has been deduced from spectroscopic information. On the other hand, single metal-metal bonded Rh^0_2 and Rh^0Rh^{II} complexes with the bridging ligand (difluorophosphine) methylamine, $(PF_2CH_3N(PF_2)$, undergo excitation into a dσ* orbital, and are expected to show bond lengthening upon excitation.[35] In none of these cases is the distortion of the ligands and the change in metal-ligand bond distances upon excitation known.

b) Photo induced electron transfer and charge-separated states

Equally intriguing is the possibility for the study of electron transfer at atomic resolution. The electronic response of systems at the molecular level has been an active area of research in recent years. It has been noted that electron transfer within a single 'triad' molecule with donor, bridging and acceptor groups, is of relevance for energy conversion and the storage of information.[36,37] Such molecules have been proposed for the design of molecular computing devices.[38] Synthetic triad and tetrad molecules, containing for example a carotenoid donor, porphyrin or metal-porphyrin bridge and quinone acceptor(s) are being studied extensively for such purposes.[39] Many of such molecules do not lend themselves easily to crystallization. On the other hand very small crystals will suffice, and have indeed have significant advantages, as argued above. It is evident that a technique that provides information on geometry changes along the reaction path of photochemical charge separation and recombination would be a fascinating tool for our understanding of photochemical processes.

4. Concluding remarks

Structure analysis by diffraction methods has become a mostly routine, though indispensable component of supramolecular chemistry. We have outline in this article two of the new frontiers in crystallography. The ability to map the charge density in solids, and the development of time-resolved techniques are now enhancing the power of diffraction methods beyond their traditional applications.

Acknowledgments

Support of the research summarized in this article by the National Science Foundation (CHE9615586), the Petroleum Research Fund administered by the American Chemical Society (PRF #32638-AC3) and the U.S. Department of Energy (DE-FG02-86ER45231) is gratefully acknowledged.

References

1. Koritsansky, T., Howard, S., Richter, T., Su, Z., Mallinson, P. R. and Hansen, N. K. (1995) *XD Program Package for Multipole Refinement and Analysis of Electron Densities from Diffraction Data*. Free University Berlin.
2. Debye, P. (1915) Dispersion of Röntgen Rays. *Ann. Phys.* **46**, 809.
3. Bragg, W. L., James, R. W. and Bosanquet, C. H. (1922) The distribution of electrons around the nucleus of sodium and chlorine atoms. *Phil. Mag.* **44**, 433; James, R. W. (1948) *The Optical Principles of the Diffraction of X-rays*, Bell & Sons, London.
4. Coppens, P. (1997) *X-ray Charge Densities and Chemical Bonding*, Oxford University Press, New York.
5. Coppens, P., Dam, J., Harkema, S., Feil, D., Feld, R., Lehmann, M. S., Goddard, R., Krüger, C., Hellner, E., Johansen, H., Larsen, F. K., Koetzle, T. F., McMullan, R. K., Maslen, E. N. and Stevens, E. D. (1984) IUCr Commission on Charge, Spin and Momentum Densities: Project on comparison of structural parameters and electron density maps of oxalic acid dihydrate, *Acta Cryst.* **A40**, 184-195.
6. Souhassou, M., Lecomte, C., Blessing, R. H., Aubry, A., Rohmer, M.-M., Wiest, R., Benard, M. and Marraud, M. (1991) Electron distributions in peptides and related molecules. 1. An experimental and theoretical study of *N*-acetyl-L-tryptophan methylamide, *Acta Cryst.* **B47**, 253-266.
7. Stevens, E. D. (1980) Comparison of theoretical and experimental electron density distributions of oxalic acid dihydrate, *Acta Cryst.* **B36**, 1876-1886.
8. Stewart, R. F. (1976) Electron population analysis with rigid pseudoatoms, *Acta Cryst.* **A32**, 565-574.
9. Hansen, N. K. and Coppens, P. (1978) Electron population analysis of accurate diffraction data. VI. Testing aspherical atom refinements on small-molecule data sets, *Acta Cryst.* **A34**, 909–921.
10. Spackman, M. A. (1992) Molecular electric moments from X-ray diffraction data, *Chem. Rev.* **92**, 1769.
11. Gao, J. (1996) in K. B. Lipkowitz and D. B. Boyd (eds.), *Reviews in Computational Chemistry*, vol. 7, VCH: New York, p. 119; Gao, J. and Xia, X. (1992) A priori evaluation of aqueous polarization effects through Monte Carlo QM-MM simulations, *Science* **258**, 631.
12. Gao, J. and Alhambra, C. (1997) Solvent effects on the bond length alternation and absorption energy of conjugated compounds, *J. Am. Chem. Soc.* **119**, 2962-2963.
13. Gatti, C., Silvi, B. and Colonna, F. (1996) Dipole moment of the water molecule in the condensed phase: A periodic Hartree-Fock estimate, *Chem. Phys. Lett.* **247**, 135-141.
14. Gatti, C., Saunders, V. R. and Roetti, C. (1994) Crystal field effects on the topological properties of the electron density in molecular crystals. The case of urea. *J. Chem. Phys.* **101**, 10686-10696.
15. Swaminathan, S. and Craven, B. M. (1984) Electrostatic properties of phosphorylethanolamine at 123K from crystal diffraction, *Acta Cryst.* **B40**, 511.
16. Destro, R., Bianchi, R. and Morosi, G. J. (1989) Electrostatic properties of L-alanine from x-ray diffraction at 23 K and ab initio calculations, *J. Phys. Chem.* **93**, 4447-4457.
17. Coppens, P., Carducci, M. and Pressprich, M. R, to be published.
18. Beachy, M. D., Chasman, D., Murphy, R. B., Halgren, T. A. and Friesner, R. A. (1997) Accurate ab initio quantum chemical determination of the relative energetics of peptide conformations and assessment of empirical force fields, *J. Am. Chem. Soc.* **119**, 5908-5920.
19. R. F. W. Bader, (1990) *Atoms in Molecules: A Quantum Theory*, Clarendon Press, Oxford Science Publications: Oxford.
20. Abramov, Y (1997). On the possibility of kinetic energy density evaluation from the experimental electron-density distribution, *Acta Cryst.* **A53**, 264-272.
21. Abramov, Y. and Coppens, P. to be published.
22. Buckingham, A. D. (1978) in B. Pullman (ed.), *Intermolecular Interactions from Diatomics to Biopolymers*, Wiley and Sons: Chichester, New York, p. 1.

23. Cohen, M. D. and Schmidt, G. M. J. (1964) Topochemistry. Part I. A survey, *J. Chem. Soc.*, 1996-2000; Cohen, M. D. (1975) The photochemistry of organic solids, *Angew. Chem.* **14**, 386-393.

24. Carducci, M. D., Pressprich, M. R. and Coppens, P. (1997) Diffraction studies of photo-excited crystals: metastable nitrosyl-linkage isomers of sodium nitroprusside, *J. Am. Chem. Soc.* **119**, 2669-2678; Fomitchev, D. V. and Coppens, P. (1996) X-ray diffraction analysis of geometry changes on excitation: The ground-state and metastable-state structures of K2[Ru(NO2)4(OH)(NO)], *Inorg. Chem.* **35**, 7021-7026.

25. Coppens, P. (1997) Time-resolved diffraction in chemistry and materials science: The developing field of photocrystallography, *Synchr. Rad. News* **10**, 26-30.

26. Mills, D. M. (1989) Feasibility study into the use of mechanical choppers to alter the natural time structure of the APS, *Rev. Sci. Instrum.*, 2338.

27. Ren, Z., Ng, K., Borgstahl, G. E. O., Getzoff, E. D. and Moffat, K. (1996) Quantitative analysis of time-resolved Laue diffraction patterns, *J. Appl. Crystallogr.* **29**, 246-260; Bourgeois, D., Ursby, T., Wulff, M., Pradervand, C., Legrand, A., Schildkamp, W., Laboure, S., Srajer, V., Teng, T. Y., Roth, M. and Moffat, K. (1996) Feasibility and realization of single-pulse Laue diffraction on macromolecular crystals at ESRF, *J. Synch. Rad.* **3**, 65-74.

28. Genick, U. K., Borgstahl, G. E. O., Ren, Z., Ng K., Pradervand, C., Burke, P. M., Srajer, V., Teng, T. Y., Schildkamp, W., Mcree, D. E., Moffat, K. and Getzoff, E. D. (1997) Structure of a protein photocycle intermediate by millisecond time-resolved crystallography, *Science* **275**, 1471-1475; Srajer, V., Teng, T. Y., Ursby, T., Pradervand, C., Ren, Z., Adachi, S., Schildkamp, W., Bourgeois, D., Wulff, M. and Moffat, K. (1996) Photolysis of the carbon monoxide complex of myoglobin: nanosecond time-resolved crystallography, *Science* **274**, 1726-1729.

29. *Photochemistry in Organized and Constrained Media*, V. Ramamurthy (ed.), VCH: New York, Weinheim, London (1991).

30. Koenig, O., Burgi, H. B., Armbruster, T., Hulliger, J. and Weber, T. (1997) A study in crystal engineering: structure, crystal growth, and physical properties of a polar perhydrotriphenylene inclusion compound, *J. Am. Chem. Soc.* **119**, 10632-10640; Hoss, R., Koenig, O., Kramerhoss, V., Berger, U., Rogin, P. and Hulliger, J. (1996) Crystallization of supramolecular materials: perhydrotriphenylene (PHTP) inclusion compounds with nonlinear optical properties, *Angew. Chem., Inter. Ed.* **35**, 1664-1666.

31. Marlow, F., Demuth, D., Stucky, G. and Schüth, F. (1995) Polarized IR spectra of p-Nitroaniline-loaded AlPO4-5 single crystals, *J. Phys. Chem.* **99**, 1306-1310.

32. Steiner, T., Koellner, G., Gessler, K. and Saenger, W. (1995) Isostructural replacement of a N-H···O by a C-H···O hydrogen bond in complex stabilization: crystal structures of β–cyclodextrin complexed with diethanolamine and with pentane-1,5-diol, *J. Chem. Soc., Chem. Commun.*, 511-512.

33. Ozawa, Y., Pressprich, M. R. and Coppens, P. (1997) On analysis of reversible lighted-induced changes in molecular crystals, *J. Appl. Cryst.* In Press.

34. Roundhill, D. M., Gray, H. B. and Che, C.-M. (1989) Pyrophosphito-bridged diplatinum chemistry, *Acc. Chem. Res.* **22**, 55-61; Nocera, D. G. (1995) Chemistry of the multielectron excited state, *Acc. Chem. Res.* **28**, 209-217.

35. Dulebohn, J. I., Ward, D. L. and Nocera, D. G. (1990) Synthetic design of a multielectron series of homologous dirhodium fluorophosphines possessing an emissive dσ* excited state, *J. Am. Chem. Soc.* **112**, 2969-2977.

36. *Topics in Current Chemistry*, vol. 156, J. Mattay (ed.), Springer-Verlag: Berlin, New York (1989).

37. Mirkin, C. A. and Ratner, M. A. (1992) Molecular electronics, *Ann. Rev. Phys. Chem.* **43**, 719.

38. Hopfield, J. J., Onuchic, J. N. and Beratan, D. N. (1988) *Science* **241**, 817; Hopfield, J. J., Onuchic, J. N. and Beratan, D. N. (1989) Electronic shift register memory based on molecular electron-transfer reactions, *J. Phys. Chem.* **93**, 6350-6357.

39. Hasharoni, K., Levanon, H., Tang, J., Bowman, M. K., Norris, J. R., Gust, D., Moore, T. A. and Moore, A. L. (1990) Singlet photochemistry in model photosynthesis: identification of charge separated intermediates by Fourier transform and CW-EPR spectroscopies, *J. Am. Chem. Soc.* **112**, 6477-6481; Collin, J.-P., Harriman, A., Heitz, V., Odobel, F. and Sauvage, J.-P. (1994) Photoinduced electron- and energy-transfer processes occurring within porphyrin-metal-bisterpyridyl conjugates, *J. Am. Chem. Soc.* **116**, 5679, and references therein.

Time resolved x-ray diffraction in solids and liquids

I. V. Tomov, P. Chen and P. M. Rentzepis
Department of Chemistry, University of California, Irvine, CA 92697

Abstract

A novel, time resolved x-ray system is described which allows for the study of transient structure changes in molecular and solid state systems. The apparatus has been used for ultrafast time resolved x-ray diffraction studies with nanosecond and picosecond resolution. Some of the processes studied include crystal lattice dynamics under short laser pulse heating and amorphous solids and liquid structures.

1. Introduction

X-ray diffraction has been used for many years to study the structural properties of crystalline and to some extent amorphous materials on the atomic scale. One of the basic advantages of x-ray diffraction is that, to a large extent, x-rays interact weakly with matter and therefore provide an ideal, non-destructive, probe for structural studies of bulk as well of surfaces. Most of the x-ray diffraction studies have been related to steady state, long lived systems. Extension of the x-ray techniques into the time domain will contribute immensely to the understanding of the mechanism of formation and transformation occurring in material systems. Ultrashort x-ray pulses provide an excellent means for basic research studies in time resolved intra- and intermolecular structural changes and also find use in several technological applications. To develop those new technologies which are based on the interaction of matter with short laser pulses it is critically important to understand better the radiation-matter coupling and mechanism of energy

G. Tsoucaris (ed.), Current Challenges on Large Supramolecular Assemblies, 287–317.

transfer. Such knowledge will make possible the evaluation of the transient properties of materials and propose the design of new materials suitable for new technological applications. Time resolved structural studies with higher resolution are expected to become more and more important as the development of more sophisticated technological materials progresses. Time-resolved x-ray diffraction studies have been demonstrated in recent years using short x-ray pulses produced by laser plasma, synchrotrons and x-ray diodes [1- 9].

In recent years we have developed an optically driven x-ray diode which generates hard x-ray pulses used for time resolved x-ray diffraction experiments [10 - 12]. This technique is especially suitable for pump/probe type x-ray diffraction experiments when accumulation of many shots are required to characterize each point. This can be achieved in practice because this technique allows for the generation of kHz repetition rate highly reproducible x-ray pulses. These pulses can be synchronized with the pumping laser pulses or different ultrafast sources with picosecond accuracy. Here we present some of the results on pico and nanosecond time resolved x-ray diffraction studies on gold and platinum crystals as well as discussion on application of the system for study of liquids.

2. Generation of nano- and picosecond x-ray pulses in an optically excited x-ray diode

The x-ray diode described here consists of a conventional x-ray diode, in which the cathode is replaced by a photocathode (Fig. 1). The use a photocathode excited by short laser pulses allows for nano and picosecond x-ray pulse generation [10-13]. The duration of the x-ray pulse produced by a diode is determined by the duration of the electron pulse impinging on the anode. Therefore in order to evaluate the operation of the x-ray photodiode we must consider the generation of short electron pulses on the cathode and their propagation between the cathode and anode. To obtain enough x-ray flux to be measured and for use in diffraction and other time resolved experiments, high peak current electron pulses have to be generated and subsequently propagated through the x-ray diode. The operation of the x-ray diode with a photocathode is similar to those found in a conventional x-ray tube where the output x-ray spectrum is determined by the anode material and electron energy. In order to evaluate the performance of a photoexcited x-ray diode we follow the analysis of a space charge limited photodiode as reported in references [14 -16]. Depending of the ratio of the duration of the exciting laser pulse and the transit time of the electrons between the cathode and the anode the x-ray diode will operate in transient or steady state mode. In a practical diode the anode cathode separation is in the 5 - 30 mm range and when excited by picosecond light pulses it normally operates in transient mode, however nanosecond excitation will drive the diode in steady state mode of operation. The results of the analysis for an x-ray diode with cathode - anode separation of 10 mm are presented in Table I and Table II.

In our study of optically driven x-ray diode, we have used a laser system which generates 193 nm picosecond and nanosecond pulses (Fig. 2). The details of the system have been reported elsewhere [17] and here only the main features related to the x-ray diode are summarized.

A cw mode locked Nd:YLF laser generating 50 ps pulses at 76 MHz is used as the master oscillator. Part of the output from the laser is converted to the second harmonic which pumps a dye laser emitting at 724 nm. The other part of the 1053 nm radiation is used to seed a regenerative amplifier operating at 300 Hz. The dye laser was operated in two ways, the first with a single jet (gain) to produce pulses with duration of 5 ps, and the second with two jets (gain and saturable absorber) to generate pulses with duration of about 1 ps. After amplification in a two stage dye

Electron energy [keV]	Max charge/pulse [pC/mm^2]	CuKα photons [/pulse Sr]
20	17.7	0.11×10^5
30	26.6	0.53×10^5
40	35.4	1.26×10^5
50	44.3	2.35×10^5
60	53.1	3.65×10^5
70	62.0	5.03×10^5
80	70.8	6.63×10^5
90	79.7	8.46×10^5
100	88.5	9.95×10^5
150	132.8	1.82×10^6
200	177.1	2.54×10^6
300	265.6	3.65×10^6

Table I. Production of Kα photons from thick copper target at different electron energies from optically driven x-ray diode in transient mode of operation.

Voltage [kV]	Current density [A/cm^2]	Electrons [pulse mm^2]	CuKα photons [pulse/Sr]
40	18.6	1.4×10^{10}	0.8×10^7
50	26.1	2.0×10^{10}	1.7×10^7
60	34.2	2.6×10^{10}	2.8×10^7
70	43.2	3.2×10^{10}	4.2×10^7
75	47.8	3.6×10^{10}	5.0×10^7
100	73.7	5.5×10^{10}	9.9×10^7
150	135.4	10.1×10^{10}	22×10^7

Table II. Production of Kα photons from thick copper target at different electron energies in x-ray diode driven by 12 ns ArF laser pulse.

Table III. Relevant parameters of gold, mica and platinum used in the calculations.

Parameter	Au [26]	mica [27]	Pt [26,28]
Reflectivity at 193 nm	0.21		0.25
Radiation penetration [nm]	13.3		10.6
Thermal conductivity - K [W/cm K]	3.15	0.0067	0.716
Specific heat - C_p [J/g K]	0.129	0.124	0.133
Density - ρ [g cm^{-3}]	18.88	2.85	21.45
Linear expansion coefficient [K^{-1}]	1.42×10^{-5}		8.8×10^{-6}
Elastic constants [x10 ^{11} dynes/cm^2]	c_{11} = 19.6		34.67
	c_{12} = 16.45		25.07
	c_{44} = 4.2		7.65
Mass abs. coeff., CuKα [cm^2/g]	208		199
Lattice constant [A]	4.0786		3.9237

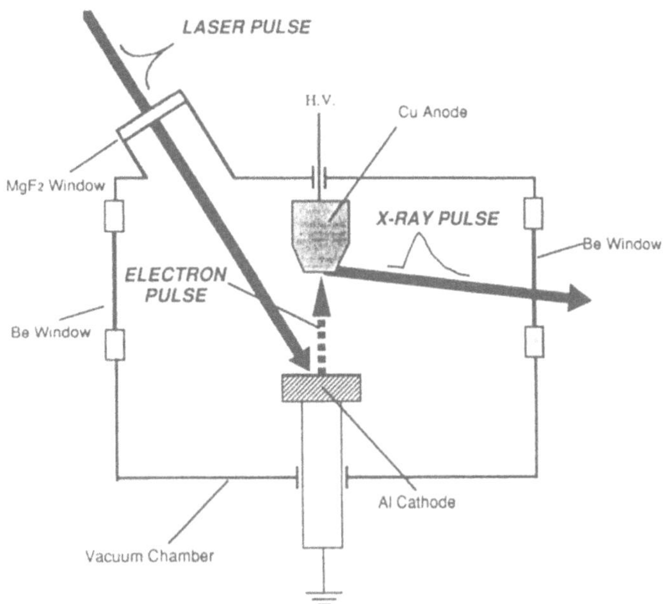

Fig. 1. Schematic diagram of the planar x-ray diode

amplifier the dye laser pulses with energy up to 10 μJ and 300 Hz repetition rate were obtained. These pulses were mixed with the forth harmonic of the regenerative amplifier in BBO crystals to produce seed pulses at 193 nm. A 6 mm thick BBO crystal was used to generate seed pulses with 10 ps duration and energy about 3 μJ and a 0.4 mm BBO crystal to generate pulses of 1-2 ps duration and energy about 200 nJ. These seed pulses passed through an ArF excimer amplifier. In the case of 10 ps seed pulses, two passes in the amplifier were used to increase the pulse energy to 3 mJ, the ASE was less than 2% of the amplifier output . In the case of shorter pulses (1-2 ps) three pass amplification was employed to increase the pulse energy to 1 mJ and about 5% ASE. In the latter case a spatial filter after the second pass was used. Synchronization of the different modules of the system was done by using the 38 MHz RF signal from the mode locker of the master oscillator.

When nanosecond operation was desired the ArF amplifier was converted in a laser generating 12 ns pulses with energy up to 30 mJ.

Part of the 193 nm radiation was used to drive the x-ray diode and the rest to excite the sample. The diode consists of two flat, parallel, electrodes made from bulk polished metal. The copper anode was 8 mm diameter separated by 10 mm from a 12 mm diameter aluminum photocathode. The pressure of the x-ray chamber was kept at about 5×10^{-9} Torr. The UV radiation was directed at about 25° relative to the normal of the cathode surface forming a spot of 2-3 mm diameter situated opposite to the anode. The x-ray radiation exits the diode via two 250 mm Be windows mounted on opposite sides of the chamber.

Temporal measurements of the x-ray output were carried out with a high repetition rate, picosecond resolution streak camera (Kentech Instruments Ltd.). The photocathode of the camera was divided into two sections: a quartz window coated with 30 nm gold film sensitive to 193 nm radiation and 1.5 μm mylar film coated with 30 nm thick, low density CsI which is sensitive to x-ray photons. The streak camera output was amplified by an electrostatic image intensifier coupled to a video camera. The output of the video camera is stored on a video processor for measurements and enhancement. Due to the low sensitivity of our streak camera photocathode, the x-ray intensity output was insufficient to record a single shot. Therefore, frame accumulation of many shots was necessary to obtain a measurable signal. Because of this accumulation the timing jitter in the system effects the pulse duration measurement. The streak camera was triggered by the second harmonic pulse of the regenerative amplifier. In order to estimate the jitter, the UV intensity incident on the streak camera slit was attenuated so the UV pulse image measured was obtained by the accumulation of the same number of frames as the x-ray image. The images, of both x-ray and UV pulses, were recorded simultaneously. By comparing of the accumulated x-ray and UV signals and knowing the duration of the UV pulse , we have determined an upper limit for the x-ray pulse duration to be in the range 20 ps to 50 ps for 10 ps UV excitation pulses and about 7 - 10 ps for 1.8 ps UV excitation pulses. For nanosecond excitation the x-ray pulse duration was equal to the UV pulse of 12 ns

Other x-ray output characteristics were measured using a large area CCD camera designed specifically for direct x-ray imaging. The camera consists of a Ford Aerospace 2048 x 2048 CCD chip (15 μm pixel), cooled by liquid nitrogen, and interfaced to a Macintosh IIfx computer. Cooling to low temperatures (-100°C) reduces the dark current to allow for single photon detection. Single and double Si crystal monochromator arrangements were used to measure the spectral properties of the output radiation. The measured ratio of intensities of the $K\alpha_1$, $K\alpha_2$ and $K\beta$ lines were found to be similar to the output from a conventional x-ray tube. When the x-ray diode was operated in the saturated mode the fluctuations of the x-ray output energy were found to be an order of magnitude less than the fluctuations in the input UV pulse energy. This system has been operated for more than 10^9 shots without any noticeable degradation in the x-ray output or the anode and cathode electrodes.

3. Time resolved x-ray diffraction studies

3.1. Theoretical consideration

General theoretical studies of x-ray diffraction are based on the approach which treats the crystal as a medium with a three dimensional periodic dielectric constant and solve Maxwell's equations for such a medium. A more complete description of the diffraction processes by crystals is given by the dynamical theory [18]. Very often the simpler kinematical theory is used, which treats the scattering from each element in the sample as being independent and therefore the extinction effects may be neglected. When the maximum reflecting power is small (a few percent), the results predicted by the dynamical diffraction theory are essentially the same as those predicted by the kinematical theory [19,20]. In the case of Bragg diffraction the diffracted intensity profiles (rocking curves) are highly sensitive to depth as well as lateral variations strain and damage distributions in the crystal. However the analysis of the information contained in the rocking curves presents some difficulties: One of the main problems is related to the lack of phase detection which precludes direct inversion of the rocking curves. The other problems are related to the complexity of the dynamical theory for x-ray diffraction from non uniform crystals. Time dependent x-ray diffraction has not been treated in detail so far. For nonhomogenious crystalline media the x-ray diffraction problem has been treated for the case where the homogeneity of the crystal varies only in one direction, i.e. depth [21-23]. In this analysis it is always assumed that the crystal properties do not change during the x-ray illumination and time dependence in the corresponding equations is omitted. However the use of ultrashort laser pulses for sample irradiation creates transient effects which vary during x-ray pulse illumination even when nanosecond and picosecond x-ray pulses are used [16]. Therefore, further extension of the stationary theory to include time dependent x-ray diffraction is necessary. Using the results presented in the above studies we outline here a treatment of the time dependent diffraction problem starting from Maxwell's equations. These equations with time dependence preserved, have to be solved simultaneously with the equations describing the time dependent properties of the material which are responsible for the changes in the diffracted x-ray radiation.

The incident x-ray wave in the vacuum is assumed to be space and time dependent:

$$D(r,t) = D_0(r,t)\exp\left[i\left(\omega t - k_0 \cdot r\right)\right] \qquad (1)$$

This wave is reflected from the lattice planes to produce a scattered wave inside the crystal given by

$$D(r,t) = D_H(r,t)\exp\left[i\left(\omega t - k_H \cdot r\right)\right] \qquad (2)$$

where $k_{0,H}$, are the wave vectors of the incident and diffracted waves and r is a point in real space.

For the case of a crystal oriented near the Bragg conditions we can consider only the incident and one reflected wave. In the x-ray region, the dielectric constant of the medium can be written as $\varepsilon = 1 + \psi$, where ψ is the susceptibility which is much smaller than unity. Therefore, the Maxwell equation for the displacement vector can be written in the form

$$\nabla \times \nabla \times (1 - \psi) D = \frac{1}{c^2}\frac{\partial^2 D}{\partial t^2} \qquad (3)$$

where D represents the sum of the incident and reflected waves.

To the first order approximation and considering a crystal with a center of symmetry, we obtain the following equations

$$i\frac{\lambda}{\pi}\frac{\partial D_0}{\partial x_0}+i\frac{\lambda}{\pi c}\frac{\partial D_0}{\partial t}=\psi_0 D_0+\psi_H D_H$$
$$i\frac{\lambda}{\pi}\frac{\partial D_H}{\partial x_H}+i\frac{\lambda}{\pi c}\frac{\partial D_H}{\partial t}=\psi_0 D_H+\psi_H D_0-\alpha_H D_H \qquad (4)$$

where $\psi_{0,H}=-(e^2/mc^2)(\lambda^2/\pi)(F_{0,H}/V_c)$; $\frac{e^2}{mc^2}$ is the classic electron radius, and $F_{0,H}$ is the structure factor for the incident and diffracted waves, V_c - unit cell volume, $\alpha_H \approx -2(\Theta-\Theta_B)\sin 2\Theta_B$, and Θ_B is the Bragg angle. These equations, (4), express the variation of the complex amplitudes of the incident and diffracted waves along the incident and diffracted directions respectively. If we neglect the time dependence in equations (4) one obtains the well known dynamical diffraction equations [21] used to calculate the time independent rocking curves of crystals. The incident wave amplitude equation can be neglected when the reflecting power is small, and the solution for the diffracted wave amplitude equation may be obtained by simply solving the second equation in (4) (kinematical limit [19]).

Equations (4) have been modified to study the diffraction from curved crystals [21, 23] and further applied to the study of time dependent diffraction from pulsed laser annealed silicon crystal [3] and shock compressed crystals [6]. For a crystal with a strained lattice in a direction perpendicular to the sample surface the change in the Bragg angle, $\Delta\Theta=\Theta-\Theta_B$, in eq. (4), is replaced by:

$$\Delta\Theta+\phi(z,t)\tan\Theta_B \qquad (5)$$

where $\phi(z,t)$ specifies the strain induced change in the Bragg angle.

If the dimensions of the heated area on the crystal surface are larger than, both absorption length for the laser radiation or the heat diffusion length, then we can consider the problem as one dimensional. The temperature distribution in the crystal may be obtained by description of the diffusion heat conductivity initiated by the laser pulse energy absorbed by the crystal. The heat conduction equation is

$$\frac{\partial T}{\partial t} = \chi \frac{\partial^2 T}{\partial z^2} + \frac{1}{\rho C_p} W(z,t) \qquad (6)$$

here χ is the thermal diffusivity, C_p is the specific heat, ρ is the crystal density and W(z,t) is the volume density of the thermalized energy. For the case of heating a semi-infinite absorbing target with a laser pulse the source term W(z,t) becomes [24]

$$W(z,t) = I_m \frac{(1-R)}{l} f(t)\exp(-z/l) \qquad (7)$$

where I_m is the maximum intensity of the incident laser pulse, l is the absorption length and the function *f(t)* describes the laser pulse envelope.

The solution of (6) gives the time evolution of the temperature distribution in the crystal. This nonhomogenious temperature distribution will generate a strain in the crystal. The strain normal to the heated surface layer as a function of depth is given by

$$\phi(z,t) = \int_{T_0}^{T(z,t)} \alpha(T)\eta(h,k,l)dT \qquad (8)$$

Where T_0 is the temperature of the unheated part of the crystal, $\alpha(T)$ is the linear thermal expansion coefficient, $\eta(h,k,l)$ is a factor that takes into account the one dimensional nature of the strain which for cubic crystal is given by [25]:

$$\eta(1,1,1) = 3(1 + 2C_{12}/C_{11})/(1+2C_{12}/C_{11} + 4C_{44}/C_{11})$$

In our calculations we neglect the temperature dependence of the linear thermal expansion coefficient and the elastic constants (C_{ij}), because in the temperature range of our experiment, they change by less than a few percent.

The time and depth dependent strain, $\phi(z,t)$, determined from the solution of equations (6) and (8) for a heating laser pulse $I_m f(t)$, is substituted in the equations (4) for a probing x-ray pulse $D_o(r,t)$ to determine the x-ray diffracted signal. Because the mathematics involved may be quite complex and an analytical solution difficult to obtain, some simplifications may be made which allow for an easier solution of a particular case. Previously [19], for the steady state case, in order to reduce the complexity of data analysis several approximations have been made, such as layer slicing and kinematical model for mosaic crystals. Additional approximations are possible for time dependent equations based on the transient properties of the crystal studied.

In our case we use such a simplified approach based on the physical considerations of our experiment. Using our experimental set up, it takes less than 20 ps for the diffracted x-ray radiation to pass through the Pt crystal and less than 1 ps to pass through Au crystal. For such short times the change of the crystal temperature due to thermal conduction can be neglected and the diffraction process may be treated as been in a steady state condition. The calculations are carried out by slicing the input nanosecond x-ray pulse into a set of micro pulses with picosecond duration and then perform the corresponding calculations for every micro pulse. For picosecond x-ray pulses only one micro pulse is used. During the passage of each micro pulse the temperature distribution and related strain in the crystal is assumed to be constant but evolves for the subsequent micro pulses. By these means we calculate first the temperature distribution $T(z,t)$ in the crystal and then the thermal strain $\phi(z,t)$ associated with specific $T(z,t)$. We have calculated the heat distribution in the experimental crystals using the results presented in [24] where the eq. (6) was solved. For a Gaussian pulse envelope, $f(t) = \exp(-t^2/\tau^2)$, assuming a flat top cross section laser beam, one dimensional temperature distribution is given by :

$$T(z,t)=\frac{I_m(1-R)\chi}{2Kl}\int_{-\infty}^{t}dt'\exp\left[-\left(\frac{t'}{\tau}\right)^2\right]$$

$$\times\{\exp\left[-z/l+\chi(t-t')/l^2\right]erfc\left(\frac{-z+2\chi(t-t')/l}{2\sqrt{\chi(t-t')}}\right)$$

$$+\exp\left[z/l+\chi(t-t')/l^2\right]erfc\left(\frac{z+2\chi(t-t')/l}{2\sqrt{\chi(t-t')}}\right)\} \tag{9}$$

here z is the direction normal to the crystal surface, K is the thermal conductivity, and erfc (x) is the complementary error function. The t = 0 time, is defined as the time when the intensity maxima of both x-ray and UV pulses overlap, the positive delay time corresponds to the case when probing x-ray pulse follows the heating UV pulse. In order to calculate the diffracted x-ray signal we solve eqs.(4) for a micro pulse. X-ray measurements have shown that the crystals used in our experiments have a mosaic structure. Therefore, the kinematical model should be applied [19]. Within the kinematical approach the crystal is divided into many thin layers. For each layer we neglect the extinction and normal absorption and assume that the strain has an average value determined by the position of the layer and temperature distribution in the crystal. The incident x-ray pulse on each layer is attenuated by the absorption losses, determined by the mass absorption coefficient, and experienced during x-ray traversal through the previous layers. In this case we have to solve only the second equation in system (4) in which the incident x-ray pulse is a constant. In addition because we are concerned with the solution for a micro pulse, we can omit the time dependent term. The result for the j-th layer is

$$D_j=i\frac{F_H}{|F_H|}\sqrt{|b|}\exp(-iAY)\frac{\sin(AY)}{Y} \tag{10}$$

where $b=\gamma_o/\gamma_H$; γ_o,γ_H are direction cosines of the incident and diffracted wave vectors.

$$A = \frac{e^2}{mc^2}\frac{\lambda|F_H|}{V}\frac{y}{(|\gamma_o\gamma_H|)^{1/2}}$$

y represents the thickness of each layer.

$$Y = \frac{[(1-b)/2]\psi_o - b(\Delta\Theta + \phi\tan\Theta_B)\sin 2\Theta_B}{\sqrt{|b|}|\psi_H|}$$

D_j is the diffracted amplitude from one layer of the crystal for a micro pulse. The total diffraction amplitude of a micro pulse is then the coherent sum of the diffracted waves from each layer adjusted for the phase lags and absorption losses during the micro pulse passage through the crystal. The total diffracted amplitude for N layers is:

$$D_H = i\frac{F_H}{|F_H|}\left(\sqrt{|b|}\right)\sum_{j=1}^{N} q_j \exp\left(-i(A_jY_j + \beta_j)\right)\frac{\sin(A_jY_J)}{Y_j} \qquad (11)$$

where

$$q_j = \exp\left[-\mu\frac{(\gamma_o + |\gamma_H|)}{|\gamma_o\gamma_H|}\sum_{i=j+1}^{N} y_j\right]; \qquad \beta_j = 2\sum_{i=1}^{j-1} A_jY_j;$$

Here variables A, Y and y with subscript j indicate their dependence on strain distribution in space and time and μ is the mass absorption coefficient. These types of calculations are carried out for each micro pulse and then the overall diffracted x-ray signal is calculated by integrating over the x-ray pulse envelope to obtain the rocking curve. In practice the calculated rocking curve needs to be convoluted with the instrumental broadening function. Thus we calculate the diffracted signal which will be recorded by the CCD detector.

3.2. Picosecond x-ray diffraction in single crystals

Time resolved x-ray diffraction can be employed to study the lattice behavior during pulsed laser illumination by means of time resolved Bragg profile measurements. When energy from a laser pulse is deposited on a material, it generates a non uniform transient temperature distribution, carrier concentration or other effects, which alter the lattice structure of the crystal. The deformed crystal lattice will change the angle of diffraction for a monochromatic x-ray beam by $\Delta\Theta = -(\Delta d/d)\cot\Theta_B$, where d is the spacing of the diffracting planes, Δd is the change of the spacing due to an outside influence. Using CCD detectors, very small changes in the position of the diffracted x-ray beam can be detected which makes it possible to measure accurately the lattice deformation. Thus the crystal transient structure changes induced by low energy short laser pulses can be directly measured. The transient temperature and strain distribution in the crystal may be studied by the use of the thermal expansion coefficients and elastic constants of the crystal.

The picosecond x-ray pulses produced by the system described above were used to study the deformation of gold (111) single crystal lattice heated by 1.8 ps, 193 nm, laser pulses. The experimental set up is a standard one used in pump and probe type experiments. The 193 nm pulses from ArF amplifier were split by a beam splitter: 70% of the pulse energy after passing a variable delay line, was focused by a 35 cm lens on the Au crystal to a 3.5 mm diameter spot size. The rest of the 193 nm pulse energy was used to drive the x-ray diode. In our experiments the 193 nm energy which impinged upon the gold crystal sample was in the range 100 - 150 μJ/pulse which gave an average energy density of up to 1.6 mJ/cm^2. The spatial distribution of the energy in the cross section of the 193 nm beam was a flat top with about 30% fluctuations. The limit of 150 μJ/pulse was due to the substantial loss of 193 nm energy in the air and lack of good quality reflectors for 193 nm, used in the delay lines and beam steering optics.

Two vertical slits 25x0.5 mm^2 and 25x0.3 mm^2 separated by 29 cm were used to define the x-ray beam to be diffracted from the crystal. The distance from the x-ray diode to the crystal was 40 cm, which gave about 10^5 CuKα photons per cm^2 per second impinging upon the crystal. Taking into account the geometry of the experiment one finds that there is an experimental set up time resolution, imposed by the slit widths and their separation, anode take off angle and crystal Bragg angle as well as distances between the experimental elements. In our experiment, the estimated geometrical resolution is about 12 ps, which is longer than the x-ray pulse duration.

The Bragg angle for 1.542 A radiation diffraction from Au(111) crystal is 19.112^o, thus the incident x-ray beam covered a 1.7 mm wide strip on the crystal surface. The beam divergence determined by the slits set up was 3 mrad. The gold crystal was L = 150 nm thick, grown on 100 μm thick mica crystal. The surface of the crystal was 10x20 mm^2. Electron diffraction patterns showed a well ordered Au(111) crystal over several mm parallel to the surface. Thus we assume that there is a mosaic structure along the surface, but along the thickness it consists mainly of a single crystal. The crystal was placed on a three axis Eulerian cradle allowing for accurate alignment. In our experiment we used both Cu Kα_1 and Cu Kα_2 radiation for which the corresponding difference in the Bragg angle is 0.85 mrad. The calculated rocking curve width determined by the crystal thickness of 150 nm is 1 mrad which is larger than the angle width of 0.85 mrad determined by the Kα doublet.

The diffracted x-ray radiation from the heated and unheated area of the crystal were detected simultaneously in a two-dimensional pattern by the CCD camera, which was set at 4x1 binning (x1 being along the Bragg angle deviation). In the measurement reported here two consecutive one hour exposures (about 1.1x10^6 shots) were carried out. The first one with the selected part of the crystal

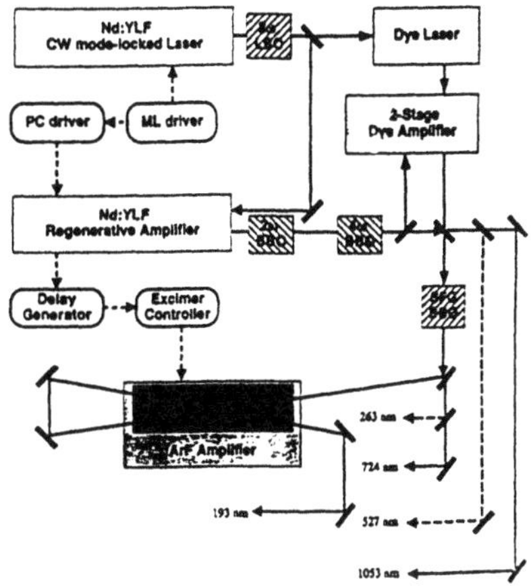

Fig. 2. High repetition rate laser system for generation and amplification of picosecond pulses at 193 nm.

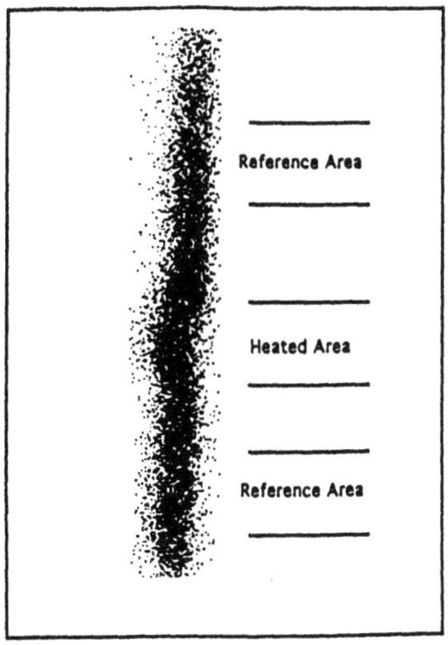

Fig. 3. CCD image of the diffracted x-rays from Au(lll) crystal. The CCD binning is xl along the horizontal (Bragg angle deviation) axis and x4 along vertical axis. The reference and heating area are marked.

under UV irradiation and the following without UV heating. Then a comparison of both exposures was made. In Fig. 3 a typical picture of the diffracted x-rays as displayed by the CCD detection system, is presented.

In the experiment described here we have studied the diffracted x-ray signal as a function of the delay between the heating UV pulse and the probe x-ray pulse. For every delay point two consecutive one hour exposures are made. The first one with UV radiation heating a selected part of the Au crystal and the second one without UV heating. Both x-ray patterns are compared to assure that there is no change due to external effects. Since the size of the heated area (3.5 mm) is smaller than the size of the crystal diffracting the x-rays (20 mm along the slit), we compare portions of the scattered signal above and below the heated area which in both exposures are without UV heating (see Fig. 3). In order to compare the experimental results with the calculated ones a fit to the experimental points is made using the a Gaussian shape for the diffraction signal. These smoothed curves from the experimental data are normalized to their peak value and compared with correspondingly normalized curves calculated using the above procedure. In the experiment the time delay was varied by 10 ps steps in the range of - 40 ps to + 100 ps and bigger steps at longer times outside this range. The experimental and calculated results for several delay times are depicted in Fig. 4. The delay time of -20 ps corresponds to a cold crystal, at t = 0 actually half of the UV pulse energy is absorbed by the crystal and the change of the rocking curve is time resolved and clearly observed. While the temperature distribution in the crystal is inhomogenious the scattered x-ray signal is a combination of the signals from the heated and cold parts of the crystal. After 100 ps, thermal equilibrium has been established and we observe a shift of the rocking curve towards bigger Bragg angles. In Fig. 5, the experimental rocking curves for the cold and hot crystal are presented. We note that for 100 μJ absorbed energy, the temperature rise of the

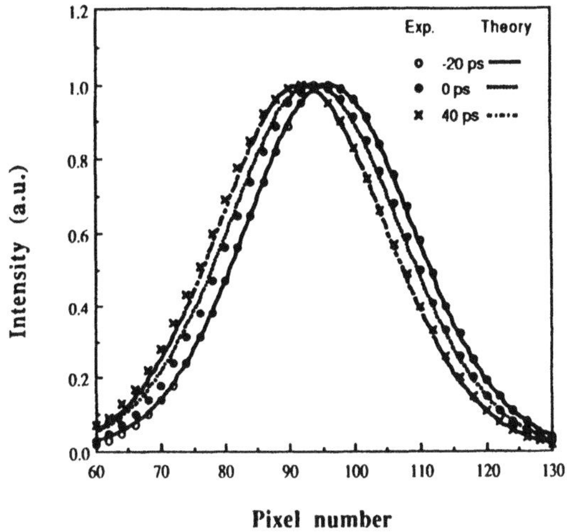

Fig. 4. Time resolved Bragg profile curves for x-ray diffraction from Au(lll) crystal. The solid lines are the calculated results. For comparison the experimental results are presented by points.

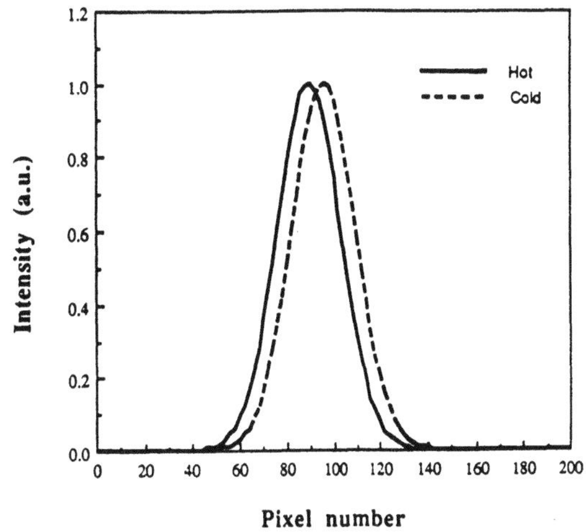

Fig. 5. Experimental time resolved Bragg profile curves for x-ray diffraction from Au(lll) crystal. Dashed line - cold crystal; solid line - hot crystal, the delay time is 120 ps.

heated part of the gold crystal is about 30°C. In Fig. 6, the shift of the peak of the rocking curve as function of delay is shown. The results shown are the average of five runs at delay times in the range -100 ps to 500 ps. The transition through a thermally nonuniform crystal lattice in the first 50 ps is clearly seen. In this transition time, the width of the rocking curve is also slightly larger than the one at equilibrium. The spread of the experimental points is partially due to the shot to shot fluctuations in the UV pulse energy which results from the jitter in the triggering of the excimer amplifier. After about 100 ps and up to 500 ps, which was the longest delay used in the experiment, no change in the shift was observed. From these measurements we conclude that our experimental system is capable of less than 10 ps time resolution and observing lattice temperature changes of about 20°C.

3.3 Nanosecond x-ray diffraction in single crystals

On the nanosecond time scale, the absorbed laser energy by the crystal surface is fully thermalized because the energy transfer from electrons to the lattice takes place in picoseconds or less. However, heat diffusion into the bulk of the crystal is much slower, therefore we have used nanosecond pump/probe techniques to study the dynamics of the heat diffusion in the crystal.

The nanosecond x-ray diffraction measurements were carried out on polished, (111), platinum crystal. The crystal, 6.5 mm diameter and 0.5 mm thickness, was mounted on a three-axis Eulerian cradle, allowing accurate alignment. Using the mass absorption coefficient data listed in Table III we find that for *1/e* attenuation of the incoming x-ray beam, at the Bragg angle, the penetration depth in the crystal is about 800 nm. From Scherrer`s formula, this thickness gives a rocking curve width of about 0.2 mrad.

The nanosecond x-ray pulses are generated in an x-ray photodiode with copper anode, pumped by 12 ns ArF excimer laser pulses at a repetition rate of 300 Hz. The radiation emitted by the ArF laser was split into two beams one used to heat the Pt crystal and the other, after the proper delay, to pump the x-ray diode. The output from the x-ray diode consisting of 12 ns (FWHM) pulses was reflected from a pyrolytic graphite monohromator to select Cu Kα characteristic radiation. This radiation, after passing a variable slit, was directed to the Pt crystal at the Bragg angle. For a divergent incoming beam the Bragg reflected beam had a divergence slightly less than 3 mrad. In this experiment the Cu Kα doublet was resolved by the Pt crystal, however, by rotating the crystal along the Bragg direction the quality of the image along the slit was changing, which, also suggests that it is a mosaic crystal. In the measurements reported here, the divergence of the input x-ray beam was kept about 1 mrad larger than the divergence of the reflected beam, allowing for the changes in the Bragg angle induced by the laser heating to be covered by the divergence of the input beam.

The diffracted x-ray radiation was measured using the CCD camera with 4x1 pixel binning (x1 being along the Bragg angle axis). A metal screen positioned in front of the crystal surface divided the image into two segments, the upper section where the heating UV radiation was directed and the lower section without UV heating. The lower section of the image was used as a reference to account for the variation in the x-ray flux. The results reported here were obtained by exposures ranging from 10 min to 1 hour. Different exposures were necessitated by the decrease of the UV flux pumping the x-ray diode due to noticeable loses in the air for 193 nm radiation at different delays. For every delay point two consecutive exposures were made , one with UV radiation heating the upper part of the crystal and the next one without UV heating. Both x-ray patterns were compared to assure that there is no change due to external effects. Since the heating is done

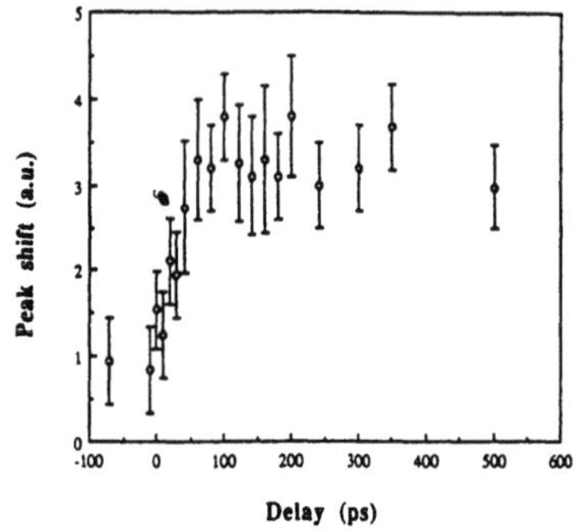

Fig. 6. The shift of the rocking curve peak as a function of delay time.

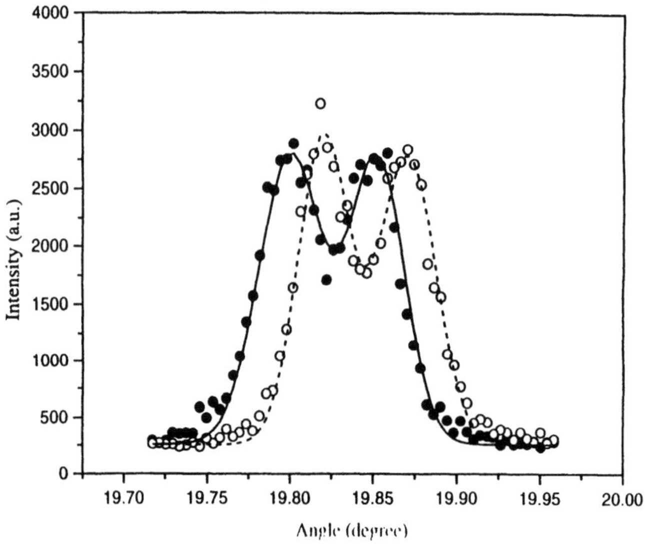

Fig. 7. Time resolved x-ray diffraction Bragg profile curves for Pt(lll) crystal.
a) dashed line - cold crystal, b) solid line - hot crystal. The points are
b) experimental data, the lines are fitted to the experimental points.
c) The delay time of the probing x-ray pulse is 5 ns.

only above the metal screen, the recorded pattern below the screen must be the same for both exposures, and any changes induced by the UV irradiation will show only on the upper part of the diffracted pattern. In Fig. 7 two traces from the upper part of the CCD patterns are shown. The signal is averaged over 30 pixels along the vertical axis and scanned along the horizontal axis. The points are the CCD records and the lines are fitted to a Gaussian line shape. The dotted line corresponds to the signal observed when the crystal was under UV irradiation and the solid line shows the same area of the crystal but without UV heating. The shift due to the laser heating is clearly seen. It was also seen that in the same CCD record there was no shift in the traces obtained from the lower, reference, part of the crystal which was not heated during either exposure.

In Table III we list the optical, thermal and other properties of platinum which are relevant to our experiment and used in the calculations. From these data we calculate that for a laser pulse, $\tau_p = 2\ (\ln 2)^{1/2}\ \tau = 12$ ns (FWHM), the diffusion length $L_d = (2K\tau_p/C_p\rho)^{1/2} = 776$ nm. This length represents the crystal depth heated during pulse illumination. From the above calculations we find that 2 mJ energy absorbed in a spot size of $S = 0.1\ cm^2$ will increase the temperature of the volume SL_d of platinum by about 70°C. The heat dissipation from this volume is mainly through the rest of the platinum crystal and the crystal holder, also there is a very small loss through the air. The hot volume will become almost twice as big ,50 ns, after the laser pulse. Calculations have shown that at least in the first few microseconds, after illumination, heat dissipation takes place mostly in the Pt crystal. The temperature change within the crystal due to the residual heat remaining between laser pulses (3.3 ms) is less then 5 degrees and therefore was neglected.

The above estimates show that for the first 100 ns after the laser pulse, the UV irradiated area of the crystal is in an nonequilibrium transient stage. In the bulk of the crystal there is a thermal strain associated with the heated surface layer. The time necessary for the strain profile to set up is roughly the heat diffusion length divided by the speed of sound in the crystal. The sound speed for Pt is 3.26×10^{5} cm/s which means that the strain profile occurs in subnanosecond time. Consequently for our experiment the time lag between the establishment of temperature and strain profile in the crystal can be neglected. Using the data listed in Table III, we find that the one-dimensionality of the strain leads to lattice expansion in the direction normal to the crystal face which is 2.204 times larger than the isotropic three dimensional thermal expansion.

In order to extract the transient lattice spacing from experimental data, we used equation (11) to calculate the Bragg profile. The instrumental broadening function was first derived from static x-ray rocking curve, and then convoluted with the calculated Bragg profile. At each time delay, the strain distribution can be obtained by the best match with the x-ray rocking curve, and this strain distribution can be further used to calculate the lattice constant and temperature distribution. Fig. 8 shows the best fit of the lattice constant within Pt crystal at various time delays. From this graph we can see that the surface layer suffers the largest lattice constant change, because the heat propagation is not fast enough to dissipate the energy into the bulk within the heating pulse duration. The lattice constant of the surface layer reaches its maximum value 2.2677 A (0.13% change) 10 nanosecond after the laser pulse illumination. The time resolved Bragg profile can also be constructed by using temperature distribution calculated from the heat diffusion equation. A laser pulse with 2 mJ energy and 12 ns pulse duration was used to calculate the temperature distribution within Pt crystal. This temperature distribution is then used to calculate the x-ray rocking curve. In order to compared the simulation result with the experimental data, the smoothed curves, from the experimental data, are normalized to their peak value and compared with the calculations. Since in our experiment both $CuK\alpha_1$ and $CuK\alpha_2$ were resolved, the calculations were done for both lines. In this experiment, the time delay was varied in the range of - 15 ns to + 50 ns and the experimental and calculated results for the induced strain in the crystal due to the heating for several delay times are depicted in Fig. 9. The overall agreement between the measured and calculated scattered x-ray intensities profiles is good, indicating that detailed time resolved x-ray diffraction measurements can be used to measure the transient properties of the crystal with nanosecond resolution for relatively small temperature changes. The system was able to detect changes in the lattice spacing of about $2x10^{-4}$ A and nanosecond time resolution. The data presented here were obtained by probing the bulk lattice changes in one dimension, however, the x-ray diffraction beam probes the changes in the lattice spacing for a particular set of planes, therefore changes in three dimensions within the bulk may be easily measured using this method by changing the relative orientation of the probing x-ray pulses.

3.4 Time resolved x-ray diffraction in amorphous solids and liquids

In order to evaluate the possible application of the ultrashort x-ray pulses, generated by our system, for time resolved x-ray diffraction studies in molecular systems, we have carried out some preliminary measurements in liquids and amorphous solids. The goal of these measurements was to evaluate the feasibility of performing time resolved x-ray diffraction experiments in various states of aggregation. Using the CCD described above for detection it is possible to achieve the simultaneous observation of diffracted x-ray patterns over a large area. In our experimental set up the CCD camera covered an angle of 20°. An additional advantage of the CCD detector is the digital nature of the recorded data, which can be easily and quickly

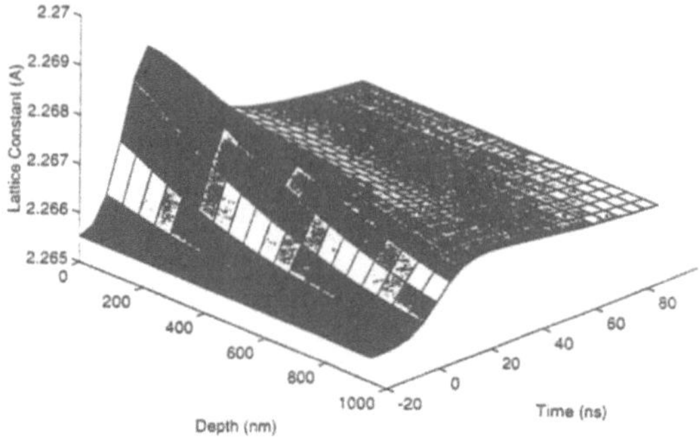

Fig. 8. Lattice constant evolution within Pt crystal.

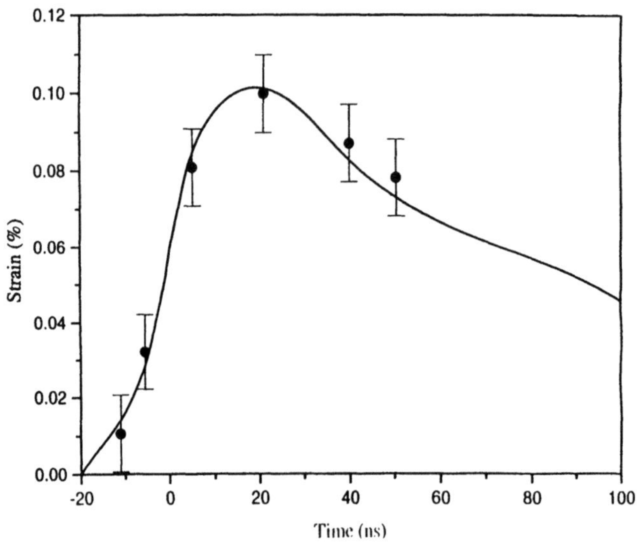

Fig. 9. Evolution of the strain as function of time calculated by kinematical model (solid curve). Points are measurement from the x-ray rocking curve.

processed by a computer. Fig. 10 shows the CCD image of the scattered x-rays from cyclohexane. The exposure time was 2 hours and the sample was in a 1 mm diameter glass capillary. The direct x-ray beam was blocked by a lead screen which is seen on the right side edge of the diffraction pattern shown in Fig. 10. Similar diffraction data were obtained from benzene, brombenzene, water and other liquids and solutions as well from liquid crystals..

The x-ray diffraction of several powder materials such as tantalum carbide and polycrystalline metals were also obtained. In Fig. 11 part of the CCD recorded image of the scattered x-rays from a 25 μm Tantalum sheet as well as its intensity trace is reproduced which is the average over twenty pixels along the vertical axis. Exposure time of 2 hours shows scattering from several planes and the split of the diffraction lines shown in Fig. 11 is the result of Cu Kα and CuKβ lines in the incoming x-radiation. These results show that this system is capable of nanosecond time resolved measurements of lattice parameters of polycrystalline and single crystal materials as well as powder materials. One possible application is the study of heat and shock wave propagation in these materials under short laser pulse irradiation and their effect on the lattice. Possibly of higher interest will be time resolved studies of the change in the macroscopic structure of liquids after laser excitation and the structural changes associated with other fast phenomena.

4. Conclusion

The experiments described, rather briefly in this paper, represent only a small part of the effort exerted in time resolved x-ray diffraction. Although there are many difficulties encountered and several problems still remaining to be solved including instrumental, theoretical, experimental and proper sample preparation, yet we believe that the knowledge to be obtained by the

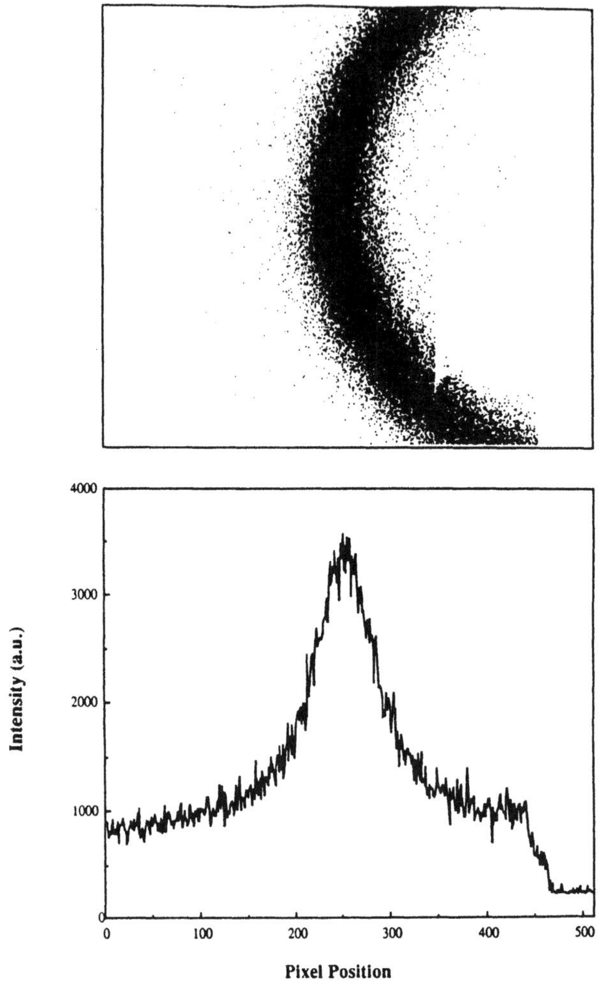

Fig. 10. CCD recorded image of the x-ray diffraction pattern of cyclohexane (upper). The diffraction pattern trace averaged over 20 pixels along the vertical axis and scanned along the horizontal axis (lower).

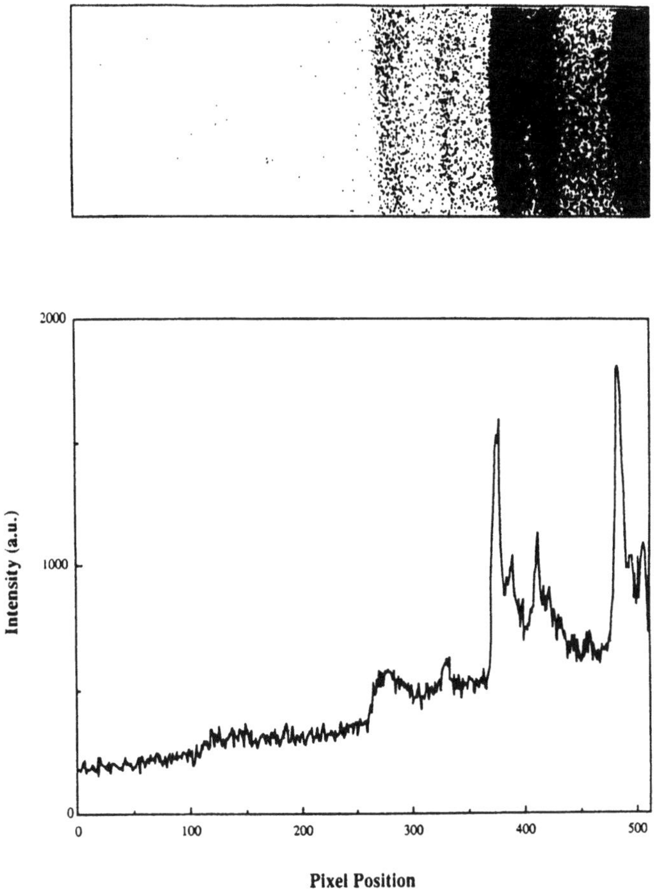

Fig. 11. Part of CCD recorded image of the diffracted x-rays from a Tantalum sheet (upper) and its trace,(Iower), averaged over 20 pixels along the vertical axis and scanned along the horizontal axis.

realization of transient structures which occur in practically every physical, chemical and biological reaction is worth the large effort and time necessary to overcome these problems.

5. Acknowledgments

This work was supported in part by the National Science Foundation grant # CHE-9501388 and the W. M. Keck Foundation.

6. References

1. J. G. Lunney, P. J. Dobson, J.D. Hares, S.D. Tabatabaei, and R. W. Eason, Opt. Commun. **58**, 269 (1986).
2. B. C. Larson and J. Z. Tischler, SPIE **1345**, 90 (1990).
3. B. C. Larson, J. Z. Tischler and D. M. Mills, J. Mater. Res. **1**, 144 (1986).
4. S. Kojima, I. Maekawa, S. Kawado, T. Takahashi, T. Ishikawa, and S. Kikuta, Rev. Sci. Instrum. **63**, 1164 (1992).
5. J. R. Bushchert, J. Z. Tischler, D. M. Mills, Q. Zhao, and R. Colella, J. Appl. Phys. **66**, 3523 (1989).
6. J. S. Wark, D. Riley, N. C. Woolsey, G. Keihnand, and R. R. Whitlock, J. Appl. Phys., **68**, 4531 (1990).
7. D. M. E. Szebenyi, D. H. Bilderback, A. LeGrand, K. Moffat, W. Schildkamp, B. S. Temple, and T. Teng, J. Appl. Cryst., **25**, 414 (1992).
8. D. Bourgeois, T. Ursby, M. Wulff, C. Paradervand, V. Srajer, A. Legrand, W. Schildkamp, S. Laboure, C. Rubin, T. Y. Teng, M. Roth, and K. Moffat, SPIE, **2521**, 178 (1995).
9. I. V. Tomov, P. Chen, P. M. Rentzepis, J. Appl. Cryst. **28**, 358 (1995)
10. B. Van Wonterghem and P. M. Rentzepis, SPIE **1204**, 784 (1990).
11. T. Anderson, I. V. Tomov and P. M. Rentzepis, J. Chem. Phys., **99**, 869 (1993).

12. I. V. Tomov, T. Anderson and P. M. Rentzepis, J. X-ray Sci. Technol., **4**, 44 (1993).

13. I. V. Tomov, P. Chen and P. M. Rentzepis, Rev. Sci. Instrum. **66**, 5214 (1995).

14. J. P. Girardeau-Montaut and C. Girardeau-Montaut, J. Appl. Phys. **65**, 2889 (1989).

15. C. Girardeau-Montaut and J. P. Girardeau-Montaut, H. Leboutet, Appl. Phys. Lett. **55**, 2556 (1989).

16. I. V. Tomov, P. Chen, S. H. Lin and P. M Rentzepis, in "Time resolved diffraction", J. R. Helliwell and P. M. Rentzepis eds. Oxford University Press, Oxford (1997).

17. I. V. Tomov, T. Anderson, and P. M. Rentzepis, Appl. Phys. Lett. **61**, 1157 (1992); **61**, 3193E (1992).

18. W. H. Zachariasen, "Theory of X-ray Diffraction in Crystals", Wiley, N Y, 1945.

19. V. S. Speriosu, J. Appl. Phys. **52**, 6094 (1981).

20. C. R. Wie, T. A. Tombrello and T. Vreeeland, Jr., J. Appl. Phys. **59**, 3743 (1986).

21. D. Taupin, Bull. Soc. Franc. Miner. Cryst., **87**, 469 (1964).

22. S. Takagi, J. Phys. Soc. Japan, **26**, 1239 (1969).

23. B. Klar and F. Rustichelli, Nuovo Cimento, **13B**, 249 (1973).

24. J. H. Bechtel, J. Appl. Phys., **46**, 1585 (1975).

25. A. Fukuhara and Y. Takano, Acta Cryst. A**33**, 137 (1977).

26. American Institute of Physics Handbook (1972), 3rd ed. (MaGraw-Hill, N.Y.)

27. Touloukian Y. S., Powell R. W., Ho C. Y. & Klemens P.G., eds. (1970), "Thermophysical properties of matter",(IFI/Plenum, N.Y.)

28. G. Simmons and H. Wang, "Single Crystal Elastic Constants and Calculated Aggregate Properties: A Handbook" MIT Press, Cambridge (1971).

SINGLE MOLECULE FORCE SPECTROSCOPY WITH INDIVIDUAL PROTEINS

Julio M. Fernandez

Department of Physiology and Biophysics,

Mayo Foundation, Rochester Minnesota 55905, USA

Mathias Gautel

Biological Structures Division, European Molecular Biology Lab .

69012 Heidelberg

Mathias Rief and Hermann E. Gaub

Lehrstuhl für Angew. Physik Ludwig-Maximilians-Univ. München,

80799 München

1. A physiological question

The passive elastic properties of muscle fibers play an important role in muscle function. For example, the passive diastolic tension of the cardiac muscle wall determines the size of the stroke volume (Brady, 1991). When two skeletal muscles are working against each other, like during the activity of asynchronus flight muscles, muscles actively contract against the passive elastic properties of the other, taking turns. In this case the storage of elastic strain in the passively elongating muscle is a crucial and large component of the mechanical performance (Alexander & Bennet-Clark, 1977).

G. Tsoucaris (ed.), Current Challenges on Large Supramolecular Assemblies, 319–336.

It is a widely accepted view that titin proteins regulate the passive elastic properties of muscle (Labeit and Kolmerer, 1995). Titin is a large, rod-like modular protein that spans the length of a half-sarcomere between the M and Z lines (see figure 1). Human cardiac titin has a molecular mass of 2993 kD and it is mostly composed of individually folded domains of the immunoglobulin (Ig) and fibronectin type III (FN-III) type. There are 112 Ig and 132 FN-III like domains in human cardiac titin. A titin molecule overlaps the sarcomere over the A band and the I band. The elastic region of titin has been identified as the region that overlaps the I band. The composition of I band titin is regulated by alternative splicing. For example, the I band of human cardiac muscle titin is composed of 37 tandem Ig modules, 6 FN-III modules and a 163 amino acid long region that is rich in Prolines (P), Glutamate (E), Valine (V) and Lysine (K), named the PEVK region. By contrast, the I band of human skeletal titin contains an additional 53 Ig domains and a PEVK region that is 2174 residues long (Labeit and Kolmerer, 1995). It has been proposed that the alternative splicing of I band titin explains the strikingly different elastic properties observed in cardiac and skeletal muscles (Wang et al, 1991; Labeit and Kolmerer, 1995).

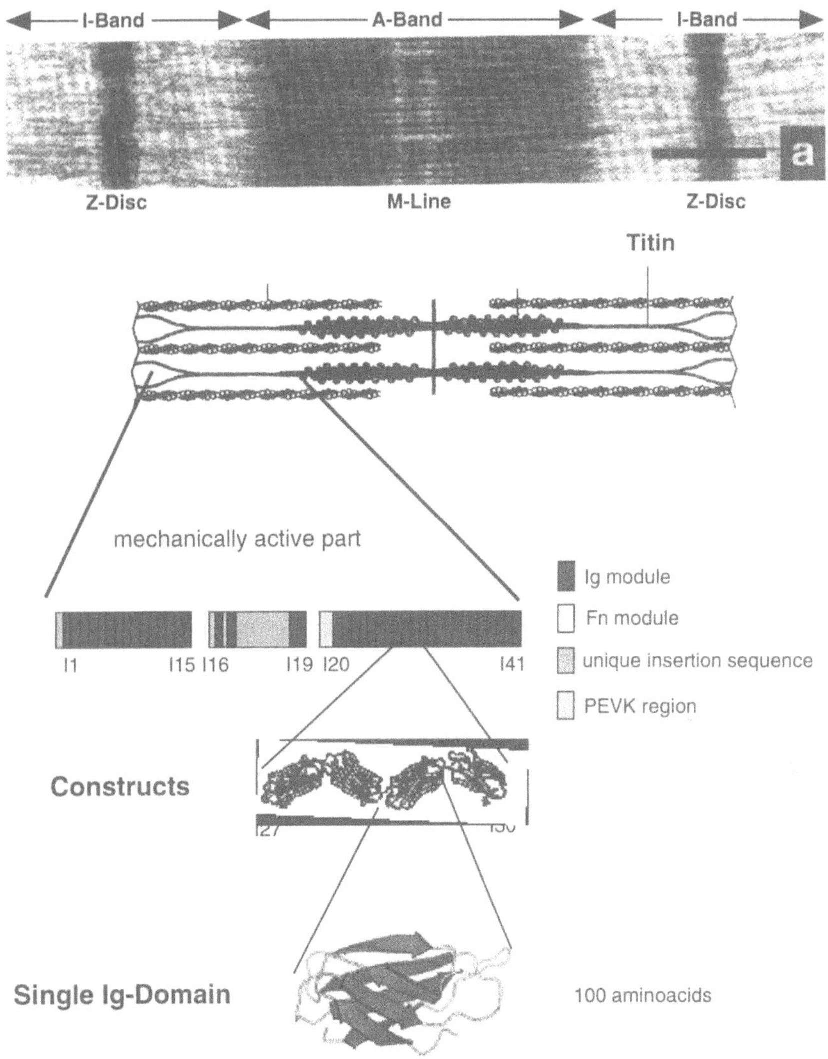

Figure 1. Molecular architecture of muscle fibers. The top panel shows a single sarcomere characterized by its band structure that results from the higly ordered arrangement of actin, myosin and titin proteins. The sequence of panels shows a schematic of the protein architecture of a sarcomere, a diagram of the modular arrangement of the I band region of titin, a model of the tandem Ig type domains found in this region and a schematic of a single titin Ig domain.

2. A hypothesis for titin elasticity

Soteriou et al, (1993) and also Harold Erickson (1994) independently proposed a molecular mechanism to explain the elasticity of I band titin. They proposed that when placed under strain, the individual immunoglobulin and fibronectin domains unfold, lengthening the contour length of the molecule and allowing for an extension of the sarcomere. Chemical denaturation experiments showed that the stability of Ig and FN-III modules falls in the range of $\Delta G = 1\text{-}10$ kcal/mol or 4.2-42 kN·m/mol (Politou et al., 1995). Since the folded size of these modules is ~ 4nm and their fully extended (denatured) size is ~35 nm, unfolding of a single module lengthens the protein's contour length by $\Delta x \sim 31$ nm. Since $F \cdot \Delta x = \Delta G$, we can calculate the force required to unfold an average single protein module as F ~ 1 pN. Hence, the force required to unfold a single Ig or FN-III domain was calculated to be very low, in the range of 1-5 pN (e.g. Erickson, 1994). Furthermore, they also proposed that upon release of the pulling force, the unfolded modules would rapidly renature to their original size. This mechanism for protein elasticity was further supported by the estimate that during stretching of a sarcomere, peak forces per titin molecule of up to ~50 pN were generated (Erickson, 1994). These considerations predicted that unfolding/refolding reactions were a major component of I band titin elasticity. Very recently, this hypothesis was put to the test in three different laboratories with results that were more complex and interesting that had been anticipated (Kellermayer et al, 1997; Tskhovrebova et al, 1997; Rief et al, 1997). Using laser tweezers, Kellermayer et al, 1997, stretched single skeletal muscle titin proteins and observed force-extension curves with a large amount of hysteresis that closely matched the passive force-extension curves observed in rabbit psoas muscles (Fig. 5 in Kellermayer et al, 1997). These experiments confirmed that the passive elastic properties of a muscle fiber could be accounted for by the molecular elasticity of titin proteins. Although not observed directly, the large hysteresis measured in the titin force-extension curves was assumed by Kellermayer et al, to be due to Ig and FN-III domain unfolding. In similar experiments, Tskhovrebova et al, 1997, measured the force-extension curves of single titin molecules and found a similar degree of hysteresis than that observed by Kellermayer et al. In addition to force-extension curves, they also did stepwise force-relaxation experiments which showed that upon stretching, titin molecules elongated in a stepwise mode with each step increasing the contour length by an average extension of ~19 nm. These data showed that stretching titin molecules caused the unfolding of titin domains as predicted earlier by Soteriou et al, 1993, and by Erickson, 1994.

3. Single molecule force spectroscopy by AFM

The atomic force microscope is a simple instrument capable of measuring forces down to piconewtons and can resolve force changes caused by the displacement of its probe by a fraction of a nanometer. Furthermore, these measurements can be done under water, allowing for the study of biological material under conditions that resemble those in vivo. The AFM functions like a miniature phonograph. A sharp tip mounted on a cantilever interacts with the sample, causing minute deflections that can be calibrated as a force. The AFM, in its force measuring mode, has been applied to the study of molecular mechanics. For example it has been used to measure the adhesion force between single avidin-biotin pairs (~160 pN; Florin et al, 1994). These measurements were independently confirmed by molecular dynamic simulations of the avidin-biotin interaction that revealed the atomic trajectory of a rupture and predicted the associated forces (Grubmuller et al, 1996). Hence, the techniques of molecular mechanics now rival the capabilities of patch-clamp techniques in describing the activity of single proteins.

More recently, single molecule AFM techniques were applied to study the elasticity of recombinant titin proteins (Rief et al, 1997). In a typical experiment, the protein sample is placed on a gold coated coverslip which is attached to the piezoelectric positioner. The recombinant proteins were engineered so that one end (C-terminus) covalently anchors to the gold substrate. Random segments of titin are then picked up by adsorption by the AFM tip and stretched for up to several hundred nanometers. When the protein is stretched, it pulls on the cantilever causing small cantilever deflections that can be converted into units of force if the cantilever spring constant, k_C, is known. As the protein is stretched the force increases in a non-linear (non-Hookean) fashion; this is the energy required to decrease the degrees of freedom of the polypeptide (an entropic-spring behavior that is a common property of polymers). If the protein is further stretched the probability for the unfolding of individual domains becomes very high. When a domain unfolds, the AFM tip travels back to its resting position and the force goes back to zero.

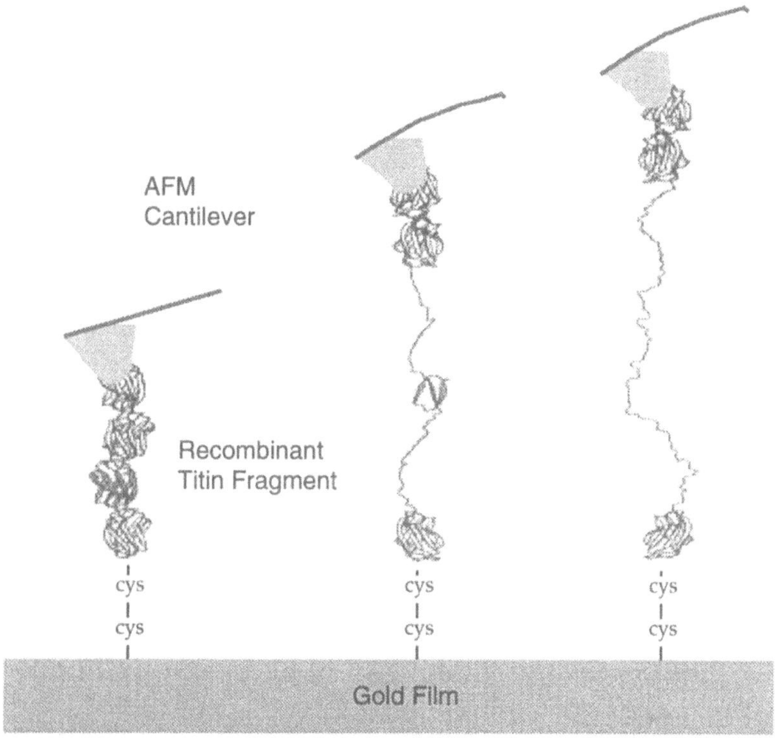

Figure 2. Schematic representation of the forced unfolding of protein domains. Recombinant titin proteins are engineered with terminal cysteines in order to provide a covalent attachment to a gold covered substrate. The other end of the protein is picked up by an AFM tip by simple adsorption. Upon stretching, the idividual Ig domains unfold abruptly in an all-or-none fashion.

4. Folding/unfolding reactions captured by AFM

Figure 2 shows a typical experiment as reported by Rief and collaborators (Rief et al, 1997). Recombinant DNA techniques were used to express a segment of 8 Ig domains spanning from I27 to I34 of the I-band region of human cardiac muscle titin (Fig. 1). The recombinant proteins were region of human cardiac muscle titin (Fig. 1). The recombinant proteins were constructed with two cystein residues added to the C terminus of the protein. These sulfhydril groups are known to interact with the gold coated coverslip forming covalent bonds and provide a firm anchor for one end of the protein. In order to pick up proteins with the AFM tip, the tip was pressed down onto the sample for 1-3 seconds and at forces of several nN creating a strong bond with the protein by adsorption. The protein is then extended by separating the AFM tip from the substrate. The results of these experiments, as reported by Rief and collaborators (Rief et al, 1997), show force-extension curves that follow a saw-tooth pattern with several regularly spaced force peaks (e.g. see figure 4) which are interpreted as follows: With increasing extension of the protein, the tension along the polymeric molecule gradually increases until one of the domains unfolds and the force drops suddenly. The unfolded domain adds contour length to the protein which needs to be stretched in order to build up enough tension again to refold another domain and so on. The last force peak corresponds to detachment of the protein from the AFM tip. The heights of the peaks reflect unfolding forces in the range of 150 to 300 pN. The distance between force-peaks, measured at ~100 pN, was ~25 nm. This distance is consistent with the length of an unfolded Ig domain containing 89 aa (Rief et al, 1997).

The elastic response of titin is non-linear and is determined by a hierarchy of processes (See Fig. 3). The low force regime is governed by entropy driven polymer elasticity. The medium force regime is determined by protein unfolding and at forces above 500 pN, we start to see bond angle deformations of the aminoacid backbone. A theoretical description of the experiments has to take theses different mechanisms into account.

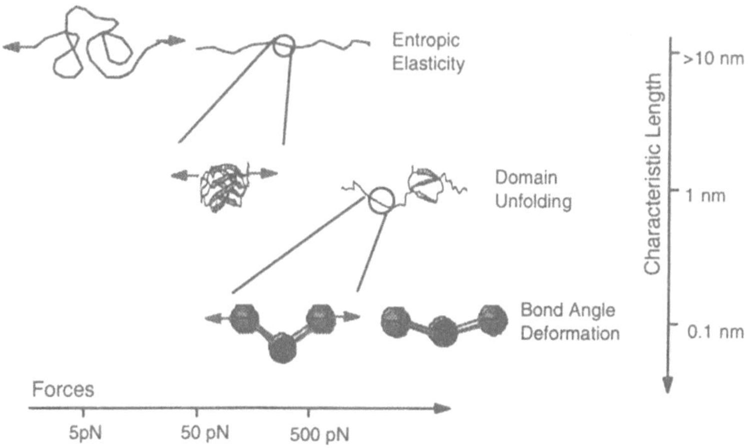

Figure 3 Elastic regimes of a polypeptide chain. The initial stages of stretching a protein (force range 0-10 pN) generates restoring forces that are mostly entropic in nature. At higher forces (10-400 pN) protein domain unfolding is observed. In the highest force range (> 1000 pN) bond deformation and bond rupture may be observed.

5. The Worm Like Chain model of polymer elasticity.

Figure 4 shows a detailed analysis of a force-extension curve obtained from a single Ig8 protein. The force-extension curve is explained by modeling the elasticity of the Ig8 polypeptide as a Worm-Like-Chain (WLC) with a persistence length of 0.4 nm and a variable contour length. Stretching the folded Ig8 protein causes a rapid increase in the force until a threshold (force peak) is reached where the first unfolding event occurs (peak to trough transition). The first unfolding event causes a large increase in the contour length of the protein (63%) rapidly relaxing the polypeptide to a zero holding force. In contrast, the 4th unfolding event increases the contour length by only 28% and cannot relax the holding force to zero, reaching a trough that represents the force obtained by stretching a polypeptide that is 28 nm longer. Another striking feature of the force-extension curves is that the left side slope of the peaks, which reflects the elasticity of the protein, decreases from peak to peak. This means that the elasticity of the stretched protein, which may be derived from the slope, is dominated by a spring that becomes softer with each Ig domain unfolding. This effect is also well described by the WLC model shown in figure 4.

An important feature of the titin force-extension traces is that the peak maximum increases with increasing extension. Since every downstroke (peak to trough) reflects the unfolding of a domain and since the domains are different, this finding means that the weakest domains unfold first and the strongest last. Thus the domains are sorted in the extension curves according to their unfolding force and not to their position in the chain. This may be an important feature of the I band organization of Ig modules.

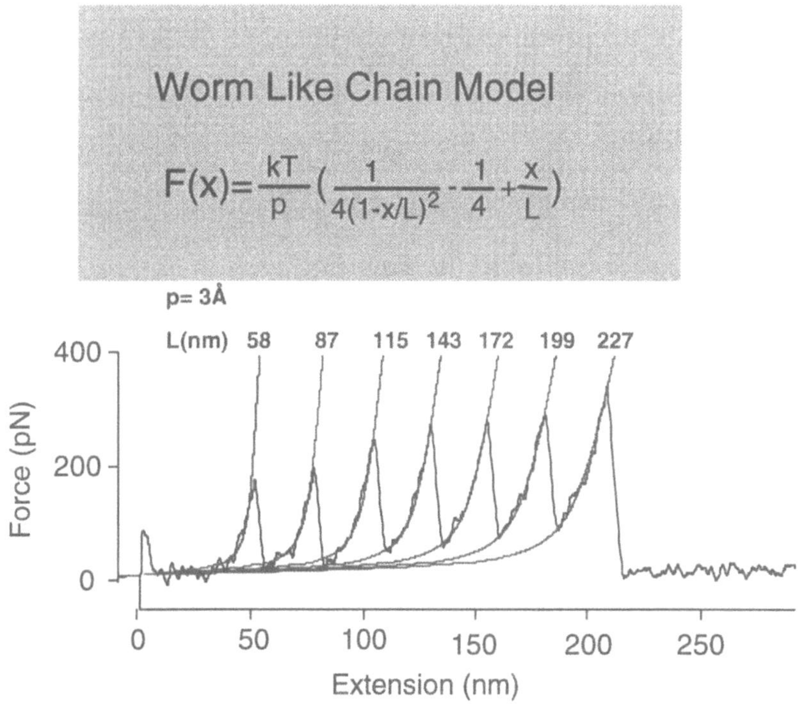

Figure 4 Force-extension curve measured for a single recombinant titin protein composed of eight Ig domains. These recordings are characterized by a sawtooth like curve where each force peak marks the unfolding of an individual Ig domain. The elasticity of the protein is described by the Worm Like Chain model of polymer elasticity (top panel and fitted lines in bottom panel) with a persistance lenght, p=0.3 nm and a contour length, L, that incremented by ~28 nm each time that a domain unfolds.

6. A Markovian model of protein unfolding.

In its most simple representation which completely neglects the complex hirarchy of the folding potential the folding/unfolding of a domain can be modeled as a two state Markovian process where the reaction coordinate is length and the rate constants of folding (β) and unfolding (α) are determined by the activation energies and reaction lengths. The free energy along the reaction coordinate is affected by an applied force. Hence, the rate constants for the folding/unfolding reaction also depend on the applied force. This description can be made similar to the well described dissociation of noncovalent bonds placed under an external force (Bell, 1978; Evans and Ritchie, 1997). In this case the rate constants are given by $\alpha(F) = \alpha_0 \exp[F\Delta x/kT]$ and $\beta(F)= \beta_0 \exp[-F\Delta x/kT]$ where F is the applied force, Δx is the distance over which the unfolding event occurs, α_0 and β_0 are the rate constants in the absence of an applied force (figure 5), and k and T have their usual meaning. It is clear from this description that we cannot speak of an "unfolding force" but rather of a domain unfolding probability. Under an applied force, the folding rate constant β becomes small whereas the unfolding rate α becomes large. As a first approximation we can neglect β and if we assume that N domains connected in series experience the same force at all times, then the probability of observing an unfolding event at a given force, F, is $P_u(F)=N\alpha(F)\Delta t$ (with $\Delta t<<1$ and $0>P<1$). These simple considerations predict that unfolding may occur at any applied force, albeit with different probabilities. Thus, in addition to the applied force, observing an unfolding event depends on the length of time that we wait at a given force or the rate of force increase during extension and on the number of available domains (Rief et al, 1997; Kellermayer et al, 1997).

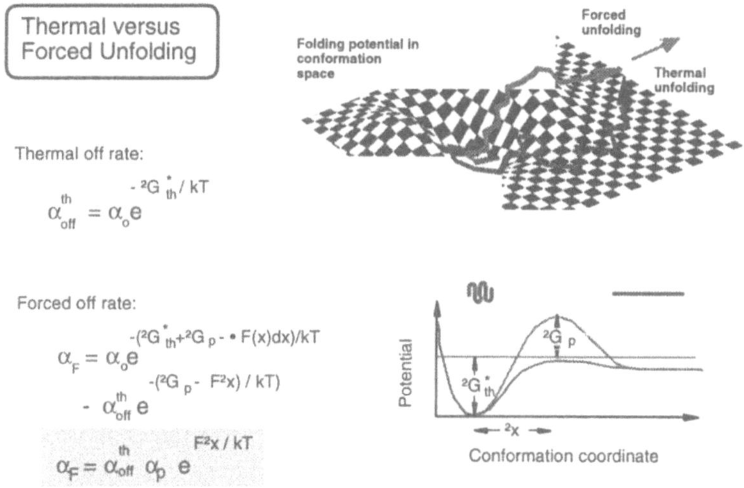

Figure 5 Forced unfolding of Ig domains as a two state Markovian process. An external force reduces the activation barrier for unfolding and thus increases the thermal unfolding-rate. The forced path may deviuate from the thermal path giving rise to an additional contribution to the activation barrier. Note that the force introduces a direction and a length scale; the width of the potential.

7. Monte Carlo simulation of the elasticity of a modular protein.

In this very simple model which more ment to educate ourselves the principles of the elastic response of modular polymers than to give a theoretical explaination of protein (un)folding, we model the extensibility of titin with an entropic elasticity such as described by the Worm Like Chain (WLC) and a force dependent all-or-none unfolding of the individual modules. The unfolding of the modules is described by Eyring rate theory with a simple two state Markovian model and a force dependent rate constant (see above). This description gives the probability of observing the unfolding of any module as $P=N\alpha\Delta t$ where N is the number of modules, α is the unfolding rate constant and Δt is the polling interval. With these simple considerations, the force-extension curves for the unfolding of a modular polymer can be predicted under a variety of experimental conditions. In order to simulate the extension of titin, the force experienced by titin modules during a ramp extension is calculated by using the WLC. Once the force is known for a given extension, the probability of unfolding of any module can be calculated and then each module is polled to determine its status, following the Monte Carlo approach.

Figure 6 shows a Monte Carlo simulation of the force-extension curve obtained by stretching, at a constant speed (4000 nm/s), 8 titin Ig domains. As shown, this simple model closely reproduces the published force extension curves of recombinant titin segments (Rief et al, 1997). In this simulation the average force at which unfolding is observed was 200 pN. However, if the extension speed is reduced to 4 nm/s the average force drops to 100 pN (Fig. 5 in Rief et al, 1997). Furthermore, if the number of domains is increased from 8 to 200 the average force at which unfolding is observed drops further to 50 pN. The same model can readily simulate the step relaxation results of Tskhovrebova et al (1997).

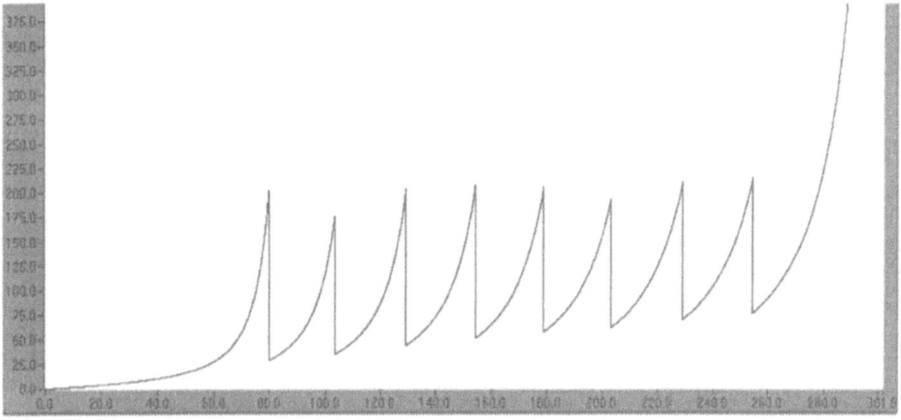

Figure 6. *Monte Carlo simulation of the extension of a recombinant titin protein with eigth Ig type domains. The axis correspond to force (pN) and extension (nm). Unfolding of individual Ig domains was modeled as a two state Markovian process. The rate constants and the worm like chain parameters were taken from Rief et al, (1997). The simulation closely reproduces the observed force extension of a recombinant titin fragment with 8 lg type modules. This simulation demonstrates that combing the purely entropic elasticity of the polymer and a Markovian extensibility that results from conformational changes induced by an applied force provides a basic description of the elasticity of a modular protein.*

8. Conclusions.

The findings produced by the combination of molecular biology, nano-scale physics and computational techniques illustrates the mechanical function of folding/unfolding reactions by Ig domains. AFM-related techniques provides a considerable advantage compared to other experimental aproaches. For example, folding and unfolding reactions involve transitions from one equilibrium state to another. However, the unfolding of isolated Ig domains is a rare event estimated by NMR to occur spontaneously at a rate of ~ $5x10^{-4}$ s^{-1} whereas refolding is typically much faster at rates of ~ 10-100 s^{-1} (Fong et al, 1996). Hence, the unfolding/refolding reactions could only be studied using chemical denaturation which forced the domains into various degrees of unfolding (e.g. Fong et al, 1996). However, it is uncertain whether chemically denaturation truly mimics the unfolding reactions that occur during a physiological extension of titin. By contrast, the AFM techniques trigger unfolding of Ig domains by applying a force which is closer to the actual extension of titin in a sarcomere (Rief et al, 1997). Furthermore, although unfolding of an unthetered domain is a rare event, application of a force increases the rate of unfolding exponentially making it easily observable without requiring denaturants such as urea. Another significant advantage of AFM techniques is that they provide real time measurement of the unfolding/refolding reactions at the level of a single protein and thus have the potential for revealing kinetic reactions which may not be apparent when measuring the steady state properties of a large number of molecules (e.g. NMR).

In addition to understanding I band titin elasticity, studies of the mechanical properties of modular domains may also have wider significance. Ig domains are found in a large number of proteins and share a similar structure characterized by a seven-stranded beta barrel fold (Bork et al, 1996; Potts and Campbell, 1996). Thus, it is likely that the mechanical roles played by these modules are not restricted to titin and may have significance for the function of a wide array of other modular proteins.

References:

Alexander, R. McN., and Bennet-Clark, H.C. (1977). Storage of elastic strain energy in muscle and other tissues. *Nature* 265:114-117.

Bell, G.I. (1978). Models for the specific adhesion of cells to cells. *Science* 200:618-627.

Bork, P., Downing, A.K., Kieffer, B., and Campbell, I.D. (1996). Structure and distribution of modules in extracellular proteins. *Quart. Rev. Biophys.* 29(2):119-167.

Brady, A.J. (1991). Mechanical properties of isolated cardiac myocytes. *Physiol. Rev.* 71(2):413-428.

Erickson, H.P. (1994). Reversible unfolding of fibronectin type III and immunoglobulin domains provides the structural basis for stretch and elasticity of titin and fibronectin. *Proc. Natl. Acad. Sci. USA* 91:10114-10118.

Evans, E., and Ritchie, K. (1997). Dynamic strength of molecular adhesion bonds. *Biophys. J.* 72:1541-1555.

Florin, E.L., Moy, V.T., and Gaub, H.E. (1994) Adhesion forces between individual ligand-receptor pairs. *Science* 264(5157):415-417.

Fong, S., et al. (1996). Structure and stability of an immunoglobulin superfamily domain from twitchin, a muscle protein of the nematode *Caenorhabditis elegans*. *J. Mol. Biol.* 264:624-639.

Grubmuller, H., Heymann, B., and Tavan, P. (1996) Ligand binding: molecular mechanics calculation of the streptavidin-biotin rupture force. *Science* 271(5251):997-999.

Kellermayer, M.S.Z., Smith, S.B., Granzier, H.L., and Bustamante, C. (1997). Folding-unfolding transitions in single titin molecules characterized with laser tweezers. *Science* 276:1112-1116.

Labeit, S., and Kolmerer, B. (1995). Titins: giant proteins in charge of muscle ultrastructure and elasticity. *Science* 270:293-296.

Politou, A.S., Thomas, D.J., and Pastore, A. (1995). The folding and stability of titin immunoglobulin-like modules, with implications for the mechanism of elasticity. *Biophys. J.* 69:2601-2610.

Potts, J.R., and Campbell, I.D. (1996). Structure and function of fibronectin modules. *Matrix Biol.* 15(5):313-320.

Rief, M., Gautel, M., Oesterhelt, F., Fernandez, J.M., and Gaub, H.E. (1997a). Reversible unfolding of individual titin immunoglobulin domains by AFM. *Science* 276:1109-1112.

Rief, M., Oseterhelt, F., Heymann, B., Gaub, H.E. (1997b). Single molecular force spectroscopy on polysaccharides by atomic force microscopy. *Science* 275(5304):1295-1297.

Soteriou, A., Clarke, A., Martin, S., and Trinick, J. (1993). Titin folding energy and elasticity. *Proc. R. Soc. Lond. B Biol. Sci.* 254(1340):83-86.

Tskhovrebova, L., Trinick, J., Sleep, J.A., and Simmons, R.M. (1997). Elasticity and unfolding of single molecules of the giant muscle protein titin. *Nature* 387:308-312.

Wang, K., McCarter, R., Wright, J., Beverly, J., and Ramirez-Mitchell, R. (1991). Regulation of skeletal muscle stiffness and elasticity by titin isoforms: a test of the segmental extension model of resting tension. *Proc. Natl. Acad. Sci. USA* 88:7101-7105.

CRYSTAL SOLVING PROCEDURES APPLIED TO PEPTIDES, FOLDAMERS AND CHANNEL FORMING ADAMANTYL/PEPTIDE HYBRID MACROCYCLES

ISABELLA L. KARLE
Laboratory for the Structure of Matter
Naval Research Laboratory
Washington, D.C. 20375-5341, U.S.A.

INTRODUCTION

Solving crystal structures by direct phase determination has become quite routine for many types of substances, providing that a sufficient number of X-ray reflections of reasonable magnitude can be measured. Resolution of 0.9 to 1.0 Å is usually desirable. Resolution of less than 1.15 to 1.2 Å causes severe problems for crystals composed of only C, H, N and O atoms. The presence of heavier atoms, S for example, makes phase determination and structure determination considerably easier. For equal atom structures, problems arise for large molecules, molecules with flexible groups, crystals with several independent molecules per asymmetric unit, and structures with large channels or cavities containing water or disordered solvents. The above conditions contribute to an attenuation of the magnitude of the scattered reflections to the point that the number of reflections that can be used in phase determination is lessened and that the probability of the correctness of individual phases becomes small. At times, it is not the resolution that is the problem, but the fact that the intensities in general may be weak and also that any particular atom has a rather small contribution to a structure factor. Several examples of structure analyses of crystals that contain one or more of the above problems will be discussed.

METHODOLOGY AND RESULTS

Characteristics of five crystals are listed in Table 1. Routine direct phase determination as contained in the SHELXTL systems of programs[1], even for 10,000 trials in the TREF mode (tangent formula[2] refinement of phases), did not result in solutions for the crystals in Table 1. The crystal of MeO(Aib)$_4$-COPyrCO-(Aib)$_4$OMe (Py8Aib)[6] (where Aib ≡ alpha aminoisobutyryl residue and Pyr = pyridyl) was solved in space group P1, rather than $P\bar{1}$. Reducing space group symmetry from $P\bar{1}$, P2$_1$ or C2 to P1 (which doubles the number of unknowns) and later locating the symmetry element (a center of symmetry, a 2-fold screw axis or a 2-fold axis, for example) has been a successful methodology for

G. Tsoucaris (ed.), Current Challenges on Large Supramolecular Assemblies, 337-348.

many years.[3] In this particular case, fragments from several molecules (of the four molecules that needed to be placed in the cell) were recognized in the E-map[4] calculated from the best combined figure of merit in 1000 trials of TREF in the SHELXTL program.[1] Phases based on the 96 atoms in the fragments were refined and phases for additional reflections were determined by a partial structure procedure[5] (corresponding to TEXP in SHELXTL[1]). The E-map after one cycle of tangent expansion contained 242 of the 254 atoms in the structure. At this point a center of symmetry relating two of the molecules to the other two was recognized. The origin of the cell was translated to correspond to the center and the remainder of the structure was solved and refined in space group $P\bar{1}$.[6] Two procedural items should be noted. First, in the TREF calculation, the SUBS (subset) command had to be designated as type 8 (in which the reflections with the highest estimated α-values are used). Second, it is prudent, in cases where the fragments found in the initial E-map are relatively small, not to recycle in the TEXP procedure (3 cycles in the default instruction) but to examine the E-map after each single cycle and to select manually the atoms for the following cycle. Automatic selection of peaks by the program often results in too many incorrect "atoms" which lead to meaningless E-maps. The helix-plane-helix conformation of the two molecules of MeO(Aib)$_4$-COPyrCO-(Aib)$_4$OMe in the asymmetric unit are shown in Figure 1. Four molecules of the hybrid peptide associate by hydrogen bonding through water bridges, as well as direct NH···OC

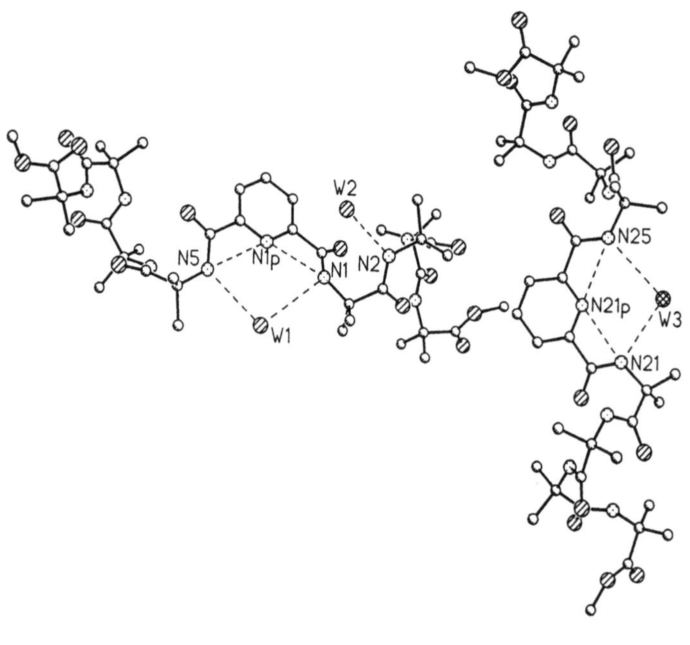

Figure 1. Crystal conformation of the two molecules of MeO(Aib)$_4$COPyrCO(Aib)$_4$OMe in an asymmetric unit in space group $P\bar{1}$.

bonds, to form a hydrophobic channel with a cross-section of at least 5.8 Å x 10.6 Å, Figure 2. Ethyl acetate solvent molecules pack into the hydrophobic channel loosely and most likely contribute to the problems of solving the structure.

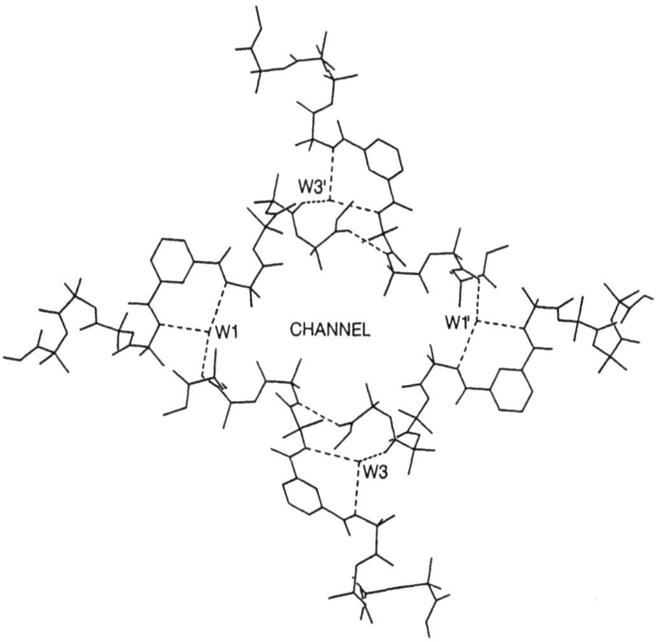

Figure 2. Hydrophobic channel (5.8 Å x 10.6 Å) formed by assembly of Py8Aib molecules shown in Figure 1. The assembly is held together mainly by hydrogen bonds with water molecules W1 and W3 acting as bridges.

Three of the structures listed in Table 1 were solved by a vector search of a Patterson function followed by a translational search (the latter not required in space group P1) of a model from a known structure (or a fragment thereof) that corresponds closely to part of the unknown structure. The PATSEE program[7] is a free-standing program that is part of the SHELXTL package. The procedures for rotation, translation and tangent expansion are quite easy to use. The selection of the initial search fragment is problematical at times. For the cyclic(Val-Ser-Adm)$_2$, Figure 3, (where Adm is 1,3-dicarbonyl-adamantane) only 12 atoms, consisting of the adamantane frame plus two attached C atoms, could be used as the known model for the search fragment. Any additional atoms attached to this fragment would have introduced unknown rotational parameters. The 12-atom model, which represented only 10% of the unknown structure (two independent molecules in P1), yielded an additional 15 recognizable atoms after rotation and tangent expansion. The successful set of rotation parameters were ranked only fourth best with respect to the CFOM. In successive cycles of E-map and further tangent expansion,.the number of atoms found increased to 38 and 93, after which difference maps were used. Two independent clam-shaped molecules form a channel that contains 9 water molecules. Almost all the carbonyl moieties and some of the NH moieties extend into the channel to form hydrogen bonds with the water[6], Figure 4.

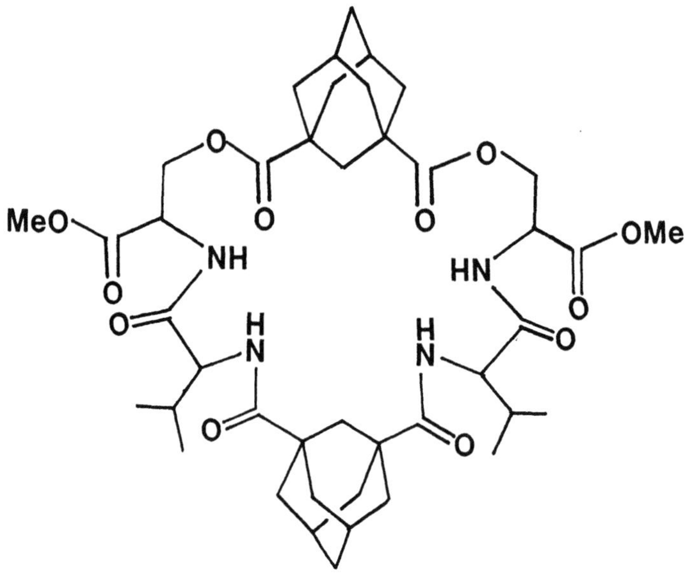

Cyclo(Val–Ser–Adm)$_2$

Figure 3. Chemical formula of (Val-Ser-Adm)$_2$.

An entirely new type of helix has been constructed with β-peptides having the formula Boc-[NH–C_6H_{10}–C(0)]$_n$-OBz. The structure of the tetramer (n=4) was solved routinely. The hexamer (n=6) had three molecules per asymmetric unit[8] and its solution with exhaustive trials of routine programs was not successful. A model consisting of 34 atoms from the tetramer (16% of the unknown atoms), Figure 5, was used more or less successfully for rotation and translation in the PATSEE program. The parameters for the second best fit, based on the CFOM according to criteria in the program, led to an E-map that showed a number of groupings of atoms that appeared to correspond to fragments of helical structures. None of the fragments in the E-map used individually would expand

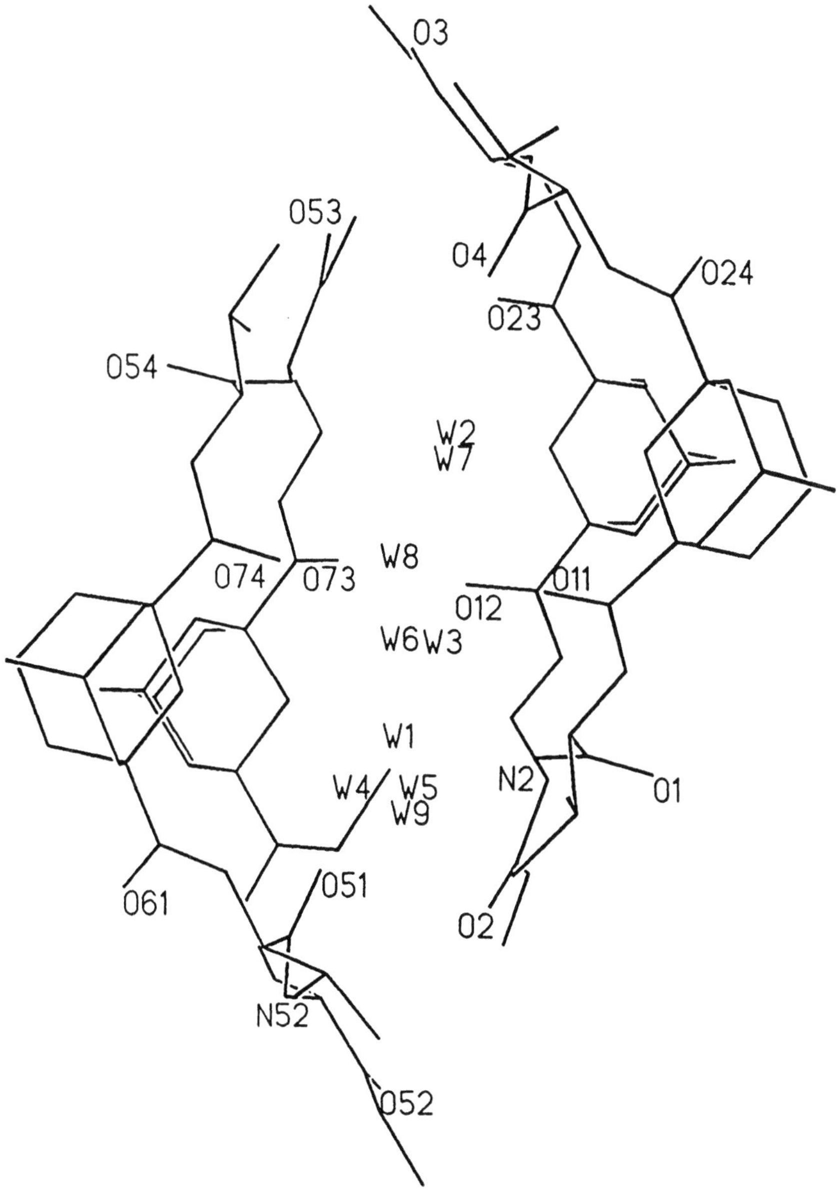

Figure 4. Two independent clam-shaped molecules of $(Val\text{-}Ser\text{-}Adm)_2$ form a channel that contains 9 water molecules. The substance has been shown to be a K^+ ionophore. There are no intramolecular NH...OC hydrogen bonds. Only two NH moieties form intermolecular hydrogen bonds that bind the [2 molecule + 9 water] complex into linear arrays. The remainder of the carbonyl moieties and some of the NH moieties are directed into the channel and form hydrogen bonds with the water molecules.[6]

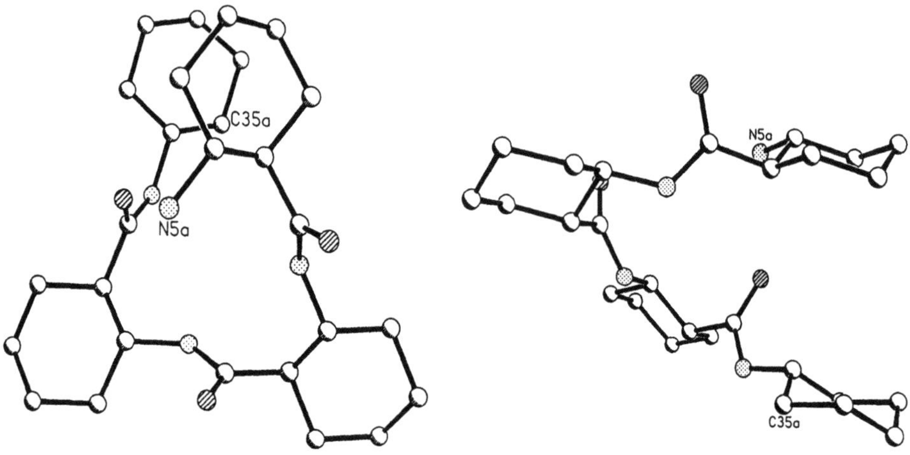

Figure 5. Two views of the search model from the tetramer (34 atoms) used for the structure solution of three independent molecules of the hexamer.

by means of the tangent formula to meaningful larger fragments. It was only after several fragments totaling 61 atoms were used together in the tangent expansion that some progress with growth on some of the fragments occurred. Repetition of the process showed very slow growth. Eventually, after the sites of 140 atoms were determined, the remainder of the 69 atoms were found by successive least-squares refinement and difference maps. The cause of the slow development of the structure is not apparent, especially since the ratio of the number of reflections observed >4s per atom was so much larger than in the usual experiment. It is true, however, that hydrophobic channels are formed which are filled with highly disordered $C_2H_4C\ell_2$ solvent molecules. The three independent molecules of the hexamer, stacked in a column, are shown in Figure 6.

The next example concerns a conventional α-helix with many Phe residues, Boc-(VAFU)$_3$-OMe.[9] The intensity of scattering dropped to a low level at a relatively small scattering angle and only 10% of the possible data were still barely measurable at a resolution of 0.9 Å. A model of 34 atoms from a known α-helix was used for the search fragment in PATSEE. The best ranked CFOM rotation parameters resulted in an E-map having a recognizable 31 atom fragment. Slowly, with manual intervention, the fragment was developed into two α-helices with the C^{β} atoms and most of the C^{γ} atoms located. The six phenyl groups are still being eked out by alternate least squares refinement and difference maps. It is not clear at this time whether the space group is P1 with two similar independent molecules or C2 with one independent molecule. In each

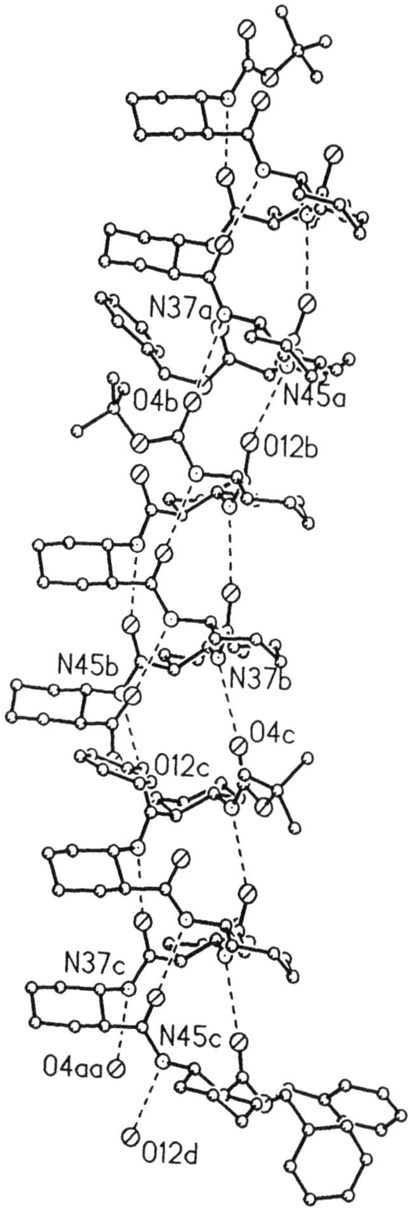

Figure 6. Three independent molecules of the hexamer of Boc-[NH–C_6H_{10}–C(O)]$_n$-OBz stack to form a column. Each foldamer is rotated by ~120° with respect to the one below. The new helix formed contains 14-membered hydrogen bond rings with the direction of the hydrogen bonding opposite to that found in alpha-helices.[8]

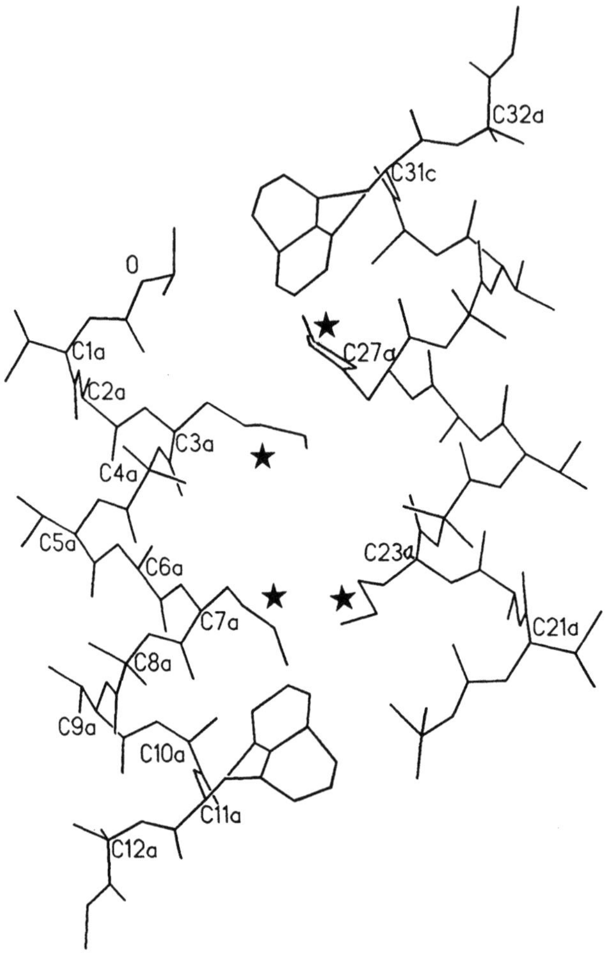

Figure 7. Present status of the structure solution of Boc(Val-Ala-Phe-Aib)$_3$-OMe. All eight phenyl groups, facing each other in the crystal, are disordered. Regions of missing atoms are denoted by asterisks.

helix, the three phenyl groups are aligned along one side of the helix. The point of interest is the packing (perhaps interdigitation) of the phenyl groups that face each other in adjacent antiparallel helices, Figure 7.[9]

The final structure analysis concerns the 16-residue antiamoebin (Ac-Phe-Aib-Aib-Aib-Iva-Gly-Leu-Aib-Aib-Hyp-Gln-Iva-Hyp-Aib-Pro-Phol).[10] The last half of the sequence is identical to that of zervamicin whose structure is known.[11] Further, both compounds crystallize in space group $P2_12_12_1$ with the a and b cell edges very similar in length and the c cell edge ~4.5 Å longer for the antiamoebin. It was suspected, correctly, that the conformation of the two molecules would be quite similar. The number of measured reflections was too meager for direct phase determination. The use of fragments of various size, taken from zervamicin, as a search model was not successful in the PATSEE procedure. The initial placement of the antiamoebin molecules in the unit cell was accomplished by using Brunger's annealing least-squares program X-PLOR[12-14] on a model based on the zervamicin molecule. Subsequently, alternate cycles of ordinary least-squares and difference maps corrected the gross misplacement of the phenyl ring in Phe^1 and located three cocrystallized octanol molecules. The particular interest in this structure is the assembly of an ion transport peptide in a membrane environment. The octanol may serve as a partial mimic for a lipid. Figure 8 shows the ion channel in antiamoebin, flanked by octanol molecules.

These different approaches for structure solution of large molecules with less than ideal X-ray data have been illustrated with actual examples and calculations.

ACKNOWLEDGEMENTS

I wish to acknowledge my collaborators who designed, purified, synthesized and grew crystals of the five peptides and peptide hybrids whose structures are discussed in this paper: Prof. P. Balaram, Indian Institute of Science, Bangalore, India; Prof. Samuel Gellman, University of Wisconsin, Madison Wisconsin; Dr. Darshan Ranganathan, Regional Research Laboratory (CSIR), Trivandrum, India; and Dr. Mary Ann Perozzo, Naval Research Laboratory, Washington, D.C., who participated in the structure solution of antiamoebin. Financial assistance from the Office of Naval Research and the National Institutes of Health, Grant #GM30902, is gratefully acknowledged.

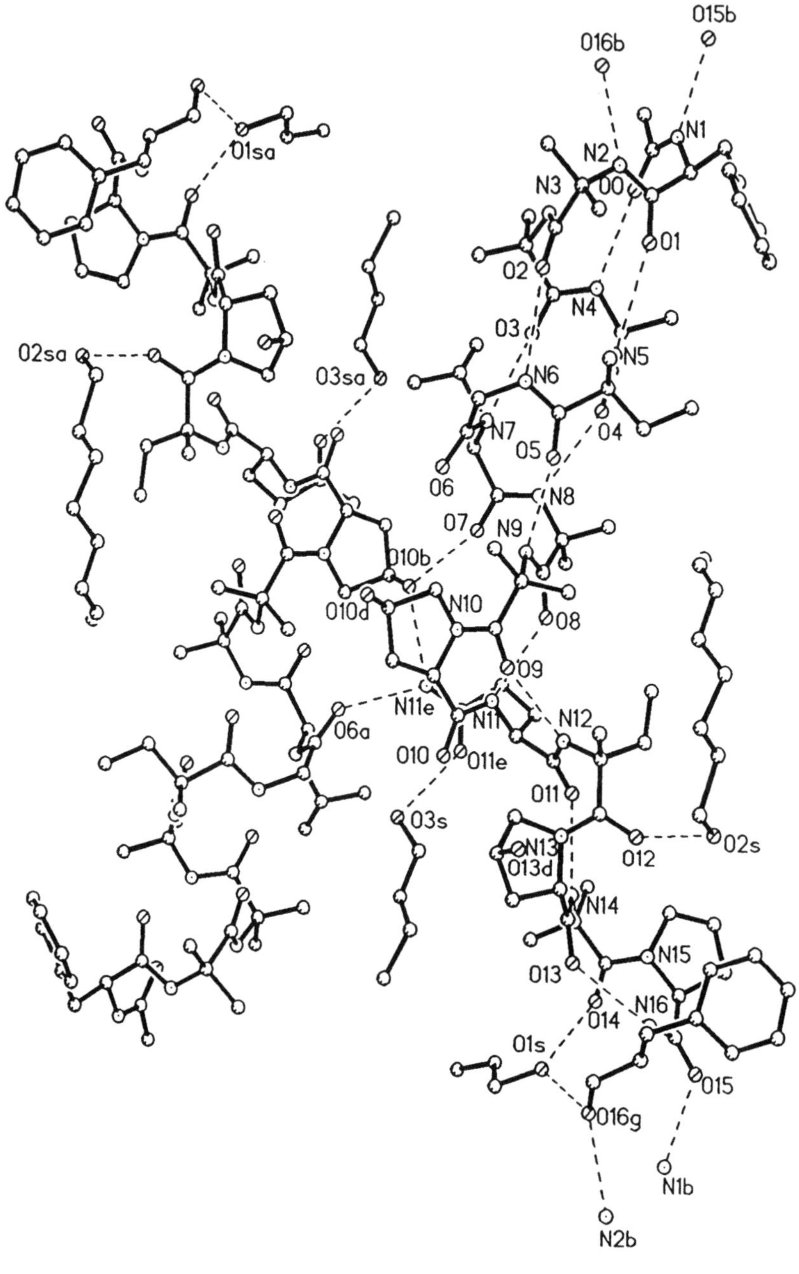

Figure 8. Hourglass-shaped ion channel formed by antiamoebin. Octanol molecules on either side may mimic the membrane into which these molecules insert themselves for K^+ ion passage. Probably due to lack of water in the crystallizing solution, octanol molecules also are inserted into the ion channels with the OH ends forming hydrogen bonds with carbonyls of the peptide that extend into the channel. The nonpolar ends of the octanols are grossly disordered.

Table 1. **CRYSTAL CHARACTERISTICS**

	Beta-- Foldamer	Py8Aib	Cyclic $(ValSerAdm)_2$	$Boc(VAFU)_3$-OMe	antiamoebin/ 3 octanol
sp. gr.	$P2_1$	$P1^*$	P1	P1	$P2_12_12_1$
#indep. mols.	3	4	2	2	1
#C, N, O atoms	209	254	122	194	196
a, Å	15.393	16.064	12.604	8.838	9.142
b	21.764	16.156	12.623	19.878	28.590
c	27.869	21.655	16.575	23.369	44.289
α(deg)	90.	90.14	68.48	75.61	90.
β	101.46	101.38	89.87	89.95	90.
γ	90.	97.07	74.50	77.27	90.
$V(g/cm^3)$	9150	5465	2350	3872	11578
#indep. data	26600	14800	6316	11444	7438
#>4σ	13300	7250	3080	weak data 3347	4133
R_1 factor	11.7	(11.0)	8.8		11.9
Comments	grossly disordered solvent	disordered EtAc *actually in $P\bar{1}$	large water channel	6 wobbly phenyls *possibly in C2	2 disordered octanols
Structure Solution	Vector Search Rot/Trans 34 atoms from tetramer	TREF 1000 SUBS 8	Vector Search Rot/Trans on Adm+2C (12 atoms)	Vector Search 34 atoms from α-helix	Brunger annealing, tangent expansion

REFERENCES

1. SHELXTL system of programs, Siemens Instruments, Madison, WI, USA.

2. Karle, J. and Hauptman, H. (1956) A Theory of Phase Determination for the Four Types of Non-Centrosymmetric Space Groups 1P222, 2P22, $3P_12$, $3P_22$, *Acta Cryst.* **9**, 635-651.

3. Karle, I. L. and Karle, J. (1971) Structure of the Chromophore from the Fluorescent Peptide Produced by Iron-deficient Azobacter Vinelandii, *Acta Cryst.* **B27**, 1891-1898; Karle, I. L., Karle, J., Mastropaolo, D., Camerman, A. & Camerman, N. (1983) [Leu[5]]Enkephalin: Four Cocrystallizing Conformers with Extended Backbones that Form an Antiparallel β-Sheet, *Acta Cryst.* **B39**, 625-637.

4. Karle, I. L., Hauptman, H., Karle, J. and Wing, A. B. (1958) Crystal and Molecular Structure of p,p'-Dimethoxybenzophenone by the Direct Probability Method, *Acta Cryst.* **11**, 257-263.

5. Karle, J. (1968) Partial Structural Information Combined with the Tangent Formula for Non-Centrosymmetric Crystals, *Acta Cryst.* **B24**, 182-186.

6. Ranganathan, D. and Karle, I. L., to be published.

7. Egert, E. and Sheldrick, G. M. (1985) Search for a Fragment of Known Geometry by Integrated Patterson and Direct Methods, *Acta Cryst.* **A41**, 262-268.

8. Appella, D. H., Christianson, L. A., Karle, I. L., Powell, D. R. and Gellman, S. H. (1996) β-Peptide Foldamers: Robust Helix Formation in a New Family of β-Amino Acid Oligomers, *J. Am. Chem. Soc.* **118**, 13071-13072.

9. Balaram, P. and Karle, I. L., to be published.

10. Balaram, P., Perozzo, M. A. and Karle, I. L., to be published.

11. Karle, I. L., Flippen-Anderson, J. L., Agarwalla, S. and Balaram, P. (1994) Conformation of the Flexible Bent Helix of Leu[1]-Zervamicin in Crystal C and a Possible Gating Action for Ion Passage, *Biopolymers* **34**, 721-735.

12. Brunger, A. T., Kuriyan, J. and Karplus, M. (1987) Crystallographic R Factor Refinement by Molecular Dynamics, *Science* **235**, 458-460.

13. Brunger, A. T., Krukowski, A. and Erickson, J. (1990) Slow-Cooling Protocols for Crystallographic Refinement by Simulated Annealing, *Acta Cryst.* **A46**, 585-593.

14. Brunger, A. T. (1992) The Free R Value: A Novel Statistical Quantity for Assessing the Accuracy of Crystal Structures, *Nature* **355**, 472-474.

TETRACYCLINE REPRESSOR ACTS AS A MOLECULAR SWITCH REGULATED BY TETRACYCLINE BINDING

W. HINRICHS[1], P. ORTH[1], C. KISKER[1], D. SCHNAPPINGER[2], W. HILLEN[2] and W. SAENGER[1]

[1] *Freie Universität Berlin, Institut für Kristallographie, Takustr. 6, D-14195 Berlin.*

[2] *Universität Erlangen-Nürnberg, Institut für Mikrobiologie, Staudtstr. 5, D-91058 Erlangen.*

1. Introduction: A protein is a supramolecular entity

Depending on their function, the biological macromolecules represented by proteins, nucleic acids (ribonucleic acid, RNA and deoxyribonucleic acid, DNA) and polysaccharides have globular or fibrous three-dimensional structures. The globular forms are mainly found for proteins and RNA (in transfer RNA and in the ribozymes) and may be considered as very complex supramolecules. Whereas proteins are composed of 20 different amino acids, RNA consists of only 4 nucleotide building blocks, and both form linear polymers several hundred to thousand units long. Proteins and RNA fold in well defined three-dimensional forms, and as such they exert their specific biological functions. The three-dimensional shape is mainly stabilized by hydrogen bonds, by van der Waals (hydrophobic) interactions and, for RNA, by stacking interactions between the 4 bases adenine, guanine, uracil, and cytosine.

G. Tsoucaris (ed.), Current Challenges on Large Supramolecular Assemblies, 349–365.

In this short review, the function of the tetracycline repressor protein, TetR, is described. TetR acts as a molecular switch by binding to a short, very specific sequence of operator DNA and prevents the genetic expression of the protein TetA which confers resistance against the broad spectrum antibiotic tetracycline. If tetracycline binds to TetR, the latter experiences conformational changes and releases the operator DNA so that now genetic expression of the protein TetA can occur and resistance is achieved. This mechanism is not only of pharmaceutical and medical interest but TetR has recently found application as a genetic switch in transgenic organisms where it allows to activate or silence genes that have been introduced into the genomes of these organisms.

2. Tetracycline and Tet repressor

Tetracyclines (Tc) are among the most commonly used broad-spectrum antibiotics. They act by binding to the small ribosomal subunit, thereby interrupting protein biosynthesis by yet unknown mechanisms. In recent years, the applicability of antibiotics against infectious deseases has been severely reduced due to the emergence and wide distribution of different resistance mechanisms [1]. They are associated with proteins encoded by genes which are located on movable genetic elements (transposons, plasmids and R-factors), and consequently can easily be exchanged among pathogenic bacteria.

Against Tc, the most abundant resistance mechanism in gram-negative bacteria is based on the active export of the drug out of the bacterial cell. This is due to the membrane embedded protein TetA, which acts as an antiporter across the cell membrane and couples the efflux of a complex between Tc and a divalent cation, $[M\ Tc]^+$, to the uptake of a proton (Fig.1) [2]. The genetic expression of TetA is regulated by the tetracycline repressor (TetR), which belongs to the family of bacterial repressors occurring as a homodimer with two identical α-helix-turn-α-helix (HTH) sequence motifs for DNA binding. In the absence of $[M\ Tc]^+$, TetR binds to two specific DNA operators (17 base pairs long DNA sequences, Fig. 2) and regulates the transcription of its own gene and of the gene encoding the protein TetA by a mechanism corresponding to the classical model developed by Jacob and Monod. The two operators have a symmetric sequence with one extra base pair inserted on their symmetry axis. Upon binding of $[M\ Tc]^+$ to TetR in the TetR-operator complexes, TetR changes its conformation and is no longer able to bind to the two operators. As a consequence, TetR and the resistance protein TetA are expressed. TetA is inserted into the cytoplasmic membrane and will export $[M\ Tc]^+$ out of the bacterial cell before it can inhibit protein biosynthesis. This is reflected by the affinity of the $[M\ Tc]^+$ complex to the repressor with an association constant $K_A \approx 10^9 M^{-1}$, which is much higher than that to the ribosome, $K_A \approx 10^6\ M^{-1}$ (For reviews see [3, 4]).

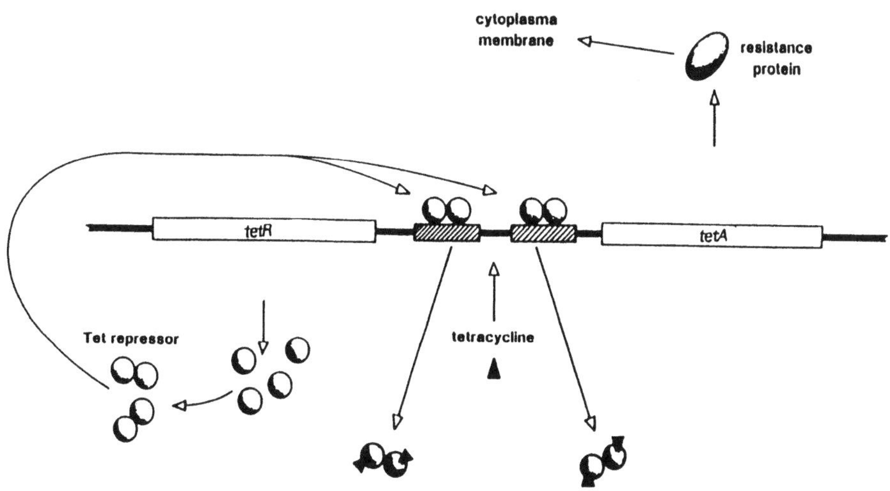

Figure 1. Schematic presentation of the resistance mechanism against Tc. The resistance protein TetA is located in the cytoplasmic membrane. It acts as an antiporter, exporting $[Mg\ Tc]^+$ and importing H^+. Homodimeric TetR, shown as double spheres, binds to operators O1, O2 thereby inhibiting expression of genes encoding for TetR and TetA. If $[Mg\ Tc]^+$ (triangles) binds to TetR, conformational changes take place and the TetR/$[Mg\ Tc]^+$ complexes release O1, O2 so that expression of TetR and TetA is initiated.

Base numbering	.... -2 -1 **0** 1 2 3 4 5 6 7 8
Operator O1	A C A C T C T A T C A **T** T G A T A G G G A T A
	T G T G A G A T A G T **A** A C T A T C C C T A T
Operator O2	C A A C T C T A T C A **A** T G A T A G G G A A T
	G T T G A G A T A G T **T** A C T A T C C C T T A

Figure 2. DNA-operator sequences of O1 and O2 with numbered base pairs. The symmetric (palindromic) sequence is underlined. Note the central base pairs (bold).

There are at least seven TetR proteins which are 207 to 219 amino acids long (Fig. 3). They have very similar amino acid sequences, and most of the amino acids which are engaged in recognition and binding to $[M\ Tc]^+$ and those in the HTH motifs binding to DNA are conserved. We focus here on the crystal structures of TetR of class D, TetR(D), in complex with tetracycline-Mg^{2+}, and cocrystallized with operator DNA-fragments. The structures have been determined at resolutions of 2.1 Å and 2.5 Å, respectively [5, 6, 7].

The crystal structure analyses confirmed the homodimeric structure of TetR. The polypeptide chain of each monomer of TetR is 207 amino acid residues long and folded into 10 α-helices (α1 - α10 and α1' - α10' for the second monomer). The N-terminal DNA binding domain is formed by a bundle of three-α-helices containing the HTH motif comprising helices α2 and α3. The long α4 connects this bundle with the main body of TetR. The helices α5 to α10 and α5' to α10' form the TetR core with the dimerization interface of the homodimeric structure. Each core subunit contains a Tc-binding pocket filled with a $[Mg\ Tc]^+$ complex.

Of interest in the context with the present paper are two questions: how is $[Mg\ Tc]^+$ recognized and bound to TetR, and why does this complex not bind to operator DNA? Both questions can be answered on the basis of the crystal structure described in Fig. 4.

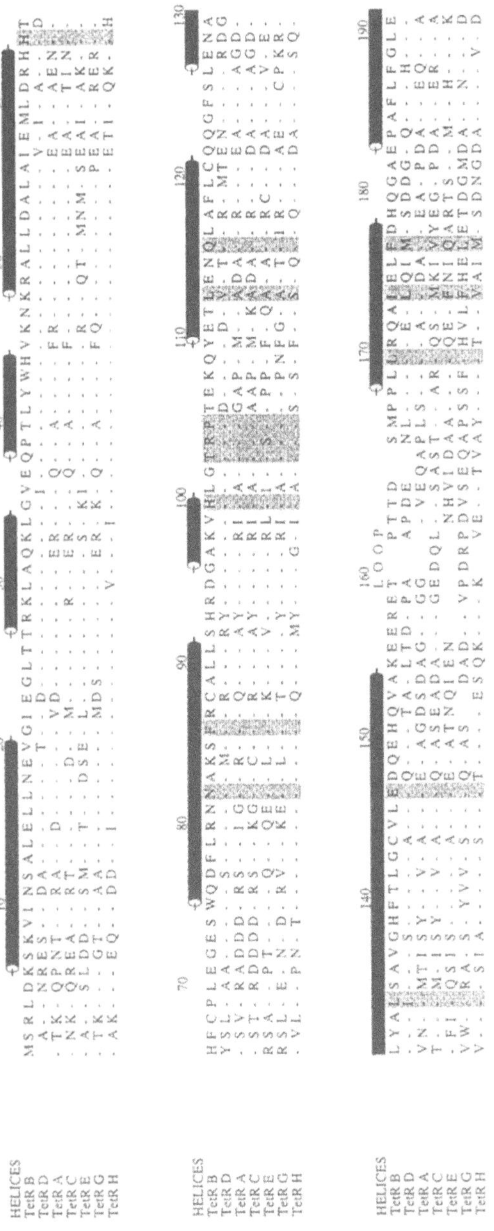

Figure 3. Amino acid sequences in one-letter code of seven known TetR, classes A -E, G and H. Identical amino acids indicated by (-). The positions of the 10 α-helices α1 to α10 are marked on the top line. Amino acids involved in binding of [Mg Tc]$^+$ are shaded.

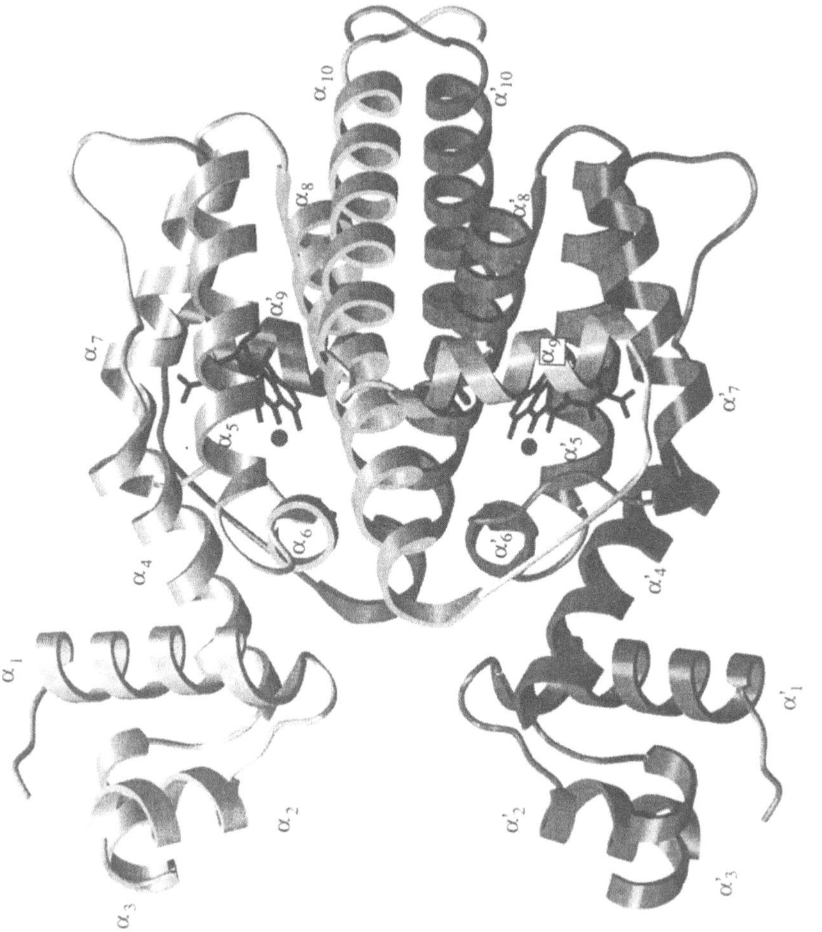

Figure 4. Crystal structure of the complex formed between TetR and $[Mg\ Tc]^+$. The twofold rotation axis relating the two monomers in the homodimer is obvious. Note clear separation into the DNA-binding domains (α1 - α3 and α1' - α3') with the α-helix-turn-α-helix motifs (α2, α3) and the core of TetR responsible for $[Mg\ Tc]^+$-binding and dimer formation. The two α-helices α3, α3' recognizing and binding to the adjacent major grooves of operator DNA are separated by 39 Å so that binding cannot occur. Primed α-helices belong to the "second" monomer in the homodimer, $[Mg\ Tc]^+$ is shown as ball-and-stick representation.

The L-shaped Tc interacts with TetR through a number of hydrogen bonds and van der Waals contacts (Fig. 5). The hydrogen bonds are to functional groups in rings A and B of Tc, and van der Waals contacts are mainly to the methyl group attached at position 6 of ring C, and to the aromatic ring D and its substituents. Of interest are the hydrogen bond between the OH group at Tc-rings A, B and the π-electrons of the phenyl group of Phe86, and the chelate coordination by the ketoenolate function of the Tc-rings B and C to the divalent cation, Mg^{2+}. The octahedral Mg^{2+}-coordination is completed by Nε of His100 and by 3 water molecules, two of which mediate hydrogen bonding to the carboxylate group of Glu147' (the ' indicates that this amino acid belongs to the second monomer in the homodimer).

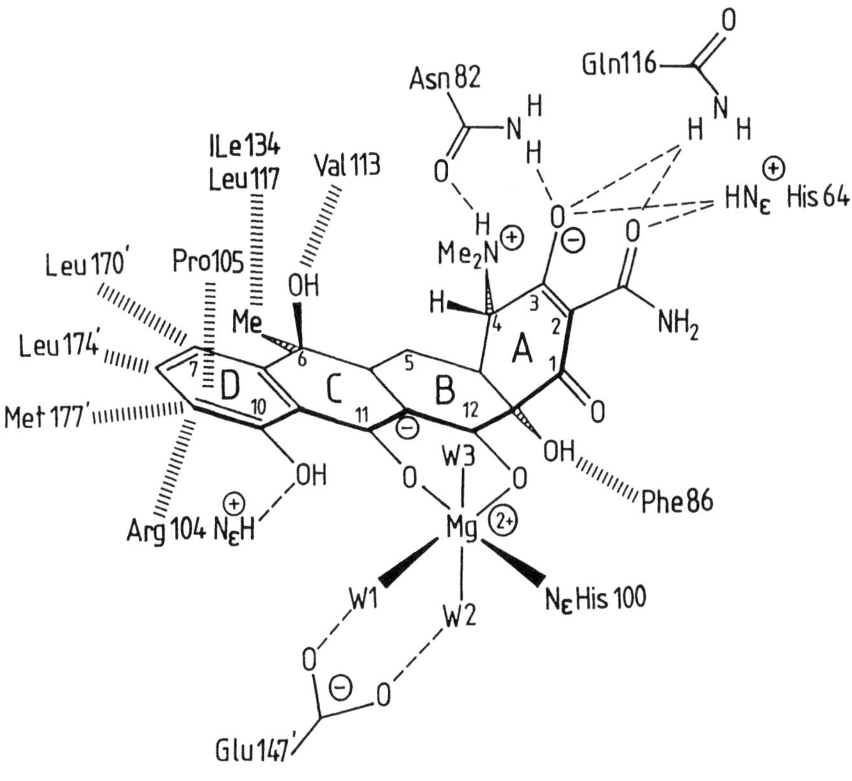

Figure 5. Contacts between [Mg Tc]$^+$ and TetR by hydrogen bonds (----) and van der Waals contacts (||||||). Note the hydrogen bond formed to the π-electron-system of Phe86 and the unfavourable interaction Val113...O(6)H. Primed amino acids belong to the "second" monomer in the homodimer.

The interactions occurring between $[Mg\ Tc]^+$ and TetR are consistent with observations made for a number of Tc derivatives. Modifications or deletions of functional groups have a severe influence on the affinity of $[MgTc]^+$ to TetR, so that the association constants for binding of different $[Mg\ Tc]^+$-derivatives to TetR may vary within 5 orders of magnitude [8]. For example, this holds for epimerisation at position 4 in Tc-ring A which largely reduces the binding affinity. On the other hand, removal of the hydroxyl group in position 6 of ring C, which forms an unfavourable contact with the hydrophobic side chain of Val113, increases the affinity to the repressor.

The question why the TetR/$[Mg\ Tc]^+$-complex does not bind to operator DNA is obvious from its overall conformation. As observed for a number of related proteins and as shown by biochemical studies, the DNA recognition helices α3 and α3 ' of the HTH motifs insert into the major groove of DNA and form specific contacts, thereby recognizing and binding to the DNA. This also the case for the complex of the TetR with operator DNA, as observed in our X-ray structure analysis (see below). In the TetR/$[Mg\ Tc]^+$-complex the two recognition helices have a distance of 39 Å which is 5 Å wider than required for the insertion into adjacent major grooves along the B-DNA double helix with 34 Å repeat length. Consequently, the complex TetR/$[Mg\ Tc]^+$ has only weak affinity to operator DNA.

3. Structure of the complex between TetR and operator DNA

Fig.2 shows the DNA operator sequences to which TetR binds very specifically. It has to be stressed that this interaction occurs only in the absence of $[Mg\ Tc]^+$. Many attempts were necessary to find the optimum length of the operator DNA which would produce TetR-complex crystals suitable for X-ray diffraction [7]. This does not only concern the length of the operator DNA but also its ends; they can be blunt, i.e. at both ends there are base pairs, or they can have overhangs of one or several nucleotides (A, T, G, C), or the operator DNA can be longer on one side than on the symmetically related other. The best operator DNA in our hands was a 15 mer without overhanging nucleotides.

In the crystals of the complex between TetR and operator DNA, the latter is stacked in the crystal lattice to form an infinite double helix to which TetR is bound. The resolution of 2.5 Å permits to see that operator DNA is essentially straight and not bent as observed for many other known protein-DNA complexes (Fig. 6). The recognition helices α3 and α3' of the two HTH motifs of TetR are inserted into the major grooves of the operator DNA and form the following specific contacts (see also Fig. 7):

TetR		*interaction*	**operator DNA**	
Arg28	Nη1	hydrogen bond to	N7 of G2	one strand
	Nη2	hydrogen bond to	O6 of G2	
Gln38	Nε	hydrogen bond to	N7 of A3	
	Oε	hydrogen bond to	N6 of A3	
Pro39	Cβ	van der Waals	T4	
	Cγ	van der Waals	T5	other strand
Thr40	Cγ	van der Waals	C6	
	Oγ	hydrogen bond	phosphate	
Trp43	indol	van der Waals	T7, T8	
His44	Nε	salt bridge	phosphate	

4. Conformational changes of TetR associated with binding of inducer [Mg Tc]$^+$

Since the binding of the inducer Tc in complex with a divalent cation M^{2+} forces TetR into a conformation which is no longer able to bind to operator DNA, it is of particular interest to analyse conformational changes associated with the induction. The crystal structure of free, not induced TetR has been compared with the induced form, the TetR/[Mg Tc]$^+$-complex. This complex formation triggers a series of conformational changes leading finally to a form of the complex TetR/[MgTc]$^+$ in which the HTH motifs are fixed in positions where they are unable to bind to operator DNA.

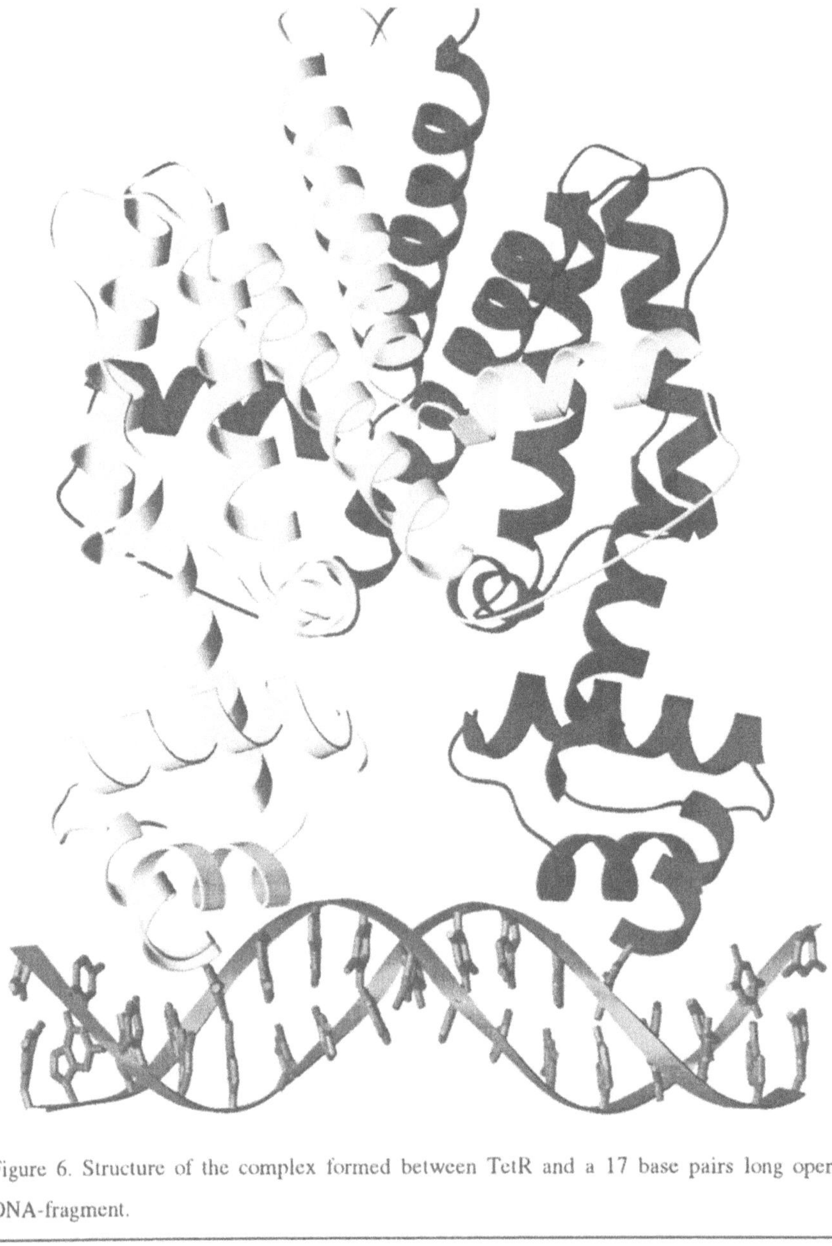

Figure 6. Structure of the complex formed between TetR and a 17 base pairs long operator DNA-fragment.

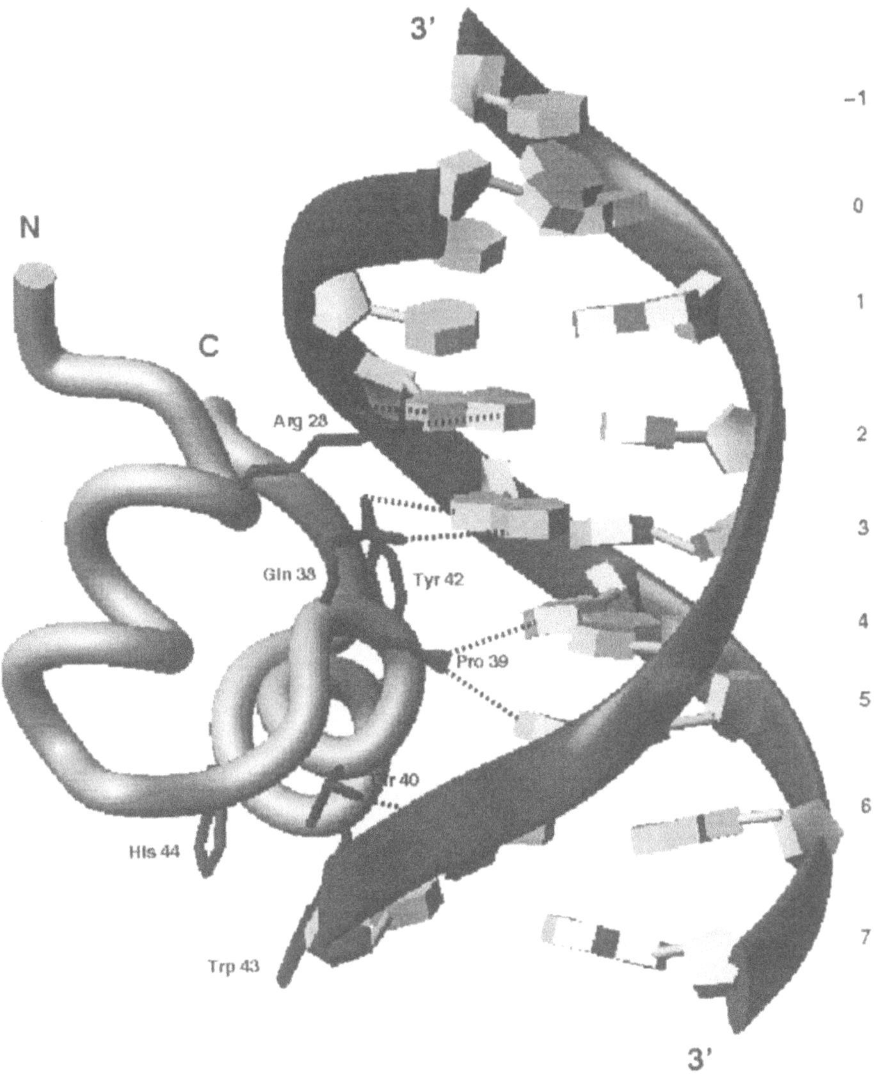

Figure 7. Details of specific interactions between recognition helix α3 and base pairs of the operator-DNA. Base pairs are numbered according to Fig. 2.

The separation between the recognition helices α3 and α3' is now 39 Å, too wide to be able to recognize operator DNA. We must stress here that these conformational changes occur in the TetR-complex with operator DNA showing a very tight association constant ($\sim 10^9$ M^{-1}). The binding of $[MgTc]^+$ to the complex between TetR and operator DNA must consequently have sufficient strength to enforce conformational changes of TetR for releasing operator DNA. The following events have to occur during induction of TetR (see Fig. 8) [9]:

a) The binding pocket in TetR for the inducer $[MgTc]^+$ has the form of a tunnel with one open and one narrow side. Structural consideration and molecular dynamic simulations have shown that the wider side is the entrance for $[MgTc]^+$. It most probably enters head-on with ring A binding tightly to His64, Asn82 and Gln116; these amino acids do not change significantly their position and orientation upon binding to Tc-ring A.

b) Binding of Mg^{2+} to the imidazole of His100 triggers formation of a type II β-turn, residues His100 to Thr103, at the expense of the C-terminal turn of helix α6. His100 remains in position, but the Cα of Gly102 und Thr103 are respectively displaced by 3.9 and 3.1 Å. In addition, hydrogen bonding of Thr103 Oγ to one of the water molecules coordinating to Mg^{2+} supports the formation of this β-turn.

c) A major structural change extends along the polypeptide chain. Residues 104-106 are reoriented to form part of the hydrophobic pocket for Tc-ring D, and the side chain of Arg104 is shifted into a position to form a salt bridge with Asp178' of helix α9'. After $[MgTc]^+$ has been inserted into the binding tunnel, this helix moves by 2.4 Å like a „sliding door“ closing the entrance.

d) The carboxylate group of Glu147' from the second monomer in the TetR homodimer rotates by 90° and forms hydrogen bonds with two aqua-ligands of the Mg^{2+} and to Gly102N of the newly formed β-turn.

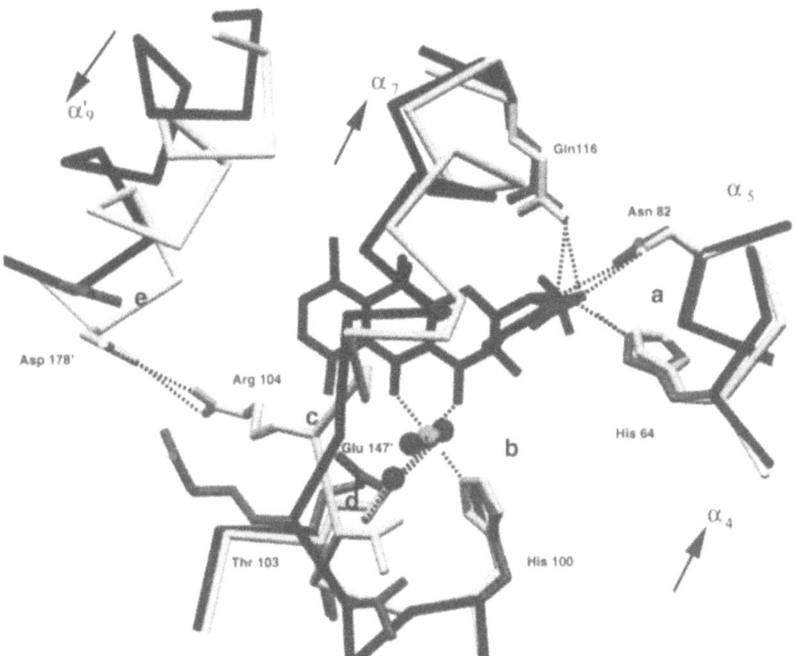

Figure 8. Schematic representation of the series of events (a) to (e) occurring upon [Mg Tc]⁺-binding to TetR. The superposition of the structures of Tc-free TetR (dark lines) and the TetR/[Mg Tc]⁺-complex (light lines) shows significant conformational changes. For details see text.

e) The hydrophobic pocket is closed around $[MgTc]^+$ by the above (c) mentioned movement of helix α9 which brings the side chains of Leu170', Leu174', Met177' in proximity to the aromatic Tc-ring D, and a salt bridge is formed between Arg104 and Asp178', the latter located on α-helix 9'.

There are two major observations associated with all these changes. First, unwinding of the C-terminal turn of α6 and its transformation into a β-turn moves the associated peptide segment 102-105 by about 3 Å, and this is coupled with movement of helix α4 and the attached HTH motif into the same direction. Second, there is formation of a remarkable water-mediated hydrogen bonding network fixing helix α4 to the Tc-recognition site. A zigzag chain of 8 water molecules is in hydrogen bonding contact on one side with the peptide oxygen atoms 53, 56, 57, 60, 64 of helix α4 and on the other side with the polypeptide segment N-terminal to helix α7, 104-109, with the side chain of Glu109 and Tc-ring A (Fig. 9).

This very tight binding of α4, which links the DNA-binding domain to the TetR-core, forces the HTH motif into an orientation unable to bind to operator DNA. This is only possible after insertion of the inducer $[Mg\ Tc]^+$ which not only triggers a number of conformational changes in TetR, but also contributes actively to the formation of the chain of water molecules, preventing TetR/operator binding.

5. TetR as a molecular switch

The ternary system operator DNA / TetR / inducer $[MgTc]^+$ allows very sensitive regulation of the expression of target genes encoding for certain proteins. If one wishes to regulate the expression of a selected gene in the genome of an organism, the gene encoding for TetR with its respective control region has to be cloned into the control region of the target gene. TetR binds to the specific operator DNA and prevents expression of the target gene. If the inducer $[MgTc]^+$ is provided to this system, it binds to TetR, induces the conformational changes described in this paper, the complex TetR/$[MgTc]^+$ releases operator DNA, and the expression of the target gene

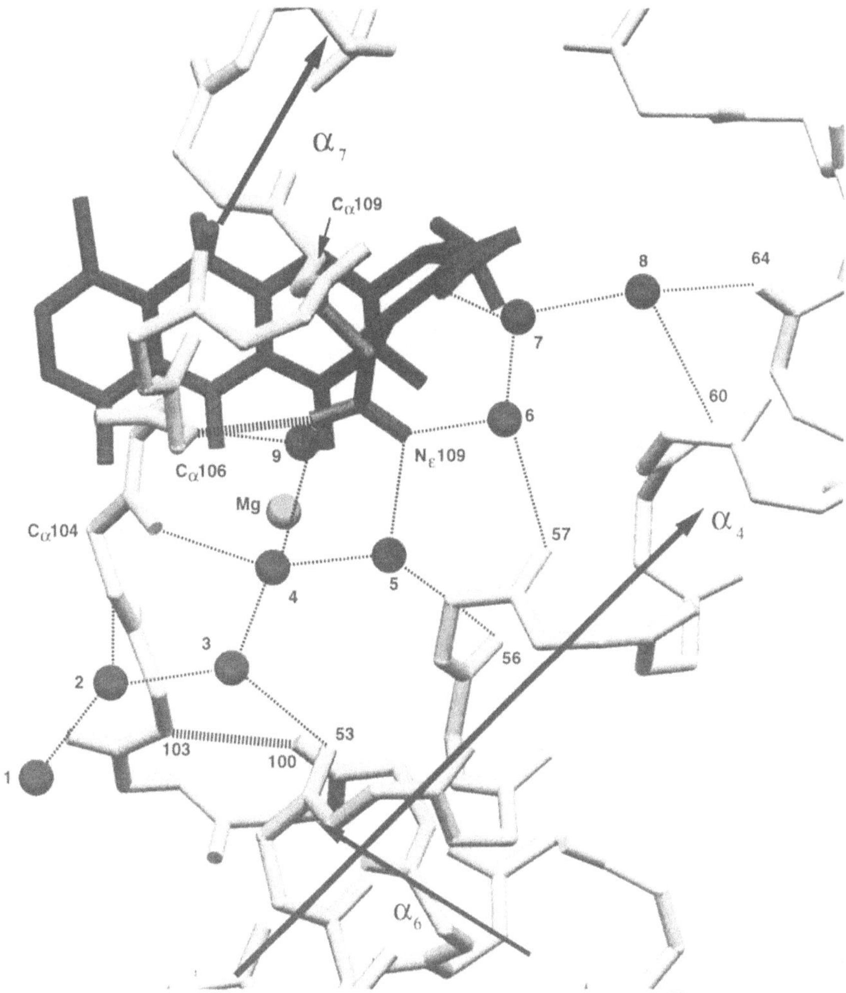

Figure 9. Schematic representation of the chain of water molecules tightening α-helix 4 in the TetR / [Mg Tc]$^+$-complex via an extended network of hydrogen bonds. This is associated with separating the HTH-motifs at 39Å (Fig. 4) to prevent binding to DNA.

can occur. This is at present actually used in the selective expression of certain genes in transgenic organisms [10].

For preparative working chemists, the TetR system is much too large and elaborated to be tackled, but there might be the possibility to develop smaller systems using chemical methods which make use of some of the ideas that nature has found to regulate sophisticated processes. This is the reason why the invitation to talk about TetR was accepted. This topic is of course on the borderline of supramolecular chemistry but might inspire colleagues to think about systems which are similarly regulated as TetR.

Acknowledgements
This work was supported by Deutsche Forschungsgemeinschaft through Sonderforschungsbereich 344 and by Fonds der Chemischen Industrie.

References

1. Neu, H.C. (1992). The Crisis in Antibiotic Resistance. *Science* **257**, 1064-1073.

2. Yamaguchi, A., Udagawa, T. and Sawai, T. (1990). Transport of divalent cations with tetracycline as mediated by the transposon *Tn10*-encoded tetracycline resistance protein. *J. Biol. Chem.* **265**, 4089-4813 (1990).

3. Hillen, W. and Berens, Ch. (1994). Mechanisms underlying expression of Tn*10*-encoded tetracycline resistance. *Annu. Rev. Microbiol.* **48**, 345 - 369.

4. Schnappinger, D. and Hillen, W. (1996). Tetracyclines: antibiotic action, uptake, and resistance mechanisms. *Arch. Microbiol.* **165**, 359-369.

5. Hinrichs, W., Kisker, C., Düvel, M., Müller, A., Tovar, K., Hillen, W. and Saenger, W. (1994). Antibiotic resistance: Structure of the Tet repressor-tetracycline complex and mechanism of induction. *Science* **264**, 418-420.

6. Kisker, C., Hinrichs, W., Tovar, K., Hillen, W. and Saenger, W. (1995). The complex formed between Tet repressor and tetracycline-Mg^{2+} reveals mechanism of antibiotic resistance. *J.Mol.Biol.* **247**, 260-280.

7. Orth, P., Alings, C., Schnappinger, D., Saenger, W. and Hinrichs, W. (1998). Crystallization and preliminary X-ray analysis of the Tet-Repressor/operator complex. *Acta Cryst. D***54**, in the press.

8. Lederer, Th., Kintrup, M., Takahashi, M., Sum, P.-E., Ellestad, G.A. and Hillen, W. (1996). Tetracycline analogs affecting binding to Tn*10*-encoded Tet repressor trigger the same mechanism of induction. *Biochemistry* **35**, 7439-7446.

9. Orth, P., Cordes, F., Schnappinger, D, Hillen, W., Saenger, W. and Hinrichs, W., Tetracycline trapping and induction mechanism of Tet repressor, submitted.

10. Shockett, P.E. and Schatz, D.G. (1996). Diverse strategies for tetracycline-regulated inducible gene expression. *Proc.Natl.Acad.Sci. USA* **93**, 5173-5176.

Esterolytic Catalytic Antibodies

Bernard S. Green[a] and Marcel Knossow[b]

[a]The Hebrew University, Faculty of Medicine - School of Pharmacy, Department of Pharmaceutical Chemistry, P.O.Box 12065, Jerusalem 91120, Israel; [b]Laboratoire d'Enzymologie et de Biochimie Structurales, CNRS, 91198 Gif sur Yvette Cedex, France

Introduction

The concept of synthesizing stable analogs of the rate-limiting transition state for a given chemical reaction and then using these substances as haptens to raise antibodies which serve as specific catalysts for that reaction, has now been demonstrated for a considerable number of reactions [1]. Although this notion dates back to Pauling [2] and was first succinctly stated by Jencks [3], the field of catalytic antibodies may still be considered to be in its developmental stage even though it represents the most successful approach for achieving tailor-made catalysts and has been applied to some 100 different chemical reactions [1]. Perhaps because of the seemingly unlimited possibilities inherent in our ability to now "program" a specific catalyst for a given chemical reaction, and of the dramatic success of catalytic antibodies relative to all other current approaches for achieving enzyme-like catalysts, there is in some quarters a sense of unfulfilled expectation: one is asked, why have catalytic antibodies not yet found applications in medicine or chemistry or biology? The recent milestones in the development of catalytic antibodies have been recorded [4] and one catalytic antibody has recently entered the marketplace as a chemical reagent [5]; this antibody catalyzes aldol condensations and acts

G. Tsoucaris (ed.), Current Challenges on Large Supramolecular Assemblies, 367–374.

via a mechanism similar to that of Nature's aldolase enzymes [6]. Here we address the present status of antibody-catalyzed ester hydrolysis reactions and relate to some of the questions that are often raised in connection with this field.

Ester hydrolysis was the first reaction to be unambiguously enhanced by a specific monoclonal antibody [7] and it is this reaction that is catalyzed by the majority of reported catalytic antibodies [1]. The mechanistic information on antibody-catalyzed reactions is therefore more abundant for this reaction and, most importantly, the largest number of x-ray structural studies are available for antibody Fab's which catalyze ester hydrolysis.

It should be noted that a number of esterolytic polyclonal catalytic antibodies have also been described and one can imagine applications where their use would be advantageous. (See [8].) Only monoclonal catalytic antibodies are discussed here because only with these homogeneous materials is it possible to accurately assess catalytic parameters and carry out unambiguous and reproducible mechanistic and, especially, structural studies.

Phosphonate transition state analog haptens

Almost all of the haptens used to mimic the transient states (the oxyanion intermediate and the flanking transition states on the path from substrate to this oxyanion and on the path from oxyanion to product) in ester hydrolysis comprise the phosphonate structure (Scheme 1). Many other tetrahedral structures may be considered as mimics for esterolytic transient states. These include phosphate and phosphorothioate [9a] and phosphonodithioate [9b] haptens; tetrafluoro ketone hydrate [10]; sulfonamides [11]; α-keto amides [12]; 1-aryl-1-hydroxy derivatives with a positive or negative charge on the aryl group [13]; β-hydroxy quaternary ammonium and phosphonoamidate haptens, used separately or together (heterologous immunization) [14]; N,O-arylphosphoramidates [15]. Not all of these structural types have been tested as haptens and some have not been studied with ester substrates even though they do catalyze the hydrolysis of, say, carbonates, carbamates, etc. Phosphonates have also been used in transition state analogs in order to elicit antibodies for reactions other than ester hydrolysis or acyl transfer (e.g. [16]).

Scheme I

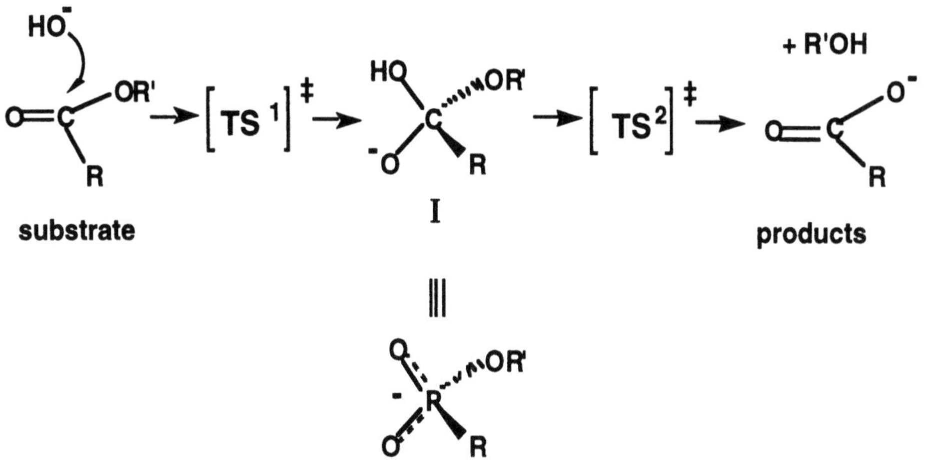

One would have wanted to be able to compare the relative effectiveness of different TSAs as haptens for eliciting catalytic antibodies but there are very few such studies [14]. Ideally, the significance of such comparisons would be enhanced by the direct screening of *all* of the hybridomas from a fusion for catalytic activity (e.g., using a method such as catELISA [17]). The reliability of such comparisons would be convincing if one also saw the same trend with monoclonals resulting from more than one fusion and/or with different mouse strains or different animal species. This represents an enormous effort but such comparisons are going to eventually be necessary in order to reliably decide which haptens to use for a given application.

Each individual transition state analog which is designed and synthesized that successfully serves as a hapten to elicit catalytic antibodies generally gives rise to a number of different catalytic antibodies. Each of the resulting antibodies, which may differ slightly or even markedly from one another, are generally studied with regard to the substrate which is most congruent with the hapten used. However, each of the antibodies may also be tested for reaction with substrates that vary in structure and one may then find unexpected, i.e., unprogrammed, reactivity patterns [18]. Especially when crystal structures are available, it may then be possible to draw new conclusions from the triangular relationship beween hapten-antibody-substrate and predict new or optimal haptenic structures for a desired substrate.

"Reactive Immunization"

A particularly interesting new approach for generating catalytic antibodies, termed "reactive immunization", refers to the use of a chemically reactive hapten, one that can form a covalent bond with a particular amino acid side chain functional group in the initial B-cell recognition event; there is then a high probability of this amino acid being present in the final isolated hybridomas. This approach was first demonstrated using a reactive phosphonate diester [19] and has recently been applied, using the same phosphonate diester function for the kinetic resolution of naproxen [20]. The ability "to program for a chemical reaction", e.g., select for a nucleophilic group promises to have wide use and has afforded highly proficient catalytic antibodies. In the former group of antibodies specific phosphonylation of the antibody combining site was observed when the catalytic antibodies were allowed to react with the hapten, whereas in the latter family of antibodies, both the phosphonate diester and monoester were inhibitors; further work is called for to uncover other appropriate haptens for reactive immunization which will combine specificity with high reactivity. (See also [6].)

Structural Studies

The recent X-ray structure analyses of a significant number of anti-phosphonate esterolytic catalytic antibodies has afforded a wealth of information. These include the nature of the antibody-phosphonate interactions, and thereby conclusions as to the mechanism of reaction [21,22]; effects of pH on structural and mechanistic properties [23]; similarities of the active sites of antibodies arising from a single immunization [24]; mechanisms of antibody inactivation [25]; detailed descriptions of reaction pathway, from substrate through transition state to product structures [26] (Figure 1); and elucidating the ways in which binding sites become catalytic sites [27].

One of the chemical surprises from these structures was the striking similarity of structure -- similar, often identical, amino acid residues of the combining site participating in catalysis in antibodies arising from the same immunization -- and the virtual identity of catalytic mechanism, oxyanion intermediate stabilization, when the transition state analog approach was used.

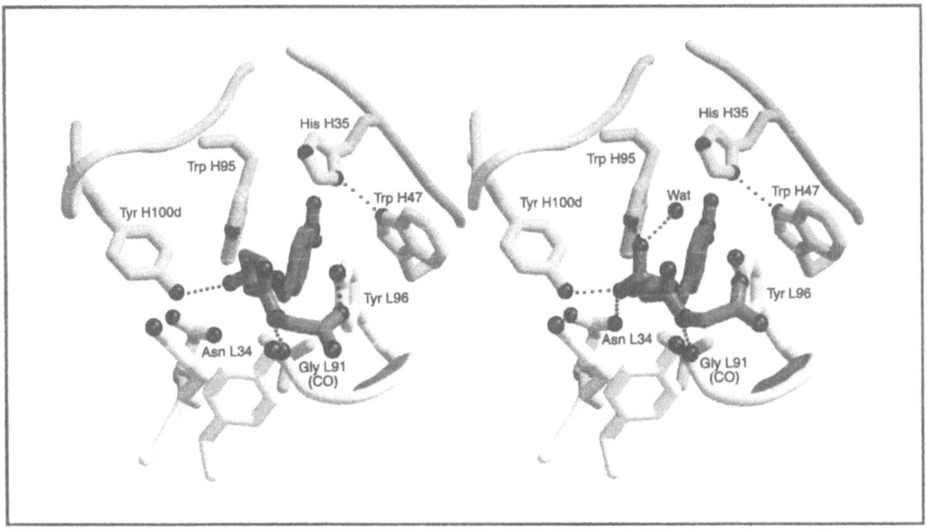

Figure 1 : Complexes of the esterase-like antibody D2.3 with an amide which is a stable substrate analogue (left) and a phosphonate TSA (right). Hydrogen bonds are shown as dotted lines. The structures demonstrate how residues H100d (a tyrosine) and L34 (an asparagine) of the antibody stabilize better the TSA than the substrate by establishing two (resp. one) hydrogen bonds with the oxygen atom that mimics the oxyanion intermediate (resp. the carbonyl oxygen of the ester bond to be hydrolysed).

Applications

In general, the available antibodies have been produced in order to demonstrate activity and not necessarily application, in spite of the intrinsic interest many of these have for chemistry, biology, or medicine. In addition, the initial catalytic activity is generally not routinely measured after large numbers of turnovers (see [28]). Nevertheless many current applications are being pursued. The widely useful, available aldolase antibody was mentioned [5,6] and antibodies may play roles in natural product synthesis [4].

Therapeutically active catalytic antibodies represent particularly attractive applications -- an extension of the antibodies' natural function -- and a number of directions are being investigated. Hydrolysis of an ester bond in cocaine renders the drug inactive and progress is being made in developing a phosphonate-elicited catalytic antibody for addiction studies [29]. A number of studies are investigating catalytic antibodies as agents to convert prodrugs to cytotoxic agents for cancer therapy [15,30].

Naturally occuring ester bonds important in human biochemistry and pathology include acetylcholine, PAF (platelet activating factor) which is involved in many body functions in health and disease, and lipid A, the causative factor of Gram-negative sepsis; all of these represent intriguing targets for esterolytic catalytic antibodies.

Conclusions

Catalytic antibody research has both obvious practical aspects (programmed, catalytic antibodies for applications in chemistry and biomedicine) and more basic ones, such as: understanding the mechanism of action of each of the novel catalysts so produced; uncovering the relation, if any, to natural enzymes; answering questions concerning the evolution of efficient catalysts, the limits of catalytic parameters and properties using the antibody scaffold, and the possible extension of the information from, and experience with, catalytic antibodies to other new, enzyme-like catalytic systems.

Acknowledgement

The research from our laboratories involved the collaboration of the following: Z. Eshhar, D.S. Tawfik, R. Arad-Yellin, R. Chap, J.-B. Charbonnier, B. Gigant, B. Golinelli-Pimpaneau, S.-H. Kim, D.G. Schindler, A. Lindner and R. R. Zemel. BSG is grateful to the US-Israel Binational Research Foundation, Grant No. 94-00397, for support.

References

[1] (a) R.A. Lerner, S.J. Benkovic and P.G. Schultz, *Science* **252**, 659-667 (1991); (b) D.S. Tawfik, Z. Eshhar and B.S. Green, *Molecular Biotechnology* **1,** 87-103 (1994); (c) N.R. Thomas, *Natural Product Reports* 479-511 (1996).

[2] L. Pauling, *Am. Sci.* **36**, 51 (1948).

[3] W.P. Jencks, *Catalysis in Chemistry and Enzymology*, McGraw-Hill, New York, (1969) p. 288.

[4] E. Keinan and R.A. Lerner, *J. Chem.* **36**, 113- 119 (1996).

[5] S. Borman, *Chem. Eng. News,* 1 Sept. 1997 issue, pp. 9-10.

[6] (a) G. Zhong, T. Hoffman, R.A. Lerner, S. Danishefsky and C.F. Barbas, *J. Am. Chem. Soc.* **119**, 8131 (1997); (b) T. Hoffman, G. Zhong, B. List, D. Shabat, J. Anderson, J. Gramatoikova, R.A. Lerner and C.F. Barbas, *J. Am. Chem. Soc.* **120**, 2768-2779 (1998).

[7] F. Kohen, J.B. Kim, H.R. Lindner, Z. Eshhar, and B.S. Green, *FEBS Letters,* **111**, 427-431 (1980).

[8] K. Shreder, R. Thomas, M. Wallace, E. Helms and B. Iverson, *Isr. J. Chem.* **36**, 215- 220 (1996) and refs. therein.

[9] (a) B.H. Wilmore and B.L. Iverson *J. Amer. Chem. Soc.* **116**, 2181-2182, (1994); (b) S.F. Martin, A.S. Wagman, G.G. Zipp, and M.K. Gratchev, *J. Org. Chem.,* **59**, 7957-7958 (1994)

[10] T. Kitazume, T. Tsukamoto, and K. Yoshimura, *J.C.S., Chem. Comm.* 1355 (1994).

[11] F. Benedetti, F. Berti, A. Colombatti,C. Ebert,P. Linda,F. Tonizzo, *J.C.S., Chem. Comm.* 1417-141! (1996)

[12] J.Yli-Kauhaluoma and K.D. Janda, *Bioorg. Med. Chem.* **2**, 521-528(1994)

[13] K.D. Janda, M.I. Weinhouse, T. Danon, K.A. Pacelli and D.M. Schloeder *J. Amer. Chem. Soc.* **113,** 5427-5434 (1991).

[14] H. Suga, O. Ersoy, S.G. Williams, T. Tsumuraya, M.N. Margolies, A.J. Sinskey, and S. Masamune *J. Amer. Chem. Soc.* **116**, 6025-6026 (1994).

[15] S.A. Taylor, M.-J. Chen, A.N. Dinaut, and R.A. Batey, *Tetrahedron* **54**, 4223-4242 (1998).

[16] C. Gonzalez-Bello, C. Abell and F. J. Leeper, *J. Chem. Soc., Perkin Trans. 1*, 1017-1024 (1997).

[17] D.S. Tawfik, B.S. Green, R. Chap, M. Sela, and Z. Eshhar, *Proc. Nat. Acad. Sci., U.S.* **90**, 373-377 (1993).

[18] (a) K.D. Janda, S.J. Benkovic, D.A. McLeod, D.M. Schloeder and R.A. Lerner, *Tetrahedron*, **47**, 2503-2506 (1991); (b) D.S. Tawfik, A. Lindner, R. Chap, Z. Eshhar, and B.S. Green *Europ. J. Biochem.* **244**, 619-626 (1997).

[19] P. Wirsching, J.A. Ashley, C.-H. L. Lo, K.D. Janda and R.A. Lerner, *Science* **270**, 1775-1782 (1995).

[20] C.-H. L. Lo, P. Wentworth, K.W. Jung, J. Yoon, J.A. Ashley, and K.D. Janda, *J. Am. Chem. Soc.* **119**, 10251-10252 (1997).

[21] (a) G. MacBeath and D. Hilvert, *Chemistry & Biology* **3**, 433-445 (1996) and references therein; (b) J.-B. Charbonnier, B. Gigant, B. Golinelli-Pimpaneau and M. Knossow, *Biochimie* **79**, 653-660 (1997).

[22] J.B. Charbonnier, E. Carpenter, B. Gigant, B. Golinelli-Pimpaneau, Z. Eshhar, B.S. Green and M. Knossow, *Proc. Nat. Acad. Sci., U.S.* **92,** 11721-11725 (1995).

[23] J.B. Charbonnier, B. Golinelli-Pimpaneau, B. Gigant, Z. Eshhar, B.S. Green and M. Knossow, *Israel J. Chem.* **36**, 143-149 (1996).

[24] J.-B. Charbonnier, B. Golinelli-Pimpaneau, B. Gigant, D.S. Tawfik, R. Chap, D.G. Schindler, S.-H. Kim, B.S. Green, Z. Eshhar and M. Knossow, *Science* **275**, 1140-1142 (1997).

[25] B. Gigant, J.-B. Charbonnier, B. Golinelli-Pimpaneau, R. R. Zemel, Z. Eshhar, B.S. Green and M. Knossow, *European J. Biochem.* **246**, 471-476 (1997).

[26] B. Gigant, J.-B. Charbonnier, Z. Eshhar, B.S. Green and M. Knossow *Proc. Nat. Acad. Sci., U.S.* **94,** 7857-7861 (1997).

[27] P.A. Patten, N.S. Gray, PL.Yang, C.B. Marks, G.J. Wedemeyer, J.J. Boniface, R.C. Stevens and P.G. Schultz, *Science*, **271**, 1086-1091 (1996).

[28] D.S. Tawfik, Z. Eshhar, and B.S. Green,*Molecular Biotechnology* **1,** 87-103 (1994).

[29] G. Yang, J. Chun, H. Arakawauramoto, X. Wang, M.A. Gawinowicz, K. Zhao, and D. W. Landry, *J. Amer. Chem. Soc.* **118**, 5881-5890 (1996).

[30] (a)P. Wentworth, A. Datta, D. Blakey, L.J. Partridge, G.M. Blackburn *Proc. Natl. Acad. Sci. USA*, **93**, 799- (1996); (b) H. Miyashita, Y. Karaki, M. Kikuchi and I. Fujii, *Proc. Natl. Acad. Sci. USA*, **90**, 5337- (1993); (c) D. A. Campbell, B. Gong, L.M. Kochersperger, S. Yonkovich, M.A. Gallop, and P.G. Schultz. *J. Amer. Chem. Soc.* **116**, 2165-2166 (1994).

WEAK HYDROGEN BONDING IN CYCLODEXTRIN COMPLEX STABILISATION

TH. STEINER* AND W. SAENGER

Institut für Kristallographie, Freie Universität Berlin,

Takustraße 6, D-14195 Berlin, Germany

Abstract. A short overview is given on crystallographic studies of weak hydrogen bond interactions in the stabilisation of cyclodextrin inclusion complexes. Most frequent are C–H···O hydrogen bonds donated from C–H groups of the cyclodextrin cavity wall to polar guest molecules such as water or alcohols. Normally, these interactions serve to satisfy hydrogen bond potentials which would otherwise be vacant. The observations on the behaviour of polar guest molecules in the relatively apolar cyclodextrin molecules are generally valid for polar molecules in partly apolar surrounding.

1. Introduction

Cyclodextrins (CD) are macrocyclic host molecules composed of α(1-4)-linked D-glucoses. Best known are the cyclodextrins composed of six to eight glucose residues, which are commonly called α-, β-, and γ-CD, respectively (or, alternatively, CA6 to CA8). They readily form inclusion complexes with various guest molecules of suitable size. The molecular cavity is lined by two C–H groups per glucose, and by the glycosidic oxygen atoms which link the glucose units. This renders the cavity relatively hydrophobic in nature. The three hydroxyl groups per glucose are placed at the rims of the donut-shaped molecules. The secondary hydroxyl groups have fixed positions and cannot form hydrogen bonds in the direction of the molecular cavity. The primary hydroxyl groups, C(6)–O(6)H, may rotate around the C(5)–C(6) bonds, and can

G. Tsoucaris (ed.), *Current Challenges on Large Supramolecular Assemblies*, 375–383.

be oriented in direction of the molecular axis so as to form hydrogen bonds with enclosed guest molecules.

If polar molecules are included in cyclodextrin cavities, they have only limited opportunity to satisfy their hydrogen bond potentials. Formerly, it was assumed that the only possible polar host-guest interactions are O–H···O/N hydrogen bonds between CD primary hydroxyl groups and the guest, and O/N–H···O(4) hydrogen bonds donated from guest molecules to glycosidic oxygen atoms. The latter are sterically poorly accessible, so that this type of hydrogen bond is formed only occasionally. A good example of this view on host-guest interactions in cyclodextrin complexes is our neutron diffraction study of β-CD–ethanol–octahydrate published in 1990 [1]. At that time, it has been common practice in analyses of intermolecular interactions, to disregard hydrogen atoms bonded to carbon, and in structure illustrations, they were frequently 'omitted for clarity'. This leads to pictures such as shown for of β-CD–ethanol–octahydrate in Figure 1: the ethanol and water molecules included in the β-CD molecular cavity appear not to form any interactions with the host molecule that deserve being drawn in an illustration. Some of the guest molecules obviously suffer from very poor hydrogen bond coordination; this is laconically attributed to steric restrictions within the cavity, and no further discussion is made. In the present paper, we summarize more advanced views on weak polar host-guest interactions that have been developed in the following years.

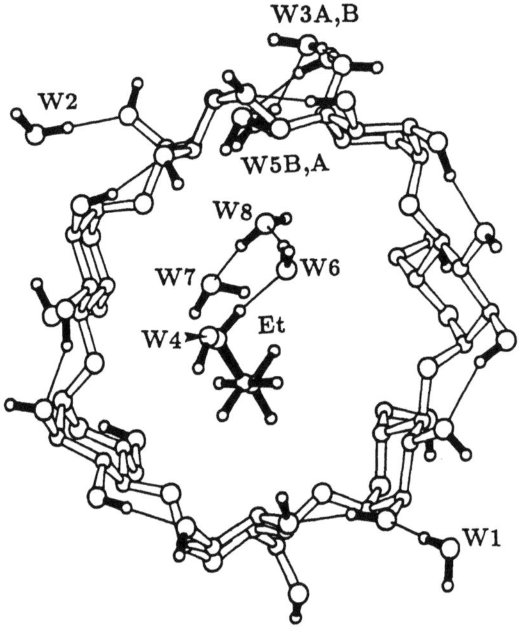

Fig. 1. Neutron crystal structure of β-CD–ethanol–octahydrate at 15 K, drawn as in the original publication [1] with 'hydrogen atoms bonded to C omitted for clarity'.

2. Water molecules and alcohols: C–H···O hydrogen bonds

In the frame of a general analysis of C–H···O hydrogen bonds in carbohydrates, the very precise neutron crystal structure of β-CD–ethanol–octahydrate at 15 K has been re-inspected [2] (background on C–H···O hydrogen bonds can be found in references [3] and [4], and the literature cited therein). At a closer look, it becomes obvious that numerous C–H···O contacts are formed between the host and guest molecules, Figure 2, which have the typical geometries of weak hydrogen bonds. In these interactions, C–H groups of the cavity wall can act as the donors and O-atoms of the guests as the acceptors, but there are also examples of C–H···O hydrogen bonds donated from guest molecules to O(4) atoms of the cavity wall. Note that the C–H groups lining the CD cavity are of the type

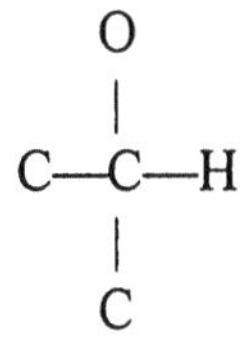

i.e. 'activated' by the adjacent electron-withdrawing O-atom. They are therefore stronger donors than those occurring in saturated pure hydrocarbons.

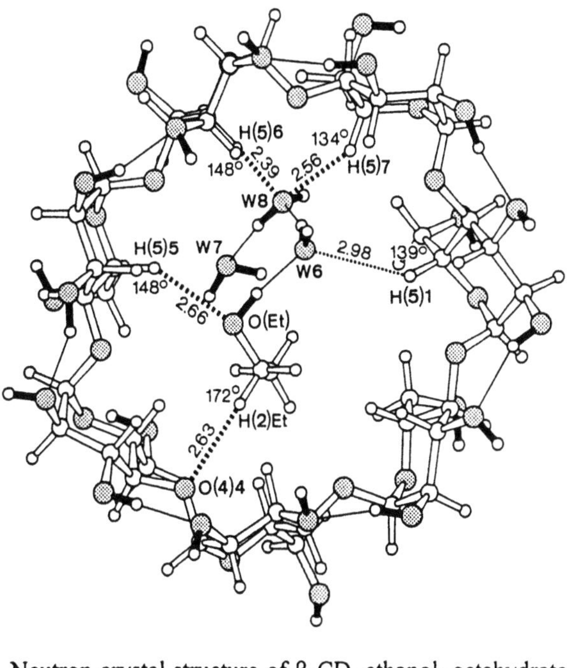

Fig. 2. Neutron crystal structure of β-CD–ethanol–octahydrate at 15 K, shown in the same projection as in Fig. 1, with host-guest C–H···O interactions drawn by dashed and dotted lines [2].

Comparison with several related crystal structures shows that the structural features of β-CD–ethanol–octahydrate are not unique, but are typical for the situation where hydrophilic guest molecules like water or alcohols are included in cyclodextrin cavities: if these molecules cannot satisfy their hydrogen bond potentials with conventional partners like hydroxyl groups or water molecules, they resort to the weaker C–H···O hydrogen bonds involving the cavity wall. This phenomenon is not restricted to low temperatures as in the crystal structure shown above, but occurs at room temperature as well: all examples presented below are from room temperature crystal structures.

In theoretical calulations on C–H···O hydrogen bonds, interaction energies are found covering a wide range [4]. For acidic C–H groups like in alkynes and haloforms, hydrogen bond energies can be > 3 kcal/mol, whereas for the weakest of the C–H donors, in particular for methyl groups, hydrogen bond energies are < 1 kcal/mol. For activated $C(sp^3)$–H donors as those in carbohydrates, intermediate enrgies in the range 1 to 2 kcal/mol can be expected. In quantum chemical *ab initio* molecular orbital calculations on the complex β-CD–but-2-yne-1,4-diol heptahydrate, Figure 3, we have obtained typical C–H···O hydrogen bond energies of slightly above 1 kcal/mol [5].

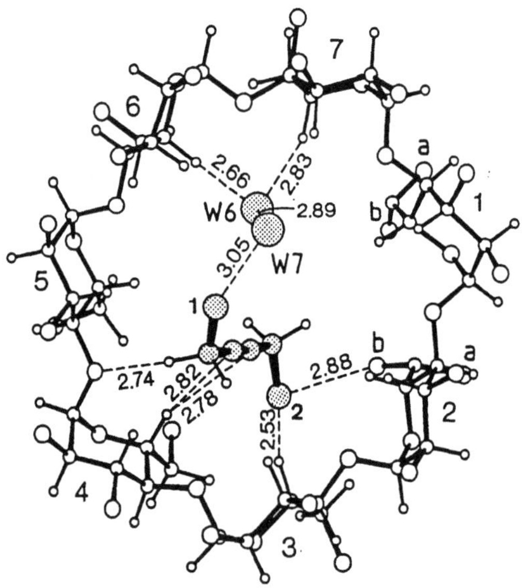

Fig. 3. Inclusion geometry in the complex β-CD–but-2-yne-1,4-diol heptahydrate, which is stabilised by an interplay of O–H···O, C–H···O and C–H···π hydrogen bonds [6].

A particularly interesting case of C–H···O hydrogen bonding has been found in the complex β-CD–1,5-pentanediole [7]. In a systematic investigation of the inclusion geometry of short chainlike molecules in β-CD, it was found that the inclusion complexes with diethanolamine and with 1,5-pentanediole crystallize isostructurally, and the guest molecules are included in virtually identical geometries. The conventional N–H···O hydrogen bond formed by the secondary amino group of the diethanolamine molecule is replaced in the 1,5-pentanediol complex by a C–H···O hydrogen bond donated to the same hydroxyl acceptor, Figure 4. The geometries of the two hydrogen bonds are, of course, different with H···O separations of 1.81 and 2.49 Å, respectively.

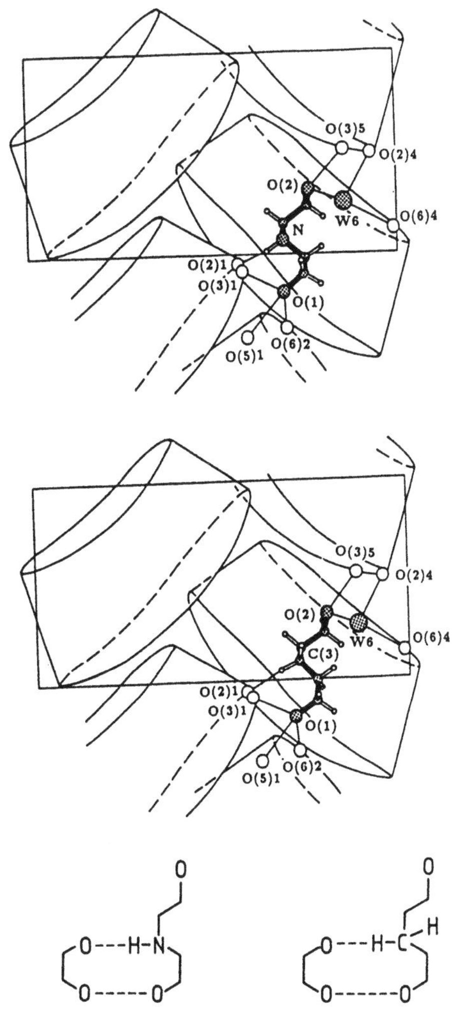

Fig. 4. Crystal structures of β-CD–diethanolamine hexahydrate (top) and β-CD–1,5-pentanediole hexahydrate (bottom) [7]. The scheme illustrates the isofunctional replacement of the N–H···O hydrogen bond formed in the diethanolamine complex by a C–H···O hydrogen bond in the 1,5-pentanediole complex.

3. Aromatic guest molecules: C(sp^2)–H···O(4) hydrogen bonds

For water molecules and alcohols, the formation of C–H···O hydrogen bonds in cyclodextrin cavities can be easily rationalised. However, this kind of interaction is formed also with orther classes of guest molecules. If small aromatic guests are included in α-CD, the aromatic plane is typically placed on the molecular axis of the host. The orientation is not random, but such that guest—host C_{ar}–H···O(4) hydrogen bonds are formed. As an example, the crystal structure of the crystalline complex α-CD–(benzylalcohol)$_2$–hexahydrate [8] is shown in Figure 5.

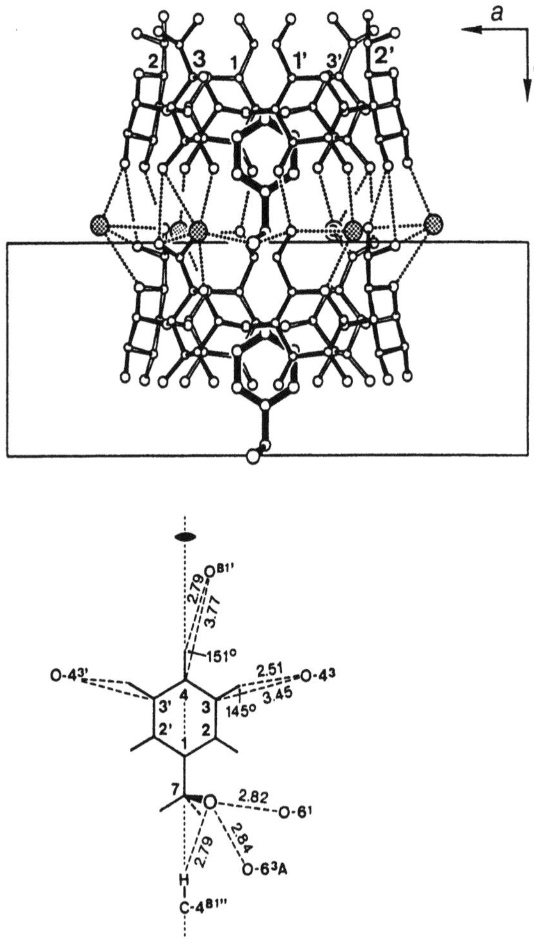

Fig. 5. Top: Crystal structure of the complex α-CD–(benzylalcohol)$_2$–hexahydrate [8]. Bottom: the directional interactions formed by the guest molecule.

The inclusion of aromatic molecules into α-CD allows more general discussion of C–H···O hydrogen bonds in complex formation from water solution. Complex formation as such is driven by the so-called hydrophobic effect, *i.e.* by the simple circumstance that aromatic groups are brought from aqueous environment into the 'hydrophobic' cyclodextrin cavities. Once the complexation has occurred, the precise orientation of the guest is determined by weak directional interactions and, of course, also by optimising van der Waals interactions. C–H···O hydrogen bonds are in this case a 'fine-tuning effect' governing the geometry of the inclusion complex.

4. C–H···π hydrogen bonds

Interplay of C–H···O and C–H···π hydrogen bonds in the stabilisation of inclusion complexes is well documented for organic hosts [9]. X–H···π hydrogen bonds can be readily accepted by phenyl–rich hosts like calixarenes and related molecules, in particular if they carry negative net charges. In cyclodextrin inclusion complexes, these interactions are of less importance because the host carries no aromatic groups. When aromatic moieties are included in CD cavities, their edges may donate C–H···O(4) hydrogen bonds, but the electron-rich faces are at relatively long separations from the cavity wall.

Short host-guest contacts of π-bonded moieties to CD host molecules are possible, however, for alkynes. These may orient their C≡C bond parallel to the CD molecular axis and then approach the C–H groups of the cavity wall to hydrogen bond distances. An example is observed in the crystal structure of the complex β-CD–but-2-yne-1,4-diol heptahydrate, Figure 3 [6]. In quantum chemical calculations, the energy of the C–H··· C≡C interaction is estimated to be around 0.7 kcal/mol, that is even smaller than those of the C–H···O hydrogen bonds occurring in the same complex [5].

5. Formation of weak hydrogen bonds is not mandatory

The weak C–H···O and C–H···π hydrogen bonds differ from the conventional O–H···O hydrogen bonds in energies and geometries. A further important difference lies in the relative frequencies of occurrence. For hydroxyl groups and water molecules, formation of hydrogen bonds is (almost) mandatory: if present in a crystal structure, these functional groups find hydrogen bond partners in almost all cases. This is not true for C–H donors and π-acceptors, which are frequently found unsatisfied. In consequence, it is difficult or even impossible to predict, whether in a given cyclodextrin inclusion complex, weak hydrogen bond effects are of importance or not. In several cases, we find C–

H···O hydrogen bonding to play only a marginal role. An example is the inclusion complex of acetone in dimethyl-α-CD [10], in which the guest molecule is found oscillating more or less freely in the sterically available cavity volume.

6. Generalisation to water molecules in partially apolar environment

The observations discussed above have been generalised to water molecules in all crystalline environments. Database studies on hydrates were performed using published neutron- [11] and X-ray [12] diffraction data. The essential results are illustrated in Figure 6.

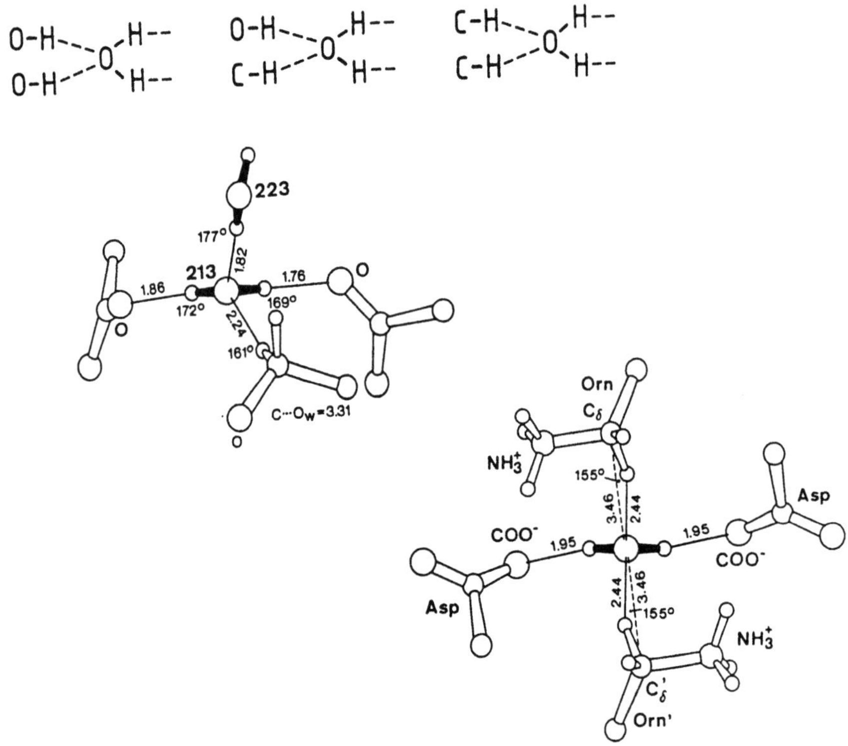

Fig. 6. Top: schemes of possible hydrogen bond configurations with tetrahedral coordination of water molecules. Center: example for a water molecule accepting one O–H···O$_W$ and one C–H···O$_W$ hydrogen bond (in vitamin B_{12} coenzyme, coordinates from Bouquiere et al., *Acta Crystallogr., Sect. B*, 1993, 49, 79). Bottom: example for a water molecule accepting two C–H···O$_W$ hydrogen bonds (in L-ornithine L-aspartate hemihydrate, coordinates from Salunke and Vijayan, *Int. J. Prot. Pept. Res.*, 1983, 22, 154). After Refs. [11, 12].

It is well known that the preferred acceptor functionality of water molecules is two-fold. In the optimal case, two hydrogen bonds from strong donors such as O–H or N–H are accepted (Fig. 6, Scheme, left). However, this requires that two such donors must be available in close proximity with a distance of less than about 5 Å between each other, *i.e.* the local density of donors must be sufficiently high. If the local density of donors is too low, a water molecule can accept only one O–H···O_W or N–H···O_W hydrogen bond at a time, necessarily leading to unsatisfied hydrogen bond acceptor potentials. As a resort, such water molecules tend to accept hydrogen bonds from the weaker C–H donors to 'fill up' the preferred two-fold acceptor functionality (Scheme, center). Occasionally, water molecules are found, which accept no hydrogen bonds from O/N–H donors at all, and still achieve overall tetrahedral coordination by accepting two C–H···O_W hydrogen bonds (Scheme, right). Two of the many examples observed in non-cyclodextrin crystal structures are shown in Figure 6.

Water molecules can also have different than two-fold acceptor functionality. If C–H groups are considered as potential donors, and contacts to metal ions M^+···O_W are included in the statistics, about 20% of the water molecules in our data sample accept only one interaction, about 60% accept two, and about 20% accept three interactions [11]. Higher acceptor functions are possible in principle, but occur only very rarely. Within the three-fold acceptor functionality of water, O/N–H···O_W and C–H···O_W hydrogen bonds, and metal contacts M^+···O_W may be combined in all possibilities, and all these variants are actually observed in crystal structures.

Acknowledgements. These studies were supported financially by the Deutsche Forschungsgemeinschaft, Bundesministerium für Bildung und Forschung, and Fonds der Chemischen Industrie.

7. References

1. Steiner, Th., Mason, S. A. and Saenger, W. (1990). Cooperative O–H···O hydrogen bonds in β-cyclodextrin ethanol octahydrate at 15K: a neutron diffraction study. *J. Am. Chem. Soc.* **112**, 6184—6190.
2. Steiner, Th. and Saenger, W. (1992). Geometry of C–H···O hydrogen bonds in carbohydrate crystal structures. Analysis of neutron diffraction data. *J. Am. Chem. Soc.* **114**, 10146—10154.
3. Desiraju, G. R. (1996). The C–H···O hydrogen bond: structural implications and supramolecular design. *Acc. Chem. Res.* **29**, 441—448.
4. Steiner, Th. (1997). Unrolling the hydrogen bond properties of C–H···O interactions (Feature Article). *Chem. Commun.* 727—734.

CYCLODEXTRINS: TOWARDS LARGE ASSEMBLIES FOR EFFICIENT SLOW RELEASE DEVICES

G. TSOUCARIS and N. RYSANEK
C.N.R.S. - Université Paris-Sud
Centre Pharmaceutique
92290, Châtenay-Malabry, France

1. Introduction

Cyclodextrins (CD) are semi-natural materials produced from starch by bacteria. They can be used in large scale applications in agriculture, being compatible with environmental constraints. A problem of great practical importance is the stabilisation and subsequent controlled release in the atmosphere of the volatile pheromones by using Supramolecular (SM) devices. The pheromones of *Bactrocera oleae* (also known as *Dacus oleae*) encapsulated into cyclodextrins are the object of a collaborative project of the European Programme Value (with the Biology and the Physical Chemistry Institutes of NCSR "Demokritos", Greece, and the Co Agrigense, UK. The main pheromone component for this insect is "spiroacetal" (1,7-Dioxaspiro [5,5] undecane).

We have pointed out in previous papers to the very strong cohesion, and therefore negligible release from crystalline β-CD spiroacetal $9H_2O$. This strong cohesion has been accounted for by the particular crystal structure of the CD-pheromone complex [1]. An important characteristic of crystalline β-CD complexes is the formation of dimers where two CD molecules are linked head-to-head by seven hydrogen bonds between the secondary hydroxyls. The dimers are linked together mostly via hydrogen bonded water molecules, forming infinite 2D layers with their seven-fold axis approximately parallel.
A particular type of a *clearly defined supermolecule* in the crystal state is observed in the so called *"channel type assembly"*[2]. In this type of assembly the layers are stacked on the top of each other so that chains perpendicular to the layers are formed by piling up CD dimers via hydrogen bonds, creating thus a continuous channel. The pheromone molecules are included within this channel. The resulting dense and strong 3D hydrogen bond network linking together CD and water molecules is likely to be a most important factor in the stability of the complex and the low release rate. We have indeed here a sort of double SM inclusion: pheromone columns included into CD channels, CD chains inserted into a strongly held hydration network.

These facts suggest the study of new assemblies that would avoid the channel SM assembly (sections 2-3). Such assemblies can be provided by chemically modified CD's. The presence of 21 hydroxyls for β-CD offers great possibilities for chemical

G. Tsoucaris (ed.), Current Challenges on Large Supramolecular Assemblies, 385–391.

modifications and subsequent construction of large SM assemblies that seem necessary to providing an efficient release control adapted to various demands and external environmental conditions.

However, the easy preparation and low cost of the now classical β-CD complex prompted us to furher examine the mechanisms of pheromone release and search for a "release accelerating" device (sections 4-5). Indeed, the slow release of a guest molecule associated with a molecular or crystalline receptor is merely a phenomenon of "**dis-assembly**" of a SM system. It is therefore useful to get more information on the mechanism of slow release at the molecular level, whereas most of current works concern *construction* of assemblies. A general way to accelerate the release rate is heating; hence the thermal decomposition of the complex was studied. But in view of a a practical device, complexes containing suitable 'helper' component were investigated.

2. A xenophilic host: strong molecular encapsulation versus induced fit

In collaboration with B.S. Green and R.Arad-Yellin

The pheromone molecule is first included into the percinnamate β-CD [3], followed by a photoclosure reaction between two cinnamoyl moieties and subsequent formation of cyclobutane derivative, trapping permanently the pheromone inside the cavity [.4]. The slow release can be controlled through the inverse photo-opening reaction by sunshine UV. Another version of the device consists in incomplete closure allowing the pheromone to escape as in the so called hemicarcerates (let us suggest that "Xenophiles" may be a more pleasant term for a SM Assembly where the guest stays with the host as long as needed for a specific function).

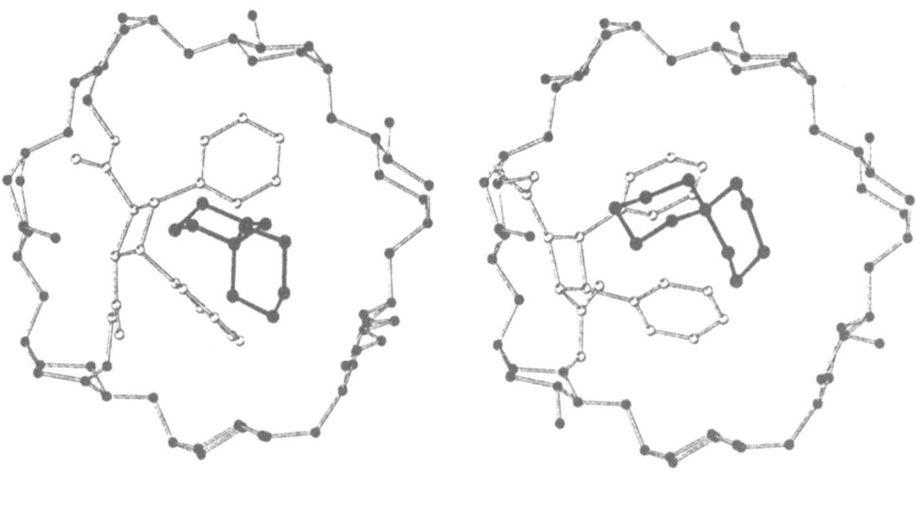

(a) (b)

Figure 1. Computer simulation for photoclosure of cinnamoylated CD
(For simplicity all non reacting cinnamates have been omitted)
(a) primary face; (b) secondary face

The partial closure can be achieved by partial esterification with cinnamic acid, or by photoclosure between only a fraction of available cinnamoyl pairs. The number of stereochemical possibilities for the photoclosure is very high, yet molecular modelling shows that a reaction between only two cinnamates is sufficient for a partial closure by capping the primary or secondary face, as shown in Fig. 1. For more than one pair of reacting cinnamates, as well for the complete reaction, we have a difficult problem of combining experimental methods to characterise the end product geometry.

The efficient closure of the cavity by a small number of covalent links can be termed as *"strong molecular encapsulation"*, as opposed to the usual inclusion achieved only by *weak forces* and enhanced, to a certain point by the induced fit phenomenon.

Induced fit is often sufficient to assure a fine tuning for the formation of a specific SM compound, but it is not sufficient to control the whole of the inclusion-delivery phenomena. An example of fine tuning by induced fit has been provided by the structure of the so called "overmethylated" dimethyl βCD [5], where an appropriate rotation of -C-O-Me group around the C-O bond signifcantly contributes to the closing of the CD cavity containing a spiroacetal molecule. However, if strong molecular encapsulation is a definitely efficient way to "solve" the slow delivery problem, the weak induced fit has still the power of a fine tuning in the usual encapsulation experiments.

This fine tuning is particularly striking in the enantiomeric discrimination. For instance, it has been found that the S-spiroacetal enantiomer alone is selected by the "overmethylated" βCD, while the R- alone is selected by trimethylated α–CD [6]. In both cases it is likely that induced fit greatly contributes to the natural enantiomeric discrimination engendered by the intrinsic chirality of different regions of the CD host. It should be noted, however, that induced fit cannot be "predisigned" in general, whereas strong molecular encapsulation based on covalent bonds results from an *a priori* planned experiment.

We will show now that induced fit may also lead to unexpectedly unfavourable effects on the inclusion process. The complex heptakis (2,6-di-O-methyl)-β-cyclodextrin methanol (dimeb, methanol) is not isomorphous to the other dimeb complexes (a = 14.139 Å, b = 23.05 Å, c = 27.53 Å; space group $P2_12_12_1$), and the structure has been solved by the molecular replacement algorithm AMoRe [7].

In this complex, helicoïdal chains of dimeb are built along the a crystallographic axis (Fig. 2) Two O-methyl group bonds, O62-C82 and O67-C87 are oriented towards the center of the cavity of the macrocycle so that the cavity is partialy closed on the primary (O6) side. On the other side (secondary O2,O3), the O-methyl group O61*-C81* of an adjacent dimeb molecule (atoms noted by *) is deeply inserted into the cavity: the methyl group C81* is close to the middle O4 plane (distance C81*-O4plane: 1,0 Å; shortest contact with the cavity: O61*-O45: 3,31 Å). This molecular packing, involving induced fit for the O-Me groups, "prevents" therefore the inclusion of the pheromone molecule, and even that of the solvent molecules. The cohesion of the structure is assured by one direct hydrogen bond between two dimeb molecules in the same helicoïdal chain (O62*-O37: 3,01. Å) and by two bond bridges between two successive dimeb molecules via solvent molecules.

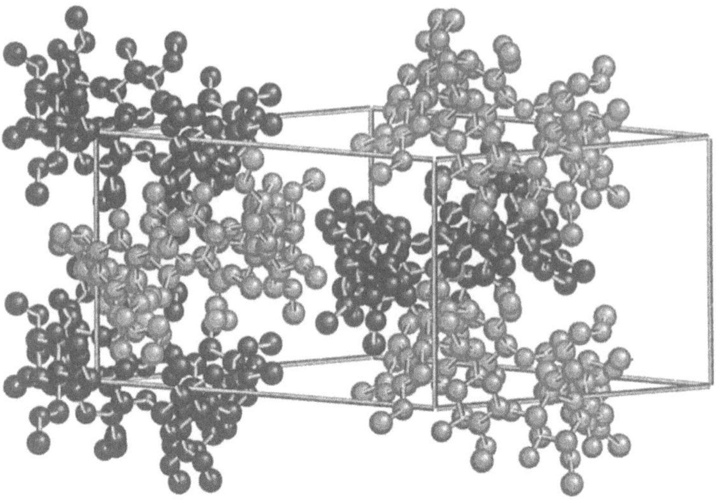

Figure 2. Helicoïdal chains of dimethyl β–cyclodextrin

3. Extremely water-soluble monosubstitued CD's

In collaboration with Y. BAHADDI and H. GALONS

An entirely different approach to the inclusion-delivery problem has been developed with extremely soluble CD derivatives. Glutamic acid is grafted at the omega carboxyl edge, so that the free amino-acid function is kept free [8]. This new CD derivative has a surprising high solubility in water: 150%. It is to be noted that the water solubility of β-CD is 1.8% and that of Lglutamic acid is of 0.9 %.

More strikingly, this solubility corresponds to 2.3 water molecules per hydroxyl, or 7 water molecules per residue or 48 water molecules per Glu-β-CD. All attempts to crystallise have failed. We could consider this extremely high solubility effect as a striking phenomenon of *non-assembly.*

Solubility is very difficult to predict in general (cf. the well known solubility of α-, β- and γ-CD, respectively equal to 14.5; 1.8 and 23.2g/100mL), yet the *non-assembly* fact may provide a useful clue. Indeed, the CD cavity has a hydrophobic character unfavourable to a zwiterion charge distribution. This is likely to prevent the quite usual phenomenon of self or polymeric inclusion where an antenna type substituent enters the next CD molecule along a crystallographic direction (Fig. 1). Speaking simply, this toroid molecule with a pending antenna has difficulties to pack in a crystal.

Moreover, the spiroacetal is well included in this CD derivative, also with a striking solubility for an hydrophobic guest: 10%. In comparison, the solubility of guest-free β-CD is 1.8%; that of the β-CD spiroacetal, of the order of 0.7%. This complex also failed to crystallise; however, the high concentration of pheromone in solution might be used in immobilisation experiments in order to provide a practical device as pheromone dispenser.

4. Thermal decomposition of the β-cyclodextrin-spiroacetal-water complex
in collaboration with M. OLLIVON

In view of the study of the release mechanism of the pheromone and the loss of solvent molecules from a dry powder of CD complex, a heating cycle has been applied. It is expected that the acceleration of the decomposition of the complex will allow to observe phenomena which otherwise would not be easily acccesible during the slow release process occuring at room temperature or field experiments.
Powder X-Ray diffraction using synchrotron radiation of LURE (Orsay) λ =1,489 Å coupled with Differential Scanning Calorimetry has been used for temperatures ranging from 30°C to 116°C (heating rate: 2°C/minute) up to θ =12°. The recorded diffraction patterns show a certain overall stability of the complex for the above range of temperatures (Fig.3a). The general features of the powder patterns remain up to the final recording temperature. The unit cell parameters do not undergo sensible changes. A general decrease of intensity of the peaks is noticed that can be probably attributed to a general loss of crystal perfection.
However, a modification of the diffraction patterns, is observed in the region from 5.46 Å to 6.80 Å).for temperatures higher than 98°C. Two new diffraction peaks appear at d = 5.91Å and d = 6.59Å , the intensity of the latter increases up to 116°C. On the contrary, the peak at d = 6.04Å steadily decreases in the same range of temperatures (Fig.3b). The experimental device allows the simultanous DSC recording [9](Fig.4) which shows an endothermic peak beginning also at about 98°C and reaching a maximum at 108°C. These modifications are reproductible and irreversible when the temperature is decreasing. This thermal event is likely to reflect the partial loss of solvent molecules and included molecules. NMR studies after heating the powder complex also show a loss of included guest molecules.
It is to be noted that the structure of the crystallites of the dry powder may be somewhat different from that reported [1] for a single crystal kept within a capillary in contact with mother liquor.

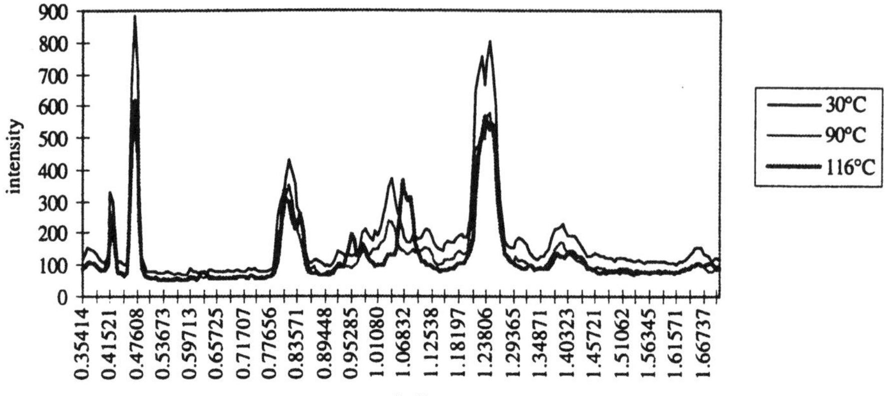

Figure 3a. X-Ray powder patterns of the complex:β-cyclodextrin, spiroacetal, water, at different temperatures.

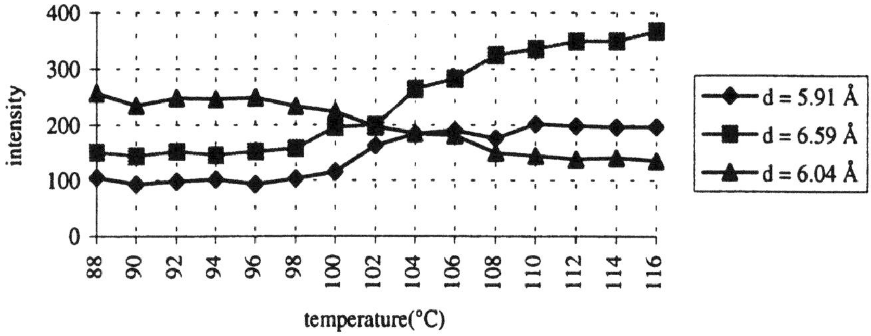

Figure 3b. Variation of the intensity of three diffraction peaks

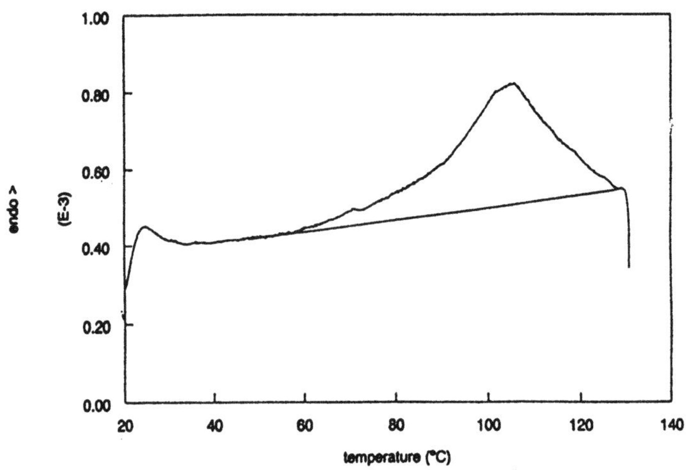

Figure 4. Differential scanning calorimetry of the complex β-cyclodextrin, spiroacetal, water

5. The helper component concept

The term is borrowed from a the sometimes called *hilf-gas* effect in hydrates. Butane fails to form a crystalline hydrate, being too long to fit into the hydrate cavities. But, inclusion into small cavities of a small "helper molecule" (H2S) allows the "crystalline assembly" where butane is now included into the larger cavities. In other words the hilf gas stabilises the otherwise unstable water-butane crystalline assembly through a subtle interaction between small and large cavities. .

Here we wish an opposite effect: find a helper molecule that «de-stabilises» the CD pheromone crystal, and, subsequently accelerates the release process. Molecules known for their affinity for the cavity such as phenylalanine have been first studied, but this solid-solid reaction is too slow. On the opposite, DMF or DMSO known for their ability to "push" the guest out of the cavity, have a too brutal effect.

Finally a suitable family is now under thorough study with promising results. Glycerol mixed in small proportions with the β-CD-spiroacetal complex induces an important accelaration of the pheromone release and offers the basis for a controlled release device.

REFERENCES

1. Rysanek N., Le Bas G., Villain F. and Tsoucaris G. (1996) Inclusion of the main pheromone component of Dacus Oleae:1,7-dioxaspiro(5,5)- undecane in β-cyclodextrin, *Acta Cryst* **C52** 2932-2936
2. a) Le Bas G. and Rysanek N.(1987) Structural aspects of cyclodextrins in Duchêne D (ed) *Cyclodextrins and their industrial uses*, Edition de santé, Paris
   b) Tsoucaris G. (1996) Stability, solvent patterns and molecular recognition in cyclodextrins in Tsoucaris and al.(eds) *Crystallography of supramolecular Compounds,* Kluwer academic publishers, Dordrecht, p 203-222.
3. Arad-Yellin R. and Green B.S. (1994) Photochemical closing and opening of the guest-binding cavity of cyclodextrins, *Nature* **371**, 320-322
4. Arad-Yellin R., Tsoucaris G. and Green B.S.(1998) Encapsulation and release of the pheromone,1,7-dioxaspiro[5,5]undecane, using β-cyclodextrin percinnamate (accepted) *Supramolecular Chemistry*
5. Rysanek N., Le Bas G., Villain F. and Tsoucaris G. (1992) Structure of the (1:1:1) complex pentadecakis (2a,2b,2c,2d,2e,2f,6a,6b,6c,6d,6e,6f,6g, 3a,3g) -O-methyl β cyclodextrin 1-7-dioxaspiro[5,5]undecane-methanol. *Acta Cryst* **C48** ,1466-1471
6. (a) Yannakopoulou, K, Mentzafos, D., Mavridis, I. M. and Dandika, K. (1996) Chiral Recognition of (R)-(-)-1,7-Dioxaspiro-[5.5]undecane by Hexakis(2,3,6-tri-O-methyl)-a-cyclodextrin. Angew. Chem. Int. Ed. Engl. 35, 2480-2481
   (b) Mentzafos D., Mavridis I.M. and Yannakopoulou K. (1998) Structure of the 1:1 complex of hexakis(2,3,6-tri-O-methyl) α-cyclodextrin with (R)-1,7-dioxaspiro[5,5],undecane *Journal of Inclusion Phenomena* (in print)
7. Navaza J. (1994) AMoRe program package *Acta Cryst* **B50** , p157
8. Bahaddi Y., Lelièvre F.,Gareil P.,Maignan J., Galons H. (1997) *Carbohydrate Research* **303,** 229-232.
9. Keller G.,Lavigne F., Loisel C., Ollivon M.and Bourgaux C. (1996) *Jnal of Thermal Analysis* **VOL47** ,1545-1565

INFLUENCE OF THE GUEST ON THE PACKING OF DIMERIC β-CYCLODEXTRIN COMPLEXES

I. M. MAVRIDIS
Institute of Physical Chemistry
N.C.S.R. "Demokritos"
153 10 Aghia Paraskevi
Athens, Greece

Abstract

Packing pattern of host:guest complexes of β-cyclodextrin (βCD) are discussed. In most of βCD inclusion complexes two βCD molecules self-assemble and form a dimer which encloses one or more molecules of guest. The whole is a supermolecule held together by intermolecular non-bonding interactions. The dimer has the shape of a cylinder whose dimensions and external surface remain unchanged irrespective of the guest. The latter influences only the top and bottom faces of the cylinder where some parts of it are exposed. As a result of this influence four packing modes are encountered. The study of the molecular packing of the βCD complexes in conjunction with the nature of the guest is a means towards a better understanding of the association of the supermolecular units that will lead to the design of crystal structures with specific characteristics.

1. Introduction

As a general idea, crystal engineering sounds as something unrealistic because it implies that all the interactions between the building blocks are known. However, certain classes of compounds have the potential to generate specific crystal packing at will. The dimeric host-guest complexes of β-cyclodextrin (βCD) is such a case.

βCD has the ability of encapsulating a plethora of compounds to form inclusion complexes, the majority of which are dimeric [1]. The βCD dimer is a supermolecule formed by self-association of the monomers through H-bonding of the O3 secondary hydroxylic groups. It exhibits 7-fold

G. Tsoucaris (ed.), Current Challenges on Large Supramolecular Assemblies, 393–403.

symmetry, its shape resembling that of a hollow cylinder. The elongated cavity is hydrophobic including the intra-dimer ring between the two monomers. The secondary hydroxylic groups are not influenced by the guest and form an invariant network of H-bonds with neighboring dimers and water molecules in the outside [2]. On the other hand, although the primary hydroxylic groups at the top and bottom of the cylinder form a number of H-bonds with adjacent groups, common among various complexes, they are influenced also by the presence of the guest, especially if the latter possesses a polar group that emerges from the primary side. Therefore, we can claim that in the dimeric complexes a semi-invariant network of H-bonds is formed that connects each dimer to other dimers either directly or through water molecules, cocrystallized around them. The result is the formation of infinite 2-D layers [1, 2] of C-centered or pseudo C-centered geometry, depending on the space group, with two orthogonal axes of dimensions close to 19 and 24Å. The layers have the primary faces of the βCD dimers and consequently the parts of guests located there exposed to the solvent. It is thus obvious that this system is an appropriate one for the study of crystal engineering since the main building blocks keep their integrity upon complexation, while a limited surface on them is possible to change by careful selection of the enclosed guest.

2. Crystal Packing

Only four packing patterns of βCD dimeric complexes are observed [1] (a) Channel (CH), (b) Screw-channel (SC), (c) Chessboard (CB) (d) Intermediate (IM). Complexes belonging to the same mode are isomorphic, an important regularity that permits to know their crystal packing just by the knowledge of their unit cell. They are all built by linking together the previously mentioned, practically invariant layers by H-bonds involving primary hydroxylic groups water and/or the guest molecule(s). Therefore, the guest emerging from the primary faces plays a crucial role in the packing.

In the CH modes the 2-D layers are placed in such a way that dimers align almost on the top of each other to form channels (Figure 1). These are hydrophobic and the guests inside them are shielded from the water environment. The dimer's 7-fold axis forms an angle of approximately 10° with the stacking axis with the result that the channel is slightly deformed in the interdimer interface. The lateral displacement between two consecutive βCD dimers along the channel is only 2.7Å.

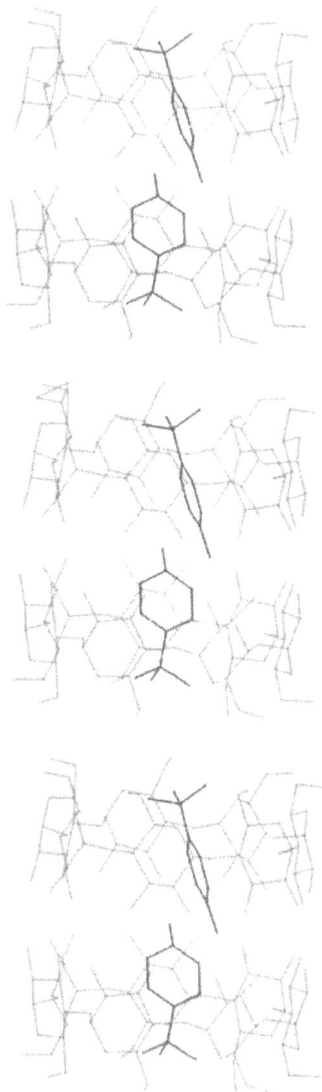

Figure 1. Channel packing. Inclusion complex of βCD -4-tert-butyltoluen [3].

On the contrary, in CB mode every dimer, immersed into the solvent, is isolated (Figure 2). The lateral distance between two dimers of successive layers is 8.9Å. The primary faces and therefore the guest are exposed to water molecules and hydroxylic groups.

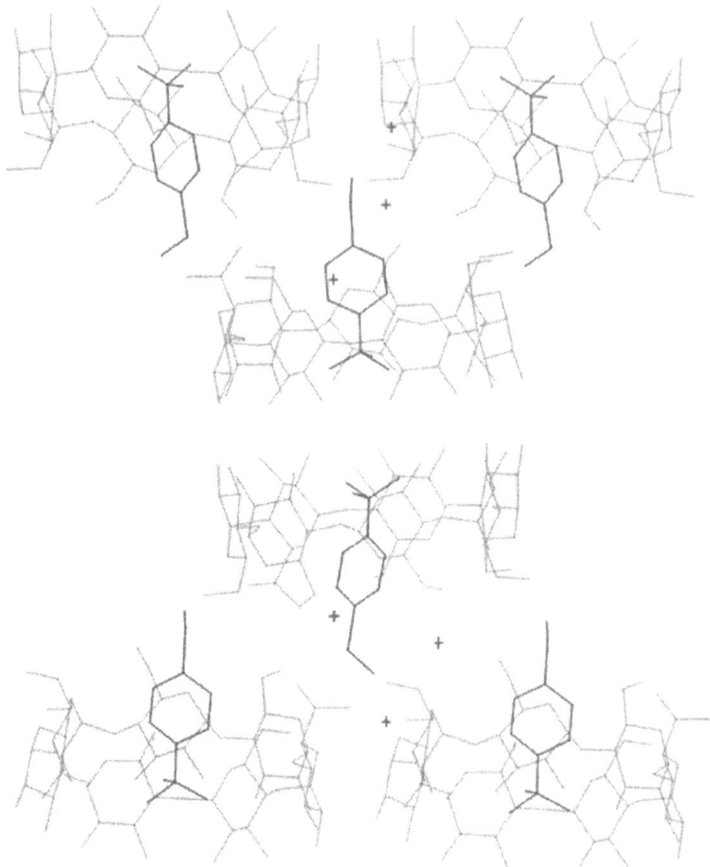

Figure 2. Chessboard packing. Inclusion complex of βCD–4-tert-butylbenzyl alcohol

Finally, in IM (Figure 3) mode, a case in-between the CH and CB modes, the dimer's 7-fold axis forms an angle of about 20° with the stacking axis and consequently adjacent layers are far from exactly aligned. The lateral displacement of dimers in consecutive layers is 6Å (the inner diameter of the primary face of βCD) thus, the dimer's cavity entrance is partially closed by the βCD atoms of adjacent layers. As a consequence, a breaking of the channel is observed that leaves parts of the guests free to interact with hydroxylic groups of the host molecules as well as with water molecules.

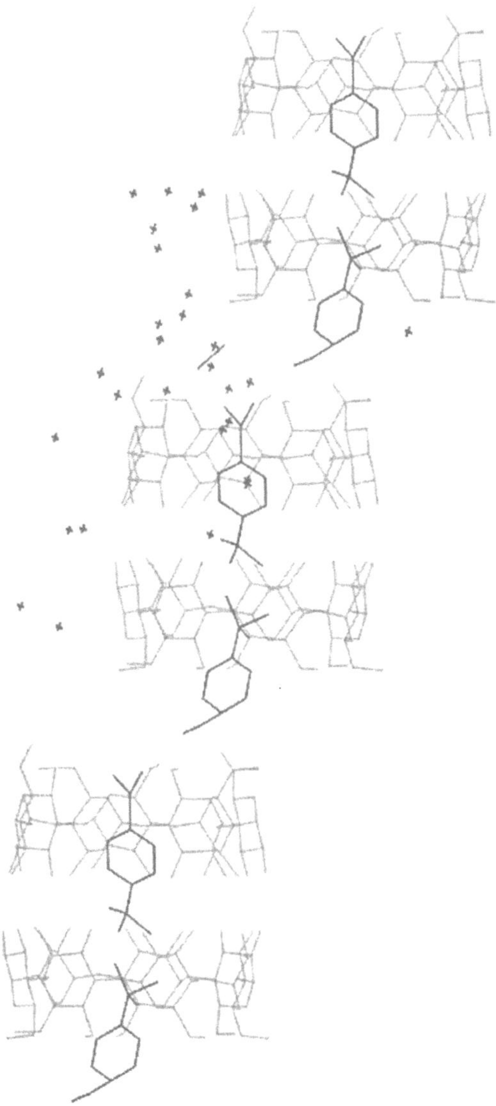

Figure 3. Intermediate mode. Inclusion complex of βCD-4-tert-butylbenzoic acid [4].

The existence of the packing independent H-bonding network around the βCD dimers in the above described layers and, in particular, the quasi-invariant primary sphere of water molecules around each dimer, point towards the hypothesis that the latter might exist in solution as a stable entity after complex formation and prior to nucleation and crystallization. The guest's nature and the mutual geometric arrangement of the host(s) and guest(s) in the complex is possible to affect the equilibrium,

monomeric complex ⇌ dimeric complex.

Indeed, dimeric complexes of permethylated [5] and natural [6] α-cyclodextrin with long aliphatic guests have been detected in aqueous solutions. The above assumption is also supported by the existence of complexes with a βCD:guest ratio of 2:1, where only one guest is included inside the cavity of two hosts, as shown below, or when a guest molecule is found in the interdimer interface [7]. Given the validity of the above hypothesis, it is easy to explain the invariability of the dimeric layers formed subsequently. The step which relates to the packing mode will result from the layers according to the geometric and energetic requirement of the dimeric complex. The guest should influence that step i.e. a guest with a hydrophilic, H-bond forming group will favor extensive hydration at the primary faces of the dimer and therefore will induce CB packing. On the contrary, a hydrophobic group at the primary faces will favor channel formation in order to be isolated from the aqueous environment. Our experience is that polar groups with a self-association tendency, like the carboxylic group [4, 8] or polar groups with intermediate tendency to be hydrated favor IM packing.

3. Dimeric complexes of β-Cyclodextrin with aliphatic acids

Based on the above observations, we have studied dimeric βCD complexes of mono- and di-aliphatic acids in order to test the transformation,

Intermediate packing mode ⇌ Channel packing mode

and *vice versa* as a function of the length of the aliphatic chain. The following cases were studied:

a) Nonanoic acid forms dimeric complexes packing in IM mode with a host:guest ratio of 2:2. Each monomer of the macrocycle hosts one molecule of acid whose carboxylic groups emerging from the corresponding primary faces form H-bonds to carboxylic groups of neighboring dimers, water molecules and βCD hydroxylic groups (Figure 4).

b) Aliphatic monoacids with 12-16 carbon atoms form βCD dimeric complexes with host:guest ratio of 2:1 that induce channel packing (Figure 5). The hydrophobic end of those amphiphilic guests influences the environment of the polar primary face. As a result, the dimers form channels in order to protect that part of the guest. The carboxylic groups found entrapped in the channel self-associate to carboxylic dimers, thus stabilizing the whole system.

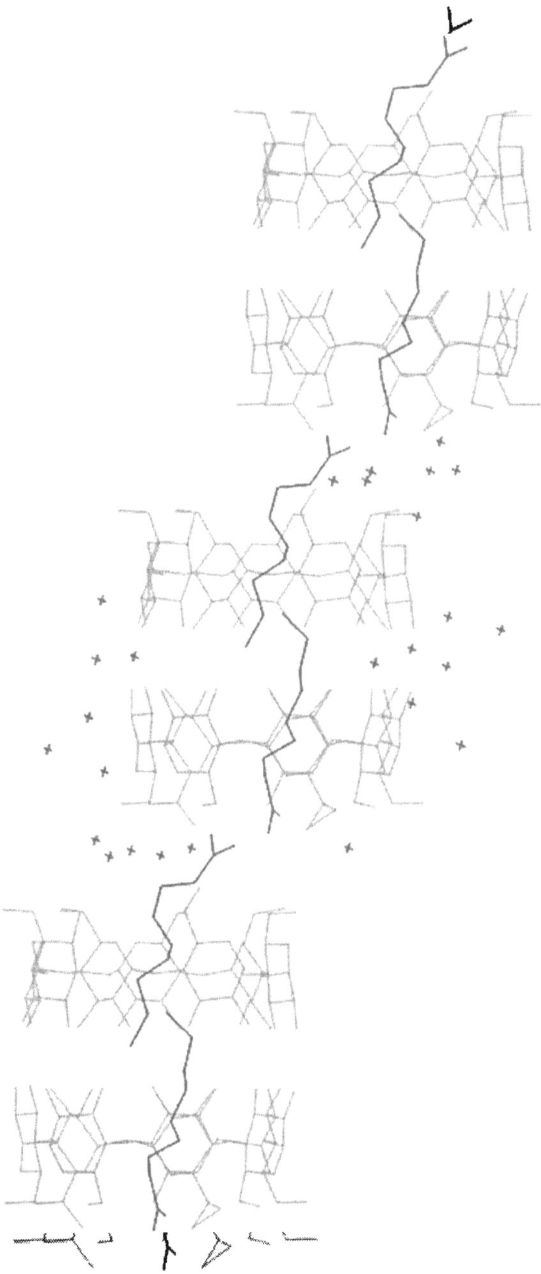

Figure 4. Inclusion complex of βCD-nonanoic acid. Intermediate packing [8].

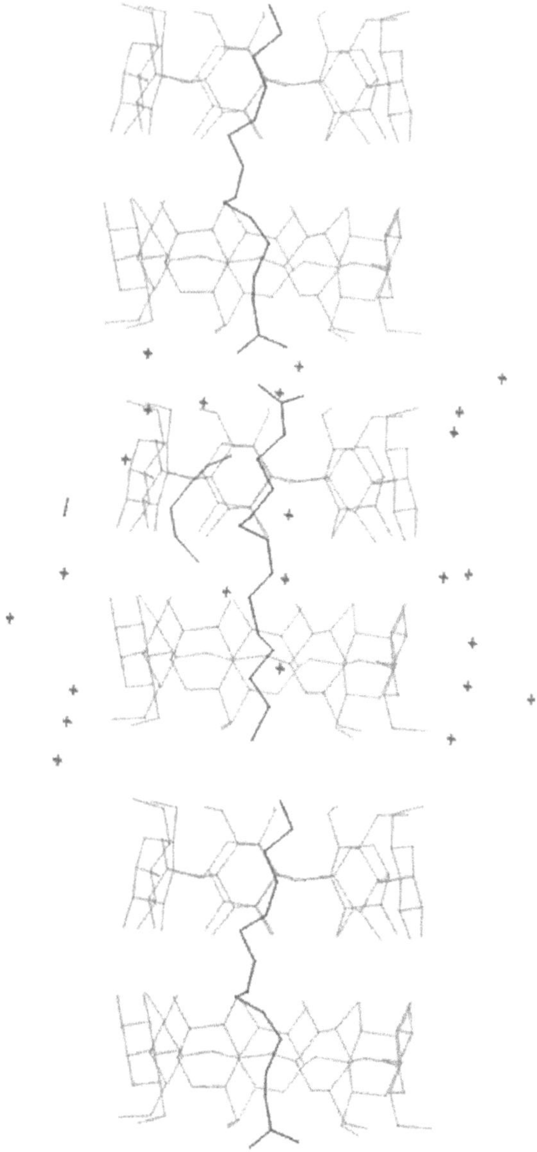

Figure 5. Inclusion complex of βCD-Z-7-tetradecenoic acid. Channel packing [9].

c) In the case of aliphatic diacids with 10-16 carbon atoms the packing is again IM (Figure 6). The hydrophylicity of the carboxylic groups overwhelms their self association tendency and favors the less densely packed IM mode. This class of complexes proves that the driving force for the channel packing in

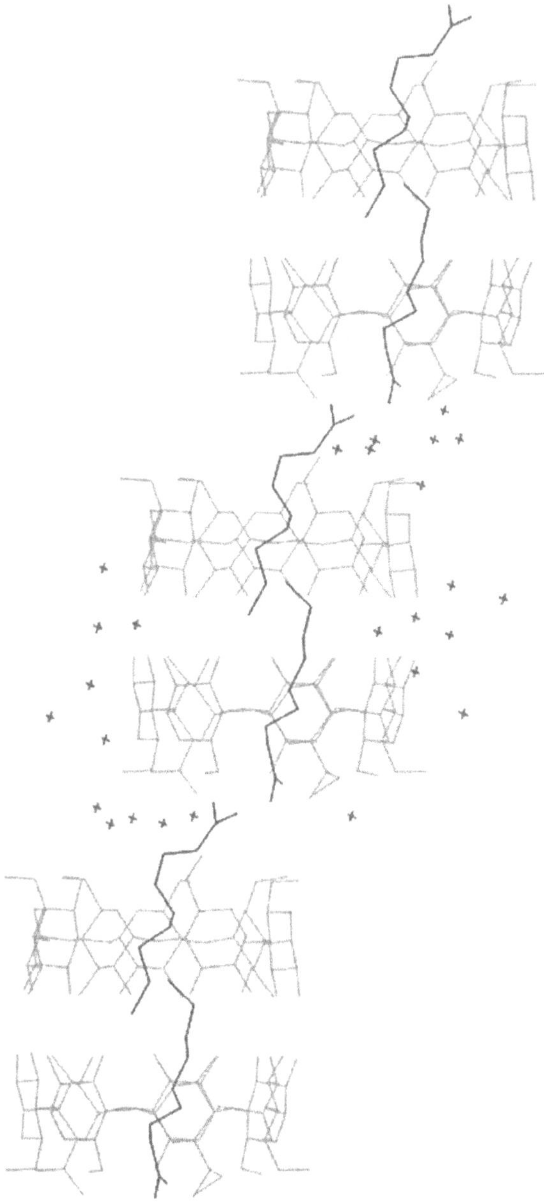

Figure 6. Inclusion complex of βCD–tridecanedioic acid. Intermediate packing [10].

the case of aliphatic mono-acids, where the alkyl chain is long enough to thread into a βCD dimer and form [3]pseudorotaxane, is the terminal methyl group. Actually, when the chain of the mono-acid is short enough to fit into the cavity of one βCD molecule, like nonanoic acid, the dimeric complex includes two molecules of the acid and the whole supermolecule resembles the complex of a diacid. It has moreover, additional advantage with respect to the latter because it is like a diacid with a variable length. Each guest molecule has the freedom to slide with respect to the other and extend more into the solvent to approach neighboring carboxylic groups in order to form carboxylic dimers, as in the case of the nonanoic acid complex (Figure 4).

4. Epilogue

In this contribution it has been discussed that the βCD dimer formed upon complexation of β-cyclodextrin is an entity that maintains its integrity irrespective of the nature of the guest This is reflected in the regularity of the packing patterns of dimeric βCD complexes which are based on the formation of semi-invariant 2D dimeric layers. Detail study of the factors that influence the packing is the means to gain insight in the way that molecules are recognized. Conversely, if we understand molecular recognition we can select appropriate building blocks that interact according to specific rules and organize into predesigned systems.

References

1. Mentzafos, D., Mavridis, I. M., Le Bas, G. and Tsoucaris, G. (1991) Structure of the 4-*tert*-butylbenzyl alcohol-β-cyclodextrin complex. Common features in the geometry of β-cyclodextrin dimeric complexes, *Acta Cryst.* **B, 47**, 746-757, and references therein.
2. Le Bas G. and Tsoucaris G. (1994) Two dimensional hydrogen bond network in β-cyclodextrin complexes, *Supramolecular Chem.*, **4,** 13-16.
3. Mavridis, I. M. and Hadjoudis, E. (1992), The crystal structure of the inclusion complex of cyclomaltoheptaose with 4-*tert*-butyltoluene, *Carbohydrate Res.*, **229**, 1-15.
4. Rontoyianni, A., Mavridis, I. M., Hadjoudis, E. and Duisenberg, A. J. M. (1994) The crystal structure of the inclusion complex of cyclomaltoheptaose (β-cyclodextrin) with 4-*tert*-butylbenzoic acid, *Carbohydrate Res.*, **252**, 19-32.
5. Botsi, A., Yannakopoulou, K., Perly, B. and Hadjoudis, E. (1995) Positive or adverse effects of methylation on the inclusion behavior of cyclodextrins. A comparative study using pheromone constituents of the olive fruit fly, *J. Org. Chem.*, **60**, 4017-4023.

6. a) Eliadou, K., Yannakopoulou, K. and Mavridis I. M. (1997) Aqueous NMR study of the inclusion complexes of linear aliphatic α-aminoacids with α-cyclodextrin in *Current Challenges on Large Supramolecular Assemblies*, 31 Oct-5 Nov 1997, Athens, Greece, Book of abstracts; b) Eliadou, K., Yannakopoulou, K., Rontoyianni, A. and Mavridis, I. M. (1998) manuscript in preparation.
7. Rontoyianni, A. and Mavridis, I. M. (1994) The crystal structure of the inclusion complex of cyclomaltoheptaose with 3,5-dimethylbenzoic acid, *J. of Inclusion Phen. and Mol. Recognition*, **18**, 211-227.
8. Rontoyianni, A. and Mavridis, I. M. (1996) β-Cyclodextrin-nonanoic acid 1:1 complex, *Acta Cryst.* **C, 52**, 2277-2281.
9. Makedonopoulou, S., Mavridis, I. M., Yannakopoulou, K. and Papaioannou, J. (1998) manuscript in preparation.
10. Makedonopoulou, S., Mavridis, I. M. and Tulinsky, A. (1998) Manuscript in preparation.

SEMICONDUCTOR NANOCRYSTAL COLLOIDS*

A. P. ALIVISATOS

Department of Chemistry
University of California, Berkeley
and Lawrence Berkeley National Laboratory

1. Semiconductor Colloids

The study of nanometer size semiconductor crystals has been advancing at a rapid pace[1,2]. Much of the interest in these materials stems form the fact that their physical and chemical properties can be systematically tuned by variation of the size, according to increasingly well established scaling laws. This paper describes colloidal semiconductor nanocrystals belonging to the II-VI and III-V families, and outlines strategies for obtaining electrical access to such dots.

If an inorganic cluster exceeds a certain size, generally in the tens of unit cells, then it will likely possess a bonding geometry characteristic of a bulk phase[3,4]. Above this critical size, the nature of the chemical bonds in the cluster remains fixed as a function of the size, but the total number of atoms, or the surface to volume ratio, change smoothly. This leads to a slow extrapolation of the properties of ideal nanocrystals towards bulk values with increasing size, according to the scaling laws. The ability to systematically control the properties of inorganic materials by variation of size and shape is an important development, with many implications for how materials should be processed and assembled[5].

Many scaling laws have been investigated, including the size variation of band gap, charging energy, magnetization reversal, melting, etc. Study of the scaling laws reveal lessons for how to make nanocrystals. This paper focuses on the properties of colloidal semiconductor nanocrystals, how to make them, and some ways of gaining electrical access to them. Advances in metal[6-8], magnetic[9], and structural[10-12] nano-materials are also occurring. Semiconductor dots produced by other processing techniques, such as self assembled semiconductor dots, and porous silicon, are reviewed elsewhere in this issue.

2. Processing on the Nanoscale

Many routes have been developed for preparing nanostructures, ranging from single atom manipulation to organic synthesis. The tremendous strengths of the colloidal preparation routes are that they yield large amounts of monodisperse nanocrystals with easy size tuning. Further, the presence of the organic molecules on the surface of a nanocrystal enable extensive chemical manipulation after the synthesis. One interesting conclusion from this work is that the preparation of nanocrystals may in many ways be far more forgiving than the fabrication of two or three dimensional materials:

* A version of this paper appeared recently in the Materials Research Society Bulletin

G. Tsoucaris (ed.), Current Challenges on Large Supramolecular Assemblies, 405–416.

2.1. NANOCRYSTALS CAN BE PREPARED AT COMPARATIVELY MODEST TEMPERATURES.

One of the most famous scaling laws relates the variation of melting temperature with size, and shows that it drops as 1/r13,14. The large drop in melting temperature in small sizes, means that it is possible to make highly crystalline, and faceted nanoparticles at temperatures that are compatible with wet chemical processing. From a kinetic perspective, a great deal of control of chemical processes can be achieved in solution, and for this reason extremely high quality inorganic nanoparticles can be prepared as colloids. The recent successes in the preparation of II-VI15,16 and III-V17-20 nanocrystals illustrate the strengths of the colloidal preparation techniques.

Nanocrystal precursor molecules are injected into a hot surfactant at 300C; nucleation and growth of the nanoparticles occurs. There is no aggregation, because the particles are terminated by a monolayer of surfactant. At the end of the preparation, nanocrystals are isolated with a monolayer surfactant coat, but this can be exchanged with other organic molecules, allowing the environment of the nanoparticles to be controlled.

2.2. NANOCRYSTALS ACT AS SINGLE STRUCTURAL DOMAINS.

Structurally, nanocrystals are at a crossover point between molecules, where the number and location of each atom can be exactly specified, and bulk matter, where we

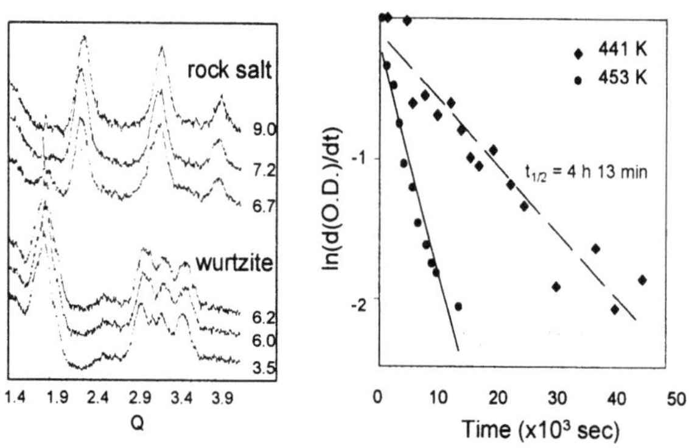

*Figure 1. A. X-ray powder diffraction patterns from 4 nm CdSe nanocrystals under hydrostatic pressure shows that they convert from four coordinate wurtzite to six coordinate rocksalt at pressures far higher than the bulk (6. 5 vs. 2.7 GPa) (Data taken as part of the UC-National Lab PRT beamline at the Stanford Synchrotron Radiation Lab.) B. When the pressure is increased abruptly, the phase transition proceeds in a characteristic relaxation time. The temperature dependence of this relaxation time yields the barrier height blocking the transition**A. X-ray powder diffraction patterns from 4 nm CdSe nanocrystals under hydrostatic pressure shows that they convert from four coordinate wurtzite to six coordinate rocksalt at pressures far higher than the bulk (6. 5 vs. 2.7 GPa) (Data taken as part of the UC-National Lab PRT beamline at the Stanford Synchrotron Radiation Lab.) B. When the pressure is increased abruptly, the phase transition proceeds in a characteristic relaxation time. The temperature dependence of this relaxation time yields the barrier height blocking the transition.*

must work with statistical descriptions. Both kinetically and thermodynamically, nanocrystals tend to exclude high energy defects, such as grain boundaries (lower energy defects, such as stacking faults are common, however). It is far easier to anneal out a defect from a very small crystal than from a large one. Assuming the barrier to defect migration is the same as the bulk solid (certainly an overestimate, since the melting temperature is lower), then the defect simply has less distance to travel and will be annealed out faster in a smaller crystal. Thermodynamically, defects are present in a large crystal of atoms at finite temperature . However, if the free energy of defect formation is relatively constant with size, then a very small solid is less likely to contain equilibrium defects.

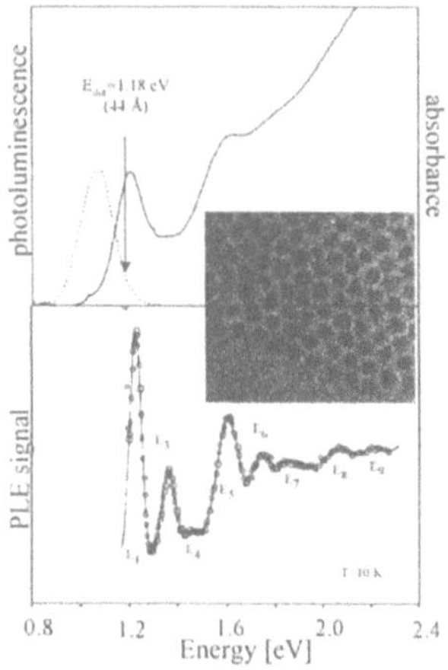

Figure 2 Optical spectra and transmission electron micrograph of InAs nanocrystals. Due to quantum size effects, the absorption and emission spectra are shifted to higher energy than in the bulk, and photoluminescence excitation spectra reveal the presence of multiple discrete transitions. The TEM shows that nanocrystals are monodisperse, and do not aggregate because of the monolayer surfactant coat on the surface.

The high quality of colloidal dots is confirmed by experiments on the kinetics of pressure induced solid-solid phase transitions in nanocrystals21,22 (Figure 1). In extended solids, pressure induced solid-solid phase transitions nucleate at high energy defects. In nanocrystals such phase transitions are kinetically suppressed because of the absence of defects. Under excess pressure beyond the thermodynamic transition point, nanocrystals are observed to undergo structural transformations on a much longer time scale than the corresponding bulk crystal. A simple scaling law, consistent with recent experiments on 2 to 5 nm dots, shows that this relaxation time for structural transformation is proportional to nanocrystal size. This is analogous to the well known scaling law for magnetization reversal time scales in single domain magnets. Larger crystals (>10 or 20 nm) inevitably contain defects, which act as nucleation sites for phase transitions, much as domain walls lower the barrier for magnetization reversal. An important conclusion from these studies of structural stability and phase transitions versus size, is that very small, nanometer size crystals may in many ways be more perfect than extended, or two dimensional solids, even though the processing is less complex1.

3. Quantum Size Effects in Electronic Spectra

The most widely investigated group of scaling laws in semiconductor nanocrystals concern the size evolution of the optical spectra. Many studies have focused on the properties of colloidal dots[23], or dots embedded in glass matrices[24]. Those studies have centered on II-VI dots of CdS and CdSe. A parallel set of studies have been directed towards the investigation of MBE grown dots, the majority of these studies being on III-V nanocrystals, such as InAs embedded in GaAs (reviewed in this issue). With the advent of colloidal synthetic routes to InAs and other III-V particles, these two parallel strains of research are now converging, and offer a fascinating opportunity to study nanostructured materials prepared by entirely different processes. Figure 2 shows the optical spectra of 5 nm diameter colloidal InAs dots. The electronic absorption spectrum is shifted to much higher energy than the bulk band gap, and the luminescence excitation reveals discrete structure in the spectra.. The systematic investigation of the evolution of the electronic structure with size remains an important topic of research, and a great deal of theoretical effort is directed at understanding this size evolution (see the article by Zunger in this issue).

4. Core-Shell Dots

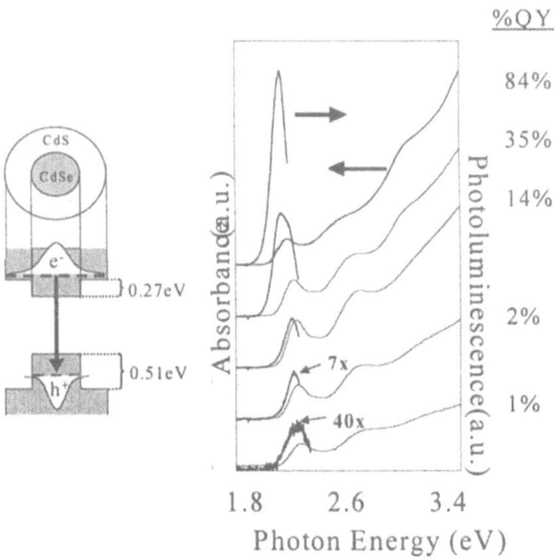

Figure 3 Core-shell nanocrystals of CdSe/CdS show greatly enhanced luminescence quantum yields. The outer shell of, CdS, is epitaxially deposited on the core of the CdSe.

[1]It is of course possible to make defective or aggregated nanoparticles. Simple scaling laws, however, suggest that the relatively modest processing conditions of colloidal routes are sufficient to prepare extremely high quality inorganic nanocrystals.)

One important issue for semiconductor nanocrystals concerns the nature of surface passivation. It is essential in a confined semiconductor structure to epitaxially passivate dangling bonds at the surface. Epitaxy and passivation in colloidal dots have now been achieved, making these dots fully comparable to MBE grown structures in surface quality[25]. As an example, CdS has been used to produce a fully epitaxial shell around CdSe[26]. That system displays band edge luminescence quantum yields as high as 80% at room temperature. (Figure 3). The outer CdS shell is still coated with a surfactant layer, so that the advantages of passivation and chemical processing are simultaneously realized.

In comparison to 2D structures, the kinetic energies in II-VI or III-V quantum dots of about 5nm diameter are of order 0.3 eV, so that the potential steps for electrons and holes at the hetero-interface need to be relatively large (order 1 eV) to be effective in confining the charges. Suitable passivation will be realized with different, and more dissimilar, pairs of materials in the 0D case, as compared to the 2D. Free standing dots also appear to be able to relax strain more readily than 2D structures, or embedded dots.

5. Nanocrystal-Polymer Composites

The electrical characteristics of quantum dots are of great interest, and we are pursuing two broad strategies for investigating them: studies of ensemble electrical characteristics, and electrical studies of individual dots. The most straightforward way to access an ensemble of colloidal nanocrystals electrically is by making a blend with a conducting or semiconducting polymer[27-29], onto which electrodes may be readily evaporated. As test cases, we have used nanocrystal/polymer blends to make photovoltaics and light emitting diodes.

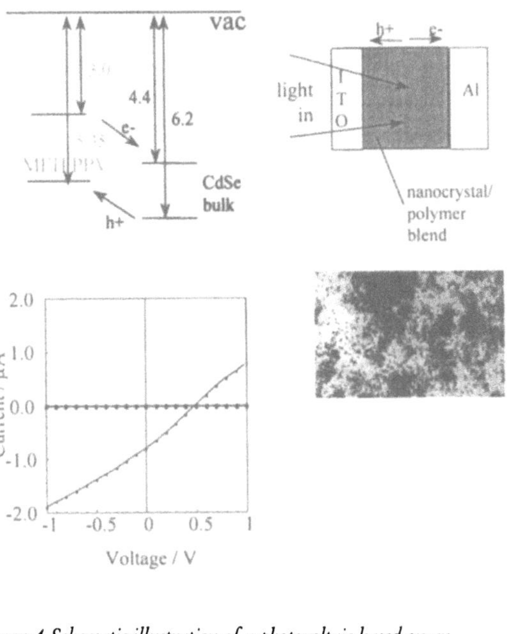

Figure 4 Schematic illustration of a photovoltaic based on an interpenetrating network of 5 nm diameter CdSe nanocrystals and the semiconductor polymer, MEH-PPV.

6. Photovoltaics

Nanocrystals are well suited to charge separation devices, since they provide a high surface area. The ability to tune the band gap readily by size variation, suggests that inexpensive multi band-gap solar cells, analogous to very high efficiency but very expensive quantum well based photovoltaics, may be realizable with nanocrystals. O'Regan and Graetzel[30] have reported an efficient (12%) and inexpensive photovoltaic based on a film of partially fused nanocrystals of TiO_2 coated with an extremely robust sensitizer molecule. That device takes advantage of the large surface area the nanoparticles provide, but does not make use of the quantum size effect to tune band gaps.

We have recently reported a photovoltaic device based on 5 nm CdSe nanocrystals and a semiconductor polymer, MEH-PPV [31](Figure 4). Both materials are semiconductors with band gaps of about 2eV, but the electron affinity of the nanocrystals is much greater than that of the polymer. As a consequence, electron-hole pairs generated with photons, either in the nanoparticles or the polymer, rapidly separate, with the hole residing on the polymer, and the electron on the nanoparticle. The nanocrystal/polymer blend morphology can be adjusted so that the nanoparticles form aggregated chains inside the polymer. This creates a pathway for holes to move along the polymer phase, and for electrons to hop (slowly) from nanocrystal to nanocrystal, until the separated charges are collected at electrodes. The efficiencies of these devices are quite low, and the fundamental physics of charge hopping and conduction in nanocrystals systems is still being studied, but it is clearly feasible to make interesting quantum dot based charge separation devices.

7. LEDs

We have also reported light emitting diodes based on nanocrystals and polymers[32-34]. Rather than making a blend, in this case we used a layered system, with a film of nanocrystals spin cast from solution onto a layer of semiconductor polymer, PPV. In this configuration holes are injected into the polymer, and electrons into the nanocrystals. Electron-hole pairs recombine by the emission of a photon in a zone near the nanocrystal-polymer interface. If the nanocrystal layer is relatively thick, then essentially all of the light comes form the nanocrystal layer only. In LEDs with moderate thickness nanocrystal layers, the recombination zone shifts with applied bias between the polymer and the nanocrystal layer, providing a form of voltage dependent color. Using a Mg/Ag electrode, we have achieved internal quantum efficiencies of 1%, and lifetimes in excess of several days. These LEDs combine the advantages of organic and inorganic LEDs. They are fabricated inexpensively in a way that can be scaled, and on flexible substrates, yet the active component, the nanocrystals, are very robust compared to most organic chromophores.

8. Single Nanocrystal Single Electron Transistor

Preliminary studies of both the electrical[35] and optical[36-39] properties of individual nanocrystals have been performed. These studies show that a single excess charge on a nanocrystal can dramatically influence its properties. We have recently reported measurements where the charge state of a single nanocrystal can be directly tuned[40]. As studies of lithographically patterned quantum dots and small metallic grains have shown, such measurements are invaluable in understanding the energy level spectra of small electronic systems. Back of the envelope estimates show that the charging energy of a 5.5 nm CdSe dot should be ca. 0.1 eV, and increases as $1/r$. The spacing of the discrete electronic levels of the dots scales as $1/r^2$, and coincidentally is also order 0.1 eV according to simple estimates.

Figure 5A shows an idealized schematic of the device. CdSe nanocrystals of a 5.5 nm diameter are bound to two closely spaced Au leads using bifunctional linker molecules. The leads are fabricated on a degenerately doped silicon wafer which is then used as a gate to tune the chemical potential, or charge state, of the nanocrystal under study.

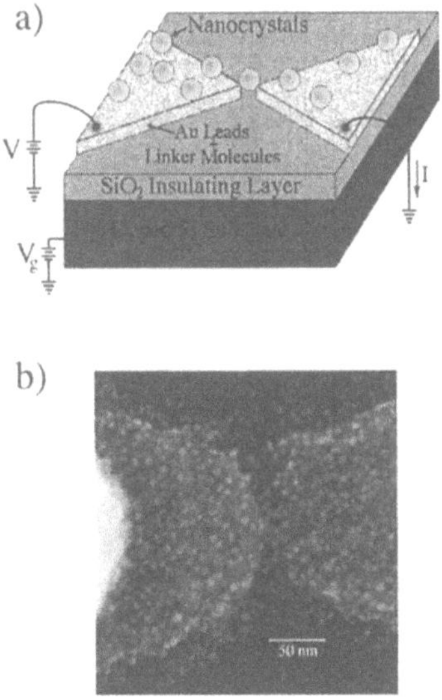

Figure 5. A. Schematic illustration of a transistor with one nanocrystal of CdSe as its active element. B. Scanning electron micrograph of such a device.

Figure 5B shows a field emission scanning electron micrograph of a completed device. A number of nanocrystals appear to be in the ~ 5nm gap between the electrodes.

Most devices, including the particular junction shown in figure 5B, have immeasurably high impedance (R > 100 GΩ). Only about one in twenty have a measurable conductance, typically in the 10 MΩ -1 GΩ range. Devices of this type typically behave as though transport is occurring through a single nanocrystal. This may initially seem surprising, since Figure 5B indicates that the number of nanocrystals in the junction region is quite large. However, tunneling through the linker molecules has an exponential decay length of less than 1 Å. As a result, only a well-placed nanocrystal (within 2 nm of each lead) can contribute to conduction.

The current-voltage characteristics of these devices at 4.2 K show characteristic single electron transistor behavior (Figure 6). At some arbitrary value of the gate voltage, no current flows at small source-drain bias. In this regime the system is blocked, until V_{sd} exceeds the charging energy of the dot. There is a value of the gate voltage at which the probability of finding an electron on the dot is ½, at which point electrons hop on and off the dot at arbitrarily small values of the source-drain bias. Figure 7 shows a map of the differential conductance of the device, plotted as a gray scale, as a function of both V_{sd} and V_g. Multiple coulomb oscillations are observed, indicating that successive charges can be added or removed from the dot.

The characteristics of this device differ substantially from what would be observed if the active element were metallic. For instance, the spacing of the successive charge states is not equal in energy. Note that on the right side of Figure 7, the Coulomb gap continues to grow with increasingly positive V_g. For even larger gate voltages than] shown in the figure (up to 40 V) the Coulomb gap exceeds 150 meV and there is no evidence for any more Coulomb oscillations. This behavior can be understood if electrons are being added to the valence band of the nanocrystal with increasingly positive V_g. In this case, for some sufficiently positive V_g, every state in the valence band is filled, and the next extended state of the nanocrystal lies in the conduction

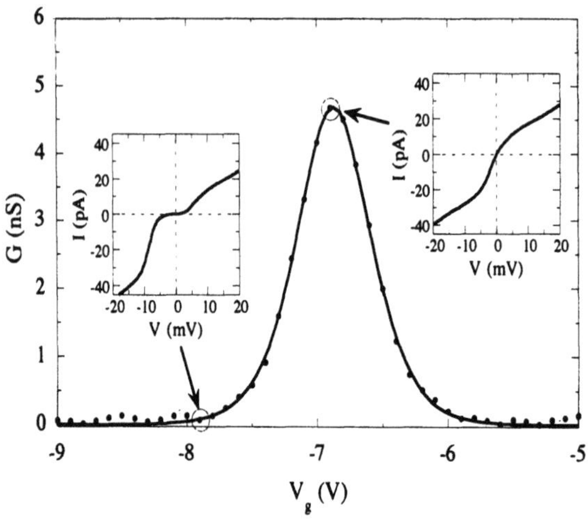

Figure 6. Current-voltage characteristics of the single nanocrystal single electron transistor device.

band, 2 eV higher in energy. In this interpretation, the large gap at positive V_g corresponds to a completely filled valence band. Starting on the right side of Figure 7c and decreasing V_g, the first Coulomb oscillation then corresponds to the removal of the first electron from, or, alternatively, the addition of the first hole to, the valence band. A further decrease in V_g adds additional holes. The addition energies for the 2nd, 3rd, and 4th holes are: $_2$ =14 ± 2 meV, $_3$ = 29 ± 3 meV, and $_4$ = 22 ± 2 meV.

This work represents a new type of spectroscopy for single nanocrystals. Unlike optical measurements, where electron-hole pairs are created, these measurements probe the energy for adding a single type of charge carrier. Future measurements will investigate how the ground state and excited state properties vary with the size, shape, and composition of nanocrystals as well as the composition of the leads. It is important to note that such devices are not yet of any practical use since their fabrication relies on the fortuitous placement of a nanocrystal fluctuations in local field modulate the gate voltage. Controlled techniques for the spatial organization of nanocrystals[41-44] are under development by a number of groups.

9. Concluding Remarks

Colloidal semiconductor nanocrystal samples have continued to advance rapidly. It is now possible to make large amounts of monodisperse II-VI and III-V nanocrystals, with a high degree of surface passivation, and with the ability to manipulate the environment. Such nanocrystals have now been incorporated into electrical devices, substantially extending the ability to understand and exploit them. In particular, composites of nanocrystals with polymers provide an excellent route for electrically

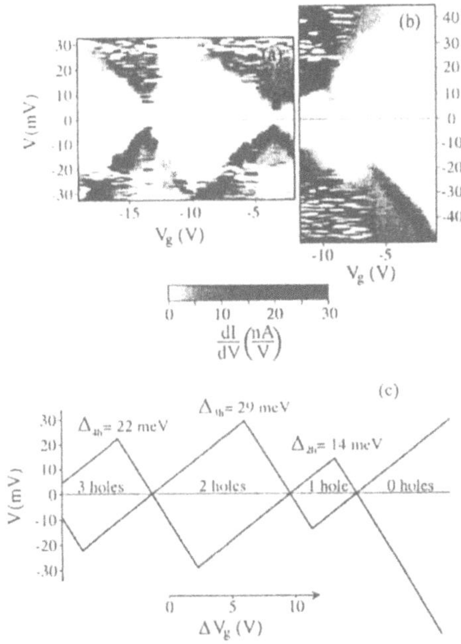

Figure 7. A plot of the differential conductance versus gate voltage and source drain bias shows the effect of adding multiple holes to the dot.

accessing ensembles of nanocrystals; single nanocrystal electrical studies are also of current interest.

Acknowledgments: Many colleagues have contributed to the work described here, including Dr. Xiaogang Peng, Dr. Chia-Chun Chen, Amy Herhold, Mike Schlamp, Dr. Neil Greenham, David Klein, Andrew Lim, and Professor Paul McEuen. We thank the Department of Energy, the National Science Foundation, and the Office of Naval research for support.

References

[1]L. E. Brus, A. L. Efros, and T. Itoh, "Special Issue On Spectroscopy Of Isolated and Assembled Semiconductor Nanocrystals - Introduction," Journal Of Luminescence 70 (V70), R7-R8 (1996).

[2]H. Weller, "Optical Properties Of Quantized Semiconductor Particles," Philosophical Transactions Of the Royal Society Of London Series a-Mathematical Physical and Engineering Sciences 354 (1708), 757-766 (1996).

[3]T. Vossmeyer, G. Reck, B. Schulz et al., "Double-Layer Superlattice Structure Built Up Of Cd32s14(Sch2ch(Oh)Ch3)(36)Center-Dot-4h(2)O Clusters," Journal Of the American Chemical Society 117 (51), 12881-12882 (1995).

[4]Y. Wang, M. Harmer, and N. Herron, "Towards Monodisperse Semiconductor Clusters - Preparation and Characterization Of Similar-to 13-Angstrom Thiophenolate-Capped Cds Clusters," Israel Journal Of Chemistry 33 (1), 31-39 (1993).

[5]A. P. Alivisatos, "Semiconductor Clusters, Nanocrystals, and Quantum Dots," Science 271 (5251), 933-937 (1996).

[6]S. A. Harfenist, Z. L. Wang, M. M. Alvarez et al., "Highly Oriented Molecular Ag Nanocrystal Arrays," Journal Of Physical Chemistry 100 (33), 13904-13910 (1996).

[7]R. L. Whetten, J. T. Khoury, M. M. Alvarez et al., "Nanocrystal Gold Molecules," Advanced Materials 8 (5), 428+ (1996).

[8]C. P. Collier, R. J. Saykally, J. J. Shiang et al., "Reversible tuning of silver quantum dot monolayers through the metal-insulator transition," Science 277 (5334), 1978-1981 (1997).

[9]Shi Jing, S. Gider, K. Babcock et al., "Magnetic clusters in molecular beams, metals, and semiconductors," Science 271 (5251), 937-41 (1996).

[10]J. M. McHale, A. Auroux, A. J. Perrotta et al., "Surface energies and thermodynamic phase stability in nanocrystalline aluminas," Science 277 (5327), 788-791 (1997).

[11]H. Gleiter, "Nanostructured materials," Advanced Materials 4 (7), 474-81 (1992).

[12]H. Gleiter, "Nanostructured materials: state of the art and perspectives," Zeitschrift fur Metallkunde 86 (2), 78-83 (1995).

[13]Ph. Buffat and J.P. Borel, " Size effect on the melting temperature of gold particles," Physical Review A 13 (6), 2287-98 (1976).

[14]A. N. Goldstein, C. M. Echer, and A. P. Alivisatos, "Melting In Semiconductor Nanocrystals," Science 256 (5062), 1425-1427 (1992).

[15]C. B. Murray, D. J. Norris, and M. G. Bawendi, "Synthesis and Characterization Of Nearly Monodisperse Cde (E = S, Se, Te) Semiconductor Nanocrystallites," Journal Of the American Chemical Society 115 (19), 8706-8715 (1993).

[16]J. E. B. Katari, V. L. Colvin, and A. P. Alivisatos, "X-Ray Photoelectron Spectroscopy Of Cdse Nanocrystals With Applications to Studies Of the Nanocrystal Surface," Journal Of Physical Chemistry 98 (15), 4109-4117 (1994).

[17]M. A. Olshavsky, A. N. Goldstein, and A. P. Alivisatos, "Organometallic Synthesis Of Gaas Crystallites Exhibiting Quantum

Confinement," Journal Of the American Chemical Society 112 (25), 9438-9439 (1990).

[18]O. I. Micic and A. J. Nozik, "Synthesis and Characterization Of Binary and Ternary Iii-V Quantum Dots," Journal Of Luminescence 70 (V70), 95-107 (1996).

[19]A. A. Guzelian, J. E. B. Katari, A. V. Kadavanich et al., "Synthesis Of Size-Selected, Surface-Passivated Inp Nanocrystals," Journal Of Physical Chemistry 100 (17), 7212-7219 (1996).

[20]A. A. Guzelian, U. Banin, A. V. Kadavanich et al., "Colloidal Chemical Synthesis and Characterization Of Inas Nanocrystal Quantum Dots," Applied Physics Letters 69 (10), 1432-1434 (1996).

[21]S. H. Tolbert and A. P. Alivisatos, "High-Pressure Structural Transformations In Semiconductor Nanocrystals," Annual Review Of Physical Chemistry 46 (V46), 595-625 (1995).

[22]Chia-Chun Chen, A. B. Herhold, C. S. Johnson et al., "Size dependence of structural metastability in semiconductor nanocrystals," Science 276, 398-401 (1997).

[23]M. G. Bawendi, M. L. Steigerwald, and L. E. Brus, "The Quantum Mechanics Of Larger Semiconductor Clusters (Quantum Dots)," Annual Review Of Physical Chemistry 41 (V41), 477-496 (1990).

[24]Ulrike Woggon, Optical Properties of Semiconductor Quantum Dots (Springer, Berlin, 1996).

[25]M. A. Hines and P. Guyotsionnest, "Synthesis and Characterization Of Strongly Luminescing Zns-Capped Cdse Nanocrystals," Journal Of Physical Chemistry 100 (2), 468-471 (1996).

[26]X. G. Peng, M. C. Schlamp, A. V. Kadavanich et al., "Epitaxial growth of highly luminescent CdSe/CdS core/shell nanocrystals with photostability and electronic accessibility," Journal Of the American Chemical Society 119 (30), 7019-7029 (1997).

[27]Y. N. C. Chan, R. R. Schrock, and R. E. Cohen, "Synthesis Of Single Silver Nanoclusters Within Spherical Microdomains In Block Copolymer Films," Journal Of the American Chemical Society 114 (18), 7295-7296 (1992).

[28]C. C. Cummins, R. R. Schrock, and R. E. Cohen, "Synthesis Of Zns and Cds Within Romp Block Copolymer Microdomains," Chemistry Of Materials 4 (1), 27-30 (1992).

[29]M. Antonietti and C. Goltner, "Superstructures of functional colloids: Chemistry on the nanometer scale," Angewandte Chemie-International Edition In English 36 (9), 910-928 (1997).

[30]B. O'Regan and M. Gratzel, "A Low-Cost, High-Efficiency Solar Cell Based On Dye-Sensitized Colloidal Tio2 Films," Nature 353 (6346), 737-740 (1991).

[31]N. C. Greenham, X. G. Peng, and A. P. Alivisatos, "Charge separation and transport in conjugated-polymer/semiconductor-nanocrystal composites studied by photoluminescence quenching and photoconductivity," Physical Review B-Condensed Matter 54 (24), 17628-17637 (1996).

[32]M. Schlamp, Xiaogang Peng, and A. P. Alivisatos, "Improved efficiencies in light emitting diodes made with CdSe (CdS) core/shell type nanocrystals and a semiconductor polymer," Journal of Applied Physics in press (1997).

PERSPECTIVES IN SUPRAMOLECULAR CHEMISTRY: FROM MOLECULAR RECOGNITION TOWARDS SELF-ORGANIZATION

Jean-Marie LEHN
Université Louis Pasteur, Strasbourg and Collège de France, Paris

Supramolecular chemistry has relied on more or less preorganised molecular receptors for effecting molecular recognition, catalysis and transport processes. A step beyond consists in the design of systems undergoing *molecular self-organisation,* i.e. systems capable of spontaneously generating a well-defined supramolecular architecture by *self assembling* from their components in a given set of conditions.

The molecular information necessary for the process to take place must be stored in the components and acts through selective molecular interactions. Thus, these *programmed supramolecular systems* operate via molecular recognition.

Several approaches to self-assembling systems have been pursued concerning supramolecular assemblies of either organic or inorganic nature:

1) the generation of *mesophases and liquid crystalline polymers* of supramolecular nature from complementary components, amounting to macroscopic expression of molecular recognition;

2) the induction of molecular recognition directed processes in *supramolecular assemblies* and in *the solid state;*

3) the *self-assembly of inorganic species* based on ligand design and on the use of suitable coordination algorithms as expressed in - the formation of helical metal complexes, *the double-stranded and triple-stranded helicates,* - the spontaneous generation of *closed cage structures,* - the self-assembling of *inorganic grids* of metal ions based on suitably designed ligands.

Molecular information controlled, *"programmed"* and functional self-organising systems offer wide perspectives in supramolecular chemistry towards the design of functional supramolecular entities for network engineering, polymolecular patterning and nanotechnology.

General reference

J.-M. LEHN, Supramolecular Chemistry: Concepts and Perspectives, VCH Weinheim, 1995.

G. Tsoucaris (ed.), Current Challenges on Large Supramolecular Assemblies, 417.

SUBJECT INDEX

MIX
Papier aus verantwortungsvollen Quellen
Paper from responsible sources
FSC® C105338

If you have any concerns about our products,
you can contact us on
ProductSafety@springernature.com

In case Publisher is established outside the EU,
the EU authorized representative is:
Springer Nature Customer Service Center GmbH
Europaplatz 3, 69115 Heidelberg, Germany

Printed by Libri Plureos GmbH
in Hamburg, Germany